Connected Vehicular Systems

Connected Vehicular Systems

Communication, Control, and Optimization

Ge Guo
The State Key Laboratory of Synthetical Automation
for Process Industries, Northeastern University
The School of Control Engineering, Northeastern
University at Qinhuangdao, China

Shixi Wen
School of Information and Engineering
The Key Laboratory of Advanced Design and Intelligent Computing,
Ministry of Education, Dalian University, China

WILEY

Published by John Wiley & Sons, Inc., Hoboken, New Jersey.
Published simultaneously in Canada.

Library of Congress Cataloging-in-Publication Data
Names: Guo, Ge (Of Dongbei da xue (1993)), author. | Wen, Shixi, author.
Title: Connected vehicular systems : communication, control, and optimization / Ge Guo, Shixi Wen.
Description: Hoboken, New Jersey : Wiley, [2024] | Includes index.
Identifiers: LCCN 2023022911 (print) | LCCN 2023022912 (ebook) | ISBN 9781394205462 (cloth) | ISBN 9781394205479 (adobe pdf) | ISBN 9781394205486 (epub)
Subjects: LCSH: Automated vehicles. | Intelligent transportation systems.
Classification: LCC TL152.8 .G86 2024 (print) | LCC TL152.8 (ebook) | DDC 629.04/6–dc23/eng/20230601
LC record available at https://lccn.loc.gov/2023022911
LC ebook record available at https://lccn.loc.gov/2023022912

Cover Design: Wiley
Cover Image(s): © kan wang/Getty Images

Set in 9.5/12.5pt STIXTwoText by Straive, Pondicherry, India

Contents

Preface *ix*
Acknowledgments *xiii*

Part I Vehicular Platoon Communication and Control *1*

1 **Control with Varying Communication Range** *3*
1.1 Introduction *3*
1.2 Problem Formulation *5*
1.3 Switching Control of Connected Vehicles *9*
1.4 Simulations and Experiments *16*
1.5 Conclusions and Future Work *23*
References *24*

2 **Control Subject to Communication Interruptions** *26*
2.1 Introduction *26*
2.2 Problem Formulation *27*
2.3 Mixed CACC-ACC Control *28*
2.4 Finite-Time Sliding-Mode Control *32*
2.5 Numerical Simulations *34*
2.6 Conclusions and Future Work *39*
References *41*

3 **Control and Communication Topology Assignment** *42*
3.1 Introduction *42*
3.2 Problem Statement *44*
3.3 Communication Topology and Control Co-Design *48*
3.4 Simulation Studies *57*
3.5 Conclusions and Future Work *70*
References *70*

4 **Control with Communication Delay and Switching Topologies** *72*
4.1 Introduction *72*
4.2 Problem Formulation *73*
4.3 Stability Analysis *77*

4.4 Controller Synthesis *82*
4.5 Simulation Studies *86*
4.6 Conclusions and Future Work *95*
References *96*

5 Control with Event-Triggered Communication *97*
5.1 Introduction *97*
5.2 Problem Formulation *99*
5.3 Event-Triggered Communication and Platoon Control *104*
5.4 Simulation Study *107*
5.5 Conclusions and Future Work *119*
References *120*

Part II Performance Guarantee Under Actuator Limitation *121*

6 Adaptive Fault-Tolerant Control with Actuator Saturation *123*
6.1 Introduction *123*
6.2 System Modeling and Problem Formulation *124*
6.3 Quadratic Spacing Policy and Adaptive PID-Type Sliding Surface *127*
6.4 Controller Design and Stability and Analysis *128*
6.5 Simulation Results *135*
6.6 Conclusions and Future Work *139*
References *142*

7 Fault-Tolerant Control with Input Quantization and Dead Zone *143*
7.1 Introduction *143*
7.2 System Modeling and Problem Formulation *144*
7.3 Improved Quadratic Spacing Policy and Adaptive PID-Type Sliding Surface *148*
7.4 Controller Design and Stability Analysis *149*
7.5 Simulation Results *155*
7.6 Conclusions and Future Work *157*
References *163*

8 Prescribed Performance Concurrent Control *165*
8.1 Introduction *165*
8.2 Problem Formulation *166*
8.3 Controller Design Guaranteed Prescribed Performance *168*
8.4 Simulation Studies *175*
8.5 Conclusions and Future Work *179*
References *179*

9 Adaptive Sliding Mode Control with Prescribed Performance *181*
9.1 Introduction *181*
9.2 Problem Formulation *181*
9.3 Model Transformation *184*
9.4 Vehicles Tracking Controller Design *185*

9.5 Simulation Studies *190*
9.6 Conclusions and Future Work *197*
References *198*

Part III Speed Trajectory Planning and Control *199*

10 Speed Planning and Tracking Control of Vehicles *201*
10.1 Introduction *201*
10.2 Problem Formulations *202*
10.3 Speed Planning *205*
10.4 Speed Tracking Controller Design *207*
10.5 Simulation and Experiments *213*
10.6 Conclusions and Future Work *221*
References *224*

11 Analytical Solution for Speed Planning and Tracking Control *225*
11.1 Introduction *225*
11.2 System Modeling and Problem Formulation *226*
11.3 Speed Optimization Based on PMP *228*
11.4 Speed Tracking Control and String Stability *232*
11.5 Simulation Studies *237*
11.6 Conclusions and Future Work *240*
References *241*

12 Speed Planning and Sliding-Mode Control to Reduce Intervehicle Spacing *242*
12.1 Introduction *242*
12.2 Problem Statement *243*
12.3 Intervehicle Spacing Optimization *246*
12.4 Sliding-Mode Controller Design *250*
12.5 Simulation Studies *253*
12.6 Conclusions and Future Work *265*
References *266*

13 Trajectory Planning and PID-Type Sliding-Mode Control to Reduce Intervehicle Spacing *268*
13.1 Introduction *268*
13.2 Problem Description *269*
13.3 Distributed Trajectory Optimization *271*
13.4 PID-Type Sliding-Mode Controller Design *275*
13.5 Simulation Results *278*
13.6 Conclusions and Future Work *288*
References *288*

14 Trajectory Planning and Fixed-Time Terminal Sliding-Mode Control *290*
14.1 Introduction *290*
14.2 Problem Formulation *291*

14.3 Vehicles Trajectory Optimization *293*
14.4 Fixed-Time Tracking Control Design *297*
14.5 Numerical Simulations *301*
14.6 Conclusions and Future Work *307*
References *307*

Index *309*

Preface

With the ubiquitous application of vehicle-to-vehicle (V2V) and vehicle-to-infrastructure (V2I) communication technologies, connected and automated vehicles (CAVs) are capable of gathering and sharing road and traffic information and even vehicle states with neighboring vehicles. In particular, enabled by the information shared, CAVs allow automated vehicle motion, car following, cooperated driving and platooning, and vehicle-traffic signal cooperative control. Therefore, CAVs are believed to be a promising technology to deliver greater safety and mobility benefits to the new generation of intelligent transportation systems (ITSs) with increased driving safety, ride comfort, traffic efficiency, and throughput, along with reduced congestion, accidents, emissions, and air pollution. However, the operation of CAVs and the associated ITSs depends heavily on timely and reliable information gathering and sharing, proper decision-making, and effective actuation of the driving decision. However, critical challenging issues are facing CAVs from all aspects including sensing, communication, control design, and command actuating, which, if not properly addressed, can result in safety risks and losses.

This book contains our research advances in the past decade in the analysis and synthesis of CAV systems from all aspects of trajectory planning, cooperative control, and communication. The focus of this book is on the development of mathematical models and methodologies for trajectory optimization and tracking control, communications conflict resolution, cooperative control subject to communication constraints, and sensor/actuator failures/faults for CAVs from different perspectives. This book is composed of 14 chapters. The contents are divided into three parts, with Chapter 1 – Chapter 5 as Part I, Chapter 6 – Chapter 9 as Part II, and Chapter 10 – Chapter 14 as Part III, respectively, concerned with cooperative vehicular communication and control, performance guarantee under actuator limitations, and speed trajectory planning and tracking control of CAVs.

Chapter 1 studies the platoon control problem subject to varying communication range with a constant-spacing policy. According to the connectivity status between the leader and each follower, connected vehicles control is modeled as a switched platoon control system with a connectivity-status-matrix-dependent controller. By using switched system theory, a series of sufficient conditions are obtained as a criterion for the stability analysis and control synthesis of the leader following platoon. Based on the obtained conditions, a useful control algorithm is proposed for connected vehicles. For each obtained connectivity-status-matrix-dependent controller, the string stability and a zero steady-state spacing error can be guaranteed by additional conditions.

Chapter 2 investigates platooning of connected vehicles considering communication interruption and latency. For a heterogeneous platoon of vehicles, a hybrid reference model with cooperative adaptive cruise control (CACC) and adaptive cruise control (ACC) is established. Then a novel CACC-ACC switching control method is suggested, which activates either a CACC scheme or an augmented ACC strategy depending on the status of communications. By introducing a platoon state tracking error system, a control algorithm is derived using finite-time sliding-mode control theory, which can robustly guarantee string stability and zero steady-state spacing error of the connected vehicles.

Chapter 3 studies the co-design problem of platoon controller and inter-vehicle communication topology (IVCT) in LTE-V2V networks. The communication assignment is achieved based on the cooperative awareness message dissemination mechanism. A sampled-data feedback controller is proposed for connected vehicles to eliminate the effect of stochastic packet dropouts and external disturbance, where the controller gain depends on the IVCT. To guarantee the stability requirement of connected vehicles with the minimized cost function, a unified control framework is established to jointly determine the optimal IVCT from all the available ones and the associated feedback controller gain. This co-design procedure is based on the optimal control and dynamic programming technique, where both fixed and periodic switching IVCTs are available. A useful algorithm is proposed to implement the established co-design framework.

Chapter 4 addresses the platoon control problem in a sampled-data setup with switching communication topology and transmission delays. A tracking error-based sampled-data control method is proposed, where the neighboring vehicle's state information is transmitted via the VANET with communication delay. By representing the switching communication topology by a Markovian chain, the connected vehicular control system is modeled as a Markovian switching time-delay system with disturbance. In the context of Markovian jumping system theory, a control methodology is obtained for connected vehicles to guarantee that the tracking errors can be stabilized mean-square exponentially with a given disturbance attenuation level. The controllers of connected vehicles with both fixed and variable gains are suggested. The results are extended to cover partially unknown transition rates of the Markov chains.

Chapter 5 studies the co-design problem considering a dynamic event-triggered communication mechanism (DECM). Under the DECM, the transmissions of sampled velocity and acceleration from a preceding vehicle to the controller can be significantly reduced. A sampled-data platoon controller is designed based on the tracking error (spacing error, velocity error, and acceleration error). Sufficient conditions for the stability of the CACC system are obtained for the DECM-based sampled-data feedback controller. According to the obtained conditions, parameter design criteria are established for the DECM to guarantee the stable performance of connected vehicles control systems.

Chapter 6 addresses a fault-tolerant control problem for connected vehicles subject to actuator faults and saturation. To compensate for the effects of actuator faults and saturation, an adaptive fault-tolerant control method is proposed based on nonlinear vehicle dynamics and a new quadratic spacing policy. The improved quadratic spacing policy is introduced to remove the assumption of zero initial spacing errors. The nonlinear vehicle dynamics is approximated by a radial basis function neural network (RBFNN). The adaptive fault-tolerant control method is developed in the context of the PID-type sliding-mode control technique and proved to be capable of guaranteeing individual vehicle stability, string stability, and traffic flow stability.

Chapter 7 revisits the fault-tolerant control problem for connected vehicles considering actuator faults, input quantization, and dead-zone nonlinearity. The occurrence of actuator faults may cause abrupt velocity and acceleration change, which may yield a violation of the spacing policy. So, an improved quadratic spacing policy reflecting the effect of actuator fault is proposed, which removes the condition of zero initial spacing errors. Then, an adaptive fault-tolerant control scheme is developed by employing RBFNN and PID-type sliding-mode control method.

Chapter 8 investigates prescribed performance concurrent control (PPCC) of connected vehicles with unknown parameters, disturbances, and actuator saturation. A closer spacing policy is introduced to achieve string stability with a virtual leader-bidirectional information flow. Based on a new transformed tracking error function and an auxiliary system introduced to deal with actuator saturation, a distributed adaptive tracking PPCC controller is designed to achieve individual vehicle stability and string stability in the sense that all the signals in the system are uniformly ultimately bounded.

Chapter 9 revisits the prescribed performance platoon control problem in the presence of actuator saturation, uncertain parameters, and unknown disturbances. Two adaptive sliding-mode control schemes based on leader-predecessor and leader-bidirectional information flows, respectively, are presented to ensure string stability and strong string stability with prescribed tracking performance. The actuator saturation nonlinearity is approximated with a smooth hyperbolic tangent function. The effects of uncertain parameters and exogenous disturbances are dealt with by introducing a set of adaptation laws.

Chapter 10 studies the speed optimization and tracking control problem for heavy-duty truck platoons. The speed planning algorithm is derived with regard to an average vehicle based on a combined fuel-time cost and receding dynamic programming. The idea of using an average vehicle instead of the leader for speed planning makes the speed profile more fuel-efficient for platooning of vehicles different in weight and size. The vehicle controller, a discrete-time back-stepping control law, is designed on the basis of a nonlinear vehicle model considering road slope and heterogeneity of vehicles. The control algorithm is strengthened by a novel string stability criterion.

Chapter 11 is concerned with speed optimization and tracking control problems for platooning of connected vehicles. A two-layered control architecture is presented: a set-point optimization layer and a vehicle tracking control layer. In the first layer, a speed-planning algorithm is derived to calculate the speed set-point for the connected vehicles by averaging the optimal speed of each vehicle, which is obtained by solving a fuel-time optimization problem based on Pontryagin's minimum principle. The second layer contains a set of distributed sliding-mode controllers for vehicle tracking control, which can guarantee string stability of the connected vehicles with the desired inter-vehicle spacing.

Chapter 12 investigates distributed trajectory optimization and adaptive platoon control of connected vehicles. A distributed hierarchical framework is proposed for trajectory optimization and tracking control. The upper layer provides an optimal trajectory for each vehicle, which is realized by minimizing the inter-vehicle spacing with regard to the desired values using convex optimization. The lower layer contains an adaptive sliding-mode controller to track the optimal trajectory. To compensate for uncertain vehicle dynamics, a parameter adaptation law based on the tracking error dynamics is involved in the controller, which guarantees both individual vehicle stability and string stability.

Chapter 13 studies distributed trajectory optimization and platoon control problems for connected vehicles with a quadratic spacing policy. The quadratic spacing policy based on the expected team speed is introduced to improve the flexibility of speed planning and regulation. The trajectory optimization problem is solved using distributed convex optimization based on spacing error minimization, resulting in an algorithm to provide the optimal trajectory for all following vehicles. Then, a PID-type sliding-mode controller with a double high-power reaching law is presented for speed-tracking control of each follower. The methodology can guarantee individual vehicle stability, string stability, and traffic flow stability with ignorable turbulence of spacing and speed.

Chapter 14 extends the distributed trajectory optimization and platooning method to achieve fixed-time tracking. The optimal trajectory for all following vehicles is obtained by minimizing the spacing errors via distributed convex optimization. The trajectory tracking controller is derived based on the tracking error dynamics in the context of fixed-time stability and terminal sliding-mode control. The method can robustly guarantee zero steady-state spacing errors and individual vehicle stability and string stability simultaneously.

In this book, special attention is given to a clear presentation of the formulations, algorithms, and their implementation in numerical simulations and experimental studies. Related works and references are given at the end of each chapter to guide the readers toward further knowledge in this area. The book can be used in courses for graduate students in modeling and control of connected automated vehicles and transportation systems. Researchers and engineers can also draw upon the book in developing mathematical models and algorithms for theoretical study and application purposes.

Ge Guo, Northeastern University, Shenyang, China
Shixi Wen, Dalian University, Dalian, China
2 March 2023

Acknowledgments

Our research work has been funded by the National Natural Science Foundation of China under Grants 60974013, 61273107, 61573077, U1808205, 61803062, and 62173079, and, in part, by the Natural Science Foundation of Liaoning Province under Grant 2022-MS-406. We are grateful to the funders for their support in the past decade. The authors acknowledge the contributions of Dandan Li, Qiong Wang, Ping Li, Hongbo Lei, Ziwei Zhao, Dongqi Yang, Jian Kang, Ren-Yong-Kang Zhang, and many others for their hard work and commitment that made this monograph come to fruition. Without their effort and time in developing quality contributions in one or two of the chapters, this book would not have been possible.

Part I

Vehicular Platoon Communication and Control

Communication issues due to bandwidth shortage are inevitable in vehicular ad hoc networks (VANET), which may induce time delay, packet dropouts, and medium access constraints that can lead to performance degradation or even instability in connected vehicular control systems. Moreover, inter-vehicle communication and control are two aspects that are strongly coupled. This part consists of five chapters. **Chapter 1** studies the platoon control problem subject to varying communication range. Based on the connectivity status between the leader and other vehicles, the platoon control system is modeled as a switching system. Then a set of sufficient conditions for stability analysis and control synthesis are given, yielding a platoon control algorithm that guarantees string stability and zero steady-state spacing errors. **Chapter 2** revisits the problem subject to communication interruption and latency. Based on a hybrid reference model of cooperative adaptive cruise control (CACC) and adaptive cruise control (ACC), a CACC-ACC switching control method is suggested. To ensure string stability and zero steady-state spacing errors with a finite time, a finite-time sliding-mode control algorithm is derived. **Chapter 3** studies the co-design problem of platoon controller and inter-vehicle communication topology. A unified co-design framework is presented to jointly determine the optimal communication topology and the sampled-data controller gains in association with the topology. The co-design algorithm is derived based on the stability requirement and minimization of the cost function via dynamic programming. **Chapter 4** studies the platoon control problem with a switching communication topology and transmission delays. The switching communication topology is described as a Markov chain, and the connected vehicular control system is modeled as a Markovian jumping system with delay. The resulting control method can guarantee exponential convergence of the tracking errors in the mean-square sense. **Chapter 5** revisits the co-design problem of the CACC controller and event-triggered communication mechanism. Sufficient conditions for the sampled-data controller and the event-triggered communication mechanism are obtained to guarantee system stability.

Connected Vehicular Systems: Communication, Control, and Optimization, First Edition. Ge Guo and Shixi Wen.

1

Control with Varying Communication Range

1.1 Introduction

Connected vehicles can share information (position, speed, etc.) with neighboring vehicles and the infrastructure via vehicular ad hoc networks (VANETs), which makes platooning or cooperative adaptive cruise control (CACC) of vehicles an important technology to alleviate traffic congestion and reduce traffic accidents [1]. In a platooning system, vehicles can travel at a small intervehicle spacing and hence lower air drag to reduce fuel consumption and improve the throughput of road traffic. Therefore, vehicular platooning has attracted much attention in recent years [2, 3]. As an extension of the ACC technology onboard most cars, CACC is a promising technique in intelligent transportation systems [4, 5] enabling cooperative control and platooning of vehicles via intervehicle communications in addition to onboard sensors. When a car is following another in the ACC mode, amplifications of acceleration/deceleration are inevitable due to the unavailability of information from the preceding vehicles. The longer the platoon, the more significant the shock wave effect in the platoon of vehicles [6], which is known as string instability. CACC relies on communication networks to exchange information between vehicles, allowing vehicles to adapt to those ahead and, hence, can effectively mitigate the shock wave effect. In addition, the time headway in CACC is smaller than ACC, leading to improved road capacity and traffic throughput [7], and reduced aerodynamic drag and fuel consumption [8].

Generally, typical spacing policies used for connected vehicles are the constant-spacing (CS) policy and the constant-time-headway-spacing (CTHS) policy, depending on whether the required spacing of the vehicle is dependent on its speed [9, 10]. CS policy has the advantage to improve road capacity, while CTHS policy can help improve road safety. From the viewpoint of string stability, connected vehicles with CS policy are harder to be guaranteed.

With the rapid development of intervehicle communication technologies, the various types of interaction topologies (see Figure 1.1) [11] are available for platoons now. Nevertheless, string stability is not always satisfied under interaction topologies when a CS policy is involved. For example, [12] proved that string stability cannot be satisfied for homogeneous platoons with predecessor following (PF) topology and CS policy owning to a complementary sensitivity integral constraint. Reference [13] showed that platoons with bidirectional topology also suffered fundamental limitations on string stability. As the leader PF (LPF) topology, the leader can periodically broadcast its information to each follower, which has been proven to yield better string stability performance [12, 14] with a CS policy.

To date, considerable research has contributed to solving the string stability problem for LPF platoons with CS policy. The important technical issues include communication constraints, optimal

Connected Vehicular Systems: Communication, Control, and Optimization, First Edition. Ge Guo and Shixi Wen.

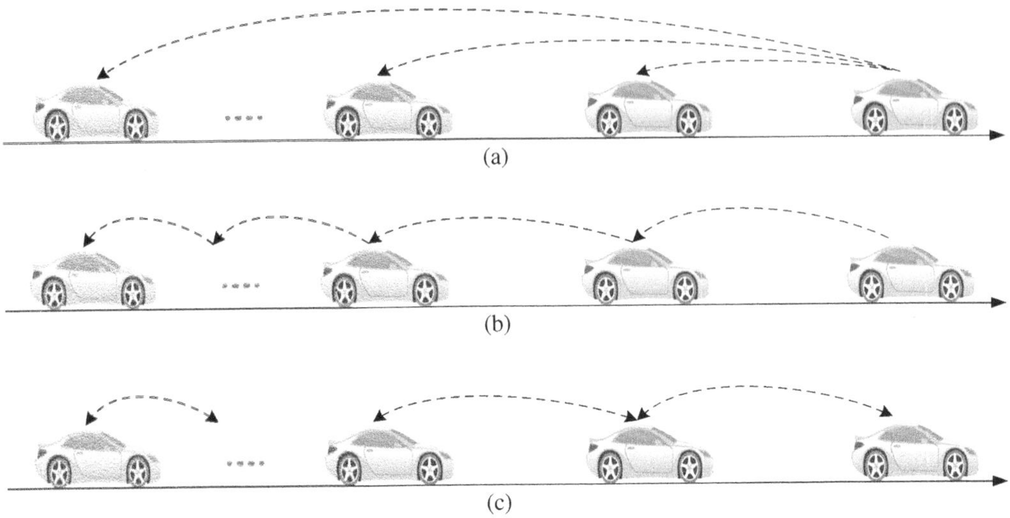

Figure 1.1 Communication topology in vehicle platoon control: (a) leader predecessor following; (b) predecessor following; (c) bidirectional topology.

control, nonlinear dynamics, range-limited sensors, uncertain vehicle models, and disturbances. To name a few studies, [15] analyzed the string stability in the presence of certain time delays in the leader state reception. Reference [16] investigated the effect of communication delays on string stability. The effects of communication limitations on LPF platoon control were studied in [17, 18] to include communication delays, packet dropouts, and quantization. Reference [19] proposed a decentralized overlapping control law by using the inclusion principle, which can preserve the string stability and steady-state behavior for the platoon controller. Reference [20] established distributed receding horizon control algorithms for platoons to deal with nonlinear vehicle dynamics. In these algorithms, the acceleration information of the leading car is not needed in the controller. The algebraic graph theory and Routh-Hurwitz stability criterion were used in [11] to analyze the stability and scalability of a platoon. Reference [21] proposed a guaranteed-cost controller for vehicle platoons to deal with actuator delays (e.g. fueling and braking delays) and the effects of sensing range limitations.

In this chapter, we are interested in stability analysis and controller synthesis for LPF platoon with a constant spacing policy, where the connectivity status between the leader and each follower in VANETs varies with time. Under such circumstances, none, part, or all of the following vehicles can receive the state information broadcast by the leader at any time. Consequently, the platoon is modeled as a switched control system [22, 23], where a compact form of the platoon controller switches according to the connectivity status of each follower. Using the average dwell time technique and piecewise Lyapunov-Krasovskii functional, the criteria for stability analysis and controller synthesis are proposed for platoons. According to the obtained conditions, a useful LPF platoon control algorithm is proposed. The contributions of this chapter are summarized as follows:

1) A novel-switched control model is established to describe the effects of varying communication ranges on LPF platoons with a CS policy.
2) An efficient platoon control structure is proposed in which the controller gain of the platoon is dependent on the connectivity status between the leader and each follower.

3) A series of new conditions are obtained as the criteria for stability analysis and controller synthesis for the switched time-delay platoon control system.
4) A useful platoon control algorithm is proposed to determine the connectivity-status-matrix-dependent feedback controller, which simultaneously guarantees the requirement of string stability and a zero steady-state space error for each obtained controller gain.

The remainder of this chapter is organized as follows. Section 1.2 analyzes the effects of varying communication ranges and establishes a switched platoon control system model. Section 1.3 gives the main result on the stability and controller synthesis for platoons in order to satisfy the string stability requirement for each obtained controller gain. Numerical simulations and experiments with laboratory-scale Arduino cars are given in Section 1.4. The concluding remarks and future research topics are given in Section 1.5.

Notation Throughout this technical note, A^T denotes the transpose of a matrix A and I is the identity matrix with appropriate dimensions. $A \geq 0$ $(A > 0)$ means that A is a positive semidefinite (positive-definite) matrix, and $A \geq B$ $(A > B)$ means that $A - B \geq 0$ $(A - B > 0)$. $\mathbb{R}^n$ and $\mathbb{R}^{n \times m}$ denote the n-dimensional real Euclidean space and the set of all $n \times m$ real matrices, respectively. $\|\cdot\|$ represents the Euclidean norm for a vector. diag$\{\cdots\}$ denotes a block diagonal matrix.

1.2 Problem Formulation

Consider a platoon of $N + 1$ vehicles in a freeway scenario. We assume that this platoon is running within a horizontal straight line under the constant spacing policy (see Figure 1.2). Denote by z_r, v_r, and a_r the r-th ($r = 0, 1, \ldots, N$) vehicle's position, velocity and acceleration, with $r = 0$ standing for the leader and the others as the followers. Each follower is equipped with an onboard sensor to measure the distance between it and the preceding vehicle. The leader periodically broadcasts its state information (velocity and acceleration) to all followers via VANETs. It is assumed that the communication range of the leader is long enough to cover all followers in a platoon. To avoid the instability of the platoon caused by saturation on the velocity, the maximum velocity of each follower is defined as $v_{\max}$ such that $v_0 \leq v_{\max}$. Moreover, it is assumed that the leader moves at a constant speed, i.e. $\dot{z}_0(t) = v_0(t)$ and $\dot{v}_0(t) = 0$ as time $t \to \infty$.

Define the spacing error of the r-th following vehicle as

$$\delta_r(t) = z_{r-1}(t) - z_r(t) - L - \delta_d \tag{1.1}$$

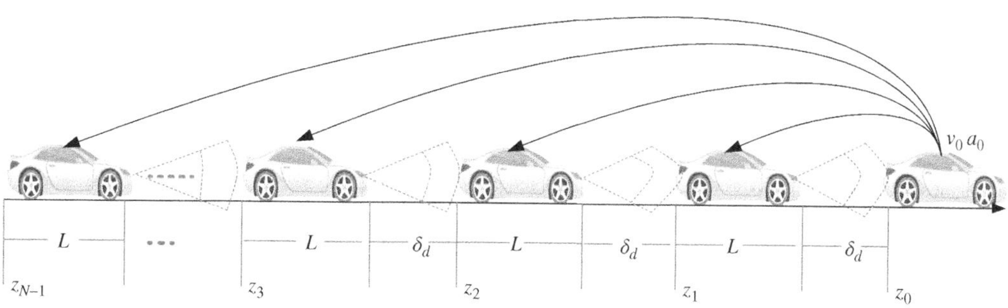

Figure 1.2 Platoon of vehicles.

where δ_d is the desired intervehicle spacing and L is the length of the vehicle. The dynamics of vehicle r, $r = 1, 2, \ldots, N$, are described by the following differential equations:

$$\dot{\delta}_r(t) = \dot{z}_{r-1}(t) - \dot{z}_r(t) = v_{r-1}(t) - v_r(t), \tag{1.2}$$

$$\ddot{\delta}_r(t) = a_{r-1}(t) - a_r(t), \tag{1.3}$$

$$\dot{a}_r(t) = -a_r(t)/\varsigma + u_r(t)/\varsigma, \tag{1.4}$$

where ς denotes the engine time constant, and $u_r(t)$ is the control input for the r-th vehicle.

In order to ensure safe braking during deployment, the desired intervehicle spacing is required to no smaller than the braking distance. According to the PATH algorithm proposed by [24], braking distance d_{brk} is adopted by the following kinematic approach

$$d_{brk} = \frac{1}{2}\left(\frac{v_r^2}{\kappa_1} - \frac{\Delta v_r}{\kappa_2}\right) + v_r\tau' + d_0, \tag{1.5}$$

where $\Delta v_r = v_r - v_{r-1}$ is the relative velocity between the successive vehicles. κ_1 is the maximum deceleration of vehicle i and κ_2 is the maximum deceleration of its preceding vehicle $i-1$. τ' is the delay time of the braking system and d_0 is the minimum distance.

Consider the existing actuator time delay τ (e.g. the fueling delay and braking delay) when a vehicle executes the control command. Then a more realistic dynamic model for (1.4) is represented as

$$\dot{a}_r(t) = -a_r(t)/\varsigma + u_r(t-\tau)/\varsigma \tag{1.6}$$

Define $x(t) = \mathrm{Col}[x_r(t)]_{r=1}^N$ and $u(t) = \mathrm{Col}[u_r(t)]_{r=1}^N$ to respectively denote the state and the control vectors, where 'Col' represents the column vector, and $x_r(t) = \left[\delta_r\ \dot{\delta}_r\ \ddot{\delta}_r\ v_0 - v_r\ a_0 - a_r\right]^T$. According to (1.2), (1.3), and (1.6), the state space equation of the entire vehicle platoon is represented as

$$\dot{x}(t) = \mathbf{A}x(t) + \mathbf{B}u(t-\tau) \tag{1.7}$$

where

$$\mathbf{A} = \begin{bmatrix} \mathbf{A}_1 & & \\ & \ddots & \\ & & \mathbf{A}_N \end{bmatrix}, \quad \boldsymbol{B} = \begin{bmatrix} -\boldsymbol{B}_{11} & & & & \\ \boldsymbol{B}_{21} & -\boldsymbol{B}_{22} & & & \\ & \ddots & \ddots & & \\ & & & \boldsymbol{B}_{N1} & -\boldsymbol{B}_{N2} \end{bmatrix},$$

$$\boldsymbol{A}_r = \begin{bmatrix} 0 & 1 & 0 & 0 & 0 \\ 0 & 0 & 1 & 0 & 0 \\ 0 & 0 & \frac{-1}{\varsigma} & 0 & 0 \\ 0 & 0 & 0 & 1 & 0 \\ 0 & 0 & 0 & 0 & \frac{-1}{\varsigma} \end{bmatrix}, \quad \boldsymbol{B}_{r1} = \begin{bmatrix} 0 \\ 0 \\ \frac{1}{\varsigma} \\ 0 \\ 0 \end{bmatrix}, \quad \boldsymbol{B}_{r2} = \begin{bmatrix} 0 \\ 0 \\ \frac{1}{\varsigma} \\ 0 \\ \frac{-1}{\varsigma} \end{bmatrix}$$

In this chapter, the long-term evolution (LTE) V2V network is utilized to support reliable V2V communication. As specified in the standard, ten subframes of 1 ms each make an LTE frame. Normally, the size of the state information of the leader varies between 50 bytes and 500 bytes. In a common LTE cell deployment (i.e. bandwidth is 10 MHZ), one subframe is sufficient for a vehicle to transmit its state information under suitable modulation and coding scheme. Here, the signal-to-interference-plus-noise (SINR) is used to characterize the reliability of V2V communication. To be specific, the SINR at the intended receiver side (the r-th follower) in a subframe is represented as [25]

$$\gamma_r = \frac{p_{v2v} h_r}{I_r + H_0}, \tag{1.8}$$

where p_{v2v} is the transmission power from the leader to follower i. h_r is the channel gain, which contains path loss, shadowing effect from the leader to follower i. I_r is the interference from cell user. H_0 is the power of additive noise in each subframe. When the SINR is above a targeted SINR threshold γ_{V2V}, the data transmission rate in the V2V communication is reliable. Due to high mobility, the condition of the wireless channel (e.g. h_r) varies dynamically. However, unreliable data transmission rates may cause communication constraints such as communication delay, packet dropouts and medium access constraints. Therefore, we assume that the velocity of the platoon can guarantee that the SINR of the V2V channel satisfies $\gamma_r > \gamma_{V2V}$. Namely, the data transmission rate in V2V communication is reliable when follower i is communicating with the leader.

The varying communication range in a VANET influences on the connectivity status between the leader and each follower. Define a binary variable $\gamma_r(t) \in \{0, 1\}$ that denotes the connectivity status between the r-th follower and the leader. Specifically, if the distance between the r-th follower and the leader is within the current communication range, then $\gamma_r(t) = 1$; otherwise $\gamma_r(t) = 0$. This is the distance between the r-th following vehicle and the leading one beyond the current communication range. Define matrix $\boldsymbol{M}_{\sigma(t)} = \mathrm{diag}\,\{\boldsymbol{\Delta}_1(t), \boldsymbol{\Delta}_2(t), \ldots, \boldsymbol{\Delta}_N(t)\}$ where $\boldsymbol{\Delta}_r(t) = \mathrm{diag}\,\{1, 1, 1, \gamma_r(t), \gamma_r(t)\}$ represents the followers' connectivity status with the leader, and $\sigma(t) \in \{0, 1, 2, \ldots N\}$ denotes the switching signal. All possible values of the connectivity status matrix $\boldsymbol{M}_{\sigma(t)}$ are taken in a finite set $\{\boldsymbol{M}_0, \boldsymbol{M}_1, \boldsymbol{M}_2, \ldots \boldsymbol{M}_N\}$, where $\boldsymbol{M}_\sigma$, $\sigma \in \{0, 1, 2, \ldots N\}$ is given as

$$\begin{aligned}
\boldsymbol{M}_0 &= \mathrm{diag}\{\boldsymbol{\Pi}_{12}, \boldsymbol{\Pi}_{22}, \ldots, \boldsymbol{\Pi}_{N2}\},\\
\boldsymbol{M}_1 &= \mathrm{diag}\{\boldsymbol{\Pi}_{11}, \boldsymbol{\Pi}_{22}, \ldots, \boldsymbol{\Pi}_{N2}\},\\
\boldsymbol{M}_2 &= \mathrm{diag}\{\boldsymbol{\Pi}_{11}, \boldsymbol{\Pi}_{21}, \boldsymbol{\Pi}_{32}, \ldots, \boldsymbol{\Pi}_{N2}\},\\
\boldsymbol{M}_3 &= \mathrm{diag}\{\boldsymbol{\Pi}_{11}, \boldsymbol{\Pi}_{21}, \boldsymbol{\Pi}_{31}, \boldsymbol{\Pi}_{42}, \ldots, \boldsymbol{\Pi}_{N2}\},\\
&\vdots\\
\boldsymbol{M}_N &= \mathrm{diag}\{\boldsymbol{\Pi}_{11}, \boldsymbol{\Pi}_{21}, \ldots, \boldsymbol{\Pi}_{N1}\}
\end{aligned}$$

where $\boldsymbol{\Pi}_{r1} = \mathrm{diag}\,\{1, 1, 1, 1, 1\}$ and $\boldsymbol{\Pi}_{r2} = \mathrm{diag}\,\{1, 1, 1, 0, 0\}$. An illustrative example of the connectivity status matrix $\boldsymbol{M}_{\sigma(t)}$ is shown in Figure 1.3.

For each follower, the r-th follower can use the received values v_0 and a_0 from the leader to compute the velocity error $v_{0r} = v_0 - v_r$ and acceleration error $a_{0r} = a_0 - a_r$ when $\gamma_r(t) = 1$; otherwise, the velocity error and acceleration error are set to $v_{0r} = 0$ and $a_{0r} = 0$ by the zero strategy. As a result, the controller of the r-th follower has the following switched form

$$u_r(t) = \begin{cases} k_\sigma^p \delta_r + k_\sigma^v \dot{\delta}_r + k_\sigma^a \ddot{\delta}_r + k_\sigma^{ve} v_{0r} + k_\sigma^{ae} a_{0r} & \gamma_r(t) = 1 \\ k_\sigma^p \delta_r + k_\sigma^v \dot{\delta}_r + k_\sigma^a \ddot{\delta}_r & \gamma_r(t) = 0 \end{cases} \tag{1.9}$$

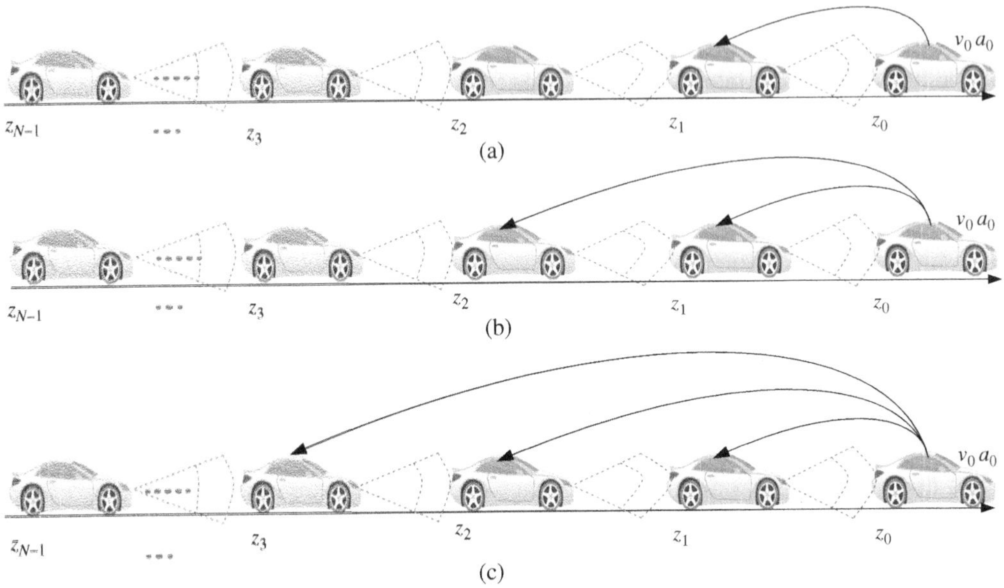

Figure 1.3 Connectivity status matrix for (a) $\boldsymbol{M}_{\sigma(t)} = \boldsymbol{M}_1$, (b) $\boldsymbol{M}_{\sigma(t)} = \boldsymbol{M}_2$, and (c) $\boldsymbol{M}_{\sigma(t)} = \boldsymbol{M}_3$.

where k_σ^p, k_σ^v, k_σ^a, k_σ^{ve}, and k_σ^{ae} are the connectivity status matrix $\boldsymbol{M}_\sigma$ dependent controller gains to be designed. According to (1.9), the compact form of a feedback controller (as shown in Figure 1.4) for a platoon can be represented as

$$u(t) = \boldsymbol{M}_{\sigma(t)}\boldsymbol{K}_{\sigma(t)}x(t) \tag{1.10}$$

where $\boldsymbol{K}_{\sigma(t)} \in \{\boldsymbol{K}_0, \boldsymbol{K}_1, \boldsymbol{K}_2, \ldots \boldsymbol{K}_N\}$, $\boldsymbol{K}_\sigma = \operatorname{diag}\left\{\overline{\boldsymbol{K}}_\sigma^1, \overline{\boldsymbol{K}}_\sigma^2, \ldots, \overline{\boldsymbol{K}}_\sigma^N\right\}$ is the connectivity-status-matrix-dependent feedback controller gain with $\overline{\boldsymbol{K}}_\sigma^r = \operatorname{diag}\{k_\sigma^p, k_\sigma^v, k_\sigma^a, k_\sigma^{ve}, k_\sigma^{ae}\}$. Let $\alpha_\sigma(t)$ denote the total time when the connectivity status matrix of the following vehicles is equal to $\boldsymbol{M}_\sigma$ in the time interval $[0, t)$. Then $\alpha_\sigma = \alpha_\sigma(t)/\sum_{\sigma=0}^{N}\alpha_\sigma(t)$ is called the mode rate.

Taking the feedback controller (1.10) into (1.7) and assuming the initial condition $x(t) = \phi(t) = \phi(t_0)$ for $t \in [-\tau, 0)$, the platoon state space equation can be rewritten as

$$\begin{cases} \dot{x}(t) = \boldsymbol{A}x(t) + \boldsymbol{B}\boldsymbol{M}_{\sigma(t)}\boldsymbol{K}_{\sigma(t)}x(t-\tau), & t \geq t_0 \\ x(t) = \phi(t) = \phi(t_0), & t \in [-\tau, 0) \end{cases} \tag{1.11}$$

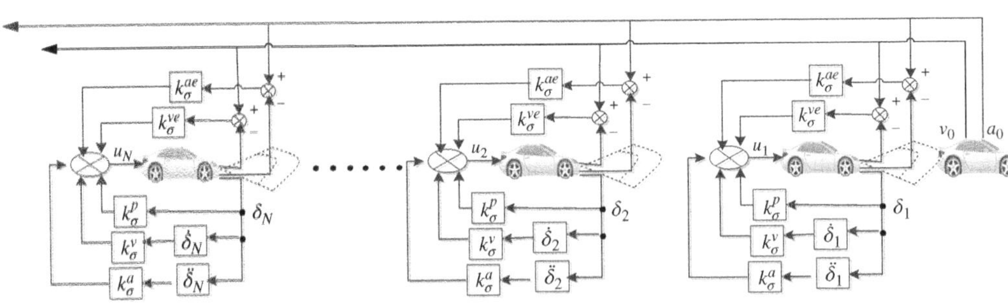

Figure 1.4 Controller structure of vehicle platoon.

The control objective of the platoon control system is to design the connectivity-status-dependent feedback controller of (1.10) for a platoon to ensure that each follower moves at the same speed with the leader while maintaining a constant desired intervehicle spacing. Namely, the platoon control system (1.11) is needed to meet the following criteria with varying communication ranges and actuator delays:

1) Individual vehicle stability: The entire closed-loop platoon control system (1.11) is mean-square exponential stable.
2) Steady-state performance: The spacing error $\delta_r(t)$ approaches to zero for all vehicles.
3) String stability: The transient errors are not amplified with the vehicle index under any maneuver of the leading vehicle for each obtained feedback controller gain, $\|G_r(s)\| \leq 1$, where $G_r(s) = \delta_r(s)/\delta_{r-1}(s)$ with $\delta_r(s)$ is the Laplace transform of the spacing error $\delta_r(t)$.

To proceed, we need the following definition and lemma.

Definition 1.1 [26] System (1.11) is said to be exponentially stabilized if there exist positive scalar c and $\rho < 0$ such that for any initial condition $\phi(t_0) \in \mathbb{R}^n$, the solution $x(t)$ satisfies

$$\|x(t)\| \leq ce^{\rho(t-t_0)}\|\phi(t_0)\|, \forall t \geq t_0 \tag{1.12}$$

where ρ is said to be the decay rate.

Definition 1.2 [27] For any switching signal $\sigma(t)$ and $t \geq t_0$, let $N(k)$ denote the number of switching times of $\sigma(t)$ over the time interval $[t_0, \ t)$. If $N(t) \leq N_0 + (t - t_0)/T_a$ holds for $N_0 > 0$ and $T_a > 0$, then T_a is called the average dwell time and N_0 the chatter bound. For simplicity, but without loss of generality, we choose $N_0 = 0$.

Lemma 1.1 [28, 29] For a scalar $\lambda \in (0, 1)$, define the positive-definite matrix $\boldsymbol{R} \in \mathbb{R}^{n\times n}$ and two real vectors $\boldsymbol{W_1} \in \mathbb{R}^n$ and $\boldsymbol{W_2} \in \mathbb{R}^n$. Define the following function $f(\lambda)$ as a reciprocally convex combination on

$$f(\lambda) = \frac{1}{\lambda}\boldsymbol{W}_1^T\boldsymbol{R}\boldsymbol{W}_1 + \frac{1}{1-\lambda}\boldsymbol{W}_2^T\boldsymbol{R}\boldsymbol{W}_2.$$

If there exists a matrix $\boldsymbol{S} \in \mathbb{R}^{n\times n}$ such that $\begin{bmatrix} \boldsymbol{R} & \boldsymbol{S} \\ \boldsymbol{S} & \boldsymbol{R} \end{bmatrix} \geq 0$, then the following inequality holds:

$$f(\lambda) \geq \begin{bmatrix} \boldsymbol{W}_1 \\ \boldsymbol{W}_2 \end{bmatrix}^T \begin{bmatrix} \boldsymbol{R} & \boldsymbol{S} \\ \boldsymbol{S} & \boldsymbol{R} \end{bmatrix} \begin{bmatrix} \boldsymbol{W}_1 \\ \boldsymbol{W}_2 \end{bmatrix}.$$

Proof: According to the lower-bounds theorem and reciprocally convex inequality given in theorem 1 of [28], it is straightforward to obtain Lemma 1.1.

1.3 Switching Control of Connected Vehicles

In this section, we first give a stability analysis and controller design criteria for the switched time-delay vehicle platoon control system (1.11). Then, we add the constraints for the obtained controller to guarantee the string stability requirement and the zero steady-state spacing error for each obtained connectivity-status-matrix-dependent controller gain.

For the switched system (1.11), introduce a piecewise quadratic Lyapunov-Krasovskii functional in the following form

$$\boldsymbol{V}_\sigma(t) = x^T(t)\boldsymbol{P}^\sigma x(t) + e^{-\lambda_\sigma t}\int_{t-\tau}^{t} e^{\lambda_\sigma s}x^T(s)\boldsymbol{Q}^\sigma x(s)ds + e^{-\lambda_\sigma t}\int_{-\tau}^{0}\int_{t+\theta}^{t} e^{\lambda_\sigma s}\dot{x}^T(s)\boldsymbol{R}^\sigma \dot{x}(s)dsd\theta \tag{1.13}$$

with $\lambda_\sigma > 0$, $\boldsymbol{P}_\sigma > 0$, $\boldsymbol{Q}_\sigma > 0$, and $\boldsymbol{R}_\sigma > 0$. The property of the defined Lyapunov-Krasovskii functional $\boldsymbol{V}_\sigma(t)$ is shown in the following lemma. For simplicity of presentation, let $\boldsymbol{\ell}_r$, $r = 1, 2, 3, 4$ be a block entry matrix with $\boldsymbol{\ell}_{rg} = \ell_r - \ell_g$. To make this clear, we give an example, $\ell_1 = [\boldsymbol{I_n}, 0, 0, 0]^T$ and $\ell_{12} = [\boldsymbol{I_n}, -\boldsymbol{I_n}, 0, 0]^T$.

Lemma 1.2 Given scalars $\lambda_\sigma > 0$, if there exist real matrices $\boldsymbol{P}_\sigma > 0$, $\boldsymbol{Q}_\sigma > 0$, $\boldsymbol{R}_\sigma > 0$, and real matrix $\boldsymbol{S}_\sigma$ of appropriate dimensions such that the following inequalities hold

$$\begin{bmatrix} \boldsymbol{R}_\sigma & \boldsymbol{S}_\sigma \\ \boldsymbol{S}_\sigma & \boldsymbol{R}_\sigma \end{bmatrix} \geq 0, \quad \boldsymbol{\Omega}_\sigma = \begin{bmatrix} \psi_\sigma & \tau \boldsymbol{G}_\sigma \boldsymbol{R}_\sigma \\ * & -\boldsymbol{R}_\sigma \end{bmatrix} < 0 \tag{1.14}$$

where

$$\begin{aligned} \psi_\sigma = {} & \ell_1^T(\lambda_\sigma \boldsymbol{P}_\sigma + \boldsymbol{Q}_\sigma)\ell_1 + G_\sigma P_\sigma \ell_1^T + \ell_1 \boldsymbol{P}_\sigma \boldsymbol{G}_\sigma^T \\ & - e^{-\tau\lambda_\sigma}\ell_3^T \boldsymbol{Q}_\sigma \ell_3 - \ell_4^T \boldsymbol{Q}_\sigma \ell_4 - e^{-\tau\lambda_\sigma}\begin{bmatrix} \ell_{23} \\ \ell_{12} \end{bmatrix}^T \begin{bmatrix} \boldsymbol{R}_\sigma & \boldsymbol{S}_\sigma \\ \boldsymbol{S}_\sigma & \boldsymbol{R}_\sigma \end{bmatrix} \begin{bmatrix} \ell_{23} \\ \ell_{12} \end{bmatrix}, \end{aligned}$$

and $\boldsymbol{G}_\sigma = \boldsymbol{A}\ell_1 + \boldsymbol{B}_\sigma \boldsymbol{M}_\sigma \boldsymbol{K}_\sigma \ell_3$, then the Lyapunov-Krasovskii functional $\boldsymbol{V}_\sigma(t)$ defined in (1.13) has the following properties

$$\dot{\boldsymbol{V}}_\sigma(t) + \lambda_\sigma \boldsymbol{V}_\sigma(t) \leq 0 \tag{1.15}$$

Proof: By $\boldsymbol{V}_\sigma(t)$ defined in (1.13), we can obtain

$$\begin{aligned} \dot{\boldsymbol{V}}_\sigma(t)\} + \lambda_\sigma \boldsymbol{V}_\sigma(t) \leq {} & 2\dot{x}_i^T(t)\boldsymbol{P}_\sigma x(t) - \lambda_\sigma \int_{t-\tau}^{t} e^{\lambda_\sigma(s-t)}x^T(s)\boldsymbol{Q}_\sigma x(s) \\ & + x^T(t)\boldsymbol{Q}_\sigma x(t) - x^T(t-\tau)\boldsymbol{Q}_\sigma x(t-\tau) - \lambda_\sigma \int_{-\tau}^{0}\int_{t+\theta}^{t} e^{\lambda_\sigma(s-t)}\dot{x}^T(s)\boldsymbol{R}_\sigma \dot{x}(s)d\theta ds \\ & + \tau \dot{x}^T(t)\boldsymbol{R}_\sigma \dot{x}(t) - e^{-\tau\lambda_\sigma}\xi_\sigma(t) + \lambda_\sigma \boldsymbol{V}_\sigma(t) \end{aligned} \tag{1.16}$$

where $\xi_\sigma(t) = \int_{t-\tau}^{t} \dot{x}^T(s)\boldsymbol{R}_\sigma \dot{x}(s)ds$.

Define $\phi(t) = \left[x^T(t) \quad x^T(t-d(t)) \quad x^T(t-\tau)\right]^T$.

Note that

$$\begin{aligned} \xi_\sigma(t) = {} & -\tau\int_{t-\tau}^{t-d(t)} \dot{x}^T(s)\boldsymbol{R}_\sigma \dot{x}(s)ds - \tau\int_{t-d(t)}^{t} \dot{x}^T(s)\boldsymbol{R}_\sigma \dot{x}(s)ds \\ \leq {} & -\frac{\tau}{\tau - d(t)}\phi^T(t)\ell_{23}^T \boldsymbol{R}_\sigma \ell_{23}\phi(t) - \frac{\tau}{d(t)}\phi^T(t)\ell_{12}^T \boldsymbol{R}_\sigma \ell_{12}\phi(t) \end{aligned}$$

Applying Lemma 1.1 to $\xi_\sigma(t)$, one can obtain

$$\xi_\sigma(t) \leq -\phi^T(t)\begin{bmatrix} \ell_{23} \\ \ell_{12} \end{bmatrix}^T \begin{bmatrix} \boldsymbol{R}_\sigma & \boldsymbol{S}_\sigma \\ \boldsymbol{S}_\sigma & \boldsymbol{R}_\sigma \end{bmatrix} \begin{bmatrix} \ell_{23} \\ \ell_{12} \end{bmatrix}\phi(t) \tag{1.17}$$

It is straightforward to know that $\dot{x}(t) = \boldsymbol{G}_\sigma \phi(t)$. Combining (1.15)–(1.17), one can obtain

$$\dot{V}_\sigma(t) + \lambda_\sigma V_\sigma(t) \leq \phi^T(t)\{\boldsymbol{\psi}_\sigma + \tau^2 \boldsymbol{G}_\sigma^T \boldsymbol{R}_\sigma \boldsymbol{G}_\sigma\}\phi(t) \tag{1.18}$$

If $\boldsymbol{\psi}_\sigma + \tau^2 \boldsymbol{G}_\sigma^T \boldsymbol{R}_i \boldsymbol{G}_\sigma < 0$, then $\dot{\boldsymbol{V}}_\sigma(t) + \lambda_\sigma \boldsymbol{V}_\sigma(t) \leq 0$ can be guaranteed. Using the Schur complement with matrix inequality $\boldsymbol{\Omega}_\sigma$ can yield the inequality $\dot{\boldsymbol{V}}_\sigma(t) + \lambda_\sigma \boldsymbol{V}_\sigma(t) \leq 0$. Hence, the proof of the lemma is completed.

We are now in a position to give the main results of this chapter.

Theorem 1.1 Consider a leader-following vehicular platoon control architecture with a varying communication range. For the scalars $\lambda_\sigma > 0$, $\kappa > 0$, and $\mu \geq 1$, if the average dwell time of the platoon systems (1.11) satisfies

$$T_a > T_a^*, T_a^* = \frac{\ln \mu}{\sum_{\sigma=0}^{N} \lambda_\sigma \alpha_\sigma} \tag{1.19}$$

and if there exist real matrices $\hat{\boldsymbol{P}}_\sigma > 0$, $\boldsymbol{Q}_\sigma > 0$, and $\boldsymbol{R}_\sigma > 0$, and real matrix $\hat{\boldsymbol{K}}_\sigma$, and $\boldsymbol{S}_\sigma$ is of appropriate dimensions such that the following inequalities hold

$$\begin{bmatrix} \boldsymbol{R}_\sigma & \boldsymbol{S}_\sigma \\ \boldsymbol{S}_\sigma & \boldsymbol{R}_\sigma \end{bmatrix} \geq 0, \hat{\boldsymbol{\Omega}}_\sigma = \begin{bmatrix} \psi_\sigma & \tau \hat{\boldsymbol{G}}_\sigma^T \\ * & 2\kappa \hat{\boldsymbol{P}}_\sigma - \kappa^2 \boldsymbol{R}_\sigma \end{bmatrix} \tag{1.20}$$

$$\boldsymbol{P}_i \leq \mu \boldsymbol{P}_j, \boldsymbol{Q}_i \leq \mu \boldsymbol{Q}_j, \boldsymbol{R}_i \leq \mu \boldsymbol{R}_j, \forall i,j \in \{0, 1, 2, 3, \ldots, N\} \tag{1.21}$$

where

$$\begin{aligned} \boldsymbol{\psi}_\sigma &= \ell_1^T(\lambda_\sigma \hat{\boldsymbol{P}}_\sigma + \boldsymbol{Q}_\sigma)\ell_1 + \hat{\boldsymbol{G}}_\sigma \ell_1^T + \ell_1 \hat{\boldsymbol{G}}_\sigma^T \\ &\quad - e^{-\tau\lambda_\sigma} \ell_3^T \boldsymbol{Q}_\sigma \ell_3 - \ell_4^T \boldsymbol{Q}_\sigma \ell_4 - e^{-\tau\lambda_\sigma} \begin{bmatrix} \ell_{23} \\ \ell_{12} \end{bmatrix}^T \begin{bmatrix} \boldsymbol{R}_\sigma & \boldsymbol{S}_\sigma \\ \boldsymbol{S}_\sigma & \boldsymbol{R}_\sigma \end{bmatrix} \begin{bmatrix} \ell_{23} \\ \ell_{12} \end{bmatrix} \\ \hat{\boldsymbol{G}}_\sigma &= \boldsymbol{A}_\sigma \ell_1 \hat{\boldsymbol{P}}_\sigma + \boldsymbol{B}_\sigma \boldsymbol{M}_\sigma \hat{\boldsymbol{K}}_\sigma \ell_3 \end{aligned}$$

then each following vehicle can be exponentially stabilized with the decay rate $\rho/2$, where $\rho = (1/T_a)\ln \mu - \sum_{\sigma=0}^{N} \lambda_\sigma \alpha_\sigma$, and the state decay estimation is given by

$$\|x(t)\| < \sqrt{b/a}\, e^{\frac{\rho(t-t_0)}{2}} \|\phi(t_0)\| \tag{1.22}$$

where

$$a = \min\{\lambda_{\min}(\boldsymbol{P}_\sigma)\}$$

and

$$b = \max\{\lambda_{\max}(\boldsymbol{P}_\sigma) + \tau \lambda_{\max}(\boldsymbol{Q}_\sigma) + \tau^2 \lambda_{\max}(\boldsymbol{R}_\sigma)/2\}.$$

Moreover, the connectivity status-dependent platoon controller can be solved by

$$\boldsymbol{K}_\sigma = \hat{\boldsymbol{P}}_\sigma^{-1} \hat{\boldsymbol{K}}_\sigma \tag{1.23}$$

Proof: First, we will prove that the controller gain obtained in (1.23) can guarantee property (1.15) for the Lyapunov-Krasovskii functional $V_\sigma(t)$ defined in (1.13) if condition (1.20) of Theorem 1.1 is satisfied. Let $\hat{P}_\sigma = P_\sigma$ in inequality (1.14) of Lemma 1.1. Define variables

$\hat{\boldsymbol{K}}_\sigma = \hat{\boldsymbol{P}}_\sigma \boldsymbol{K}_\sigma$ and $\boldsymbol{S} = \text{diag}\{\boldsymbol{I}_n, \boldsymbol{R}_\sigma^{-1}\hat{\boldsymbol{P}}_\sigma\}$. By pre- and post-multiplying both sides of (1.14) by S, one can obtain (1.20), where the term $\hat{\boldsymbol{P}}_\sigma \boldsymbol{R}_\sigma^{-1}\hat{\boldsymbol{P}}_\sigma$ deals with inequality $\hat{\boldsymbol{P}}_\sigma \boldsymbol{R}_\sigma^{-1}\hat{\boldsymbol{P}}_\sigma \leq \kappa^2 \boldsymbol{R}_\sigma - 2\kappa \hat{\boldsymbol{P}}_\sigma$. Since the conditions in Lemma 1.1 are satisfied, the inequality (1.15) holds.

Next, we prove that the switched platoon control system (1.11) is exponentially stable. Let $0 < t_1 < t_2, \ldots, < t_k < t$ denote the time switching instant of the switching signal $\sigma(t)$. According to (1.15), we can obtain

$$\boldsymbol{V}_{\sigma(t)}(t) \leq e^{-\lambda_{\sigma(t_k)}(t - t_k)} \boldsymbol{V}_{\sigma(t_k)}(t_k) \tag{1.24}$$

Combined with conditions (1.21) of Theorem 1.1 and the Lyapunov-Krasovskii functional $\boldsymbol{V}_{\sigma(t)}(t)$ defined in (1.13), one can obtain

$$\boldsymbol{V}_i(t) \leq \mu \boldsymbol{V}_j(t), \quad \forall i,j \in \{0, 1, 2, \cdots N\} \tag{1.25}$$

for any $x_i(t)$.

According to (1.25), we have

$$\boldsymbol{V}_{\sigma(t)}(t) \leq \mu e^{-\lambda_{\sigma(t_k)}(t - t_k)} \boldsymbol{V}_{\sigma(\bar{t}_k)}(t_k)$$

where $\bar{t}_k$ denotes the time instant is immediately before the time instant t_k.

It is easy to know that

$$\boldsymbol{V}_{\sigma(t)}(t) \leq \mu^2 e^{-\lambda_{\sigma(t_k)}(t - t_k)} e^{-\lambda_{\sigma(t_{k-1})}(t_k - t_{k-1})} \boldsymbol{V}_{\sigma(t_{k-1})}(t_{k-1}) \tag{1.26}$$

Applying (1.26) recursively, one obtains

$$\begin{aligned}
\boldsymbol{V}_{\sigma(t)}(t) &\leq \mu^{N(t)} e^{-\lambda_{\sigma(t_k)}(t - t_k)} e^{-\lambda_{\sigma(t_{k-1})}(t_k - t_{k-1})} e^{-\lambda_{\sigma(t_{k-2})}(t_{k-1} - t_{k-2})} \cdots e^{-\lambda_{\sigma(t_0)}(t_1 - t_0)} \boldsymbol{V}(t_0) \\
&\leq \mu^{(t - t_0)/T_a} e^{\sum_{\sigma=0}^{N} - \lambda_\sigma \alpha_\sigma (t - t_0)} \boldsymbol{V}(t_0) = \left(\mu^{1/T_a} e^{\sum_{\sigma=0}^{N} - \lambda_\sigma \alpha_\sigma}\right)^{t - t_0} \boldsymbol{V}(t_0) \\
&\leq \left(e^{(1/T_a)\ln \mu - \sum_{\sigma=0}^{N} \lambda_l \alpha_l}\right)^{t - t_0} \boldsymbol{V}(t_0) = e^{\rho(t - t_0)} \boldsymbol{V}(t_0)
\end{aligned} \tag{1.27}$$

where $\rho = (1/T_a)\ln \mu - \sum_{\sigma=0}^{N} \lambda_\sigma \alpha_\sigma$. The condition (1.19) of Theorem 1.1 guarantees that $\rho \leq 0$.

If a quadratic piecewise Lyapunov-Krasovskii functional $\boldsymbol{V}_{\sigma(t)}(t)$ is considered, then there exist $a > 0$ and $b > 0$ such that

$$a\|x(t)\|^2 \leq \boldsymbol{V}_{\sigma(t)}(t) < b e^{\rho(t - t_0)} \|\phi(t_0)\|^2, \tag{1.28}$$

where $a\|x(t)\|^2 \leq \boldsymbol{V}_{\sigma(t)}(t)$, $\boldsymbol{V}_{\sigma(t)}(t) < b\|\phi(t_0)\|^2$.

By inequality (1.28), we further obtain

$$\|x(t)\| < \sqrt{b/a} e^{((\rho(t - t_0))/2)} \|\phi(t_0)\|, \tag{1.29}$$

which means that each vehicle can be mean-square exponentially stabilized with a decay rate of $\rho/2$.

The controller obtained in the previous section can ensure the intervehicle stability of each following vehicle. However, the intervehicle stability does not guarantee the string stability for the dynamically coupled platoon control system. Therefore, additional conditions are needed to complement the obtained controller in order to satisfy the string stability and zero steady-state error requirements. Since string stability is highly dependent on the interaction topology of platoon, how to analyze string stability for a switched platoon control system still remains an open and challenging problem [12, 30]. Consequently, the discussions on string stability and the problem of zero steady spacing error have mainly focused on each obtained connectivity-status-matrix-dependent controller gain.

According to (1.3), we have

$$\dddot{\delta}_r(t) = \dot{a}_{r-1}(t) - \dot{a}_r(t) \tag{1.30}$$

Taking Laplace transforms to (1.30), the transform function $\delta_r(s)/\delta_{r-1}(s)$ can be obtained. For the varying communication range, we consider the following case for obtaining transform function $\delta_r(s)/\delta_{r-1}(s)$.

Case 1.1

Both follower i and its preceding vehicle (follower $i-1$) are within the communication range of the leader. Then, we have

$$G_1(s) = \frac{k_\sigma^p + k_\sigma^v s + k_\sigma^a s^2}{(\varsigma s^3 + s^2)e^{\tau s} + k_\sigma^p + k_\sigma^v s + k_\sigma^a s^2 + k_\sigma^{ve} s + k_\sigma^{ae} s^2} \tag{1.31a}$$

Case 1.2

Neither follower i nor its preceding vehicle (follower $i-1$) are within the communication range of the leader. Under such circumstance, $v_{0r} = v_{0(r-1)} = 0$ and $a_{0r} = a_{0(r-1)} = 0$, then one can derive

$$G_2(s) = \frac{k_\sigma^p + k_\sigma^v s + k_\sigma^a s^2}{(\varsigma s^3 + s^2)e^{\tau s} + k_\sigma^p + k_\sigma^v s + k_\sigma^a s^2} \tag{1.31b}$$

Case 1.3

Follower $i-1$ are within the communication range of the leader, but follower i is out, then one can obtain

$$G_3(s) = \frac{k_\sigma^p + k_\sigma^v s + k_\sigma^a s^2}{(\varsigma s^3 + s^2)e^{\tau s} + k_\sigma^p + k_\sigma^v s + k_\sigma^a s^2 + k_\sigma^{ve} s \sum_{j=1}^{r-1} \delta_j(s) + k_\sigma^{ae} s^2 \sum_{j=1}^{r-1} \delta_j(s)} \tag{1.31c}$$

where $v_0(t) - v_{r-1}(t) = \sum_{j-1}^{r-1} \dot{\delta}_j(t)$ and $a_0(t) - a_{r-1}(t) = \sum_{j-1}^{r-1} \ddot{\delta}_j(t)$ is used in (1.31c).

In the next, we want to illustrate that (1.31c) can be seen as a special case of (1.31b). Considering the term $\sum_{j=1}^{r-1} s\delta_j(s)$ in the denominator of transform function $G_3(s)$, one can obtain that

$$\sum_{j=1}^{r-1} s\delta_j(s) = \frac{sG_1(s)\delta_{r-1}(s) - sG_1^{1-r}(s)\delta_{r-1}(s)}{G_1(s)-1},$$

where $(\delta_j(s))/(\delta_{j-1}(s)) = G_1(s)$ is used. According to the final theorem, we can derive that $\lim_{s\to 0} sG_1(s)\delta_{r-1}(s) = 0$, which means $\lim_{s\to 0}\sum_{j=1}^{r-1} s\delta_j(s) = 0$. Similarly, it is easy to know that $\lim_{s\to 0}\sum_{j=1}^{r-1} s^2\delta_j(s) = 0$. Therefore, transform function (1.31c) is equal to (1.31b).

As previously mentioned, we have the following sufficient conditions about string stability condition $|\delta_r(jw)/\delta_{r-1}(jw)| \le 1$.

Theorem 1.2 For the spacing error transform function (1.31), if the connectivity-status-dependent feedback controller satisfies the following conditions:

$$\begin{aligned}
&\text{a)}\quad k_\sigma^p > 0\\
&\text{b)}\quad \left(1 + k_\sigma^a + d^* k_\sigma^{ae}\right)^2 - \left(k_\sigma^a\right)^2 > 0\\
&\text{c)}\quad k_\sigma^a + d^* k_\sigma^{ae} < 0\\
&\text{d)}\quad \varsigma\left(k_\sigma^v + d^* k_\sigma^{ve}\right) - k_\sigma^p = 0\\
&\text{e)}\quad 1 + \left(k_\sigma^v + d^* k_\sigma^{ve}\right) - \left(k_\sigma^v\right)^2 - 2d^* k_\sigma^{ae} k_\sigma^p > 0
\end{aligned} \tag{1.32}$$

then condition $|\delta_r(jw)/\delta_{r-1}(jw)| \le 1$ always holds true for each obtained feedback controller gain, where $d^* = 1$ if $\gamma_i(t) = 1$; otherwise $d^* = 0$.

Proof: $|\delta_r(jw)/\delta_{r-1}(jw)| \le 1$ is rewritten as:

$$|G(jw)| = \left|\frac{\delta_r(jw)}{\delta_{r-1}(jw)}\right| = \sqrt{\frac{a}{a+b}}$$

where

$$a = \left[\left(k_\sigma^p - k_\sigma^a w^2\right)^2 + \left(k_\sigma^v w\right)^2\right]$$

$$\begin{aligned}
b = {} & w^4\left[1 + 2\left(k_\sigma^a + d^* k_\sigma^{ae}\right)\cos(w\tau) + \left(d^* k_\sigma^{ae}\right)^2 + 2d^* k_\sigma^a k_\sigma^{ae}\right]\\
& + w^3 2\sin(w\tau)\left[\left(k_\sigma^v + d^* k_\sigma^{ve}\right) - \varsigma\left(k_\sigma^a + d^* k_\sigma^{ae}\right)\right]\\
& + w^2\left\{1 - \left(k_\sigma^v\right)^2 + \left(k_i^v + d^* k_\sigma^{ve}\right)[1 + 2\varsigma\cos(w\tau)] - 2\left[\cos(w\tau) + d^* k_\sigma^{ae}\right]k_\sigma^p\right\} + 2\varsigma k_\sigma^p w\sin(w\tau)
\end{aligned}$$

It can be clearly seen that $a > 0$. In order to guarantee $|G(jw)|$ holds true, we need to prove that the condition in Theorem 1.2 can make $b \ge 0$ hold true. According to condition (a) of (1.32) and using $\cos(w\tau) \le 1$ and $\sin(w\tau) \le w\tau \le 1$, for $w > 0$, we have

$$b \geq w^4\left[1 + 2\left(k_\sigma^a + d^*k_\sigma^{ae}\right) + \left(d^*k_\sigma^{ae}\right)^2 + 2k_\sigma^a d^* k_\sigma^{ae}\right] + 2w^4\tau\left[\left(k_\sigma^v + d^*k_\sigma^{ve}\right)\right.$$
$$\left. - \varsigma\left(k_\sigma^a + d^*k_\sigma^{ae}\right)\right] + w^2 2\cos(w\tau)\left[\varsigma\left(k_\sigma^v + d^*k_\sigma^{ve}\right) - k_\sigma^p\right] \tag{1.33}$$
$$w^2\left\{1 + \left(k_\sigma^v + d^*k_\sigma^{ve}\right) - \left(k_\sigma^v\right)^2 - 2d^*k_\sigma^{ae}k_\sigma^p\right\} + 2\varsigma k_\sigma^p w^2\tau$$

Then, according to (b)–(e) of (1.32), we known that $b \geq 0$, which completes the proof.

Next, we will show that the zero steady-state spacing error can be guaranteed by the obtained feedback controller. Define ξ_0 to denote the deviation of the lead vehicle's velocity from its steady-state value at time t, i.e. $\xi_0(t) = v_0(t) - v_0$. Based on (1.1), we have

$$\dddot{\delta}_1(t) = \ddot{\xi}_0(t) - \dot{a}_1(t) \tag{1.34}$$

Substituting differential equations (1.4) and feedback controller (1.10) into (1.34) with $r = 1$, we have the following spacing error equation of the first following vehicle:

$$\varsigma\dddot{\delta}_1(t) = \varsigma\ddot{\xi}_0(t) + \dot{\xi}_0(t) - \ddot{\delta}_1(t) - \left[k_\sigma^p\delta_1(t-\tau) + \left(k_\sigma^v + k_\sigma^{ve}\right)\dot{\delta}_1(t-\tau) + \left(k_\sigma^a + k_\sigma^{ae}\right)\ddot{\delta}_1(t-\tau)\right] \tag{1.35}$$

Taking the Laplace transform at both sides of (1.35), we can describe the spacing error for the fist follower with respect to the lead vehicle's speed change as follows

$$H(s) = \frac{\delta_1(s)}{\xi_0(s)} = \frac{\varsigma s^2 + s}{\varsigma s^3 + s^2 + \left[\left(k_\sigma^a + k_\sigma^{ae}\right)s^2 + \left(k_\sigma^v + k_\sigma^{ve}\right)s + k_\sigma^p\right]e^{-\tau s}} \tag{1.36}$$

Suppose that the lead vehicle can reach its final steady-state value in a finite time t_f. Therefore, we have $\xi_0(t) - \xi_0(t_f) = 0$ for all $t \geq t_f$, which means

$$\xi_0(s) = \frac{\xi_0(t_f)}{s} \tag{1.37}$$

Based on the final theorem and (1.35)–(1.37), we obtain

$$\begin{aligned}\lim_{t\to\infty}\delta_r(t) &= \lim_{s\to 0} s\frac{\delta_r(s)}{\delta_{r-1}(s)}\frac{\delta_{r-1}(s)}{\delta_{r-2}(s)}\cdots\frac{\delta_r(s)}{\xi_0(s)}\xi_0(s) \\ &= \lim_{s\to 0} sG^{i-1}(s)H(s)\xi_0(s) = 0\end{aligned} \tag{1.38}$$

Therefore, a zero steady-state error is achieved for vehicle r.

Remark 1.1 These discussions on string stability and zero steady spacing error are not equal to the case of practical switched connected vehicular control system. The main reason is that the switching of different mode of feedback controller is avoided. As a result, the relationship among mode rate, average dwell time, and the string stability is not established. Moreover, we cannot determine the number of followers in the LPF topology that can guarantee string stability of entire platoon when several of the followers are in the PF topology. According to [12], this number has the effect on the string stability performance of the platoon. Those problems will be investigated in our future research work.

Algorithm 1.1

Platoon control algorithm

1) Design a spacing, velocity, and acceleration-error-based feedback controller in (1.9) for platoon control.
2) According to Theorems 1.1 and 1.2, the linear matrix inequality technique is used to obtain the connectivity-status-dependent feedback controller gain K_σ that satisfies condition (d) of Theorem 1.2, $\sigma \in \{0, 1, 2, \ldots N\}$.
3) Using conditions (a), (b), (c), and (e) are given in Theorem 1.2 to constraint the obtained controller gain $\boldsymbol{K}_\sigma$. If this is feasible, then the resulting controller gain can be used for string stabilizing control. Otherwise, reset the matrices $\boldsymbol{P}_\sigma$, $\boldsymbol{Q}_\sigma$, $\boldsymbol{R}_\sigma$ and other related parameters, and return to step 2).

1.4 Simulations and Experiments

Through a numerical example, we first show how we can apply the proposed controller to a five-vehicle platoon control system with a varying communication range in the LF topology. In the simulation, the length of the vehicle is chosen as $L = 4.46$ m. The engine time constant is $\varsigma = 0.25$. During time intervals (0, 90 s), the prescribed acceleration of the leader in the simulation is given as

$$a_0 = \begin{cases} 0\ \text{m/s}^2 & 0\ \text{s} \le t < 25\ \text{s} \\ -0.25\ \text{m/s}^2 & 25\ \text{s} \le t < 30\ \text{s} \\ 0.25\ \text{m/s}^2 & 30\ \text{s} \le t < 40\ \text{s} \\ 0.35\ \text{m/s}^2 & 40\ \text{s} \le t < 50\ \text{s} \\ 0\ \text{m/s}^2 & t \ge 50\ \text{s} \end{cases}.$$

According to the leader's acceleration a_0, the maximum velocity of platoon is 7.5 m/s, approximately. With maximum deceleration $\kappa_1 = \kappa_2 = 7$ m/s^2, minimum relative velocity $\Delta v_r = 0$ m/s, delay time $\tau' = 0.1$ s, and minimum distance $d_0 = 5$ m, one can derive the maximum braking distance $d_{brk} = 9.8$ m by (1.5). Therefore, the desired intervehicle spacing can be chosen as $\delta_d = 10\,m$.

On the other hand, it is assumed that the LTE-V2V network has a carrier frequency of 2 GHz and bandwidth of 10 MHz. Usually, the effective communication range of the LTE-V2V network is 320 m with an absolute speed of 70–140 km/h and a relative speed 140–280 km/h for the platoon at the motorway. The transmission power is $p_{V2V} = 10$ dBm and the targeted SINR threshold is $\gamma_{V2V} = 10$ dB. According to (1.8) with the parameters of the V2V channel model specified in [31], we know that $\gamma_i > \gamma_{V2V}$. So, the data transmission rate in the communication channel is reliable under the velocity of the leader.

The parameters needed for the controller design are given as $\lambda_1 = \lambda_2 = \lambda_3 = \lambda_4 = 0.8$ and $\mu = 1.2$. Using the proposed platoon control algorithm, we obtain a series of connectivity-state-matrix-dependent controllers as follows:

$$\overline{K}_0^r = \text{diag}\{4, 2, -3, 30, 2\},\ \overline{K}_1^r = \text{diag}\{3, 5, -1, 18, -2\},\ \overline{K}_2^r = \text{diag}\{1, 1.5, -3, 6.5, 2\},$$
$$\overline{K}_3^r = \text{diag}\{0.5, 2, 1, 2, -5\},\ \text{and}\ \overline{K}_4^r = \text{diag}\{0.25, 1, 4, 1, -6\}.$$

The connectivity status of the follower with the leading one is shown in Figures 1.5a–1.7a, where $\sigma(t) = i$, $i = 0, 1, 2, 3, 4$, means that there are i successive following vehicles that can communicate with the leading vehicle. The mode rate α_σ of Figures 1.5a–1.7a is shown in Table 1.1. The average dwell time of the successive switching in Figures 1.5a–1.7a are, respectively, 1.5, 1.67, and 2.25 s, which are longer than the value $T_a^* = 0.23$ s. By using the obtained platoon controller, the spacing error, velocity, and acceleration of each follower with different mode rates α_σ are given in Figures 1.5b–d, 1.6b–d, and 1.7b–d. From these figures, it is clear that the proposed connectivity-status-dependent feedback controller can effectively stabilize the tracking error. Compared with the results in Figures 1.5–1.7, we found that the different values of mode rate α_σ may affect the stability performance of the platoon with respect to the maximum spacing error, the convergence rate of the spacing error, performance in tracking the velocity, and acceleration of the leader. To be specific,

1) For the same mode rate α_0, a larger mode rate α_4 has better string stability performance (see Figures 1.5 and 1.6).

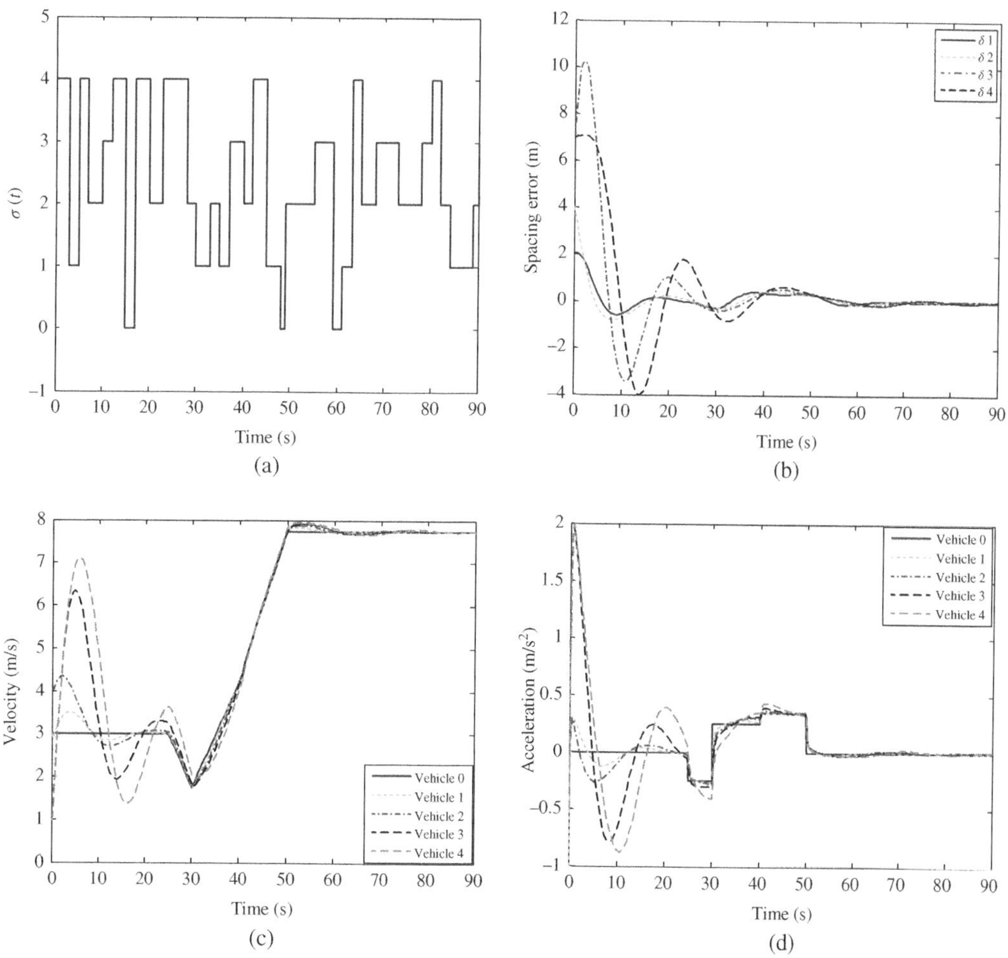

Figure 1.5 Five-vehicle platoon with varying communication range in Case 1.1: (a) connectivity status of vehicle platoon, (b) spacing error, (c) velocity, and (d) acceleration.

Table 1.1 Mode rate α_i in platoon control.

Case	α_0	α_1	α_2	α_3	α_4
1	5/90	17/90	29/90	16/90	23/90
2	5/90	13/90	12/90	11/90	49/90
3	41/90	0	0	0	49/90

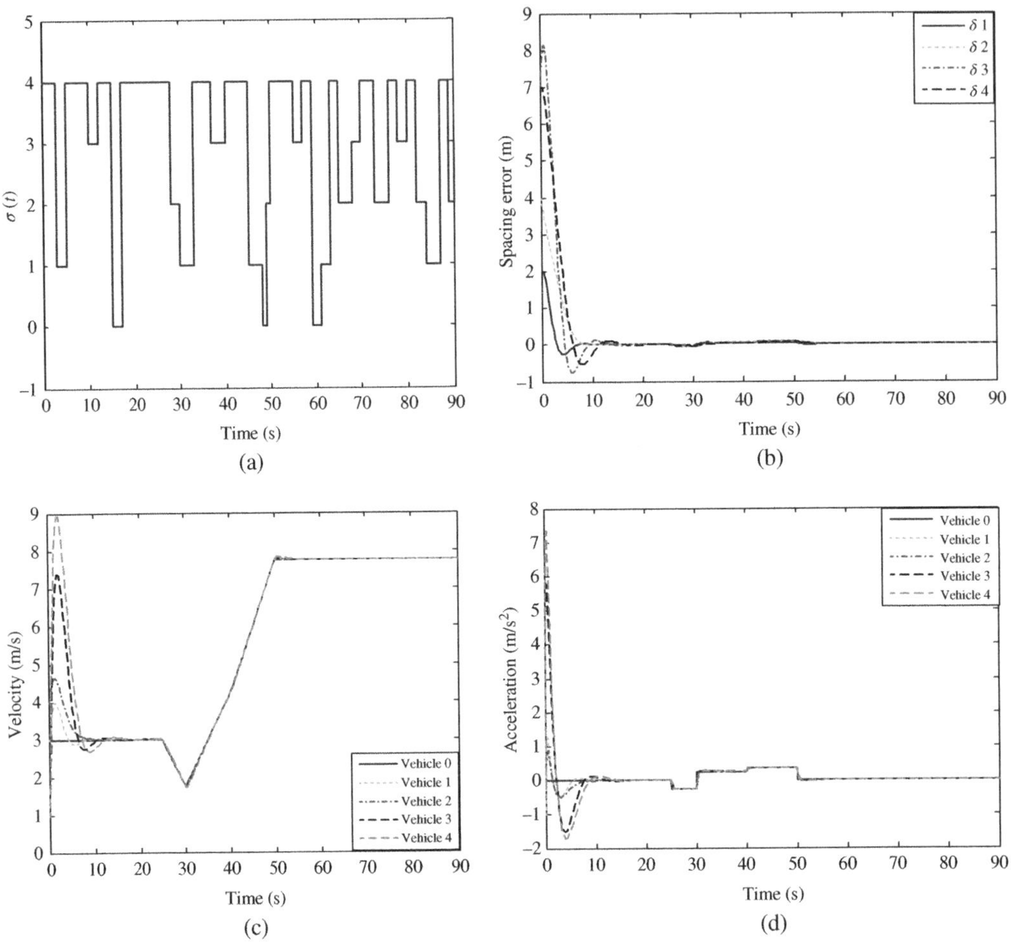

Figure 1.6 Five-vehicle platoon with varying communication range in Case 1.2: (a) connectivity status of the vehicle platoon, (b) spacing error, (c) velocity, and (d) acceleration.

2) For the same mode rate α_4, a smaller mode rate α_0 has better string stability performance (see Figures 1.6 and 1.7).
3) For a different mode rate α_0 and α_4, a larger value of $\sum_{j=1}^{4} \alpha_j$ has better string stability performance (see Figures 1.5 and 1.7).

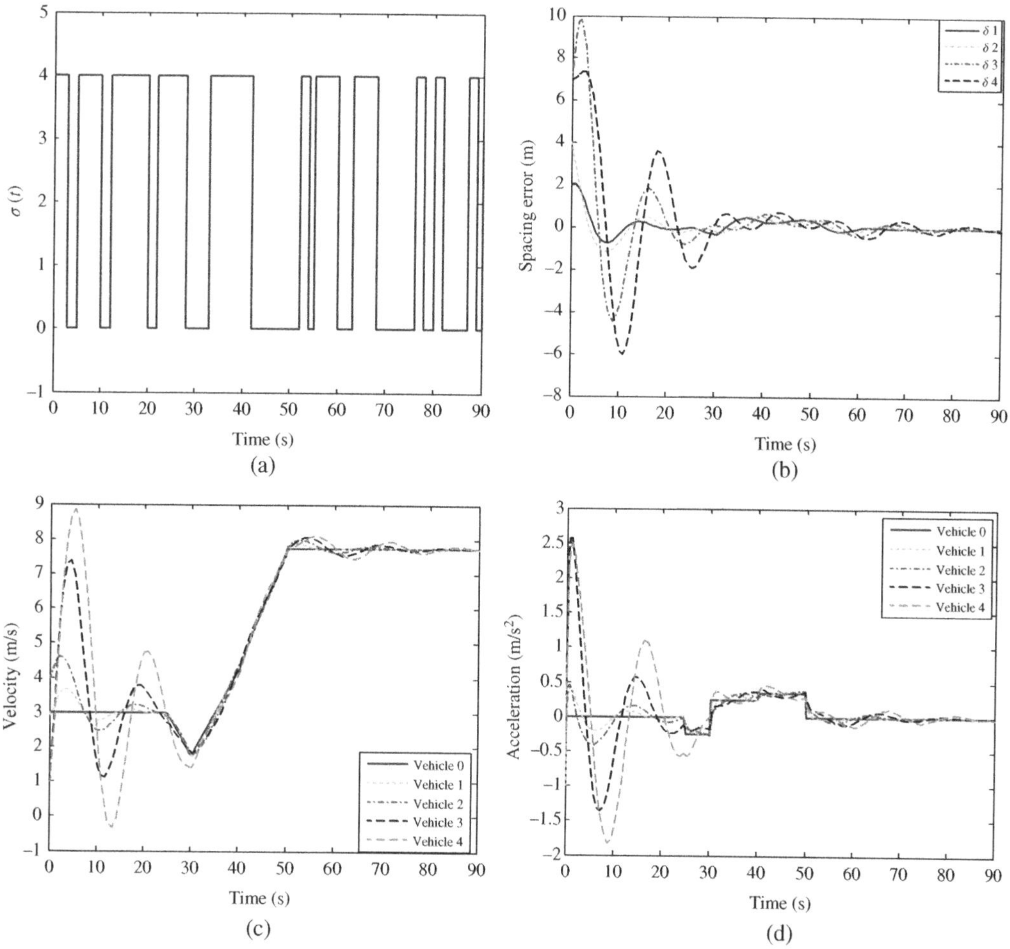

Figure 1.7 Five-vehicle platoon with varying communication range in Case 1.3: (a) connectivity status of the vehicle platoon, (b) spacing error, (c) velocity, and (d) acceleration.

Therefore, the longer the total time in which all followers in the platoon obtain information from the leader, the better the stability performance of the platoon.

In the following, we give the comparative simulation study with the sliding-model controller designed in [32]. The connectivity status of the followers is given in Figure 1.8a, in which the connectivity status of the followers is in the normal case during the time interval (0, 30 s) but changes over the rest of time. For the same situation, when adopting the sliding-model controller in [32], we can see that the platoon can only guarantee the stability requirement during the time interval (0, 30 s), as shown in Figure 1.8d, e. After 30 s, the follower cannot accurately track the velocity and acceleration of the leader. Moreover, the spacing error is unstable, as shown in Figure 1.8a–c. Although the sliding-model controller in [32] is good at its robustness, the stability performance of the platoon is degraded under the effect of the varying communication range of the leader. Consequently, it is necessary to consider the effect of the varying communication range for the LF platoon.

We use an Arduino car's experiment (in Figure 1.9) to show that the proposed controller design method is effective for a practical connected vehicular control system with a varying

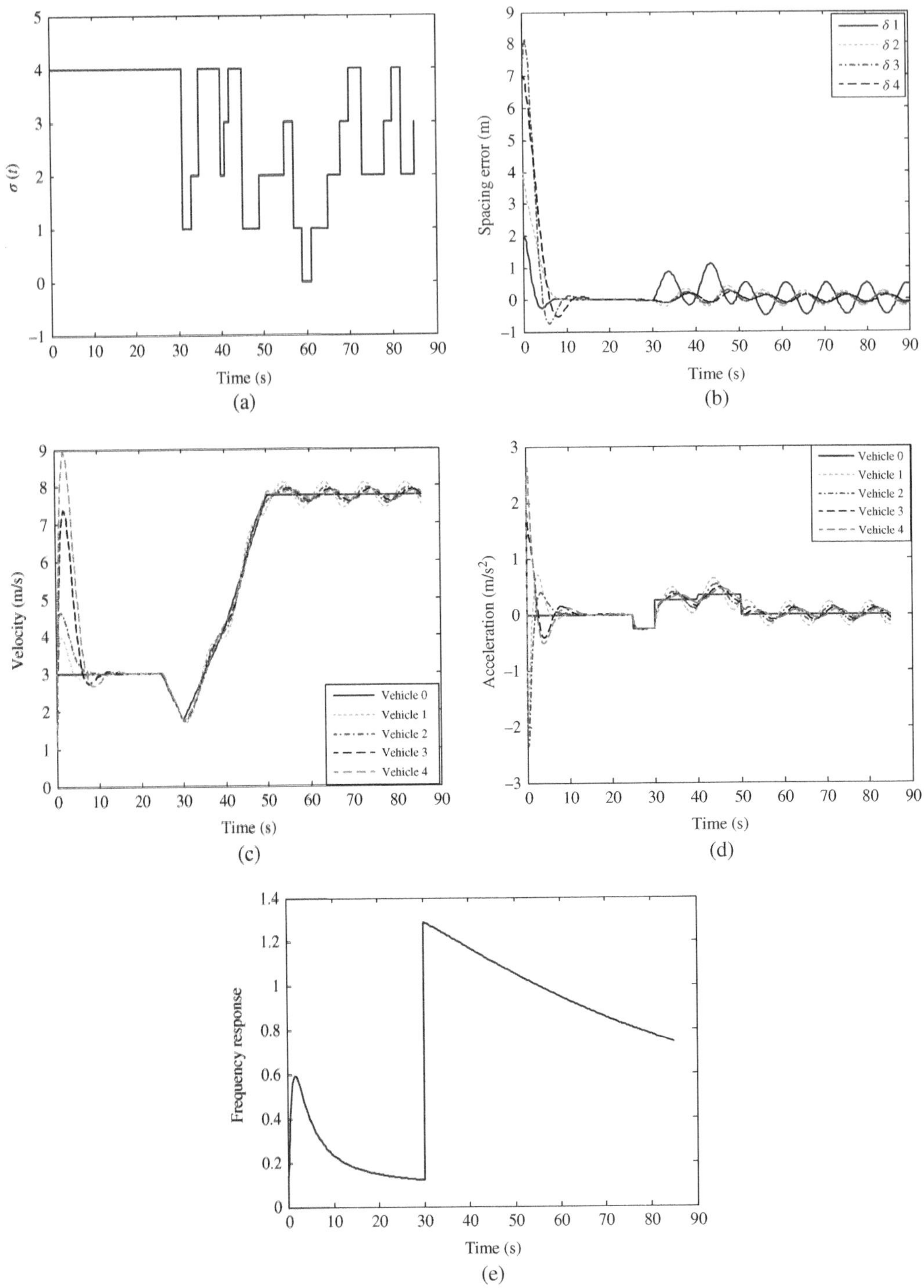

Figure 1.8 Five-vehicle platoon under sliding-model controller in [32]: (a) connectivity status of the vehicle platoon, (b) spacing error, (c) velocity, (d) acceleration, and (e) frequency response $\|G(s)\| = \|\delta_r(s)/\delta_{r-1}(s)\|$. *Source:* Adapted from Yanakiev and Kanellakopoulos [32].

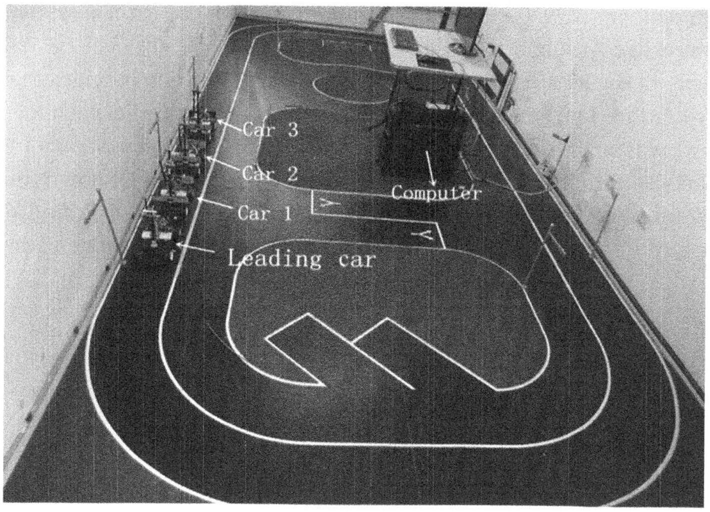

Figure 1.9 Platoon of Arduino cars on the traffic control platform.

communication range when using the PF topology. Each Arduino car uses infrared sensors GP2D12 produced by Sharp Corporation for distance measurement. The speed and acceleration are measured by an incremental encoder installed on the shaft of the rear wheels and an acceleration sensor installed on the top of the vehicle, respectively. The leading Arduino car transmits the velocity and acceleration to each following car via the wireless communication module APC220. The data transmission rate of APC220 is 1200–19,200 bps with a maximum transmission range of 1000 m. The onboard processor generates the control command to steer the two wheels of the Arduino car. In the experiment, the transmission rate of Arduino car to Arduino car communication is 7800–8900 bps, where the packet dropout rate during the data transmission is given in Figure 1.10. As shown in Figure 1.10, Arduino car-to-Arduino car communication has a reliable data transmission when they run at 15 cm/s approximately. Therefore, the velocity of the leader car for Arduino car platoons is chosen as 15 cm/s. On the other hand, the maximum deceleration of the Arduino car is 5 cm/s, delay time in its braking system is 0.15 s. According to (1.5), the

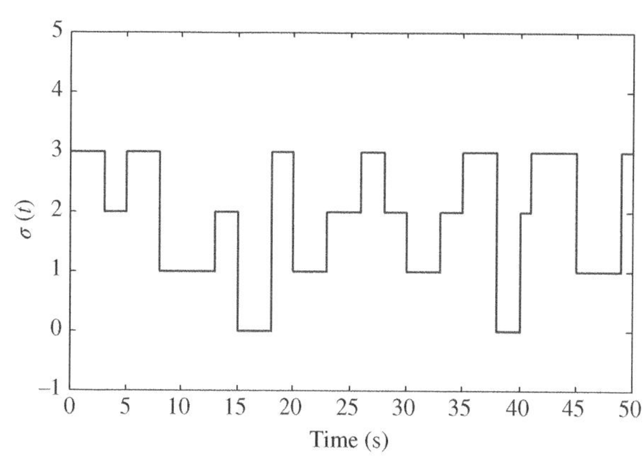

Figure 1.10 Packet dropout rate of the data transmission during Arduino car to Arduino car communication within 30–40 cm.

braking distance for Arduino car platoons is 29.75 cm with a minimum distance of 5 cm. Therefore, the desired spacing between successive Arduino cars is set as 30 cm.

In the experiments, the velocity of the leading car is given in Figure 1.12b, which includes the acceleration and deceleration situations for the Arduino cars. The connectivity status of each following vehicle with the leading car is simulated by a program in a computer, which is shown in Figure 1.11. The distances between the Arduino cars and their velocities are shown in Figure 1.12a, b, respectively. From these figures, we observe that each Arduino car can rapidly and accurately track the velocity of the leading car at the desired constant distance. The

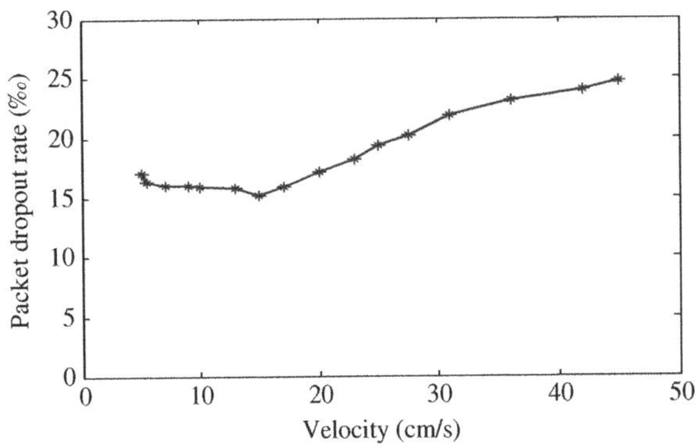

Figure 1.11 Connectivity status of leader-following Arduino cars in an experiment.

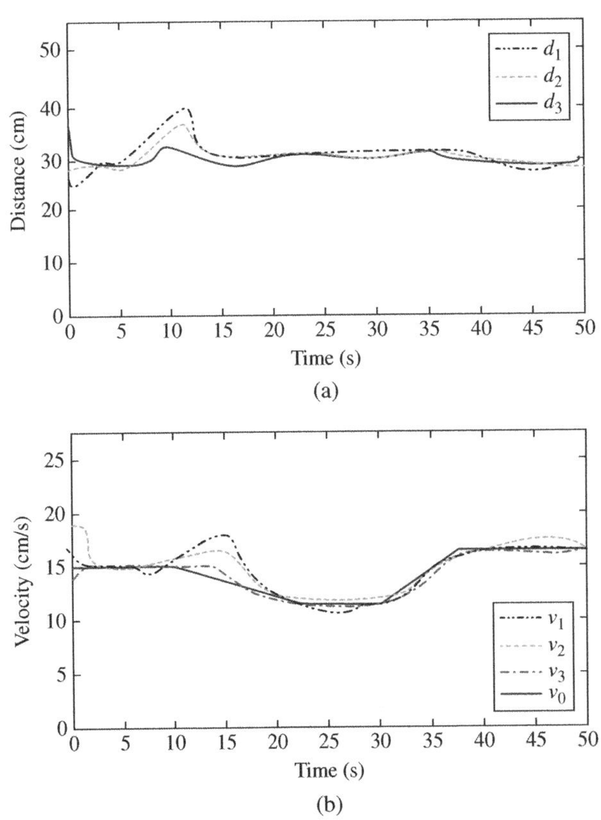

Figure 1.12 (a) Distance and (b) velocity of Arduino cars with the proposed controller.

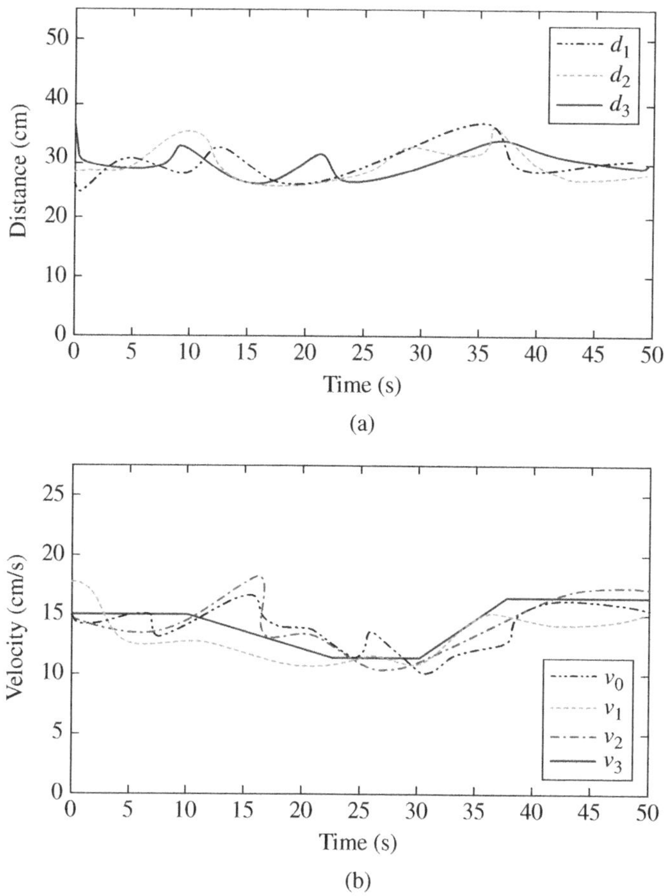

Figure 1.13 (a) Distance and (b) velocity of Arduino cars with the sliding-model controller. *Source:* Adapted from Guo and Yue [21].

proposed controller design method is effective. Under the same situation, when the sliding-model controller [32] is used, the distances and velocities of the Arduino cars are shown in Figure 1.13. Under the effect of a varying communication range, the spacing error is amplified downstream and the velocity response is unstable.

1.5 Conclusions and Future Work

In this chapter, platoon control with a varying communication range was addressed using the LF topology, where a prescribed intervehicle spacing policy is adopted. A switched platoon control model was established that reflected the influence of following vehicles' connectivity status on the platoon controller. A series of conditions on stability analysis and feedback controller design was obtained by using the switched system method. Then, a practical platoon control algorithm was proposed based on the obtained conditions. By using the proposed algorithm, the gain of the connectivity-status-matrix-dependent controller was obtained. This can guarantee string stability and zero steady-state spacing error for each obtained controller. The effectiveness of the established algorithm was shown by a numerical experiment and laboratory-scale Arduino car experiment.

References

1 Wen, S., Guo, G., Chen, B., and Gao, X. (2019). Cooperative adaptive cruise control of vehicles using a resource-efficient communication mechanism. *IEEE Trans. Intell. Veh.*, 4 (1): 127–140.

2 Kwon, J.-W. and Chwa, D. (2014). Adaptive bidirectional platoon control using a coupled sliding mode control method. *IEEE Trans. Intell. Transp. Syst.* 15 (5): 2040–2048.

3 Dolk, V.S., Ploeg, J., and Heemels, W.P.M.H. (2017). Event-triggered control for string-stable vehicle platooning. *IEEE Trans. Intell. Transp. Syst.* 18 (12): 3486–3500.

4 Vahidi, A. and Eskandarian, A. (2003). Research advances in intelligent collision avoidance and adaptive cruise control. *IEEE Trans. Intell. Transp. Syst.* 4 (3): 143–153.

5 Li, S., Li, K., Rajamani, R., and Wang, J. (2011). Model predictive multi-objective vehicular adaptive cruise control. *IEEE Trans. Control Syst. Technol.* 19 (3): 556–566.

6 Nieuwenhuijze, M.R.I., van Keulen, T., Öncü, S. et al. (2012). Cooperative driving with a heavy-duty truck in mixed traffic: experimental results. *IEEE Trans. Intell. Transp. Syst.* 13 (3): 1026–1032.

7 van Arem, B., van Driel, C.J.G., and Visser, R. (2006). The impact of cooperative adaptive cruise control on traffic-flow characteristics. *IEEE Trans. Intell. Transp. Syst.* 7 (4): 429–436.

8 Jia, D., Lu, K., Wang, J. et al. (2016). A survey on platoon-based vehicular cyber-physical systems. *IEEE Commun. Surveys Tuts.* 18 (1): 263–284.

9 Ghasemi, A., Kazemi, R., and Azadi, S. (2013). Stable decentralized control of platoon of vehicles with heterogeneous information feedback. *IEEE Trans. Veh. Tech.* 62 (9): 4299–4308.

10 Milanes, V., Shladover, S.E., Spring, J. et al. (2014). Cooperative adaptive cruise control in real traffic situations. *IEEE Trans. Intell. Transp. Syst.* 15 (1): 296–305.

11 Zheng, Y., Li, S.E., and Wang, J. (2016). Stability and scalability of homogeneous vehicular platoon: Study on the influence of information flow topologies. *IEEE Trans. Intell. Transp. Syst.* 17 (1): 14–26.

12 Seiler, P., Pant, A., and Hedrick, K. (2004). Disturbance propagation in vehicle strings. *IEEE Trans. Autom. Control* 49 (10): 1835–1842.

13 Barooah, P. and Hespanha, J.P. (2005). Error amplification and disturbance propagation in vehicle strings with decentralized linear control. *Proc. 44th IEEE Conf. Decision Control, Eur. Control Conf. (CDC-ECC)*, Seville, Spain (15 December 2005), pp. 4964–4969.

14 Sheikholeslam, S. and Desoer, C.A. (1993). Longitudinal control of a platoon of vehicles with no communication of lead vehicle information: a system level study. *IEEE Trans. Veh. Technol.* 42 (4): 546–554.

15 Peters, A., Richard, H.M., and Oliver, M. (2014). Leader tracking in homogeneous vehicle platoons with broadcast delays. *Automatica* 50 (1): 64–74.

16 Liu, X., Goldsmith, A., Mahal, S., and Hedrick, J. (2000). Effects of communication delay on string stability of vehicle platoons. *Proc. IEEE Intell. Transport. Syst.* 625–630.

17 Seiler, P. and Sengupta, R. (2005). An H_∞ approach to networked control. *IEEE Trans. Autom. Control* 50 (3): 356–364.

18 Guo, G. and Yue, W. (2011). Hierarchical platoon control with heterogeneous information feedback. *IET Control Theory Appl.* 5 (15): 1766–1781.

19 Stankovic, S., Stranojevic, M., and Siljak, D. (2000). Decentralized overlapping control of a platoon of vehicles. *IEEE Trans. Contr. Syst. Technol.* 8 (5): 816–831.

20 Dunbar, W. and Caveney, D. (2012). Distributed receding horizon control of vehicles platoons: stability and string stability. *IEEE Trans. Autom. Control* 57 (3): 620–633.

21 Guo, G. and Yue, W. (2012). Autonomous platoon control allowing range-limited sensors. *IEEE Trans. Veh. Technol.* 61 (7): 2901–2912.

22 Baldi, S., Rosa, M.R., Frasca, P., and Kosmatopoulos, E.B. (2018). Platooning merging maneuvers in the presence of parametric uncertainty. *IFAC-Papers On Line* 51 (23): 148–153.

23 Acciani, F., Frasca, P., Stoorvogel, A. et al. (2018). Cooperative adaptive cruise control over unreliable networks: an observer-based approach to increase robustness to packet loss. *European Control Conference (ECC)*, Limassol, Cyprus (12–15 June 2018). http://doi.org/10.23919/ECC.2018.8550585.

24 Seiler, P., Song, B., and Hedrick, J.K. (1998). Development of a collision avoidance system. *Neurosurg.* 46 (2): 492–492.

25 Mei, J., Zheng, K., Zhao, L. et al. (2018). Joint radio resource allocation and control for vehicle platooning in LTE-V2V network. *IEEE Trans. Veh. Technol.* 67 (12): 12218–12230.

26 Liberzon, D. (2013). *Switching in Systems and Control*. Boston: Birkhauser.

27 Zhai, G.S., Hu, B., Yasuda, K., and Michel, A.N. (2001). Disturbance attenuation properties of time-controlled switched systems. *J. Frankl. Inst.* 338: 765–779.

28 Park, P., Ko, J.W., and Jeong, C. (2011). Reciprocally convex approach to stability of systems with time-varying delays. *Automatica* 47 (1): 235–238.

29 Zhang, X., Han, Q.L., Seure, A. et al. (2019). Overview of recent advances in stability of linear systems with time-varying delays. *IET Control Theory Appl.* 13 (1): 1–16.

30 Middleton, R.H. and Braslavsky, J.H. (2010). String instability in classes of linear time invariant formation control with limited communication range. *IEEE Trans. Autom. Control* 55 (7): 1519–1530.

31 Fu, F. and Mihaela, S. (2012). Structure-aware stochastic control for transmission scheduling. *IEEE Trans. Veh. Technol.* 61 (9): 3931–3945.

32 Yanakiev, D. and Kanellakopoulos, L. (2001). Longitudinal control of automated CHVs with significant actuator delays. *IEEE Trans. Veh. Technol.* 50 (5): 1289–1297.

2

Control Subject to Communication Interruptions

2.1 Introduction

As in Chapter 1, the homogeneous connected vehicles (i.e. all vehicles have the same size and weight) are involved. However, in practice, the connected vehicles are often heterogeneous (e.g. different driveline dynamics, vehicle parameters, and network-induced uncertainties). For heterogeneous platoons, the vehicle dynamics and the spacing dynamics are different, and hence, the bounded velocity of the vehicles does not necessarily imply bounded spacing errors between all the vehicles and vice versa. Therefore, the string stability of heterogeneous platoons must guarantee simultaneous boundedness of velocities and spacing errors of all the vehicles. This is clearly more challenging and complicated than the case of homogeneous vehicular platoons [1].

For a practical communication network, the data packets in transmission between vehicles may be dropped out, delayed, or interrupted due to limited communication channels, which has a significant impact on platoon control. The problem of packet dropouts was investigated in [2] for a vehicular platoon, where a bursty packet generation algorithm was given. Reference [3] gave a scheduling and control codesign method to resolve communication conflicts and random packet dropouts in vehicular platooning systems. Reference [4] modeled network topologies by Markov chains to resolve the effects of random packet dropouts and stabilize the vehicular platooning systems. A graph-theoretic method was proposed in [5] to guarantee string stability and robustness in the presence of time delays and disturbances. Reference [6] studied the practical string stability of a cooperative adaptive cruise control (CACC) system with the constant-time-headway-spacing policy, considering the limitation of the communication link.

On the other hand, communication interruption is inevitable for large-scale connected vehicles due to limited communication channels. The issue of communication interruption in heterogeneous platooning systems has not been investigated yet except for the cooperative adaptive cruise control-adaptive cruise control (CACC-ACC) switching control method in [7]. In [7], an adaptive CACC-ACC switching control strategy was given based on the average dwell-time technique for heterogeneous platoons. This method makes the reference model tracking error bounded, and the string stability cannot be guaranteed exactly.

There is a definite need for a new control method for heterogeneous platoons that can guarantee string stability while adapting to changing conditions and communication interruptions. In this chapter, taking advantage of the finite-time stabilization technique and sliding-mode control

Connected Vehicular Systems: Communication, Control, and Optimization, First Edition. Ge Guo and Shixi Wen.

[8], we give a finite-time switching control method for heterogeneous platoons with communication latency under communication interruption. Intervehicle communication interruptions are addressed by switching the vehicle control strategy from a CACC scheme to an ACC scheme. Using a model control augmentation method, we prove analytically the convergence of the heterogeneous platoons with communication latency to an appropriately defined string stable platoons in finite time. Furthermore, the string stability of the platooning system with external disturbances is demonstrated analytically. The contributions in comparison with the current state of the art are given as follows:

1) Based on a mixed CACC-ACC reference model with communication latency, a design method for CACC-ACC baseline controllers is given, using the Routh-Hurwitz stability criterion and frequency-domain analysis, to guarantee string stability and zero steady-state spacing errors.
2) A vehicular tracking error system is established for the heterogeneous platoon considering communication interruptions, based on which the convergence of the tracking errors of the vehicular platoons is proved analytically.
3) A finite-time switching control algorithm based on sliding-mode control theory is proposed, which can robustly stabilize the vehicular platoons subject to communication interruptions with reduced fluctuation of spacing errors and improved safety.

The remainder of this chapter is organized as follows. Section 2.1 gives the heterogeneous platoon dynamics and the objective. Section 2.2 contains a string-stable CACC-ACC mixed control design for homogeneous platoon dynamics with communication latency. In Section 2.3, the finite-time switching control of heterogeneous platoons with communication latency subject to communication interruptions is given. Section 2.4 shows the simulations, which is followed by the conclusions in Section 2.5.

Notation The notations used in this chapter are quite standard. N represents the set of natural numbers. The superscript T stands for matrix transposition. sup|·| represents the least upper bound of a function. I_n indicates the identity matrix. The sgn (·) denotes the symbol function.

2.2 Problem Formulation

Consider a heterogeneous platoon with M vehicles driving on a horizontal road (see Figure 2.1) in a VANET environment. $p_i(t)$, $v_i(t)$, and $a_i(t)$ denote the ith ($i = 0, 1, \ldots, M-1$) vehicle's position (m), velocity (m/s), and acceleration (m/s^2), respectively, where $i = 0$ represents the leading vehicle.

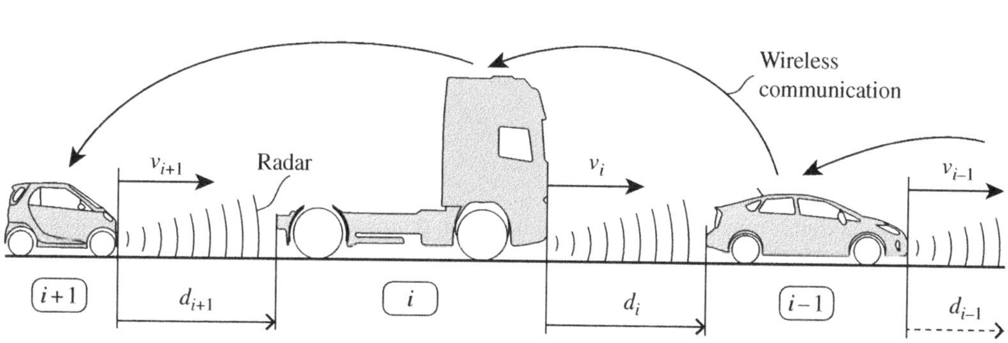

Figure 2.1 Predecessor-following communication topology of a heterogeneous vehicle platoon.

Each vehicle can only communicate with its preceding vehicle and transmit its control information via wireless communication.

All followers are equipped with onboard sensors to measure the distance between them and their preceding vehicles. In what follows, we will describe the vehicle model for a heterogeneous platoon and our objectives one by one.

Define the spacing error of the i th following vehicle as:

$$e_i(t) = d_i(t) - d_{r,t}(t), \quad \forall i \in S_M \tag{2.1}$$

where $d_i(t) = p_{i-1}(t) - p_i(t) - L_i$ is the real-time distance (m) between successive vehicles, L_i is the length (m) of the vehicle i, $d_{r,\ i}$ is the desired distance, and $S_M = \{i \in N \mid 1 \leq i \leq M-1\}$. Here, a constant time headway (CTH) spacing policy is adopted, which is defined as follows:

$$d_{r,i} = h_i v_i(t) + r_i, \quad \forall i \in S_M \tag{2.2}$$

where h_i represents the set time headway (seconds) (or time gap), r_i is a given standstill distance. Then, the dynamics of vehicle i can be represented by:

$$\begin{pmatrix} \dot{p}_i \\ \dot{v}_i \\ \dot{a}_i \end{pmatrix} = \begin{pmatrix} 0 & 1 & 0 \\ 0 & 0 & 1 \\ 0 & 0 & -\dfrac{1}{\tau_i} \end{pmatrix} \begin{pmatrix} p_i \\ v_i \\ a_i \end{pmatrix} + \begin{pmatrix} 0 \\ 0 \\ 1 \end{pmatrix} u_i \tag{2.3}$$

where u_i represents the control input of the vehicle i, and τ_i is the engine time constant of the vehicle i. Note that the heterogeneity of the connected vehicles in this chapter means that the engine time constant is different.

The objective is to design a switched control input $u_i(t)$, $\forall i \in S_M$, such that the heterogeneous vehicular platoon (2.3) can track the behavior of an asymptotic stable and string stable platoon in finite time under communication interruptions. That is, to meet the following goals:

1) *Steady-state performance*: The spacing error $e_i(t)$ approaches zero for all vehicles.
2) *String stability*: The oscillations are not amplified with the vehicle index due to any maneuver of the leading vehicle, namely $|\Gamma_i(j\omega)| \leq 1$ for any $\omega > 0$, where $\Gamma_i(j\omega) = (a_i(j\omega))/(a_{i-1}(j\omega))$, and $a_i(j\omega)$ is the expression in the frequency domain of $a_i(t)$.

2.3 Mixed CACC-ACC Control

In this section, we establish a mixed CACC-ACC reference dynamics and design CACC-ACC baseline controllers that can stabilize the system.

Define the leading vehicle control input as

$$h_0 \dot{u}_0 = -u_0 + u_r \tag{2.4}$$

where u_r is the system's input and the physical meaning represents the desired acceleration of the leader, h_0 is the time headway set by the leader. Considering the power limitation of the engine and safety, u_r is constrained by

$$a_{\min} \leq u_r \leq a_{\max} \tag{2.5}$$

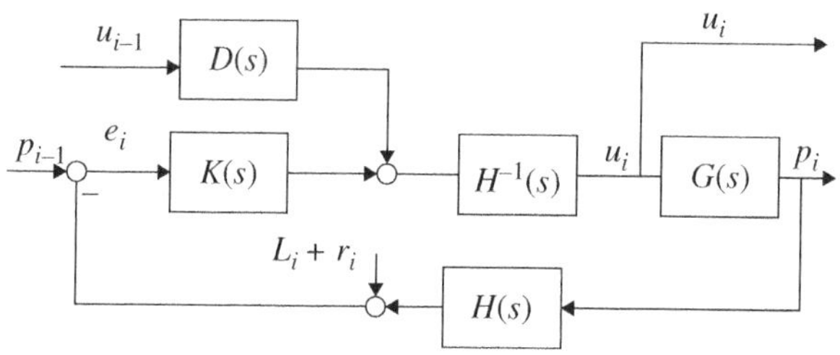

Figure 2.2 Block scheme of the CACC system.

Figure 2.2 shows the block scheme of the CACC system, and the CACC baseline controller is defined as follows:

$$h^C \dot{u}^C_{bl,i} = -u^C_{bl,i} + K^C_p e_i + K^C_d \dot{e}_i + u^C_{bl,i-1}(t-\theta_i) \tag{2.6}$$

where the time headway of CACC is set as $h_i = h^C, \forall i \in S_M$, K^C_p, K^C_d are the controller parameters to be determined, i is the communication time latency, and $u^C_{bl,0} = u_0$. Without loss of generality, the initial condition of the controller is set to zero, that is $u^C_{bl,i}(0) = 0, \forall i \in S_M$. Note that θ_i can be different for different vehicles, especially for heterogeneous connected vehicles; it can be even time-varying due to unstable communication conditions. However, as the main concern of this chapter is communication interruptions, to deal with this issue intrinsically and in more detail, we consider platoons with uniform constant time latency for simplicity, i.e. $\theta_i = \theta$.

Using (2.1)–(2.3) and (2.6) with $\tau_i = \tau_0$, and $u_{i,m} = u^C_{bl,i}$, the homogeneous CACC model can be described by

$$\begin{pmatrix} \dot{e}_{i,m} \\ \dot{v}_{i,m} \\ \dot{a}_{i,m} \\ \dot{u}_{i,m} \end{pmatrix} = \underbrace{\begin{pmatrix} 0 & -1 & -h^C & 0 \\ 0 & 0 & 1 & 0 \\ 0 & 0 & -\dfrac{1}{\tau_0} & \dfrac{1}{\tau_0} \\ \dfrac{K^C_p}{h^C} & -\dfrac{K^C_d}{h^C} & -K^C_d & -\dfrac{1}{h^C} \end{pmatrix}}_{A^C_m} \underbrace{\begin{pmatrix} e_{i,m} \\ v_{i,m} \\ a_{i,m} \\ u_{i,m} \end{pmatrix}}_{x_{i,m}} + \underbrace{\begin{pmatrix} 1 & 0 \\ 0 & 0 \\ 0 & 0 \\ \dfrac{K^C_d}{h^C} & \dfrac{1}{h^C} \end{pmatrix}}_{B^C_w} \underbrace{\begin{pmatrix} v_{i-1} \\ u^C_{bl,i-1}(t-\theta) \end{pmatrix}}_{w_i}, \quad \forall i \in S_M \tag{2.7}$$

where $x_{i,\ m}$ and w_i represent the state vector and the external input vector of the vehicle i, respectively.

Then, we give the following theorem to ensure the platoon stability of the homogeneous CACC system.

Theorem 2.1 Consider the homogeneous CACC dynamics (2.7). If the CACC baseline controller parameters satisfy the following Routh-Hurwitz conditions:

$$\begin{aligned} & h^C > 0, K^C_p > 0, K^C_d > 0, K^C_d > \tau_0 K^C_p \left(h^C\right)^2 - 2K^C_d \tau_0 \left(h^C\right)^2 \geq 0 \\ & -2K^C_p \left(h^C\right)^2 + \left(h^C K^C_d\right)^2 - 4K^C_d \tau_0 - 2K^C_d \theta - 2\tau_0 K^C_p \theta \geq 0 \\ & \left(h^C K^C_p\right)^2 - 4K^C_p \geq 0 \end{aligned} \tag{2.8}$$

then the dynamics (2.7) will be asymptotically stable around the equilibrium point $x_{i,m,eq} = (0, \bar{v}_0, 0, 0)^T$ where $\bar{v}_0$ is the equilibrium state of v_0 for $u_r = 0$, and the platoon is string stable.

Proof: The eigenvalues of the matrix A_m^C can be calculated as follows:

$$|\lambda I_4 - A_m^C| = \tau_0 \lambda^4 + \left(\frac{\tau_0}{h^C} + 1\right)\lambda^3 + \left(\frac{1}{h^C} + K_d^C\right)\lambda^2 + \left(\frac{K_d^C}{h^C} + K_p^C\right)\lambda + \frac{K_p^C}{h^C} = 0 \tag{2.9}$$

If the controller parameters of (2.6) satisfy $h^C > 0$, $K_p^C > 0$, $K_d^C > 0$, and $K_d^C > \tau_0 K_p^C$, the Routh-Hurwitz stability criterion will be satisfied. According to the input-to-state stability theory [10], the dynamics (2.7) can converge to the equilibrium point asymptotically.

Furthermore, we prove the platoon is string stable. According to the block scheme of the CACC system (see in Figure 2.2), $\Gamma_i(s)$ can be equal to

$$\Gamma_i(s) = \frac{a_i(s)}{a_{i-1}(s)} = \frac{1}{H(s)} \frac{D(s) + G(s)K(s)}{1 + G(s)K(s)} \tag{2.10}$$

where the vehicle transfer function is

$$G(s) = \frac{p_i(s)}{u_i(s)} = \frac{1}{s^2(\tau_0 s + 1)} \tag{2.11}$$

which follows from $\dddot{p} = -(1/\tau_0)\ddot{p}_i + (1/\tau_0)u_i$. The spacing policy transfer function $H(s)$ can be denoted by

$$H(s) = h^C s + 1 \tag{2.12}$$

and the feedback law can be denoted by

$$K(s) = K_p^C + K_d^C s \tag{2.13}$$

It follows from (2.10) that with communication latency ($D(s) = e^{-\theta s}$), $\Gamma_i(s)$ can be equal to

$$\Gamma_i(s) = \frac{e^{-\theta s}(s^3\tau_0 + s^2) + K_p^C + K_d^C s}{(h^C s + 1)\left(\tau_0 s^3 + s^2 + K_p^C + K_d^C s\right)} \tag{2.14}$$

Let $s = j\omega$, it gives

$$|\Gamma_i(j\omega)| = \frac{\sqrt{b_1}}{\sqrt{(h^C\omega)^2 + 1}\sqrt{\left(K_p^C - \omega^2\right)^2 + b_2}} \tag{2.15}$$

where $b_1 = \omega^4 + \tau_0{}^2\omega^6 + (K_d^C\omega)^2 + 2K_d^C\omega^3\sin(\theta\omega) - 2K_d^C\tau_0\omega^4\cos(\theta\omega) + \left(K_p^C\right)^2 - 2\omega^2 K_p^C\cos(\theta\omega) - 2\tau_0\omega^3 K_p^C\sin(\theta\omega)$ $b_2 = (K_d^C\omega - \tau_0\omega^3)^2$.

Taking into account the fact that $\sin(\theta\omega) \le \theta\omega$ and $|\cos(\theta\omega)| \le 1$ for $\forall \omega > 0$, if the following conditions are satisfied:

$$(h^C\tau_0)^2\omega^6 + \left((h^C)^2 - 2K_d^C\tau_0(h^C)^2\right)\omega^4 + \left(-2K_p^C(h^C)^2 + (h^C K_d^C)^2 - 4K_d^C\tau_0 - 2K_d^C\theta - 2\tau_0 K_p^C\theta\right)$$
$$\omega^2 + \left(h^C K_p^C\right)^2 - 4K_p^C \ge 0 \tag{2.16}$$

Then, $\sup_{\omega}|\Gamma_i(j\omega)| \leq 1$, $\forall\omega > 0$ substituting the control parameters (2.8) into (2.16), we can conclude that $|\Gamma_i(j\omega)| \leq 1$ for any $\omega > 0$, i.e. the platoon is string stable.

Remark 2.1 It follows from (2.10) that without communication latency ($D(s) = 1$), $\Gamma_i(s)$ can be simplified as follows:

$$\Gamma_i(s) = \frac{a_i(s)}{a_{i-1}(s)} = \frac{1}{H(s)} = \frac{1}{h^C s + 1} \tag{2.17}$$

Therefore, we can conclude that (2.17) satisfies the string stability condition for any $h^C > 0$. Moreover, it should be mentioned that the string stability transfer functions defined on acceleration, position, velocity, and spacing error are proved to be equal due to the assumption of homogeneous connected vehicles here. For the homogeneous connected vehicles, we have the following equalities:

$$a_i(s)/a_{i-1}(s) = p_i(s)/p_{i-1}(s) = v_i(s)/v_{i-1}(s)$$

Let $L_i = r_i = 0$, then the Laplace transform $E_i(s)$ of the spacing error equation (2.1) can be written as: $E_i(s) = p_{i-1}(s) - H(s)p_i(s)$ through some mathematical transformation, one can finally obtain that: $E_i(s)/E_{i-1}(s) = a_i(s)/a_{i-1}(s)$.

Remark 2.2 It should be noted that when the delay issue is concerned, it is more reasonable and practical to consider heterogeneous and time-varying delays. For a connected vehicles system with heterogeneous time-varying delays, one can use the Lyapunov-Razumikhin method [11] or the Lyapunov-Krasovskii method [12] to design the vehicle controller guaranteeing zero steady-state error in the context of consensus control of time-delay systems. In both cases, the string stability analysis part needs to be addressed carefully, which is not a trivial problem for amendments and is worthy of investigation. As the main concern of this chapter is not communication delay but communication interruptions, we consider homogeneous connected vehicles where the vehicle model involves a uniform delay in the input of the preceding vehicle. This assumption is reasonable since, on the one hand, the communication delay is small relative to the sampling period in the connected vehicles control system, and on the other, its impact can be neglected in comparison with other factors such as acceleration jerks.

Remark 2.3 In fact, the design of the CACC baseline controller is not unique. Involving the current acceleration/deceleration in the controller can indeed help improve the performance. Some existing works (e.g. [9, 13]) do include the acceleration of the follower and its preceding vehicle in the control law,

$$h^C \dot{u}^C_{bl,i} = -u^C_{bl,i} + K^C_p e_i + K^C_d \dot{e}_i + K^C_{dd} \ddot{e}_i + u^C_{bl,i-1} \tag{2.18}$$

where $\ddot{e}_i$ includes the acceleration of related vehicles. However, as shown in Theorem 2.1, using the lower order controller (2.6) without acceleration input, which involves only the feedback of spacing error and velocity error, is enough to achieve similar control objectives of zero steady-state error and string stability. The introduction of the acceleration information not only is redundant but also implies an extra cost for measurement and communications.

In this section, we establish the string stable mixed homogeneous CACC-ACC dynamics serving as the reference model for platooning. In the presence of inter-vehicle communication interruptions, the system in (2.7) may not be string stable since the control information $u^C_{bl,i-1}$ can no longer be present. Each vehicle in the ACC can obtain the state of its preceding vehicle through onboard sensors. The system selects the CACC spacing policy (indicated with h^C) or the ACC spacing policy (indicated with h^L) depending on its own communication reliability, where the superscript L stands for communication interruption. According to the results of California PATH [14], the time headway in ACC is generally larger than that of CACC. To do so, we adopt the ACC baseline controller in [7] as follows:

$$h^L \dot{u}^L_{bl,i} = -u^L_{bl,i} + K^L_p e_i + K^L_d \dot{e}_i \tag{2.19}$$

where K^L_p and K^L_d are the controller parameters to be designed, and $u^L_{bl,i}(0) = 0$, $\forall i \in S^L_M$, S^L_M is a subset of the set S_M containing the indices of the vehicles that lose communication, S^C_M is the subset of S_M containing the indices of the vehicles that maintain communication with their preceding vehicle. Similar to the CACC case, the homogeneous ACC model can be described by

$$\begin{pmatrix} \dot{e}_{i,m} \\ \dot{v}_{i,m} \\ \dot{a}_{i,m} \\ \dot{u}_{i,m} \end{pmatrix} = \underbrace{\begin{pmatrix} 0 & -1 & -h^L & 0 \\ 0 & 0 & 1 & 0 \\ 0 & 0 & -\frac{1}{\tau_0} & \frac{1}{\tau_0} \\ \frac{K^L_p}{h^L} & \frac{K^L_d}{h^L} & -K^L_d & -\frac{1}{h^L} \end{pmatrix}}_{A^L_m} \underbrace{\begin{pmatrix} e_{i,m} \\ v_{i,m} \\ a_{i,m} \\ u_{i,m} \end{pmatrix}}_{x_{i,m}} + \underbrace{\begin{pmatrix} 1 & 0 \\ 0 & 0 \\ 0 & 0 \\ \frac{K^L_d}{h^L} & 0 \end{pmatrix}}_{B^L_w} \underbrace{\begin{pmatrix} v_{i-1} \\ u^C_{bl,i-1} \end{pmatrix}}_{w_i}, \quad \forall i \in S^L_M \tag{2.20}$$

Furthermore, we give the following lemma to ensure the string stability of a homogeneous ACC system.

Lemma 2.1 [7] Consider the homogeneous ACC dynamics (2.20). If the ACC baseline controller parameters satisfy the following Routh-Hurwitz conditions:

$$\begin{aligned} & h^L > 0, K^L_p > 0, K^L_d > 0, K^L_d > \tau_0 K^L_p \\ & \left(h^L\right)^2 - 2K^L_d \tau_0 \left(h^L\right)^2 + \tau_0^2 \geq 0 \\ & 1 - 2K^L_p \left(h^L\right)^2 + \left(h^L K^L_d\right)^2 - 2K^L_d \tau_0 \geq 0 \\ & \left(h^L K^L_p\right)^2 - 2K^L_p \geq 0 \end{aligned} \tag{2.21}$$

then the dynamics (2.20) will be asymptotically stable around the equilibrium point $x_{i,\,m,\,eq}$ and the connected vehicles is string stable.

Thus, for homogeneous connected vehicles in the CACC strategy, when the communication link is interrupted, one can switch, from a string-stable CACC strategy designed via (2.7) to a string-stable ACC strategy designed via (2.20). When the communication is resumed, one can switch, from a string-stable ACC strategy to a string-stable CACC strategy.

2.4 Finite-Time Sliding-Mode Control

This section gives a vehicular state tracking error system model, based on which a piecewise continuous controller is designed to stabilize the heterogeneous connected vehicles system (2.3) infinite time.

The tracking error is defined as follows:

$$\begin{aligned} e_{1i} &= \tilde{p}_i = p_i - p_{i,m} \\ e_{2i} &= \tilde{v}_i = v_i - v_{i,m} \quad, \forall i \in S_M \\ e_{3i} &= \tilde{a}_i = a_i - a_{i,m} \end{aligned} \tag{2.22}$$

Substituting (2.3) into (2.22), the heterogeneous connected vehicles state tracking error system is established as follows:

$$\begin{aligned} \dot{e}_{1i} &= e_{2i} \\ \dot{e}_{2i} &= e_{3i} \\ \dot{e}_{3i} &= -\frac{1}{\tau_i}a_i + \frac{1}{\tau_i}u_i + \frac{1}{\tau_0}a_{i,m} - \frac{1}{\tau_0}u_{i,m} \end{aligned} \tag{2.23}$$

where

$$u_{i,m} = \begin{cases} u_{bl,i}^C, & \text{if} \quad i \in S_M^C \\ u_{bl,i}^L, & \text{if} \quad i \in S_M^L \end{cases}$$

The finite-time sliding-mode controller is designed in the following form:

$$u_i = -\frac{\tau_0}{\Lambda_i^*}\left(c_3\,\text{sgn}(e_{3i})|e_{3i}|^{\alpha_3} + c_2\,\text{sgn}(e_{2i})|e_{2i}|^{\alpha_2} + c_1\,\text{sgn}(e_{1i})|e_{1i}|^{\alpha_1}\right) + a_i + \frac{u_{i,m} - a_{i,m}}{\Lambda_i^*} \tag{2.24}$$

where $\Lambda_i^* = \tau_0/\tau_i$, and $c_j, j = 1, 2, 3$ are constants such that the polynomial $p^3 + c_3p^2 + c_2p + c_1$ is Hurwitz. α_j can be determined based on the following conditions:

$$\alpha_{j-1} = \frac{\alpha_j\alpha_{j+1}}{2\alpha_{j+1} - \alpha_j}, \; j = 2, 3 \tag{2.25}$$

where $\alpha_4 = 1$, $\alpha_3 = \alpha$, $\alpha \in (1 - \varepsilon, 1)$, $\varepsilon \in (0, 1)$.

Theorem 2.2 The finite-time sliding-mode controller (2.24) can make the state tracking error system (2.23) converge to zero within a finite time from any initial condition.

Proof: Substituting the control input (2.24) into (2.23) gives the following ideal sliding-mode manifold:

$$\begin{aligned} \dot{e}_{1i} &= e_{2i} \\ \dot{e}_{2i} &= e_{3i} \\ \dot{e}_{3i} &= -c_3\,\text{sgn}(e_{3i})|e_{3i}|^{\alpha_3} - c_2\,\text{sgn}(e_{2i})|e_{2i}|^{\alpha_2} - c_1\,\text{sgn}(e_{1i})|e_{1i}|^{\alpha_1} \end{aligned} \tag{2.26}$$

According to [15], the tracking error e_{ji} can converge to zero along the sliding-mode manifold in finite time.

Remark 2.4 In [7], the CACC problem with communication interruptions was considered under an average dwell-time framework, and an adaptive switching controller was proposed to compensate for vehicular heterogeneity. In this method, the state tracking error between the reference model and the actual system can only be bounded. Therefore, the control objective cannot be reached exactly. In this chapter, the sliding-mode control law (2.24) is designed to make the state of the heterogeneous error system (2.23) converge to zero in finite time, which makes the heterogeneous connected vehicular system achieve the desired performance.

Considering the practical application, external disturbances are inevitable. So, we design a robust finite-time sliding-mode controller based on [8] for the vehicle dynamics described by

$$\begin{pmatrix} \dot{p}_i \\ \dot{v}_i \\ \dot{a}_i \end{pmatrix} = \begin{pmatrix} 0 & 1 & 0 \\ 0 & 0 & 1 \\ 0 & 0 & -\frac{1}{\tau_i} \end{pmatrix} \begin{pmatrix} p_i \\ v_i \\ a_i \end{pmatrix} + \begin{pmatrix} 0 \\ 0 \\ \frac{1}{\tau_i} \end{pmatrix} u_i + \begin{pmatrix} 0 \\ 0 \\ \mu_i \end{pmatrix} \tag{2.27}$$

where μ_i is the external disturbances caused by wind gust, road condition, modeling error, etc.

Assumption 2.1 The disturbance term μ_i and its derivative are bounded:

$$|\mu_i| \leq C_{1i}, |\dot{\mu}_i| \leq C_{2i} \tag{2.28}$$

where C_{1i}, $C_{2i} > 0$ are known normal constants.

Substituting (2.27) into (2.22), the state tracking error system of connected vehicles system with the disturbances is written as

$$\begin{aligned} \dot{e}_{1i} &= e_{2i} \\ \dot{e}_{2i} &= e_{3i} \\ \dot{e}_{3i} &= -\frac{1}{\tau_i} a_i + \frac{1}{\tau_i} u_i + \frac{1}{\tau_0} a_{i,m} - \frac{1}{\tau_0} u_{i,m} + \mu_i \end{aligned} \tag{2.29}$$

A sliding-mode function is selected as

$$s_i = \dot{e}_{3i} + c_3 \operatorname{sgn}(e_{3i})|e_{3i}|^{\alpha_3} + c_2 \operatorname{sgn}(e_{2i})|e_{2i}|^{\alpha_2} + c_1 \operatorname{sgn}(e_{1i})|e_{1i}|^{\alpha_1} \tag{2.30}$$

The robust finite-time sliding-mode controller is designed as

$$u_i = \frac{\tau_0}{\Lambda_i^*} \left(u_{i,eq} + u_{i,n}\right) \tag{2.31}$$

$$u_{i,eq} = \frac{\Lambda_i^* a_i + u_{i,m} - a_{i,m}}{\tau_0} - c_3 \operatorname{sgn}(e_{3i})|e_{3i}|^{\alpha_3} - c_2 \operatorname{sgn}(e_{2i})|e_{2i}|^{\alpha_2} - c_1 \operatorname{sgn}(e_{1i})|e_{1i}|^{\alpha_1} \tag{2.32}$$

$$\dot{u}_{i,n} + \beta_i u_{i,n} = \kappa_i \tag{2.33}$$

$$\kappa_i = -(C_{2i} + \epsilon_i + \eta_i) \operatorname{sgn}(s_i) \tag{2.34}$$

where $u_{i,n}(0) = 0$, β_i and η_i are two positive constants, and ϵ_i is selected to satisfy $\epsilon_i \geq \beta_i C_{1i}$.

Theorem 2.3 The robust finite-time sliding-mode controller (2.31)–(2.34) can make the state tracking error system (2.29) converge to zero within finite time from any initial condition.

Proof: The conclusion can be easily obtained according to [8].

Remark 2.5 Note that in (2.34), s_i is not available since $\dot{e}_{3i}$ is not available in (2.30). For calculating $\operatorname{sgn}(s_i)$ in (2.34), the algorithm in [8] can be used.

2.5 Numerical Simulations

In order to verify the effectiveness of this method, we use the MATLAB/Simulink software to compare the results with the adaptive control strategy in [7] in this section.

Consider a heterogeneous platoon of 3 + 1 connected vehicles (including the leading vehicle). The length of each vehicle is 5 m. The starting positions of the four vehicles are 0, −30, −60,

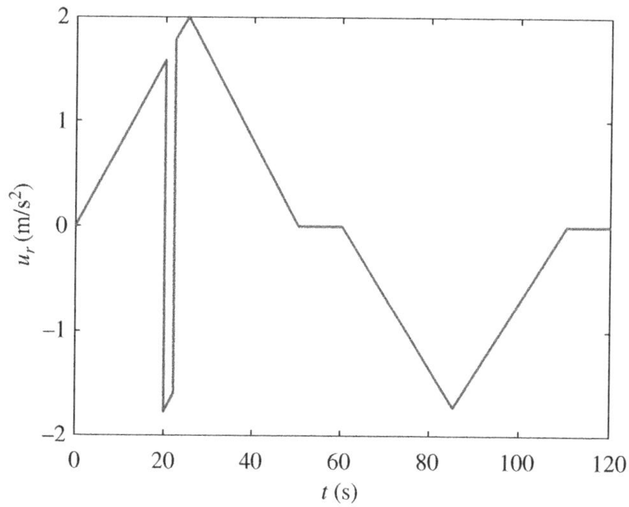

Figure 2.3 Desired platoon acceleration u_r.

and −90 m. The engine time constants are $\tau_0 = 0.1$, $\tau_1 = 0.15$, $\tau_2 = 0.2$, $\tau_3 = 0.25$. The remaining simulation parameters are $h^L = 1.1$, $h^C = 1$, $r = 15$, $K_p^C = 4.1$, $K_d^C = 4.5$, $K_p^L = 1.66$, $K_d^L = 1.62$, and $\theta = 0.1$, which satisfy the Routh-Hurwitz conditions. To approach reality, the desired platoon acceleration u_r is given in Figure 2.3 that represents a stop-and-go scenario with a sudden deceleration. The parameters of the finite-time sliding-mode controller (2.24) are selected $\alpha_1 = 7/16$, $\alpha_2 = 7/13$, $\alpha_3 = 7/10$, $c_1 = 80$, $c_2 = 66$, and $c_3 = 15$. The adaptive controller proposed in [7] is designed as

$$\begin{aligned} u_i &= u_{bl,i} + u_{ad,i} \\ u_{ad,i} &= -\Theta_i^T \Phi_i \\ \dot{\Theta}_i^T &= [0, 0, 100/\tau_0, 0] P_k \tilde{x}_i \Phi_i^T + F_i^T \end{aligned} \tag{2.35}$$

where

$$\Phi_i = [u_{bl,i}, -a_i]^T \tag{2.36}$$

$$\tilde{x}_i = [e_i - e_{i,m}, v_i - v_{i,m}, a_i - a_{i,m}, 0]^T \tag{2.37}$$

$P_k > 0$ is a symmetric matrix such that

$$A_m^T P_k + P_k A_m + 0.6 P_k \leq 0 \tag{2.38}$$

$A_m = A_m^C$ if the communication is present, $A_m = A_m^L$ if the communication is lost, and F_i is a projection term defined in [7].

Without loss of generality, four groups of comparative experiments are made, respectively, under the random communication interruptions as shown in Figure 2.4, where $\sigma_i = 0$ means the communication is present, and $\sigma_i = 1$ means the communication is lost.

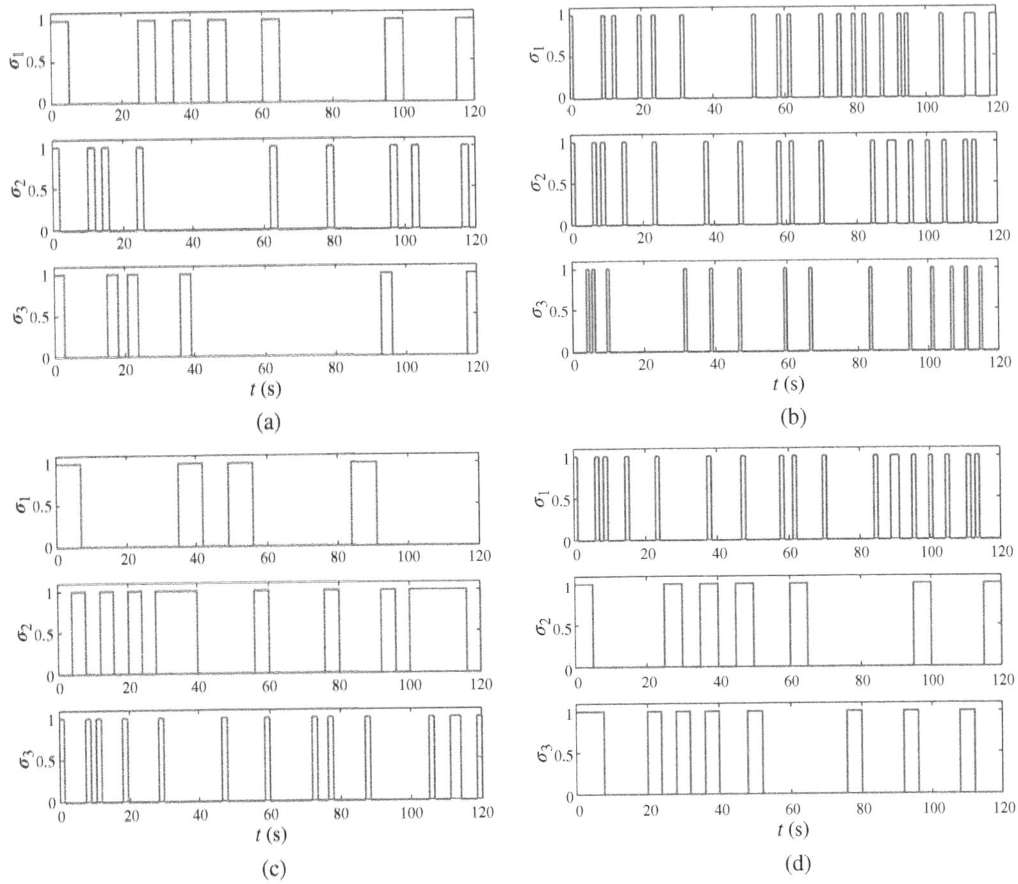

Figure 2.4 Communication interruption signals. (a) Signal 1, (b) Signal 2, (c) Signal 3, (d) Signal 4.

The simulation results are displayed in Figures 2.5 and 2.6. The position profiles show that collisions are avoided since there are no cross points. However, from the spacing error profiles, it can be seen obviously that the amplitude under the adaptive controller increases along the platoon. This is because the state tracking error under the adaptive controller is only bounded without guaranteeing the string stability exactly. The designed finite-time switching scheme can eliminate the amplification effectively.

For further performance comparison, we define a spacing error sum E_a (or E_f) as $\sum_{i=1}^{3}|e_i|$ corresponding to the adaptive controller (or the finite-time sliding-mode controller). The profiles of the variable $E_a - E_f$ are shown in Figure 2.7. It can be seen that E_a is generally larger than E_f in each case. Therefore, it can be concluded that the designed finite-time switching scheme is superior to the existing method in a statistical sense.

Finally, we study the performance of connected vehicles subject to external disturbances and communication interruptions. To this end, a disturbance $\mu_1 = 60\sin(5t)$ is involved in vehicle 1 and communication interruption 1 is used. The simulation results under the adaptive controller

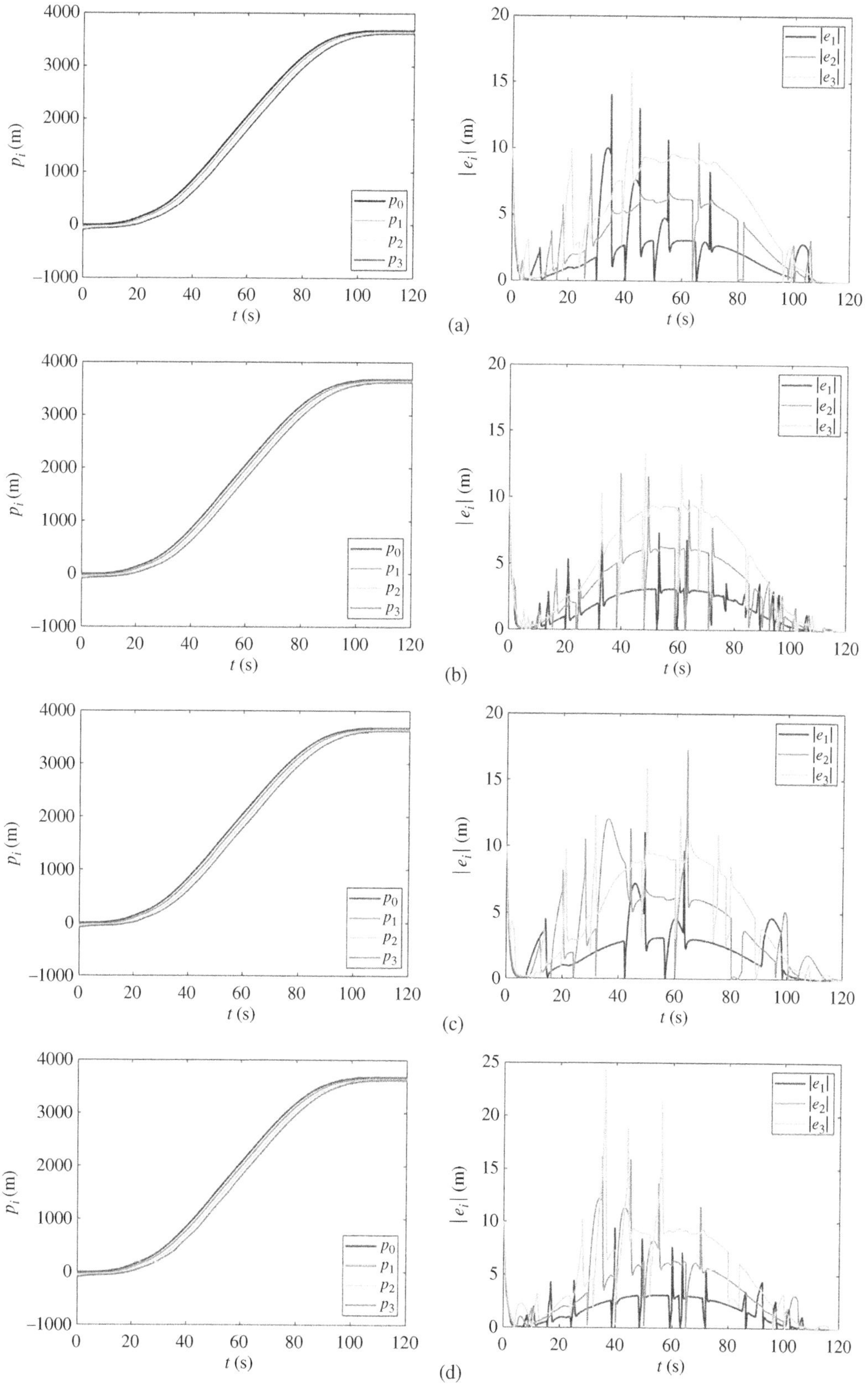

Figure 2.5 Profiles of position p_i and spacing error $|e_i|$ under the adaptive controller in [4]. (a) Profiles under signal 1, (b) Profiles under signal 2, (c) Profiles under signal 3, (d) Profiles under signal 4. *Source:* Adapted from Nguyen et al. [4].

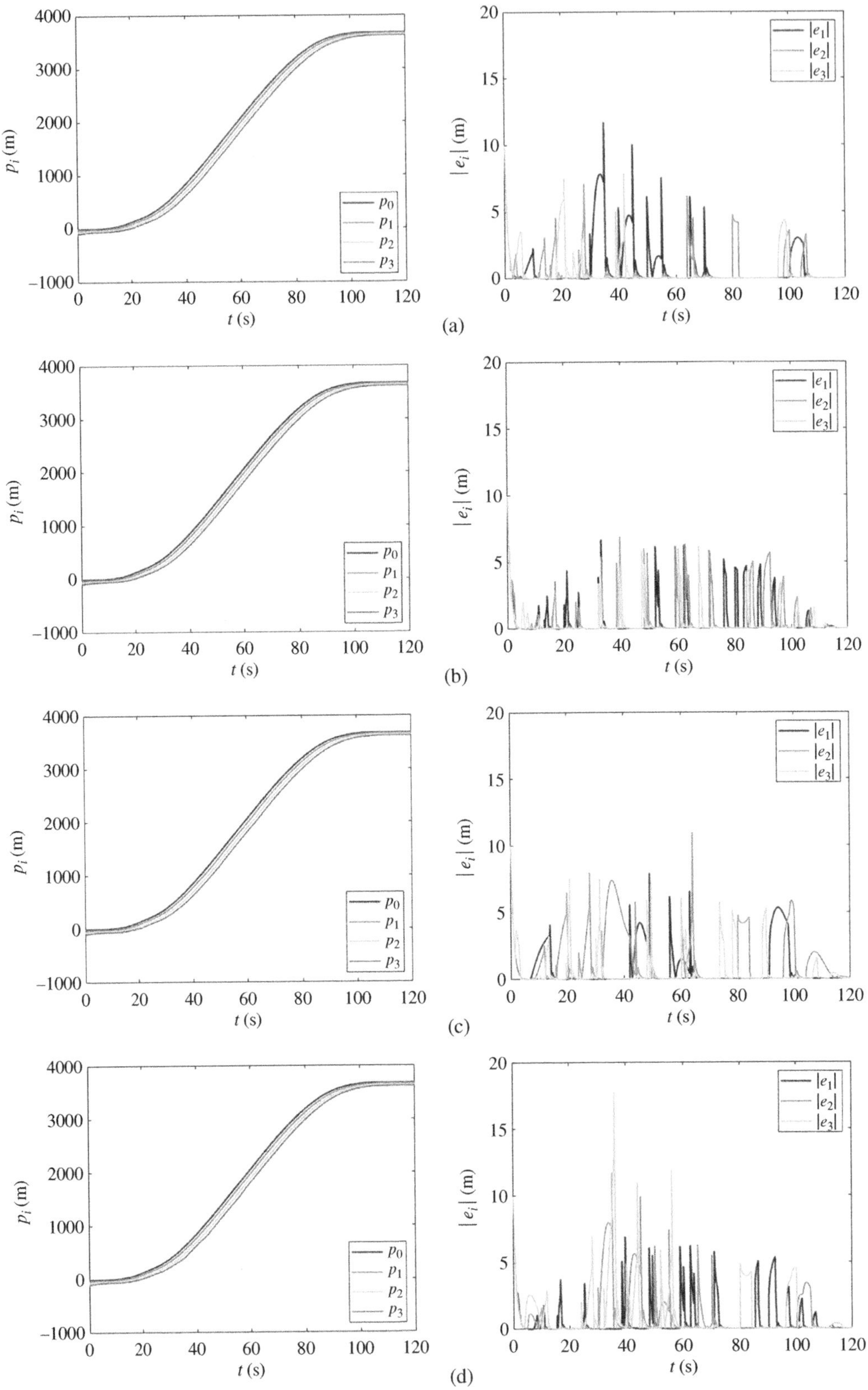

Figure 2.6 Profiles of position pi and spacing error $|e_i|$ under the finite-time sliding-mode controller (2.24). (a) Profiles under signal 1, (b) Profiles under signal 2, (c) Profiles under signal 3, (d) Profiles under signal 4.

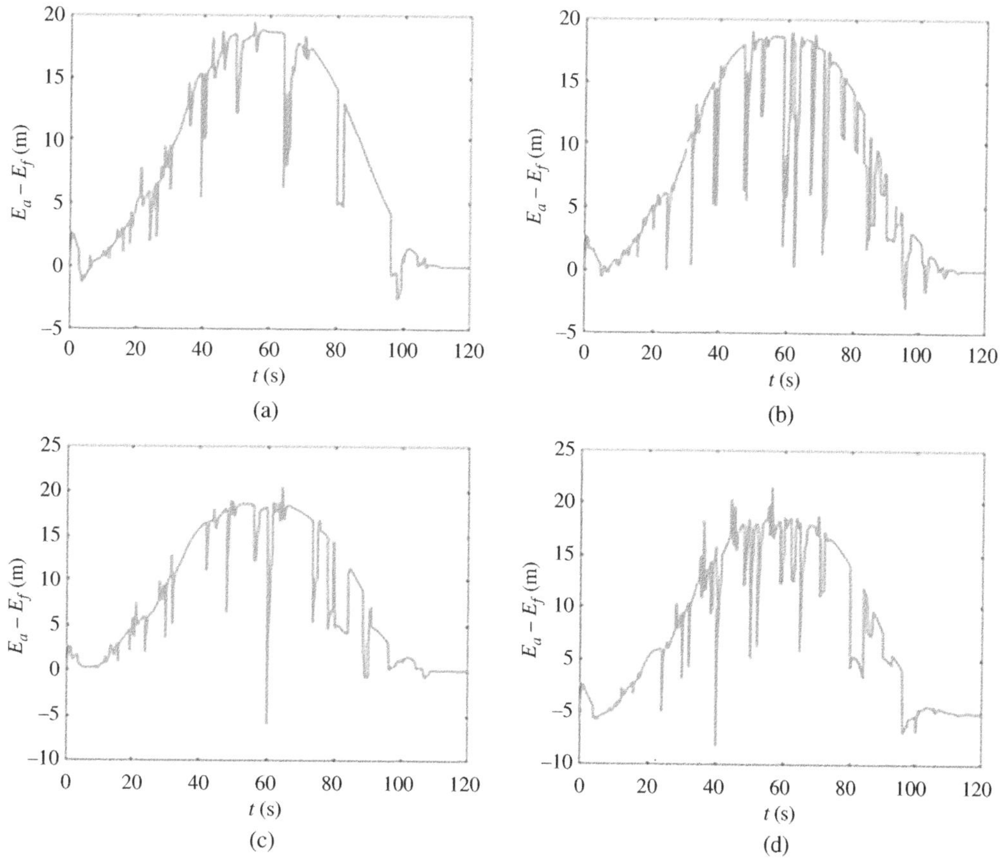

Figure 2.7 Profiles of $E_a - E_f$. (a) Profiles under signal 1, (b) Profiles under signal 2, (c) Profiles under signal 3, (d) Profiles under signal 4.

in [7] and the finite-time sliding-mode controller (2.24) are shown in Figure 2.8a and Figure 2.8b, respectively. It can be seen that the velocity profiles have serious vibrations; especially in Figure 2.8a, an overlapping phenomenon frequently occurs in the position profiles. Next, the robust finite-time sliding-mode controller (2.31)–(2.34) is adopted for vehicle 1, with parameters in (2.33)–(2.34) selected as $\beta_1 = 0.1$ and $C_{21} + \epsilon_1 + \eta_1 = 310$ and the other parameters the same. The results are shown in Figure 2.8c, where the vibrations are rejected effectively.

2.6 Conclusions and Future Work

This chapter has investigated a finite-time switching control algorithm for platooning of heterogeneous connected vehicles with communication interruptions and latency. The proposed control scheme comprises a switching CACC-ACC baseline controller and a finite-time sliding-mode controller to compensate for heterogeneous dynamics and external disturbances. Both steady-state performance and string stability are ensured.

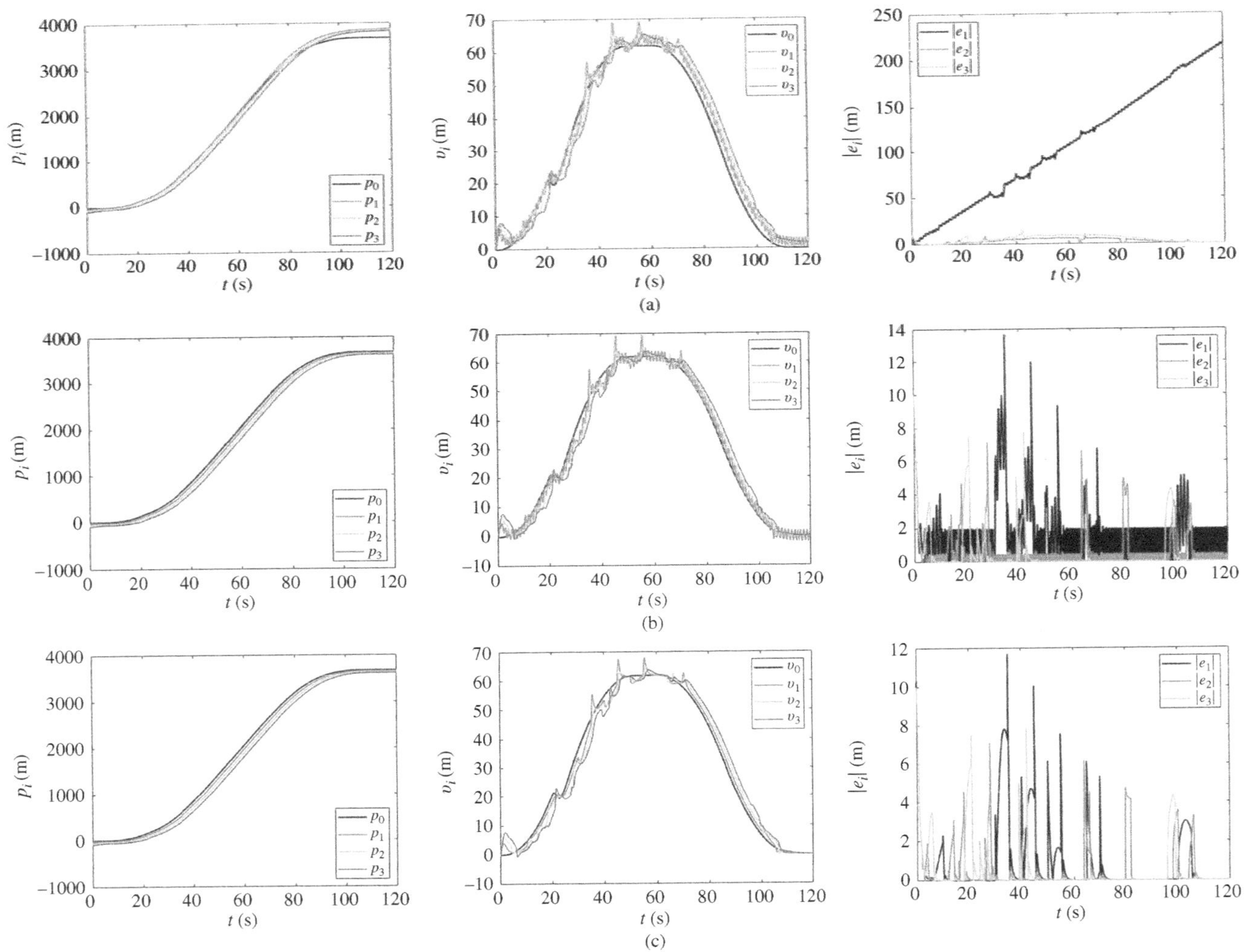

Figure 2.8 Profiles of position p_i, velocity v_i, and spacing error $|e_i|$ under the interruption signal 1 with external disturbances. (a) Profiles under the adaptive controller in [7], (b) Profiles under the finite-time sliding-mode controller (2.24), (c) Profiles under the robust finite-time sliding-model controller (2.31).

References

1 Xiao, L. and Gao, F. (2011). Practical string stability of platoon of adaptive cruise control vehicles. *IEEE Trans. Intell. Transp. Syst.* 12 (4): 1184–1194.

2 Carpenter, S.E. and Sichitiu, M.L. (2018). BUR-GEN: a bursty packetgenerator for vehicular communication channels. *IEEE Trans. Veh. Technol.* 67 (11): 10232–10242.

3 Guo, G. and Wen, S. (2016). Communication scheduling and control of a platoon of vehicles in V ANETs. *IEEE Trans. Intell. Transp. Syst.* 17 (6): 1551–1563.

4 Nguyen, T., Wang, L.Y., Yin, G. et al. (2018). Impact of communication erasure channels on control performance of connected and automated vehicles. *IEEE Trans. Veh. Technol.* 67 (1): 29–43.

5 Pirani, M., Hashemi, E., Simpson-Porco, J.W. et al. (2017). Graph theoretic approach to the robustness of k-nearest neighbor vehicle platoons. *IEEE Trans. Intell. Transp. Syst.* 18 (11): 3218–3224.

6 Naus, G.J.L., Vugts, R.P.A., Ploeg, J. et al. (2010). String-stable CACC design and experimental validation: a frequency-domain approach. *IEEE Trans. Veh. Technol.* 59 (9): 4268–4279.

7 Harfouch, Y.A., Yuan, S., and Baldi, S. (2018). An adaptive switched control approach to heterogeneous platooning with intervehicle communication losses. *IEEE Trans. Control Netw. Syst.* 5 (3): 1434–1444.

8 Feng, Y., Han, F., and Yu, X. (2014). Chattering free full-order sliding-mode control. *Automatica* 50 (4): 1310–1314.

9 Ploeg, J., van de Wouw, N., and Nijmeijer, H. (2014). Lp string stability of cascaded systems: application to vehicle platooning. *IEEE Trans. Control Syst. Technol.* 22 (2): 786–793.

10 Khalil, H.K. (2002). *Nonlinear Systems*, 3e. Upper Saddle River, NJ: Prentice-Hall.

11 di Bernardo, M., Salvi, A., and Santini, S. (2015). Distributed consensus strategy for platooning of vehicles in the presence of time-varying heterogeneous communication delays. *IEEE Trans. Intell. Transp. Syst.* 16 (1): 102–112.

12 Li, Y., Tang, C., Peeta, S., and Wang, Y. (2019). Nonlinear consensus-based connected vehicle platoon control incorporating car-following interactions and heterogeneous time delays. *IEEE Trans. Intell. Transp. Syst.* 20 (6): 2209–2219.

13 Ploeg, J., Scheepers, B.T.M., van Nunen, E. et al. (2011). Design and experimental evaluation of cooperative adaptive cruise control. *Proc. 14th Int. IEEE Conf. Intell. Transp. Syst. (ITSC)*, Washington, DC, USA (October 2011), pp. 260–265.

14 Milanés, V., Shladover, S.E., Spring, J. et al. (2014). Cooperative adaptive cruise control in real traffic situations. *IEEE Trans. Intell. Transp. Syst.* 15 (1): 296–305.

15 Bhat, S.P. and Bernstein, D.S. (2005). Geometric homogeneity with applications to finite-time stability. *Math. Control Signals Syst.* 17 (2): 101–127.

3

Control and Communication Topology Assignment

3.1 Introduction

Note that intervehicle communication topology (IVCT) is a key important factor for connected vehicle control. The stability analysis and controller design presented in Chapters 1 and 2 are based on typical IVCT such as the predecessor following (PF) model and the leader PF (LPF) model, respectively. The existing results have been extended to various types of complex information topologies (directed or undirected) in connected vehicles, e.g. bidirectional PF (BPF), two PF (TPF), multiple-PF, and mixed topology [1]. For example, Zheng et al. [2] investigated the effect of network topology on the stability and scalability of vehicular platoons. Zheng et al. [3] introduced topology selection and enlargement to improve the stability margin via asymmetric control. Zheng et al. [4] investigated distributed H_∞ control of a platoon of connected vehicles with undirected topologies. Li et al. [5] gave a distributed control method for vehicular platoons considering the interaction of topologies like PF, TPF, BPF, and LPF. In the aforementioned studies, the obtained results are available on a priori-defined IVCT.

Since IVCT and controller two aspects are strongly coupled, the most efficient way to improve the control performance of connected vehicles is to jointly design the IVCT and controller in a unified control framework. The co-design of IVCT and controller has attracted increasing research attention in recent years. By using the Routh-Hurwitz stability criterion and Rayleigh-Ritz theorem, undirected topology selection and control adjustment were implemented in [3] from a unified viewpoint to improve the stability margin of platoons. In [6], a spatially invariant system approach was used to jointly design the n vehicles' look-ahead/look-behind IVCT and controller so that string stability is satisfied. In the procedure of the proposed co-design, the solution to the algebraic Riccati equation (ARE) is approximated using a least-squares approach, where n is the order of approximation that determines the communication topology and feedback controller. Note that the results of the aforementioned papers are available for platoons under limited types of IVCTs with perfect intervehicle communication such as undirected topology in [3] and multiple vehicle look-ahead/behind and all-to-all topology in [6]. Considering the status of the practical intervehicle communication, however, it is possible that the determined IVCTs in Refs. [3, 6] are not applicable to platoons. Furthermore, the IVCT in both Refs. [3, 6] requires to be fixed. Nevertheless, it is common for practical platoons to switch to IVCT [7, 8]. A case in point is that the IVCT periodically switches between all available topologies. The possible reasons for this may be: (i) the joining and leaving of vehicles from a platoon;

Connected Vehicular Systems: Communication, Control, and Optimization, First Edition. Ge Guo and Shixi Wen.

(ii) the time-invariant connectivity of the moving vehicles to the network channels and inevitable frequent handoffs; (iii) resource limitation (e.g. bandwidth of wireless network, transmission power, and radio resource of communication network). When an IVCT is required to be assigned from the set of available topologies, it is not clear how to establish IVCT assignment and control co-design framework for platoons, especially for the optimal IVCT and associated controller. This is the motivation for this chapter.

Different from the work in Refs. [3, 6], in this chapter, we consider the case that the IVCT is assigned to a platoon from a set of available topologies. Here, we aim to establish a unified framework to jointly assign the IVCT and design the associated controller for vehicular platoons subject to unreliable IVC and stochastic external disturbances. We adopt the LTE-V2V network because it can provide more reliable intervehicle communication with lower latency and higher data transmission [12]. In addition, under the centralized communication mode LTE-V, the IVCT assignment for platoons can be governed by the control center via the CAM dissemination mechanism. A sampled-data feedback controller is proposed for platoons, wherein the controller gain is dependent on the IVCT. According to the optimal control and dynamic programming technique, from all the available IVCTs, the optimal IVCT and associated controller are jointly determined in the established control framework, where both fixed and periodic switching IVCTs are available. A useful algorithm is proposed to implement the established co-design framework, which not only guarantees individual vehicle stability and string stability but also minimizes the defined cost function. This algorithm can be easily applied with low computational cost because it can be calculated offline and only needs to be executed once. The contributions of this chapter are summarized as follows:

1) A novel IVCT-dependent sampled-data feedback controller is designed for homogeneous platoons, using which each follower can reject the effect of stochastic packet dropouts and disturbances to guarantee the stability requirements with minimized cost function.
2) A combinatorial search problem is defined and solved to determine the optimal IVCT of a set and the associated feedback controller gain for platoons. Consequently, for a predefined IVCT, one can derive the optimal controller and vice versa.
3) Stochastic string stability is analyzed for platoons, which include more generic IVCTs. The obtained results reveal the relationship between string stability and several key factors such as the probability of packet dropouts, the number of neighbors of the i-th follower, and the connection status with the leader.

The remainder of this chapter is organized as follows. Section 3.2 introduces an IVCT-dependent sampled-data platoon control model, which considers the effect of packet dropouts and disturbances. Section 3.3 presents the co-design of the optimal IVCT and the associated feedback controller. Section 3.4 describes the effectiveness of the proposed control method through numerical simulations. Section 3.5 concludes the paper.

Notations The symbol $\mathbb{R}^n$ denotes the n-dimensional Euclidean space. diag$\{\cdot\}$ denotes a block-diagonal matrix. I_N represents an $N \times N$ identity matrix and $\mathbf{1}_m$ the vector of length m containing one in each entry. For definite matrices P, matrix $P = P^T > 0$. For vector x, $\|x\|_Q^2 = x^T Qx$. Superscript T and T represent the matrix transpose and conjugate transpose of a matrix, respectively. $\|x\|_p$ denotes the p-norm of vector x. $|\cdot|$ denotes the modulus of the complex number.

3.2 Problem Statement

Consider a homogeneous platoon of $N+1$ vehicles driving on a horizontal road (see Figure 3.1). We denote z_i, v_i, and a_i as the position, velocity, and acceleration, respectively, of the i-th vehicle, $i = 0, 1, \ldots, N$. Each follower ($i = 1, \ldots, N$) has a tracking error-based feedback controller that updates its control input by collecting CAMs from its neighbors via the LTE-V2V network. Here, the IVCT is assigned by the control center, which is served by the Evolved Node B (eNB). In this section, the model for IVCT assignment and control is established, based on the CAM dissemination mechanism.

According to the feedback linearization technique, the following linear vehicle dynamic model is adopted [3, 5]

$$\begin{aligned} \dot{z}_i(t) &= v_i(t) \\ \dot{v}_i(t) &= a_i(t) \\ \dot{a}_i(t) &= -a_i(t)/\varsigma + u_i'(t)/\varsigma \end{aligned} \tag{3.1}$$

where ς is the time constant of the first-order acceleration dynamics. $u_i'(t)$ denotes the control input that represents the desired acceleration. In practice, the vehicle may experience external disturbances, for example, from wind gusts, air resistance, and road slope. We use $\omega_i(t)$ to denote the integrated external disturbance in vehicle dynamics. Using all these, we get:

$$\dot{a}_i(t) = -a_i(t)/\varsigma + u_i'(t)/\varsigma + \omega_i(t) \tag{3.2}$$

It is assumed that the disturbance $\omega_i(t) \sim \mathcal{N}(0, \nu_\omega)$ is a zero-mean Gaussian with covariances ν_ω.

CAM dissemination within a platoon can be divided into two phases. During the first phase, the control center transmits the CAM to vehicles according to the intervehicle communication topology. This CAM includes the control information (controller structure and parameters) and the uplink subframe (U subframe) allocation scheme for vehicles, that is, it decides which U subframe to use on which link of the IVC. During the second phase, according to the allocated U subframes, each vehicle transmits its CAMs (e.g. position, velocity, and acceleration) to its neighbors through the established V2V communication link. Normally, the size of the CAMs varies between 50 and 500 bytes [9]. In a common LTE cell deployment (i.e. bandwidth is 10 MHz), one subframe is sufficient for a vehicle to transmit its CAMs under suitable modulation and coding schemes. Here, it is assumed that the number of available U subframes is sufficient to support the required V2V communication simultaneously. Figure 3.2 presents an example to illustrate the PF communication topology assignment for a 4-vehicles platoon based on the CAM dissemination mechanism.

The IVCT of the platoon of vehicles is represented by a directional and connected graph, $\mathcal{G}_N = (\mathcal{V}_N, \mathcal{E}_N)$, where $\mathcal{V}_N = \{1, 2, \ldots, N\}$ is the set of vehicles, and $\mathcal{E}_N \subseteq \mathcal{V}_N \times \mathcal{V}_N$ is the set of edges. An edge is defined as $\varepsilon_{ij} \triangleq (i, j) \in \mathcal{E}_N$, which implies that there exists a directed communication link from vehicle j to vehicle i, i.e. $(i, j) \in \mathcal{E}_N$. $\mathbf{A} = [a_{ij}]_{n \times n}$ represents the adjacency matrix of graph $\mathcal{G}_N$,

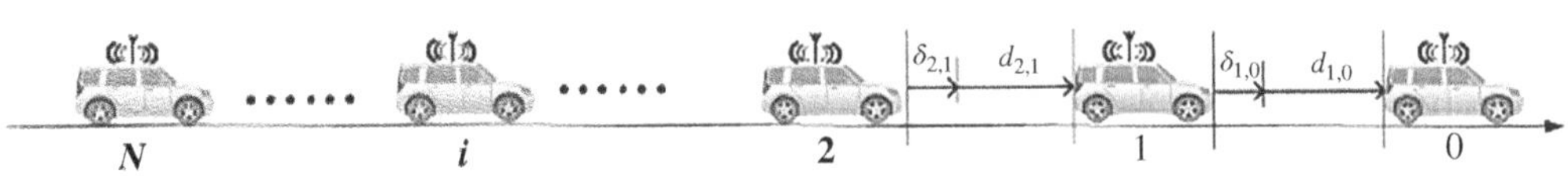

Figure 3.1 Platoon control.

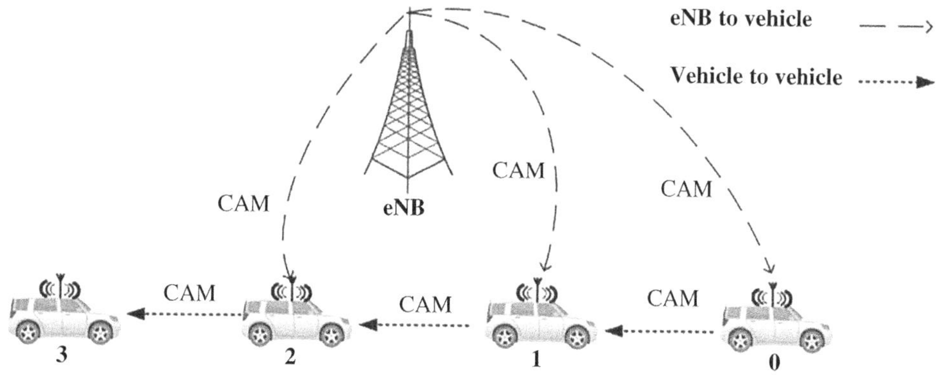

Figure 3.2 PF communication topology assignment by eNB based on CAM dissemination.

where $a_{ii} = 0$ for all i, and $a_{ij} = 1$ if $\varepsilon_{ij} \in \mathcal{E}_N$; otherwise, $a_{ij} = 0$. The degree matrix of graph $\mathcal{G}_N$ is $\mathcal{D} = \text{diag}\{d_1, d_2, ..., d_n\}$, where the diagonal elements are represented as $d_i = \sum_{j=1}^{N} a_{ij}$. Correspondingly, the Laplacian matrix, $\mathcal{L}$, describes the directional connection among followers, defined as $\mathcal{L} = \mathcal{D} - \mathbf{A}$. A path is a sequence of connected edges in a graph. Graph $\mathcal{G}$ is said to be connected if there exists a path between any nodes. A neighbor set of vehicle i in the followers is defined as

$$\mathbf{N}_i = \{j \in \mathcal{V}_N \mid a_{ij} = 1\} \tag{3.3}$$

To model the communication from the leader to the followers, we define an augmented directed graph, $\mathcal{G}_{N+1} = (\mathcal{V}_{N+1}, \mathcal{E}_{N+1})$, with a set of nodes $N+1$ vertices $\mathcal{V}_{N+1} = \{1, 2, ..., N+1\}$, which includes both the leader and the followers in a platoon. Similar to [3], a pinning matrix, $\mathcal{P} = \text{diag}\{p_1, p_2, ..., p_N\}$, is used to denote how each follower connects to the leader: $p_i = 1$ if $(i, 0) \in \mathcal{E}_{N+1}$; otherwise, $p_i = 0$. Note that $(i, 0) \in \mathcal{E}_{N+1}$ means that node i is pinned to the leader. The leader-accessible set of node i is defined as

$$\mathbf{P}_i = \begin{cases} \{0\} & \text{if } p_i = 1 \\ \emptyset & \text{if } p_i = 0 \end{cases} \tag{3.4}$$

where symbol $\emptyset$ denotes the empty set.

In this paper, we define matrix $\mathcal{H} = \mathcal{L} + \mathcal{P}$ as the IVCT in a platoon. Moreover, matrix $\mathcal{H}$ comes from set $\mathcal{M} = \{\mathcal{H}_1, \mathcal{H}_2, ..., \mathcal{H}_q\}$. For a larger platoon with limited network bandwidth, we assume that set $\mathcal{M}$ includes a limited number of available IVCTs. Furthermore, we assume the following:

Assumption 3.1 The augmented graph $\mathcal{G}_{N+1}$ contains at least one spanning tree rooted at the leader.

This assumption implies that the leader is globally reachable in $\mathcal{G}_{N+1}$. Each follower can obtain the leader information directly or indirectly, which is a prerequisite to guarantee the individual vehicle stability of a platoon.

According to the information exchange with its neighbor j, $j \in \mathbf{I}_i = \mathbf{N}_i \cup \mathbf{P}_i$, we define the tracking errors for the i-th follower

$$\begin{aligned} \delta_{i,j}(t) &= z_i(t) - z_j(t) - d_{i,j} \\ e_{i,j}^{v}(t) &= v_i(t) - v_j(t) \\ e_{i,j}^{a}(t) &= a_i(t) - a_j(t) \end{aligned} \tag{3.5}$$

where $\delta_{i,j}(t)$, $e^{v}_{i,j}(t)$, and $e^{a}_{i,j}(t)$ denote the spacing error, velocity error, and acceleration error between vehicle i and its neighbor j, respectively. $d_{i,j}$ denotes the desired headway which is given by [11]

$$d_{i,j} = (i-j)h_v v_i + (i-j)d \tag{3.6}$$

where constant $h_v > 0$ denotes the time gap and d is the standstill distance. Here, the constant time headway (CTH) spacing policy is adopted for a platoon.

As the CAMs are periodically updated with a time interval of h seconds, the controller of each follower requires the regular collection of the CAMs about the kinematics status of its neighbors. For this reason, continuous time t is divided into time slots indexed by $k \in \{0, 1, 2, ...\}$, where the time length of each time slot is equal to constant $h > 0$. At the k-th time slot, the local sampled-data feedback controller for the i-th follower is given as

$$u'_i(t) \equiv u'_i(kh) \text{ for } t \in [kh, (k+1)h) \tag{3.7}$$

with

$$u'_i(kh) = -\sum_{j\in \mathbf{I}_i} k^p_i(\mathcal{H})\bar{\delta}_{i,j}(kh) + k^v_i(\mathcal{H})\bar{e}^v_{i,j}(kh) + k^a_i(\mathcal{H})\bar{e}^a_{i,j}(kh)$$

where $k^p_i(\mathcal{H})$, $k^v_i(\mathcal{H})$, and $k^a_i(\mathcal{H})$ are the feedback controller gains to be designed, which depend on the IVCT assigned by the control center. $\bar{\delta}_{i,j}(kh)$, $\bar{e}^v_{i,j}(kh)$, and $\bar{e}^a_{i,j}(kh)$ are the actual tracking errors used to update controller $u'_i(kh)$.

As a U subframe is reused by both a cellular user and a follower, interference is induced between the V2V communication link and the cellular transmission link. At the k-th time slot, the transmission power of the ij-th V2V link is denoted by $\Xi_{ij}(kh)$. The channel gain is denoted by $\eta_{ij}(kh)$, which contains path loss, shadowing effect in the ij-th V2V link, and $I_{ij}(kh)$ as the interference from the cellular user occupying the same U subframe. Based on these definitions, the signal-to-interference-plus-noise ratio (SINR) of the ij-th V2V communication link is represented as [12]:

$$\gamma_{ij}(kh) = \frac{\Xi_{ij}(kh)\eta_{ij}(kh)}{I_{ij}(t) + N_0}, \quad i \neq j \in \{0, 1, 2, ..., N\} \tag{3.8}$$

where N_0 is the power of the additive noise in each U subframe. We use SINR to characterize the reliability of the V2V communication link. Specifically, if the SINR of the ijth V2V link is below a prescribed SINR threshold γ_{V2V}, that is, $\gamma_{ij}(kh) < \gamma_{V2V}$, the data flow of the ijth V2V link is unreliable for data transmission, thus probably leading to data-packet dropouts. For this reason, a binary-valued Bernoulli stochastic variable $\theta_{ij}(\cdot)$ is introduced to describe the status of the packet dropouts in the ijth $(i \neq j)$ V2V communication link. At the kth time slot, if $\theta_{ij}(kh) = 1$, the kinematic status of the jth neighboring vehicle is successfully transmitted to follower i via the ij-th V2V communication link, and then $\bar{\delta}_{i,j}(kh) = \delta_{i,j}(kh)$, $\bar{e}^v_{i,j}(kh) = e^v_{i,j}(kh)$, and $\bar{e}^a_{i,j}(kh) = e^a_{i,j}(kh)$; otherwise $\theta_{ij}(kh) = 0$, packet dropouts occur and the tracking errors are set to $\bar{\delta}_{i,j}(kh) = 0$, $\bar{e}^v_{i,j}(kh) = 0$, and $\bar{e}^a_{i,j}(kh) = 0$ by using the set-to-zero strategy [10, 12]. Here, the packet-arriving probabilities of the ij-th V2V communication link is

$$\bar{\theta}_{ij} = \mathbf{E}\{\theta_{ij}(kh)\} = \mathbf{Pr}\{\theta_{ij}(kh) = 1\} \tag{3.9}$$

Assumption 3.2 Each stochastic Bernoulli variable $\theta_{ij}(kh)$ has the same expectation, i.e. $\mathbf{E}\{\theta_{ij}(kh)\} = \bar{\theta}_{ij} = \bar{\theta}_i$, $0 \leq \bar{\theta}_i \leq 1$.

Taking into account the effect of stochastic packet dropouts, the sampled-data feedback controller in (3.7) is derived as in (3.10).

$$u_i'(t) = -\sum_{j\in \mathbf{I}_i}\theta_{ij}(kh)\left[k_i^p(\mathcal{H})\delta_{i,j}(kh) + k_i^v(\mathcal{H})e_{i,j}^v(kh) + k_i^a(\mathcal{H})e_{i,j}^a(kh)\right] \quad t \in [kh, (k+1)h) \tag{3.10}$$

We define the state vector $\hat{x}_i(t) = \left[\delta_{i,0}(t) \quad e_{i,0}^v(t) \quad e_{i,0}^a(t)\right]^T$; then, the tracking error dynamics of the follower $\tau'(t)$ are represented by

$$\dot{\hat{x}}_i(t) = \Phi_i\hat{x}_i(t) + \Gamma u_i'(t) + \Gamma\omega_i(t) \tag{3.11}$$

where

$$\Phi_i = \begin{bmatrix} 0 & 1 & h_{iv} \\ 0 & 0 & 1 \\ 0 & 0 & -\frac{1}{\varsigma} \end{bmatrix}, \quad \Gamma = \begin{bmatrix} 0 \\ 0 \\ \frac{1}{\varsigma} \end{bmatrix}, \quad \text{and } h_{iv} = i^* h_v$$

Using a state vector $\hat{x}_i(t)$, for $t \in [kh, (k+1)h)$, we can rewrite the sampled-data controller (3.10) in a compact form as

$$u_i'(t) = -\overline{K}_i(\mathcal{H})\sum_{j\in \mathbf{I}_i}\theta_{ij}(kh)\left(\hat{x}_i(kh) - \hat{x}_j(kh)\right) \tag{3.12}$$

where $\overline{K}_i(\mathcal{H}) = \left[k_i^p(\mathcal{H}) \quad k_i^v(\mathcal{H}) \quad k_i^a(\mathcal{H})\right]$ is the vector of feedback gains. Furthermore, by defining the state vector $x(t) = \left[\hat{x}_1^T(t),\ \hat{x}_2^T(t),\ \dots,\ \hat{x}_N^T(t)\right]^T$ and control input vector $u'(t) = [u_1'(t),\ u_2'(t),\ \dots,\ u_N'(t)]^T$, the control input for the entire platoon is derived as

$$u'(t) = -\left(\mathcal{H}(\theta(kh)) \otimes \overline{K}_i(\mathcal{H})\right)x(kh), \quad t \in [kh, (k+1)h) \tag{3.13}$$

where stochastic matrix

$$\mathcal{H}(\theta(kh)) = \mathcal{L}(\theta(kh)) + \mathcal{P}(\theta(kh))$$

with $\mathcal{L}(\theta(kh)) = [l_{ij}(\theta_{ij}(kh))]_{N\times N}$, $l_{ii}(\theta_{ii}(kh)) = \sum_{j=1, j\neq i}^{N} - l_{ij}\left(\theta_{ij}(kh)\right)$ and $l_{ij}(\theta_{ij}(kh)) = l_{ij}\theta_{ij}(kh)$. $\mathcal{P}(\theta(kh)) = \text{diag}\{\theta_{10}(kh)p_1, \theta_{20}(kh)p_2, \dots, \theta_{N0}(kh)p_N\}$. As in controller (3.13), it is clear that the two aspects, namely IVCT and controller, are strongly coupled. For a perfect IVC, that is, $\theta_{ij}(kh) = 1$ for $i \neq j$, controller (3.12) becomes

$$u'(t) = -\left(\mathcal{H} \otimes \overline{K}_i(\mathcal{H})\right)x(kh), \quad t \in [kh, (k+1)h) \tag{3.14}$$

which is the feedback controller proposed in [3, 4].

Considering the packet dropouts and stochastic external disturbance, the control objectives of this study are to jointly determine the optimal topology $\mathcal{H}$ from the set $\mathcal{M}$ and the associated controller such that all followers can track the lead's trajectory with the desired spacing and the following criteria are met:

1) *Individual vehicle stability*: The defined spacing and velocity error for each follower in (3.5) is mean-square stabilized, namely

$$\begin{cases} \lim_{t\to\infty} \mathbf{E}\left\{\left\|e_{i,0}^v(t)\right\|^2\right\} = 0 \\ \lim_{t\to\infty} \mathbf{E}\left\{\left\|\delta_{i,i-1}(t)\right\|^2\right\} = 0 \end{cases} \tag{3.15}$$

Here, we assume that the leader moves at a constant speed, that is, $a_0 = 0$.

2) *Stochastic string stability*: Under zero initial condition, the transfer function

$$G(s) = \frac{\delta_{i,0}(s)}{\delta_{i-1,0}(s)}, \quad i = 2, 3, \ldots, N \tag{3.16}$$

satisfies

$$\mathbf{E}\{|G(jw)|\} \leq 1 \tag{3.17}$$

for any $w > 0$, where $\delta_{i,0}(s)$ is the Laplace transformation of the spacing error $\delta_{i,0}(t)$.

3) From any time slot k, minimize the following LQ cost function defined for a platoon

$$\lim_{F \to \infty} \mathbf{E}\left\{\sum_{\upsilon=0}^{F} \|x((k+\upsilon)h)\|_Q^2 + \|u((k+\upsilon)h)\|_R^2\right\} \tag{3.18}$$

where F denotes the horizon length. $Q \in \mathbb{R}^{3N \times 3N}$ and $R \in \mathbb{R}^{N \times N}$ are symmetric positive definite matrices that penalize the cost of the state and the cost of the control, respectively.

3.3 Communication Topology and Control Co-Design

In this section, a unified framework is established for the platoon to jointly determine the optimal IVCT and feedback controller gain, in which both fixed and periodically switched IVCT are involved. First, we derive the optimal feedback controller gain for each available IVCT using the LQ control and dynamics programming technique. Then, by solving the defined combinatorial search problem, the optimal IVCT and the associated controller gain are determined such that the defined LQ cost function is minimized. Third, string stability analysis is carried out for the platoon to further complement the obtained feedback controller gain. Finally, an algorithm is proposed to implement the co-design procedure.

As the sampled-data controller is adopted, we discretize the continuous-time-tracking error system (3.11) with sampling period h and zero-order holder. As a result, the discrete-time version of the tracking error systems is obtained as

$$\hat{x}_i(k+1) = A_i\hat{x}_i(k) + B_1 u_i'(k) + B_1\omega_i(k) \tag{3.19}$$

where

$$A_i = \begin{bmatrix} 1 & h & \varsigma^2 e^{-h/\varsigma} + \varsigma h - \varsigma^2 + h_{iv}\varsigma(1 - e^{-h/\varsigma}) \\ 0 & 1 & -\varsigma e^{-h/\varsigma} + \varsigma \\ 0 & 0 & e^{-h/\varsigma} \end{bmatrix}$$

$$B_1 = \begin{bmatrix} \varsigma^2(-e^{-h/\varsigma} + 1) + h(-\varsigma + h) \\ \varsigma(e^{-h/\varsigma + 1} - 1) + h \\ -e^{-h/\varsigma} + 1 \end{bmatrix}$$

According to (3.19) with controller (3.13), the closed-loop connected vehicular control system can be described in the following compact form:

$$x(k+1) = \mathcal{A}x(k) - \mathcal{B}(\theta(k))K_{\mathcal{H}}x(k) + \mathcal{B}_1\omega(k) \tag{3.20}$$

where $\mathcal{A} = \text{diag}\{A_1, A_2, \ldots, A_N\}$, $\mathcal{B}(\theta(k)) = \mathcal{H}(\theta(k)) \otimes \overline{B}_1$, $\overline{B}_1 = B_1 \times \mathbf{1}_3^T$, and $\mathcal{B}_1 = I_N \otimes B_1$. External disturbance vector $\omega(k) = [\omega_1(k), \quad \omega_2(k), \quad \ldots, \quad \omega_N(k)]^T$ obeys $\omega(k) \sim \mathcal{N}(0_N, \nu_\omega)$. Controller gains $K_{\mathcal{H}} = \text{diag}\{\Delta_K^1, ..., \Delta_K^N\} \in \mathbb{R}^{3N \times 3N}$ with $\Delta_K^i = \text{diag}\{\overline{K}_i(\mathcal{H})\}$. Define $u(k) = -K_{\mathcal{H}}x(k)$ as the control aspect to be designed, then the closed-loop platoon control system in (3.20) can be represented as

$$x(k+1) = \mathcal{A}x(k) + \mathcal{B}(\theta(k))u(k) + \mathcal{B}_1\omega(k) \tag{3.21}$$

In this section, we present the results of the communication topology assignment and control co-design for the platoon under fixed communication topology $\mathcal{H}, \mathcal{H} \in \mathcal{M}$. For this purpose, we define the following LQ cost function for the platoon control system (3.21)

$$J_\infty(x(k), \mathcal{H}, \mu(x(k), \mathcal{H})) = \lim_{F \to \infty} \mathbf{E}\left\{\sum_{\upsilon=0}^{F}\left(\|x(k+\upsilon)\|_Q^2 + \|u(k+\upsilon)\|_R^2\right)x(k), \mathcal{H}\right\} \tag{3.22}$$

where $\mu(x(k), \mathcal{H})$ is the control law that maps the tracking error vector $x(k)$ to control signal $u(k)$, such that $u(k) = \mu(x(k), \mathcal{H})$.

Theorem 3.1 Consider a platoon control system with a fixed communication topology, $\mathcal{H} \in \mathcal{M}$. Then, at any time slot k, the defined LQ cost function in (3.22) satisfies

$$\min_{\mu(x(k),\mathcal{H})} J_\infty(x(k), \mathcal{H}, \mu(x(k), \mathcal{H})) = \lim_{F \to \infty} x^T(k)P_{\mathcal{H}}x(k) + \sum_{\upsilon=1}^{F} \alpha^\upsilon \text{trace}\left(\nu_\omega \mathcal{B}_1^T P_{\mathcal{H}} \mathcal{B}_1\right) \tag{3.23}$$

where $\alpha \in (0, 1)$ is the discounting factor and the positive definite matrix $P_{\mathcal{H}} \in \mathbb{R}^{3N \times 3N}$ is given for

$$P_{\mathcal{H}} = Q + \alpha\mathcal{A}^T P_{\mathcal{H}}\mathcal{A} - \alpha\mathcal{A}^T P_{\mathcal{H}}(\overline{\mathcal{H}} \otimes \overline{B}_1)K_{\mathcal{H}} \tag{3.24}$$

with

$$K_{\mathcal{H}} = \alpha\left(\overline{R} + \alpha(\overline{\mathcal{H}} \otimes \overline{B}_1)^T P_{\mathcal{H}}(\overline{\mathcal{H}} \otimes \overline{B}_1)\right)^{-1}(\overline{\mathcal{H}} \otimes \overline{B}_1)^T P_{\mathcal{H}}\mathcal{A} \tag{3.25}$$

where $\overline{R} = R \otimes I_3$, $\overline{\mathcal{H}} = \overline{\mathcal{L}} + \overline{\mathcal{P}}$ with $\overline{\mathcal{L}} = [\overline{l}_{ij}]_{N \times N}$, $\overline{l}_{ii} = -\sum_{j=1, j \neq i}^{N} \overline{l}_{ij}$ with $\overline{l}_{ij} = l_{ij}\overline{\theta}_{ij}$. $\overline{\mathcal{P}} = \text{diag}\{\overline{\theta}_{10}p_1, \overline{\theta}_{20}p_2, ..., \overline{\theta}_{N0}p_N\}$. Moreover, the optimal feedback controller for the platoon associated with communication topology $\mathcal{H}$ is obtained as

$$\mu^*(x(k), \mathcal{H}) \triangleq -K_{\mathcal{H}}x(k) \tag{3.26}$$

Proof: For a finite time length F, at time slot k, the LQ cost function in (3.22) for the platoon is given as

$$J_k(x(k), \mathcal{H}, \mu(x(k), \mathcal{H})) = \mathbf{E}\left\{\sum_{\upsilon=0}^{F}\left(\|x(k+\upsilon)\|_Q^2 + \|u(k+\upsilon)\|_R^2\right)x(k), \mathcal{H}\right\}$$

Because the pair (A_i, B_1) of the tracking error system (3.19) is controllable, it is evident that pair $(\mathcal{A}, \mathcal{B})$ with $\mathcal{B} = \mathcal{H} \otimes \overline{B}_1$ is controllable. According to the dynamic programming algorithm [11], at the time slot k, the LQ cost function for the platoon under communication topology $\mathcal{H}$ is given by (3.27).

$$\begin{aligned} &J_{k+F}(x(k+F), \mathcal{H}, \mu(x(k+F), \mathcal{H})) \\ &= x^T(k+F)Qx(k+F)J_k(x(k), \mathcal{H}, \mu(x(k), \mathcal{H})) \\ &= \mathbf{E}\left\{\|x(k)\|_Q^2 + \|u(k)\|_R^2 + \alpha J_{k+1}(x(k+1), \mathcal{H}, \mu(x(k+1), \mathcal{H}))x(k), \mathcal{H}\right\} \end{aligned} \tag{3.27}$$

Performing a straightforward induction from J_{k+F-1} as in [13], we can obtain Equation (3.28).

$$\begin{aligned} J_k(x(k),\mathcal{H},\mu(x(k),\mathcal{H})) = \min{}_{\mu(x(k),\mathcal{H})}\mathbf{E}\{x^T(k)\left(Q+\alpha\mathcal{A}^T P_{\mathcal{H}}(k+1)\mathcal{A}\right)x(k) \\ + u^T(k)\left(R+\alpha\mathcal{B}^T(\theta(k))P_{\mathcal{H}}(k+1)\mathcal{B}(\theta(k))\right)u(k) \\ + 2\alpha u^T(k)\mathcal{B}^T(\theta(k))P_{\mathcal{H}}(k+1)\mathcal{A}x(k) \\ + \sum_{v=0}^{F-1}\alpha^{v+1}\omega^T(k+v)\mathcal{B}_1^T P_{\mathcal{H}}(k+v+1)\mathcal{B}_1\omega(k+v)x(k),\mathcal{H}\} \end{aligned} \tag{3.28}$$

Because the disturbance $\omega(t)\sim\mathcal{N}(0_N,\nu_\omega)$, (3.28) can be rewritten as

$$\begin{aligned} J_k(x(k),\mathcal{H},\mu(x(k),\mathcal{H})) = \min{}_{\mu(x(k),\mathcal{H})}x^T(k)\left(Q+\alpha\mathcal{A}^T P_{\mathcal{H}}(k+1)\mathcal{A}\right)x(k) \\ + u^T(k)\left(R+\alpha\mathbf{E}\{\mathcal{B}^T(\theta(k))P_{\mathcal{H}}(k+1)\times\mathcal{B}(\theta(k))\mathcal{H}\}u(k)\right. \\ + 2\alpha u^T(k)\mathbf{E}\{\mathcal{B}^T(\theta(k))\mathcal{H}\}P_{\mathcal{H}}(k+1)\mathcal{A}x(k) \\ + \sum_{v=1}^{F}\alpha^{v}\mathrm{trace}\left(\nu_\omega\mathcal{B}_1^T P_{\mathcal{H}}(k+v)\mathcal{B}_1\right) \end{aligned} \tag{3.29}$$

Taking the derivative with respect to $u(k)$ and setting the resulting expression equal to zero result in the minimizer

$$\mu^*(x(k),\mathcal{H}) = -K_{\mathcal{H}}(k)x(k) \tag{3.30}$$

where

$$K_{\mathcal{H}}(k) = \alpha\left(R+\alpha\mathbf{E}\{\mathcal{B}^T(\theta(k))P_{\mathcal{H}}(k+1)\mathcal{B}(\theta(k))\mathcal{H}\}\right)^{-1}\times\mathbf{E}\{\mathcal{B}^T(\theta(k))\mathcal{H}\}P_{\mathcal{H}}(k+1)\mathcal{A} \tag{3.31}$$

with

$$P_{\mathcal{H}}(k) = Q+\alpha\mathcal{A}^T P_{\mathcal{H}}(k+1)\mathcal{A}-\alpha\mathcal{A}^T P_{\mathcal{H}}(k+1)\mathbf{E}\{\mathcal{B}^T(\theta(k))\mathcal{H}\}K_{\mathcal{H}}(k) \tag{3.32}$$

By inserting this into (3.29), we can obtain

$$J_k(x(k),\mathcal{H},\mu(x(k),\mathcal{H})) = x^T(k)P_{\mathcal{H}}(k)x(k)+\sum_{v=1}^{F}\alpha^{v}\mathrm{trace}\left(\nu_\omega\mathcal{B}_1^T P_{\mathcal{H}}(k+v)\mathcal{B}_1\right) \tag{3.33}$$

By computing the expectation in Equations (3.31) and (3.32), we get

$$K_{\mathcal{H}}(k) = \alpha\left(R+\alpha\left(\overline{\mathcal{H}}\otimes\overline{B}_1\right)^T P_{\mathcal{H}}(k+1)\left(\overline{\mathcal{H}}\otimes\overline{B}_1\right)\right)^{-1}\times\left(\overline{\mathcal{H}}\otimes\overline{B}_1\right)^T P_{\mathcal{H}}(k+1)\mathcal{A} \tag{3.34}$$

and

$$P_{\mathcal{H}}(k) = Q+\alpha\mathcal{A}^T P_{\mathcal{H}}(k+1)\mathcal{A}-\alpha\mathcal{A}^T P_{\mathcal{H}}(k+1)\left(\overline{\mathcal{H}}\otimes\overline{B}_1\right)^T K_{\mathcal{H}}(k) \tag{3.35}$$

According to proposition 4.4.1 of [13], it is clear that the solution of the discrete-time Riccati Equation (3.35) converges to a positive definite symmetric matrix $P_{\mathcal{H}}$, i.e.

$$\lim_{k\to\infty} P_{\mathcal{H}}(k) = P_{\mathcal{H}} \tag{3.36}$$

Furthermore, $P_{\mathcal{H}}$ is the unique solution of the algebraic matrix equation (3.24) in Theorem 3.1. As horizon time length F approaches infinity, it is clear that (3.23) of Theorem 3.1 holds.

Based on Theorem 3.1, we can obtain the minimized cost function (3.23) for each IVCT in the set $\mathcal{M}$. This allows us to solve the optimal IVCT and controller co-design problem by jointly minimizing (3.23) for the optimal IVCT

$$\mathcal{H}^* = \arg\min_{\mathcal{H}\in\mathcal{M}}\left[x^T(k)P_{\mathcal{H}}x(k) + \sum_{\upsilon=1}^{F}\alpha^{\upsilon}\text{trace}\left(\nu_{\omega}\mathcal{B}_1^T P_{\mathcal{H}}\mathcal{B}_1\right)\right] \tag{3.37}$$

with the associated feedback controller

$$\mu^*(x(k), \mathcal{H}^*) = -K_{\mathcal{H}^*}x(k) \tag{3.38}$$

The combinatorial search problem (3.37) can be solved using Algorithm 3.1.

In this section, we consider the communication topology assignment and control co-design for vehicle platoons under periodic switching IVCT. Specifically, the control center periodically assigns IVCT from set $\mathcal{M}$ to the platoon according to the prescribed order in set $\vec{\mathcal{H}}_T = \{\mathcal{H}(0), \mathcal{H}(1), ..., \mathcal{H}(T-1)\}$ with period T. At the k-th time slot, the communication topology $\mathcal{H}(\tau_k)$ is assigned to the platoon with

$$\tau_k = \text{rem}\{k/T\} \tag{3.39}$$

where rem{} is the remainder after division. In this case, the LQ cost function for the platoon is given as

$$J_\infty\left(x(k), \vec{\mathcal{H}}_T, \mu(x(k), \mathcal{H}(\tau_k))\right) = \lim_{F\to\infty}\mathbf{E}\left\{\sum_{\upsilon=0}^{F}\left(\|x(k+\upsilon)\|_Q^2 + \|u(k+\upsilon)\|_R^2\right)x(k), \mathcal{H}(\tau_k)\right\} \tag{3.40}$$

Remark 3.1 It should be noted that for a structured communication topology such as PF or bidirectional PF (BPF), when vehicles join or leave a platoon, the IVCT may be re-established so as to maintain the same structure again. In the switching communication topology assignment problem studied here, there are no specific assumptions regarding the type of topologies. In other words, the switching topologies can be unstructured ones.

Before describing the main results of this section, we provide the necessary conditions that need to be fulfilled for a stabilizing solution to exist.

Lemma 3.1 The platoon control system $(\mathcal{A}, \mathcal{B}(\theta(k)))$ is a T-step controllable with periodic switching IVCT from set $\vec{\mathcal{H}}_T$ over time interval $(k, k+T)$ if

$$rank\left\{\Upsilon\left(T, \vec{\mathcal{H}}_T\right)\right\} = 3N \tag{3.41}$$

where

$$\Upsilon\left(T, \vec{\mathcal{H}}_T\right) \triangleq \left[\vec{\mathcal{B}}(k+T-1), \mathcal{A}\vec{\mathcal{B}}(k+T-2), ..., \mathcal{A}^{T-1}\vec{\mathcal{B}}(k)\right] \tag{3.42}$$

with matrix $\vec{\mathcal{B}}(k) = -\mathcal{H}(\tau_k)\otimes\overline{B}_1$.

Proof: It is easy to obtain the results according to the proof for controllability of the periodic system given in [12].

Let the set $\mathcal{M}_T$ contain all the available communication topology sets that satisfy Lemma 3.1 with period T. Namely $\vec{\mathcal{H}}_T$ comes from set $\mathcal{M}_T = \left\{\vec{\mathcal{H}}_T^1, \vec{\mathcal{H}}_T^2, ..., \vec{\mathcal{H}}_T^q\right\}$.

Theorem 3.2 Consider a platoon control system with periodic switching communication topologies from set $\vec{\mathcal{H}}_T$, $\vec{\mathcal{H}}_T \in \mathcal{M}_T$. Then, at any time slot k, the LQ cost function defined in (3.40) satisfies

$$\begin{aligned}\min_{\mu(x(k),\mathcal{H}(\tau_k))} J_{\infty}\left(x(k), \vec{\mathcal{H}}_T, \mu(x(k), \mathcal{H}(\tau_k))\right) \\ = x^T(k)P_T(\tau_k)x(k) + \sum_{\upsilon=1}^{\infty} \alpha^{\upsilon}\text{trace}\left(\nu_{\omega}\mathcal{B}_1^T P_T(\tau_{k+\upsilon})\mathcal{B}_1\right)\end{aligned} \tag{3.43}$$

where $P_T(\tau_k) \in \mathbb{R}^{3N \times 3N}$ is the periodic positive definite solution to the following periodic Riccati equation

$$P_T(\tau_k) = Q + \alpha\mathcal{A}^T P_T(\tau_{k+1})\mathcal{A} - \alpha\mathcal{A}^T P_T(\tau_{k+1})\left(\overline{\mathcal{H}}(\tau_k) \otimes \overline{B}_1\right)K_{\mathcal{H}}(\tau_k) \tag{3.44}$$

with

$$K_{\mathcal{H}}(\tau_k) = \alpha\left(\overline{R} + \alpha\left(\overline{\mathcal{H}}(\tau_k) \otimes \overline{B}_1\right)^T P_T(\tau_{k+1})\left(\overline{\mathcal{H}}(\tau_k) \otimes \overline{B}_1\right)\right)^{-1} \times \left(\overline{\mathcal{H}}(\tau_k) \otimes \overline{B}_1\right)^T P_T(\tau_{k+1})\mathcal{A} \tag{3.45}$$

where matrix $\overline{\mathcal{H}}(\tau_k)$ is the corresponding matrix for the IVCT $\mathcal{H}(\tau_k)$ in set $\vec{\mathcal{H}}_T$, which has the same definition as matrix $\overline{\mathcal{H}}$ in (3.25). Moreover, the minimizing control policy is given by

$$\mu^*(x(k), \mathcal{H}(\tau_k)) \triangleq -K_{\mathcal{H}}(\tau_k)x(k) \tag{3.46}$$

Proof: Following the procedure as in the proof of Theorem 3.1, it is easy to obtain the Riccati equation (3.44). Moreover, (3.44) is the periodic Riccati equation [10].

Similar to the combinatorial search problem defined in (3.37), based on Theorem 3.2, the control center can determine a group of optimal IVCTs, that is, $\vec{\mathcal{H}}_T^* = \{\mathcal{H}^*(0), \mathcal{H}^*(1), \ldots, \mathcal{H}^*(T-1)\}$ from set $\mathcal{M}_T$ by computing

$$\arg\min_{\vec{\mathcal{H}}_T \in \mathcal{M}_T}\left[x^T(k)P_T(\tau_k)x(k) + \sum_{\upsilon=1}^{F} \alpha^{\upsilon}\text{trace}\left(\nu_{\omega}\overset{T}{\underset{1}{\mathcal{B}}} P_T(\tau_{k+\upsilon})\mathcal{B}_1\right)\right] \tag{3.47}$$

then, the associated minimizing control policy is derived as

$$\mu^*(x(k), \mathcal{H}^*(\tau_k)) \triangleq -K_{\mathcal{H}^*}(\tau_k)x(k) \tag{3.48}$$

To simplify the solution for $P_T(\tau_k)$, we transform the time-variant Riccati equation (3.44) into a time-invariant one by using a cyclic system description [14]. As a result, the associated cyclic matrices $\hat{\mathcal{A}} \in \mathbb{R}^{3TN \times 3TN}$ and $\hat{\mathcal{B}}(\hat{\theta}_T) \in \mathbb{R}^{3TN \times 3TN}$ for the platoon control system (3.21) are given as

$$\hat{\mathcal{A}} = \begin{bmatrix} 0 & 0 & \ldots & 0 & \mathcal{A} \\ \mathcal{A} & 0 & \ldots & 0 & 0 \\ 0 & \mathcal{A} & \ldots & 0 & 0 \\ \vdots & \vdots & \ddots & \vdots & \vdots \\ 0 & 0 & \ldots & \mathcal{A} & 0 \end{bmatrix}$$

$$\hat{\mathcal{B}} = \begin{bmatrix} \mathbf{0} & \mathcal{B}(\theta(T-1)) \\ \tilde{\mathcal{B}}(\theta) & \mathbf{0}^T \end{bmatrix}$$

where $\mathbf{0}$ is a $3N \times 3(T-1)N$ matrix containing only zeros. $\tilde{\mathcal{B}}(\theta) = \text{diag}\{\mathcal{B}(\theta(0)), (\theta(1)), \ldots, \mathcal{B}(\theta(T-2))\}$.

We define $\hat{Q} = \text{diag}\{Q, Q, \ldots, Q\} \in \mathbb{R}^{3TN \times 3TN}$ and $\hat{R} = \text{diag}\{\overline{R}, \overline{R}, \ldots, \overline{R}\} \in \mathbb{R}^{3TN \times 3TN}$. This then results in the time-invariant Riccati equation

$$\hat{P}_T = \hat{Q} + \alpha\hat{\mathcal{A}}^T\hat{P}_T\hat{\mathcal{A}} - \alpha\hat{\mathcal{A}}^T\hat{P}_T\mathbf{E}\left\{\hat{\mathcal{B}} \mid \vec{\mathcal{H}}_T\right\}\hat{K}_{\vec{\mathcal{H}}} \tag{3.49}$$

with

$$\hat{K}_{\vec{\mathcal{H}}_T} = \alpha\left(\hat{R} + \alpha\mathbf{E}\left\{\hat{\mathcal{B}}^T\hat{P}_T\hat{\mathcal{B}} \mid \vec{\mathcal{H}}_T\right\}\right)^{-1}\mathbf{E}\left\{\hat{\mathcal{B}}^T \mid \vec{\mathcal{H}}_T\right\}\hat{P}_T\hat{\mathcal{A}} \tag{3.50}$$

where $\hat{P}_T = \text{diag}\{P_T(0), P_T(1), ..., P_T(T-1)\}$ and $\hat{K}_{\vec{\mathcal{H}}_T} = \text{diag}\{K_{\mathcal{H}}(0), K_{\mathcal{H}}(1)..., K_{\mathcal{H}}(T-1)\}$.

By computing the expectations in (3.49) and (3.50), the Riccati equation can be written as

$$\hat{P}_T = \hat{Q} + \alpha\hat{\mathcal{A}}^T\hat{P}_T\hat{\mathcal{A}} - \alpha\hat{\mathcal{A}}^T\hat{P}_T\breve{\mathcal{B}}\hat{K}_{\vec{\mathcal{H}}_T} \tag{3.51}$$

with

$$\hat{K}_{\vec{\mathcal{H}}_T} = \alpha\left(\hat{R} + \alpha\breve{\mathcal{B}}^T\hat{P}_T\breve{\mathcal{B}}\right)^{-1}\breve{\mathcal{B}}^T\hat{P}_T\hat{\mathcal{A}} \tag{3.52}$$

where $\breve{\mathcal{B}} = \begin{bmatrix} \mathbf{0} & \overline{\mathcal{H}}(T-1)\otimes\overline{B}_1 \\ \tilde{\mathcal{B}} & \mathbf{0}^T \end{bmatrix}$ with $\tilde{\mathcal{B}} = \text{diag}\left\{\overline{\mathcal{H}}(0)\otimes\overline{B}_1, ..., \overline{\mathcal{H}}(T-2)\otimes\overline{B}_1\right\}$.

As a result, matrix $P_T(\tau_k)$ and $K_{\mathcal{H}}(\tau_k)$ in (3.47) can be extracted as the $(\tau_k + 1)$th $3N \times 3N$ block on the diagonal of $\hat{P}_T$ and $\hat{K}_{\vec{\mathcal{H}}_T}$, which are solved using the Riccati equations (3.51) and (3.52), respectively. The optimal controller gain $K_{\mathcal{H}^*}(\tau_k)$ in (3.48) can be extracted as the $(\tau_k + 1)$th $3N \times 3N$ block on the diagonal of the gain matrix $\hat{K}_{\vec{\mathcal{H}}_T^*}$ associated with communication topology $\mathcal{H}^*(\tau_k)$ in set $\vec{\mathcal{H}}_T^*$.

In this section, additional conditions are given to complement the controller gain obtained in (3.38) and (3.48) such that the stochastic string stability defined in (3.17) is satisfied for platooning.

On substituting the controller (3.7) into the dynamic equation (3.2), by using the Laplace transformation under zero initial condition, we obtain

$$\begin{aligned}\left(s + \frac{1}{\varsigma}\right)a_i(s) = &-\frac{1}{\varsigma}\sum_{j\in\mathbf{I}_i}\theta_{ij}(s)\left[k_i^p\left(\delta_{i,0}(s) - \delta_{j,0}(s)\right) + k_i^v\left(e_{i,0}^v(s) - e_{j,0}^v(s)\right) + k_i^a\left(e_{i,0}^a(s) - e_{j,0}^a(s)\right)\right] \\ &+ \frac{1}{\varsigma}\omega_i(s).\end{aligned} \tag{3.53}$$

where $\theta_{ij}(s)$ and $\omega_i(s)$ are the Laplace transformations of the stochastic variable $\theta_{ij}(t)$ and disturbance $\omega_i(t)$, respectively. For simplicity, k_i^p, k_i^v, and k_i^a denote the controller gains of the gain matrix $K_{\mathcal{H}^*}$ or $K_{\mathcal{H}^*}(\tau)$, $\tau \in \{0, 1, ..., T-1\}$.

According to tracking errors defined in (3.5), it is obvious that

$$\ddot{\delta}_{i,0}(t) = a_i(t) - h_{iv}\dot{a}_i(t) \tag{3.54}$$

Under zero initial condition, using the Laplace transformation for (3.54), yields

$$a_i(s) = \frac{s^2\delta_{i,0}(s)}{1 - sh_{iv}} \tag{3.55}$$

On substituting (3.55) into (3.53) with $e_{i,0}^v(s) = (a_i(s))/s$ and $e_{i,0}^a(s) = a_i(s)$, using basic algebraic calculations, we have

$$\begin{aligned}&\varsigma s^3\delta_{i,0}(s) + s^2\delta_{i,0}(s) - \omega_i(s) - \sum_{j\in\mathbf{I}_i}\theta_{ij}(s)\left[k_i^p(1 - sh_{iv}) + sk_i^v + k_i^a s^2\right]\delta_{i,0}(s) \\ &= \sum_{j\in\mathbf{I}_i}\theta_{ij}(s)\left[k_i^p(1 - sh_{jv}) + k_i^v s + k_i^a s^2\right]\delta_{j,0}(s)\end{aligned} \tag{3.56}$$

Taking the expectation on both sides of (3.56) and using the fact that $\omega_i(s)$ is zero mean, we have

$$\mathbf{E}\{\delta_{i,0}(s)\} = \mathbf{E}\left\{\sum_{j\in\mathbf{I}_i} H_{ij}(s)\delta_{j,0}(s)\right\} \tag{3.57}$$

with transfer function

$$H_{ij}(s) = \frac{\theta_{ij}(s)\left[k_i^p(1-sh_{jv}) + sk_i^v + s^2k_i^a\right]}{\varsigma_i s^3 + s^2 + \sum_{j\in\mathbf{I}_i}\theta_{ij}(s)\left[k_i^p(1-sh_{iv}) + sk_i^v + s^2k_i^a\right]} \tag{3.58}$$

Then, we give the following lemma about the string stability specification.

Lemma 3.2 A vehicular platoon is stochastic string stable (i.e. $\mathbf{E}\{|G(jw)|\} \le 1$) if the transfer function $H_{ij}(s)$ in (3.58) satisfies:

$$\mathbf{E}\left\{\left\|H_{ij}(s)\right\|_\infty\right\} \le \frac{1}{\mu_i}, \quad j \in \mathbf{I}_i \tag{3.59}$$

where μ_i denotes the number of neighbors in the set $\mathbf{I}_i$.

Proof: Based on the Parseval's theorem [15], we know that

$$\begin{aligned}\mathbf{E}\{\|\delta_{i,0}(t)\|_2^2\} &= \mathbf{E}\left\{\int_{-\infty}^{\infty}|\delta_{i,0}(t)|^2dt\right\} = \frac{1}{2\pi}\mathbf{E}\left\{\int_{-\infty}^{\infty}|\delta_{i,0}(jw)|^2dw\right\}\\ &= \frac{1}{2\pi}\mathbf{E}\left\{\int_{-\infty}^{\infty}\left|\sum_{j\in\mathbf{I}_i}H_{ij}(jw)\delta_{j,0}(jw)\right|^2dw\right\}\end{aligned} \tag{3.60}$$

where $\|\delta_{i,0}(t)\|_2^2 = \int_{-\infty}^{+\infty}|\delta_{i,0}(t)|^2dt$ denotes the square of the L_2 norm of $\delta_{i,0}(t)$.

According to the Cauchy-Schwarz inequality [16], we have

$$\begin{aligned}\mathbf{E}\{\|\delta_{i,0}(t)\|_2^2\} &\le \frac{1}{2\pi}\mathbf{E}\left\{\sum_{j\in\mathbf{I}_i}\delta_{j,0}(jw)^T H_{ij}^T(jw)H_{ij}(jw)\delta_{j,0}(jw)\right\}\\ &\le \mathbf{E}\left\{\sum_{j\in\mathbf{I}_i}\sup_w|H_{ij}(jw)|^2\frac{1}{2\pi}{}^*\int_{-\infty}^{\infty}\left(\sum_{j\in\mathbf{I}_i}\delta_{j,0}^2(jw)\right)dw\right\}\\ &\le \mathbf{E}\left\{\sum_{j\in\mathbf{I}_i}\left(\|H_{ij}(jw)\|_\infty^2\cdot\|\delta_{j,0}(t)\|_2^2\right)\right\}\end{aligned} \tag{3.61}$$

By the definition of the transfer function $G(s)$ in (3.17), it is obvious that

$$\delta_{j,0}(s) = G^{j-i}(s)\delta_{i,0}(s) \tag{3.62}$$

On substituting (3.62) into (3.61), we can obtain

$$\mathbf{E}\{\|\delta_{i,0}(t)\|_2^2\} \le \mu_i\mathbf{E}\left\{\|\delta_{i,0}(t)\|_2^2\sum_{j\in\mathbf{I}_i}\left(\|H_{ij}(jw)G^{j-i}(jw)\|_\infty^2\right)\right\} \tag{3.63}$$

which means

$$\mathbf{E}\left\{\|H_{ij}(jw)G^{j-i}(jw)\|_\infty^2\right\} \le \frac{1}{\mu_i} \tag{3.64}$$

According to condition (3.59) in Lemma 3.2, it is obvious that $\mathbf{E}\left\{\|G^{j-i}(jw)\|_\infty^2\right\} = \sup_{w>0}$ $\mathbf{E}\left\{|G^{j-i}(jw)|^2\right\} \le 1$. Therefore, $\mathbf{E}\{\sup_{w>0}|G(jw)|\} \le 1$, that is, stochastic string stability is guaranteed. This completes the proof.

Based on Lemma 3.2, we give the following theorem to find the conditions for controller gain such that the defined stochastic string stability is guaranteed.

Theorem 3.3 For the platoon spacing-error system (3.57), stochastic string stability (3.17) is guaranteed if and only if either one of the following conditions holds for each follower i

$$\begin{cases} \tau_{i2} \geq 0 \\ \tau_{i1} \geq 0 \end{cases} \tag{3.65}$$

or

$$\begin{cases} \tau_{i2} \geq 0 \\ \tau_{i1} \leq 0 \\ \Delta d_i \leq 0 \end{cases} \tag{3.66}$$

where $\tilde{\theta}_i = \sum_{j\in \mathbf{I}_i} \bar{\theta}_{ij}$ and

$$\tau_{i1} = \left(\tilde{\theta}_i^2 - \mu_i^2 \bar{\theta}_{ij}^2\right)\left(k_i^a\right)^2 - 2\tilde{\theta}_i \varsigma\left(k_i^v - k_i^p h_{iv}\right) + 1 + 2\tilde{\theta}_i k_i^a$$

$$\tau_{i2} = -2\left(\tilde{\theta}_i^2 - \mu_i^2 \bar{\theta}_{ij}^2\right) k_i^p k_i^a - 2\tilde{\theta}_i k_i^p + \left(\tilde{\theta}_i^2 - \mu_i^2\right)\left(k_i^v - k_i^p h_{iv}\right)^2$$

$$\Delta d_i = \tau_{i1}^2 - 4\varsigma^2 \tau_{i2}$$

Proof: Substituting $s = jw$ into $H_{ij}(s)$ in (3.58), we can derive

$$\mathbf{E}\left\{\left|H_{ij}(jw)\right|^2\right\} = \frac{N_{ij}}{D_{ij}} = \frac{\bar{\theta}_{ij}^2\left[\left(k_i^a w^2 - k_i^p\right)^2 + w^2\left(k_i^v - k_i^p h_{jv}\right)^2\right]}{\left[\tilde{\theta}_i\left(k_i^p - k_i^a w^2\right) - w^2\right]^2 + \left[\tilde{\theta}_i\left(k_i^v - k_i^p h_{iv}\right) w - \varsigma w^3\right]^2} \tag{3.67}$$

Consequently, we obtain

$$D_{ij} - \mu_i^2 N_{ij} = \Delta_i(w) + \left(\tilde{\theta}_i^2 - \mu_i^2 \bar{\theta}_{ij}^2\right)\left(k_i^p\right)^2 \tag{3.68}$$

where $\Delta_i(w) = \varsigma w^6 + \tau_{i1} w^4 + \tau_{i2} w^2$.

By using the conditions in Theorem 3.3 with $\tilde{\theta}_i^2 - \mu_i^2 \bar{\theta}_{ij}^2 \geq 0$, it is clear that $\Delta_i(w) \geq 0$. Consequently, $D_{ij} - \mu_i^2 N_{ij} \geq 0$, which means

$$\mathbf{E}\left\{\left\|H_{ij}(s)\right\|_\infty\right\} = \mathbf{E}\left\{\sup_{w>0}\left|H_{ij}(jw)\right|^2\right\} \leq \frac{1}{\mu_i}$$

Then, according to Lemma 3.2, we know that stochastic string stability is satisfied.

Combined with the results of the string stability analysis, Algorithms 3.1 and 3.2 are proposed to solve the combinatorial search problem defined in (3.37) and (3.47), respectively. As a result, the optimal IVCT and associated feedback controller can be jointly determined for platoons with fixed and periodic switching IVCT, respectively. Algorithms 3.1 and 3.2 have a low computational cost because they only need to be executed once at time slot $k = 0$. In addition, it is worth noting that the sum term in (3.37) and (3.47) for the solution to the recursion (3.24) and (3.51) can be calculated offline.

Algorithm 3.1 Platoon with Fixed IVCT

Initialization: Choose initial values for parameters $\alpha(0)$, $Q(0)$, $R(0)$, $\Delta\alpha$, and horizon time length F, where $\Delta\alpha$ denotes a certain step. Let $\alpha = \alpha(0)$, $Q = Q(0)$, and $R = R(0)$.

Step 1: According to (3.37) and (3.38), at time $k = 0$, the control center determines the optimal communication topology $\mathcal{H}^*$ and the associated feedback controller gain $K_{\mathcal{H}^*}$ by the following steps:

1) Solve $P_{\mathcal{H}_1}$ using (3.24) with topology $\mathcal{H}_1$ in set $\mathcal{M}$ and compute $J_F(x(0), \mathcal{H}_1, \mu(x(0), \mathcal{H}_1))$ with initial tracking errors $x(0)$. Let $J_{\min} = J_F(x(0), \mathcal{H}_1, \mu(x(0), \mathcal{H}_1))$ and set $m = 2$.
2) Solve $P_{\mathcal{H}_m}$ using (3.24) with topology $\mathcal{H}_m$ and compute $J_F(x(0), \mathcal{H}_m, \mu(x(0), \mathcal{H}_m))$.
3) If $J_{\min} > J_F(x(0), \mathcal{H}_m, \mu(x(0), \mathcal{H}_m))$, then $J_{\min} = J_F(x(0), \mathcal{H}_m, \mu(x(0), \mathcal{H}_m))$, $\mathcal{H}^* = \mathcal{H}_m$ and $K_{\mathcal{H}^*} = K_{\mathcal{H}_m}$. If index number $m = q$, then find the optimal IVCT $\mathcal{H}^*$ and associated control gain $K_{\mathcal{H}^*}$ and go to **Step 2**.
4) Increase m and return to Item 2).

Step 2: The control center verifies the string stability conditions of Theorem 3.3 for the obtained control gain $K_{\mathcal{H}^*}$. If $K_{\mathcal{H}^*}$ is not feasible, then set $\alpha = \alpha + \Delta\alpha$ and return to **Step 1** to recompute the $\mathcal{H}^*$ and controller gain $K_{\mathcal{H}^*}$. If $\alpha \geq 1$, then set a new group of parameters $\alpha(0)$, $Q(0)$, and v and return to **Step 1**.

Step 3: The control center allocates the U-subframes for intervehicle communication in platoon according to $\mathcal{H}^*$. Then, each follower updates its control input by using $K_{\mathcal{H}^*}$ and the kinematic status of the CAM collected from its neighbors.

Algorithm 3.2 Platoon with Periodic Switching IVCT

Initialization: Choose initial values for parameters $\alpha(0)$, $Q(0)$, $R(0)$, and $\Delta\alpha$ and horizon time length F, where $\Delta\alpha$ denotes a certain step. Let $\alpha = \alpha(0)$, $Q = Q(0)$, and $R = R(0)$.

Step 1: According to (3.47) and (3.48), at time $k = 0$, the control center determines the optimal communication topology set $\vec{\mathcal{H}}_T^*$ and the associated feedback controller gain $\hat{K}_{\vec{\mathcal{H}}_T^*}$ by the following steps:

1) Solve $\hat{P}_T^1$ using (3.49) with topology set $\vec{\mathcal{H}}_T^1$ in $\mathcal{M}_T$ and compute $J_F(x(0), \mathcal{H}_1, \mu(x(0), \mathcal{H}_1))$ with initial tracking errors $x(0)$. Let $J_{\min} = J_F\left(x(0), \vec{\mathcal{H}}_T^1, \mu(x(0), \mathcal{H}_m)\right)$, and set $m = 2$.
2) Solve $\hat{P}_T^m$ using (3.49) with topology set $\vec{\mathcal{H}}_T^m$ and compute $J_F(x(0), \mathcal{H}_m, \mu(x(0), \mathcal{H}_m))$.
3) If $J_{\min} > J_F\left(x(0), \vec{\mathcal{H}}_T^m, \mu(x(0), \mathcal{H}_m(\tau_0))\right)$, then set $J_{\min} = J_F\left(x(0), \vec{\mathcal{H}}_T^m, \mu(x(0), \mathcal{H}_m(\tau_0))\right)$, $\vec{\mathcal{H}}_T^* = \vec{\mathcal{H}}_T^m$ and $\hat{K}_{\mathcal{H}^*} = \hat{K}_{\vec{\mathcal{H}}_T^m}$. If the index number $m = q$, then find the optimal IVCT set $\vec{\mathcal{H}}_T^*$ with associated control gain $\hat{K}_{\vec{\mathcal{H}}_T^*}$ and go to **Step 2**.
4) Increase m and return to Item 2).

Step 2: The control center verifies the string stability condition of Theorem 3.3 for each $K_{\mathcal{H}^*}(\tau)$, $\tau \in \{0, 1, 2, ..., T-1\}$ of $\hat{K}_{\vec{\mathcal{H}}_T^*}$. If $\hat{K}_{\vec{\mathcal{H}}_T^*}$ is not feasible, then update $\alpha = \alpha + \Delta\alpha$ and return to **Step 1** to recompute the $\vec{\mathcal{H}}_T^*$ and controller gain matrix $\hat{K}_{\vec{\mathcal{H}}_T^*}$. If $\alpha \geq 1$, then set a new group of parameters $\alpha(0)$, $Q(0)$, v and return to **Step 1**.

Step 3: The control center allocates the U-subframes for intervehicle communication in the platoon according to the communication topology $\mathcal{H}^*(\tau_k)$ from the set $\vec{\mathcal{H}}_T^*$. Then, each follower updates its control input by using $K_{\mathcal{H}^*}(\tau_k)$ of $\hat{K}_{\vec{\mathcal{H}}_T^*}$ and kinematic status of the CAMs collected from its neighbors.

Remark 3.2 Algorithms 3.1 and 3.2 can also be executed at time slot $k \neq 0$. Under such circumstances, the control center requires obtaining the tracking error vector $x(k)$ for solving the combinatorial search problem defined in (3.37) and (3.47). Considering the CAM dissemination mechanism, it is possible that not all vehicles can communicate with the control center directly to transmit their state information (e.g. position, velocity, and acceleration). Because the spacing error $\delta_{i,0}(t)$ of each follower is computed more easily by the control center according to the position information provided by the global positioning systems, tracking error vector $x(k)$ can be estimated by using the following observer.

$$\hat{x}(k) = \mathcal{A}\hat{x}(k-1) + \mathcal{B}(\theta(k-1))u(k-1) + L(k)(y(k) - C\hat{x}(k-1)) + \mathcal{B}_1\omega(k-1) \tag{3.69}$$

where $L(k)$ is the observer gain. $y(k) = [y_1(k), \quad y_2(k), \quad ..., \quad y_N(k)]^T = Cx(k)$ denotes the output vector with $y_i(k) = \overline{C}_i x_i(k)$, $\overline{C}_i = [1 \quad 0 \quad 0]$ and $C = \mathrm{diag}\{\overline{C}_1, \overline{C}_2, ..., \overline{C}_N\}$. The design of the observer (3.69) is performed as described in [17].

3.4 Simulation Studies

In this section, numerical simulations are presented to verify the effectiveness of the proposed co-design algorithm for IVCT assignment and control. It is assumed that the vehicular platoon has one leader and four followers. The total simulation time is 100 seconds. In the simulation, a Monte Carlo method was used to evaluate the tracking errors in probability. In particular, all the simulation examples are repeated 100 times from 0 to 100 seconds. The average of these 100 samples is used to approximate the expectation of the tracking errors.

In the simulation, the following realistic model of nonlinear vehicle dynamics is used.

$$\begin{aligned}
\dot{z}_i(t) &= v_i(t) \\
\dot{v}_i(t) &= \frac{1}{m_i}\left(\varpi_{T,i}\frac{T_i(t)}{R_i} - C_{A,i}v_i^2(t) - m_i g f\right) \\
\varsigma\dot{T}_i(t) + T_i(t) &= T_{i,des}(t)
\end{aligned} \tag{3.70}$$

Table 3.1 Vehicle parameters used in simulation.

Parameter	Notation	Value
Aerodynamic drag coefficient	$C_{A,i}$	0.492
Acceleration of gravity	g	0.98 m/s^2
Coefficient of rolling resistance	f_i	0.01
Mechanical efficiency of the driveline	$\varpi_{T,i}$	0.9
Mass of the vehicle	m_i	2.9×10^3 kg
Inertial lag	ς	0.20 s
Time gap	h_v	0.7 s
Standstill distance	d	3 m
Tire radius	R_i	0.3 m

where m_i is the mass of the vehicle, $C_{A,i}$ is the combined aerodynamic drag coefficient, g is the acceleration due to gravity, f is the coefficient of rolling resistance, T_i denotes the actual driving/breaking torque, $T_{i,des}$ is the desired driving/breaking torque, R_i denotes the tire radius, and $\varpi_{T,i}$ is the mechanical efficiency of the driveline. The values of these parameters are listed in Table 3.1.

As shown in Figure 3.3, the platoon control system has a hierarchical architecture, which consists of an upper layer and lower layer controller. The upper layer determines the desired acceleration a_{ides}. The lower layer generates the throttle torque α_{thrdes} or brake torque b_{brkdes} to track a_{ides} by using the following inverse model.

$$T_{i,des}(t) = \frac{1}{\varpi_{T,i}}\left(C_{A,i}v_i(t)(2\varsigma\dot{v}_i + v_i) + m_i g f + m_i u_i'(t)\right)R_i \tag{3.71}$$

where $u_i'(t)$ is the designed control input by the sampled-data feedback controller (3.7). Taking the desired driving/breaking torque (3.71) into the nonlinear vehicle dynamics (3.70), one can derive the linear vehicle dynamic model in (3.1).

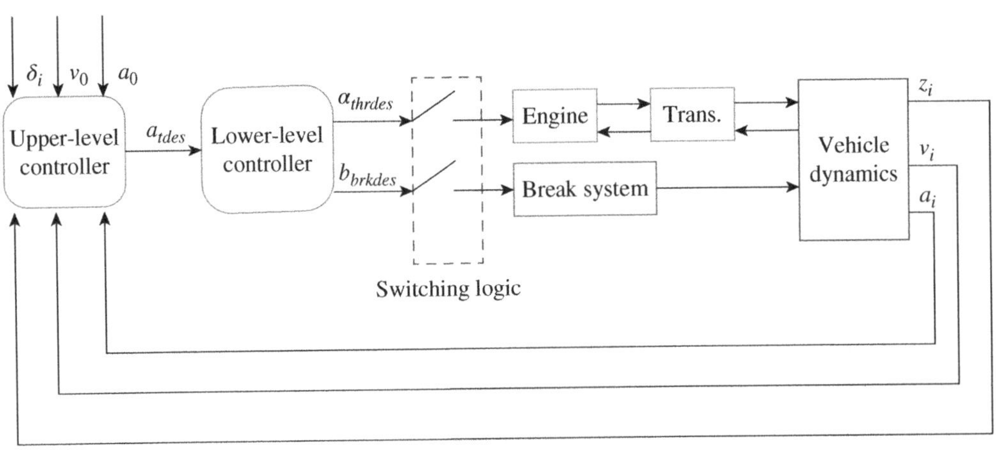

Figure 3.3 Controller structure for vehicle platoons.

At each sampling time instant, the acceleration $a_0(kh)$ of the leader is set as follows:

$$a_0(kh) = \begin{cases} 0\,\mathrm{m/s^2}, & k \leq 10 \\ 2\,\mathrm{m/s^2}, & 10 < k \leq 15 \\ -1\,\mathrm{m/s^2}, & 15 < k \leq 20 \\ 0\,\mathrm{m/s^2}, & k > 20 \end{cases}$$

As shown in Figure 3.4, the desired velocity and acceleration simulate the scenarios when vehicle platoon operates the acceleration and deceleration. The initial tracking errors of the platoon are zero and disturbance $\omega_i(t)$ is simulated by white noise with an expectation of zero and the variance.

On the other hand, it is assumed that the LTE-V2V network has a carrier frequency of 2 GHz and bandwidth of 10 MHz. As specified in the stander, an LTE frame is composed of 10 subframes of 1 ms each. The transmission power is $\Xi_{ij}(kh) = 10\,\mathrm{dBm}$ and the targeted SINR threshold is $\gamma_{V2V} = 10\,\mathrm{dB}$. According to (3.8) with the parameters of the V2V channel model specified in [9], the packet dropouts' probabilities of each V2V communication link are approximately equal to 0.1. Therefore, the packet-arriving probabilities are $\overline{\theta}_{ij} = 0.9$. The update period of the CAM within the platoon is set at $h = 10$ ms, which is equal to the length of an LTE framework.

In this section, we present the comparative simulation results for a five-vehicle platoon under the fixed IVCT using the optimal controller (3.26) and controller in [6], respectively. To evaluate the control performance, during the time interval $[0, Fh)$, the LQ cost function is computed by

$$J(Fh) = \sum_{\upsilon=0}^{F} \left(\|x(\upsilon h)\|_Q^2 + \|u(\upsilon h)\|_R^2 \right) \tag{3.72}$$

with matrix $Q = I_{12}$ and $R = I_4$.

Firstly, we present the comparative simulation for a platoon with fixed IVCT without considering the packet dropouts and external disturbance, since the controller designed in [6] ignores these negative effects. Here, we consider the case that the zero initial condition does not hold, i.e. the initial

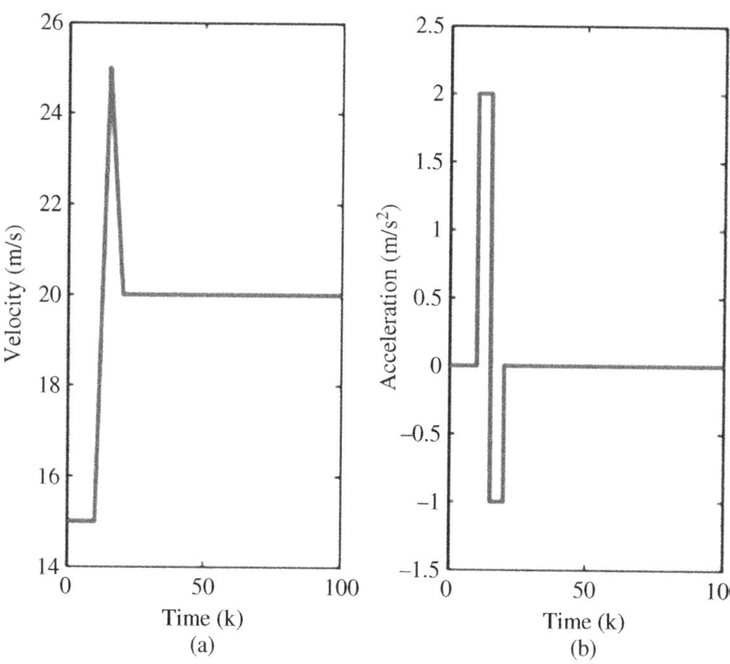

Figure 3.4 (a) Velocity and (b) acceleration of leader.

tracking errors of each follower are not equal to zero. In [6], the feedback controller and communication topology were jointly designed in the z-domain. For convenience, the order of approximation for the solution to ARE is set at $n = 1$. According to the simulation parameters listed in Table 3.1, using the method in [6], we obtain the results on the co-design of one vehicle look-ahead/look-behind (i.e. BPF) topology and feedback controller. Owing to the space limit, the obtained results on the solution $\breve{P}(z)$ of ARE and the controller gain $\breve{L}(z)$ are omitted. In contrast, by using Algorithm 3.1 for platoon under BPF topology, the gains of the optimal feedback controller (3.26) for each follower are $\overline{K}_1(\mathcal{H}) = [0.3 \quad 7 \quad 9]$, $\overline{K}_2(\mathcal{H}) = [0.25 \quad 6 \quad 8.1]$, $\overline{K}_3(\mathcal{H}) = [0.32 \quad 5.7 \quad 9.2]$, and $\overline{K}_4(\mathcal{H}) = [0.41 \quad 6.8 \quad 8.9]$, where the discounting factor is $\alpha = 0.07$ and the packet-arriving probability of the ijth communication link is $\overline{\theta}_{ij} = 1$. By using the controller obtained in our study and [6], the tracking performance of the platoon is shown in Figures 3.5 and 3.6. In Figure 3.6, the equivalent spatial of the tracking errors is obtained by means of inverse bilateral Z-transformation [6]. During the time interval [0, 20 seconds], it is observed that string stability is not satisfied for the platoon by using the designed controller in (3.26). Since the string stability in our study is based on the frequency-domain definition, the zero initial condition is required. In contrast, vehicular platoon can guarantee string stability with the controller designed in [6] because that string stability is defined in the z-domain, which can mitigate any introduced initial condition perturbation. Under the non-zero initial tracking errors, the platoon with the controller designed in [6] has a better string stability performance than ours. However, when tracking errors go into stable, string stability can be guaranteed with the feedback controller designed in this study.

Then, we consider the simulation scenario for a platoon subject to packet dropouts and disturbance. Here, the initial tracking errors of each follower are set to zero. Under such circumstances, by using Algorithm 3.1, the gain of the optimal feedback controller in (3.26) is recomputed as $\overline{K}_1(\mathcal{H}) = [1.3 \quad 5.2 \quad 8.9]$, $\overline{K}_2(\mathcal{H}) = [1.5 \quad 6.6 \quad 9.2]$, $\overline{K}_3(\mathcal{H}) = [1.7 \quad 5.6 \quad 7.8]$, and $\overline{K}_4(\mathcal{H}) = [1.6 \quad 5.4 \quad 8.3]$. By using the controller obtained in (3.26) and [6], the tracking performance of the platoon is shown in Figures 3.7 and 3.8. The LQ cost function in Figures 3.7 and 3.8 are 46 519 and 79 620 dB, respectively. From these figures, it is clear that the platoon has better control performance with a smaller LQ cost function by using the proposed controller in our study. Because packet dropouts and external disturbances are ignored in the co-design of [6]. Note that the velocity and acceleration shown in Figure 3.7a and b are not satisfied string stability requirement since the obtained conditions of Theorem 3.3 to complement the feedback gain is only available for the string stability defined by spacing error δ_{i0}.

Instead of the BPF topology, here we consider the simulation scenario for a platoon with an all-to-all topology, where each follower can communicate with other vehicles in the platoon to exchange information. By using Algorithm 3.1 with the same simulation parameters, the feedback controller for the platoon under the all-to-all topology is obtained as $\overline{K}_1(\mathcal{H}) = [0.8 \quad 9 \quad 10]$, $\overline{K}_2(\mathcal{H}) = [0.64 \quad 7.1 \quad 7]$, $\overline{K}_3(\mathcal{H}) = [0.71 \quad 8.3 \quad 9.6]$ and $\overline{K}_4(\mathcal{H}) = [0.93 \quad 8.7 \quad 10.4]$. Using the obtained optimal feedback controller, the tracking performance of the vehicular platoon is shown in Figure 3.9. Compared with the tracking performance in Figure 3.7, it is clear that platoon control with all-to-all topology has better control performance than the BPF topology (see the overshoot, convergence speed, and maximum spacing error). Moreover, the value of the LQ cost function in Figure 3.9 is 34 599 dB, which is smaller than the value in Figure 3.7. According to the analysis results in [3, 4], it is true that the more followers obtain information from the leader, the better the control performance. Different from BPF topology, each follower can directly obtain the information from the leader under the all-to-all topology. Note that the eigenvalues of the topology Laplacian matrix can influence the control performance of the platoon. A platoon with the BPF topology has larger tracking errors since the smaller eigenvalue of the topology Laplacian matrix reduces the effect of feedback control.

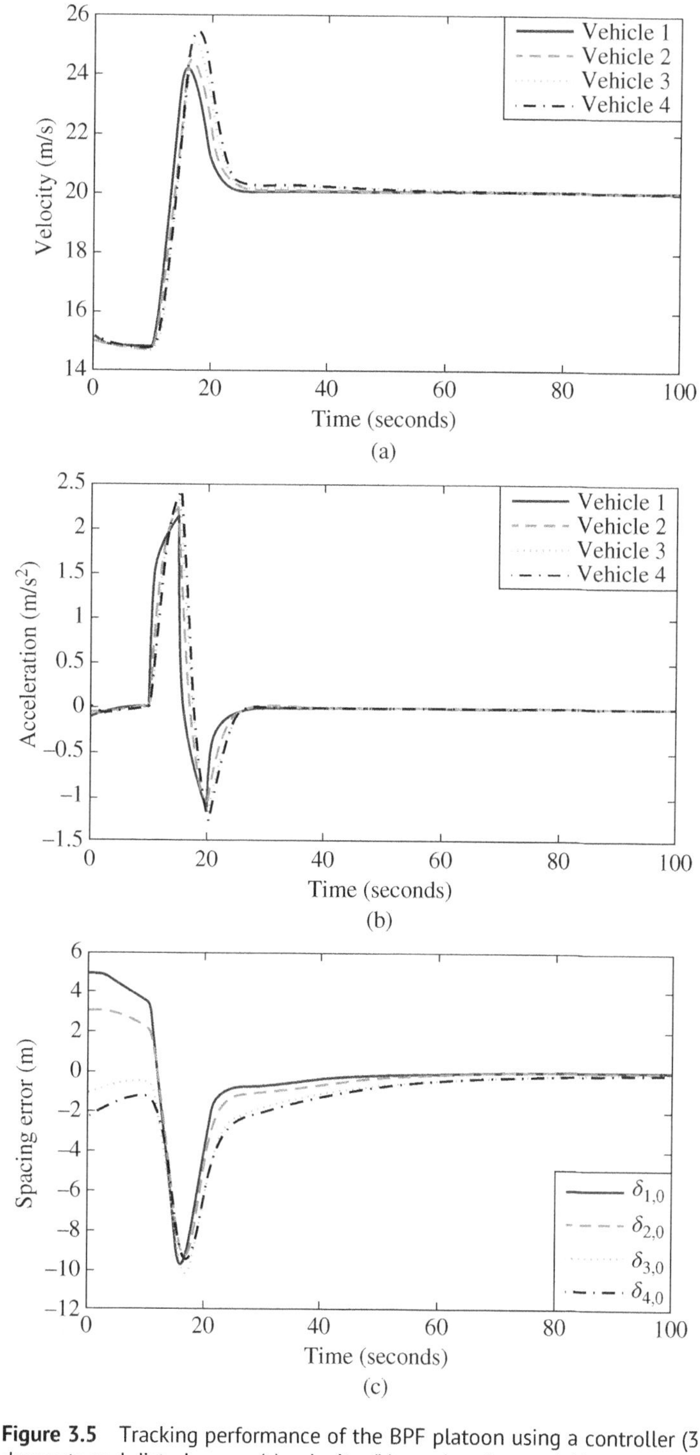

Figure 3.5 Tracking performance of the BPF platoon using a controller (3.26) without considering packet dropouts and disturbances: (a) velocity, (b) acceleration, and (c) spacing error.

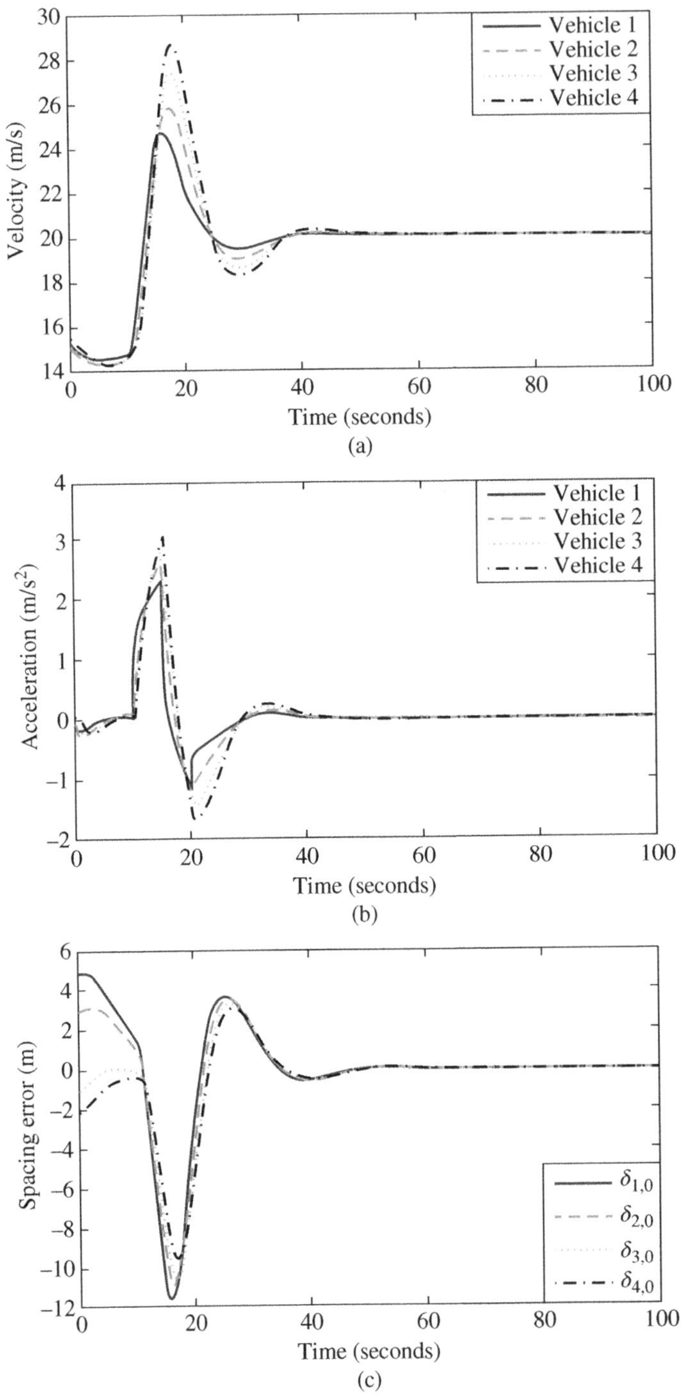

Figure 3.6 Tracking performance of the BPF platoon using the co-design method in [6] without considering packet dropouts and disturbance: (a) velocity, (b) acceleration and (c) spacing error. *Source:* Adapted from Firooznia et al. [6].

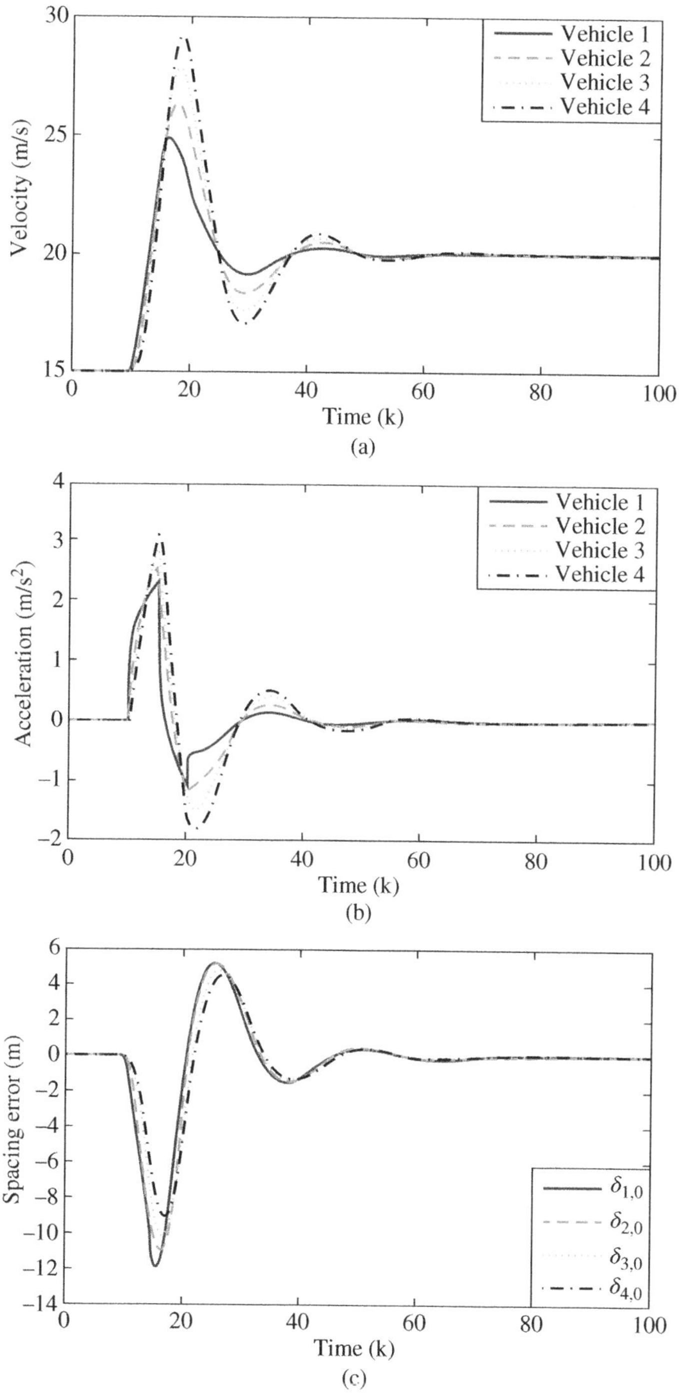

Figure 3.7 Tracking performance of the BPF platoon subject to packet dropouts and disturbance with controller (3.26): (a) velocity, (b) acceleration, and (c) spacing error.

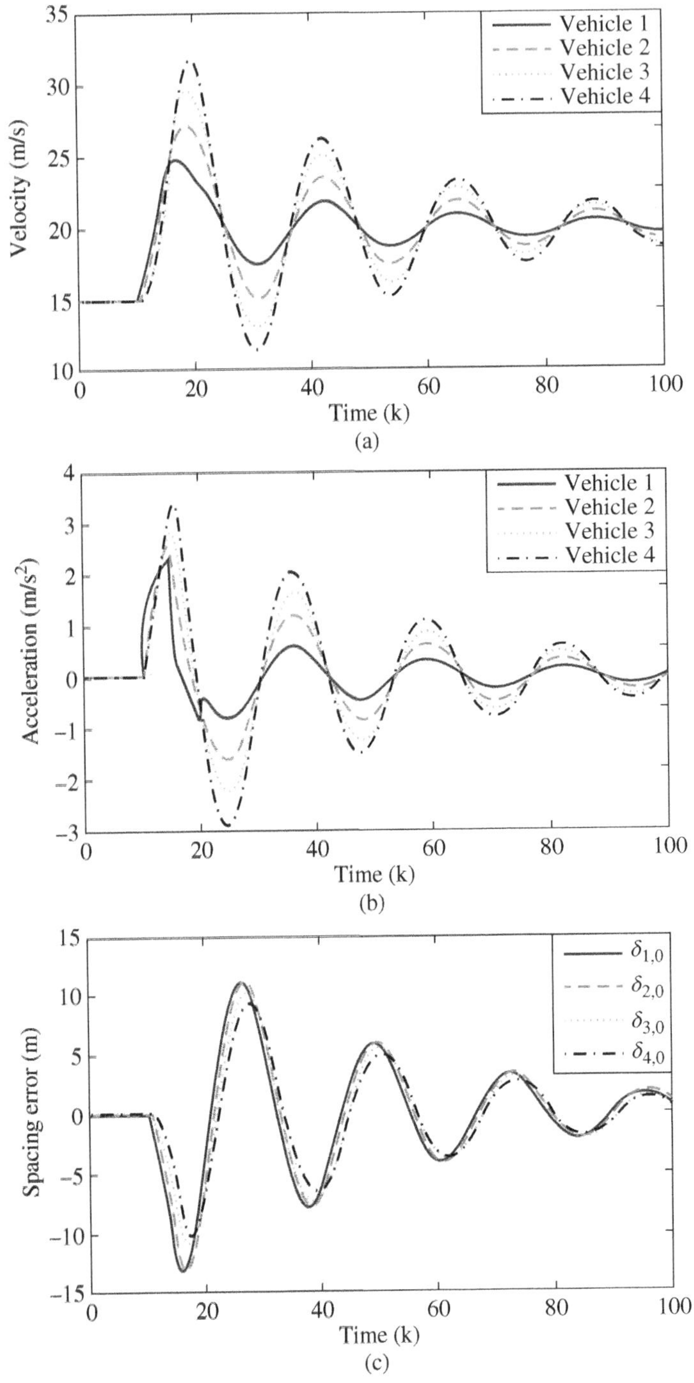

Figure 3.8 Tracking performance of the BPF platoon subject to packet dropouts and disturbance with controller designed in [6]: (a) velocity, (b) acceleration, and (c) spacing error. *Source:* Adapted from Firooznia et al. [6].

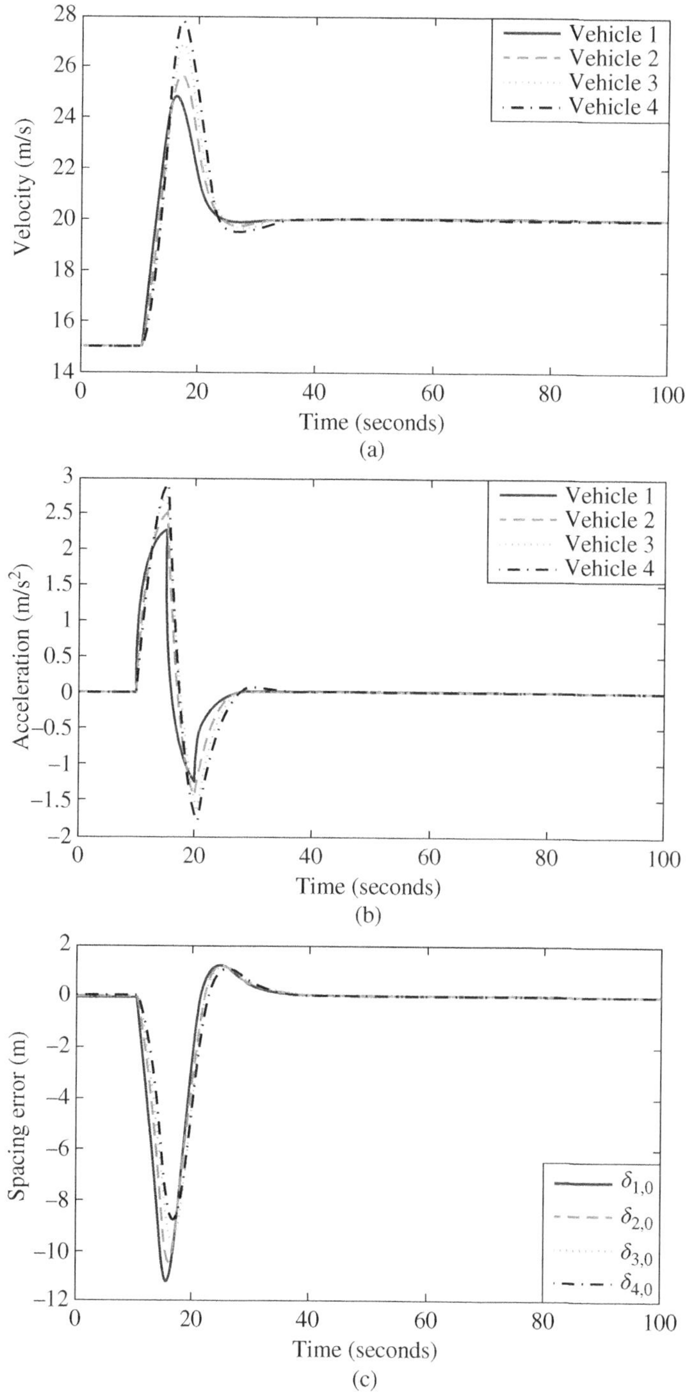

Figure 3.9 Tracking performance of the platoon with all-to-all topology using the co-design method in our study: (a) velocity, (b) acceleration, and (c) spacing error.

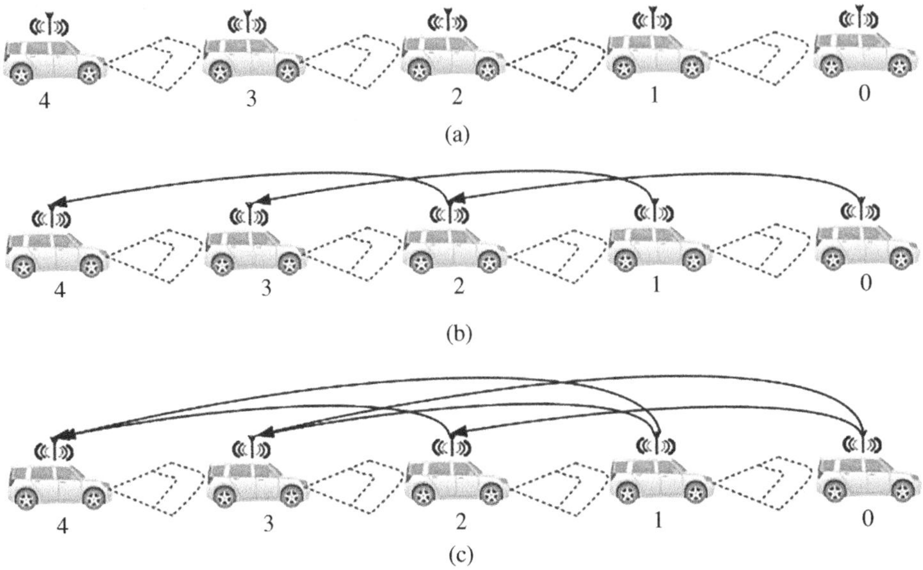

Figure 3.10 Intervehicle communication topologies of vehicle platoons in simulation. (a) 1-VLA, (b) 2-VLA, (c) 3-VLA.

Similar to the simulation scenario in Figures 3.8 and 3.9, in this section, we present the comparative simulation results for a platoon with periodic switching IVCT which is caused by the dynamically changing status of the V2V communication link. As shown in Figure 3.10, n vehicles' look-ahead (n-VLA) topology is used for platoon with $n = 1$, $n = 2$, and $n = 3$, respectively. In the 1-VLA topology, the distance between the i-th follower and its preceding vehicle is directly measured by the onboard sensors. As a result, 1-VLA topology does not require intervehicle communication because the velocity error $e^{v}_{i,i-1}$ and acceleration error $e^{a}_{i,i-1}$ can be estimated by using the computed spacing error $\delta_{i,i-1}$, for example, $e^{v}_{i,i-1}(k) = (\delta_{i,i-1}(k) - \delta_{i,i-1}(k-1))/h$ and $e^{a}_{i,i-1}(k) = (e^{v}_{i,i-1}(k) - e^{v}_{i,i-1}(k-1))/h$ [18]. In 2-VLA and 3-VLA topologies, the velocity and acceleration of the corresponding vehicles are transmitted via the LTE-V2V network. According to the status of the IVC link, it is assumed that the IVCTs are periodically switched between 1-VLA, 2-VLA, and 3-VLA with a period of 100 ms, namely period $T = 10$. Then, the control center periodically assigns the IVCT for the platoon according to the order given in the set $\overrightarrow{\mathcal{H}}_T^{m}$, $m = 1, 2, \ldots, 6$, where all the available $\overrightarrow{\mathcal{H}}_T^{m}$ in set $\mathcal{M}_T$ are shown in Table 3.2. Here, we aim to jointly determine the IVCT-assigning strategy and the associated controller gain such that individual vehicle stability and string stability are guaranteed with the minimized cost function $J(Fh)$ in (3.72). By using Algorithm 3.2, we found that the cost function of the platoon is minimized with the communication set $\overrightarrow{\mathcal{H}}_T^{6}$, where the associated feedback controller gains and the value of the cost function are listed in Tables 3.3 and 3.4, respectively.

By periodically assigning the optimal communication topology $\mathcal{H}(\tau)$ from set $\overrightarrow{\mathcal{H}}_T^{6}$ with the associated controller gain $\overline{K}_i(\mathcal{H}(\tau))$, $\tau = 0, 1, \ldots, 9$, the control performance of the platoon is shown in Figure 3.11. It is clear that both individual vehicle stability and string stability are satisfied. In the same simulation scenario, Figure 3.12 gives the control performance of the platoon by using the

Table 3.2 Periodic switching communication topologies from set $\vec{\mathcal{H}}_T^m$.

$\mathcal{H}(\tau)$ / $\vec{\mathcal{H}}_T^m$	$\tau = 0$	$\tau = 1$	$\tau = 2$	$\tau = 3$	$\tau = 4$	$\tau = 5$	$\tau = 6$	$\tau = 7$	$\tau = 8$	$\tau = 9$
$m = 1$	1-VLA	1-VLA	1-VLA	1-VLA	2-VLA	2-VLA	2-VLA	3-VLA	3-VLA	3-VLA
$m = 2$	1-VLA	1-VLA	1-VLA	3-VLA	3-VLA	3-VLA	2-VLA	2-VLA	2-VLA	2-VLA
$m = 3$	3-VLA	3-VLA	2-VLA	2-VLA	2-VLA	2-VLA	2-VLA	1-VLA	1-VLA	1-VLA
$m = 4$	1-VLA	1-VLA	1-VLA	1-VLA	1-VLA	2-VLA	2-VLA	1-VLA	2-VLA	3-VLA
$m = 5$	1-VLA	1-VLA	2-VLA	3-VLA	1-VLA	2-VLA	3-VLA	1-VLA	2-VLA	3-VLA
$m = 6$	3-VLA	3-VLA	3-VLA	3-VLA	3-VLA	3-VLA	1-VLA	2-VLA	2-VLA	2-VLA

Table 3.3 Controller gain for each follower associated with communication topology from set $\vec{\mathcal{H}}_T^6$.

	$\overline{K}_i(\mathcal{H}(\mathbf{0}))$	$\overline{K}_i(\mathcal{H}(\mathbf{1}))$	$\overline{K}_i(\mathcal{H}(\mathbf{2}))$	$\overline{K}_i(\mathcal{H}(\mathbf{3}))$	$\overline{K}_i(\mathcal{H}(\mathbf{4}))$	$\overline{K}_i(\mathcal{H}(\mathbf{5}))$	$\overline{K}_i(\mathcal{H}(\mathbf{6}))$	$\overline{K}_i(\mathcal{H}(\mathbf{7}))$	$\overline{K}_i(\mathcal{H}(\mathbf{8}))$	$\overline{K}_i(\mathcal{H}(\mathbf{9}))$
$i = 1$	[1.6 5.2 6.1]	[1.4 4.6 5.3]	[1.3 2.8 5.1]	[1.1 3.4 5.2]	[1.0 4.8 5.35]	[1.3 2.4 6.2]	[1.2 5.1 4.4]	[1.0 2.8 6.1]	[1.5 4.8 6.2]	[1.8 5.8 7.0]
$i = 2$	[1.7 6.4 7.2]	[1.8 5.6 8.7]	[1.6 5.2 6.1]	[0.9 3.6 7.1]	[1.3 2.8 6.7]	[1.2 2.6 5.3]	[0.8 1.6 6.0]	[0.9 4.2 7.9]	[0.8 2.4 6.7]	[1.2 3.7 6.2]
$i = 3$	[1.5 6.2 7.3]	[1.9 6.1 6.8]	[1.6 5.7 6.9]	[1.4 4.5 5.8]	[0.6 5.4 6.2]	[0.7 5.1 7.2]	[0.9 5.7 7.2]	[0.8 6.3 7.4]	[0.9 5.1 8.2]	[1.6 5.1 7.2]
$i = 4$	[1.3 6.5 7.0]	[1.2 5.2 8.1]	[1.1 4.2 7.3]	[2.0 4.1 5.0]	[1.2 3.1 7.3]	[1.4 4.2 8.1]	[1.1 4.9 7.8]	[1.4 6.8 8.3]	[1.3 6.6 7.1]	[1.1 5.9 6.4]

Table 3.4 Cost function of the platoon under periodic switching communication topologies.

	$\vec{\mathcal{H}}_T^1$	$\vec{\mathcal{H}}_T^2$	$\vec{\mathcal{H}}_T^3$	$\vec{\mathcal{H}}_T^4$	$\vec{\mathcal{H}}_T^5$	$\vec{\mathcal{H}}_T^6$
$J(Fh)$ (dB)	53 393	52 673	49 214	47 719	44 669	42 708

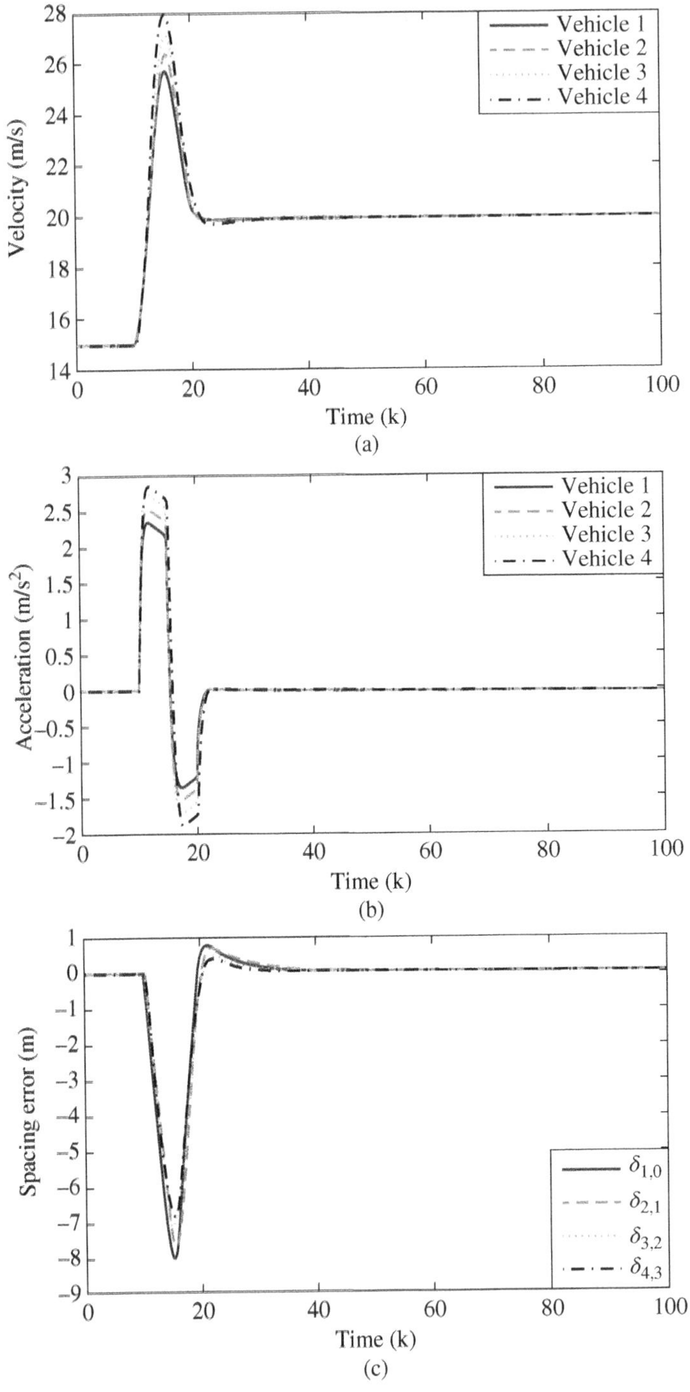

Figure 3.11 Tracking performance of the platoon with periodic switching topologies from set $\vec{\mathcal{H}}_T^6$ using the proposed co-design method: (a) velocity, (b) acceleration, and (c) spacing error.

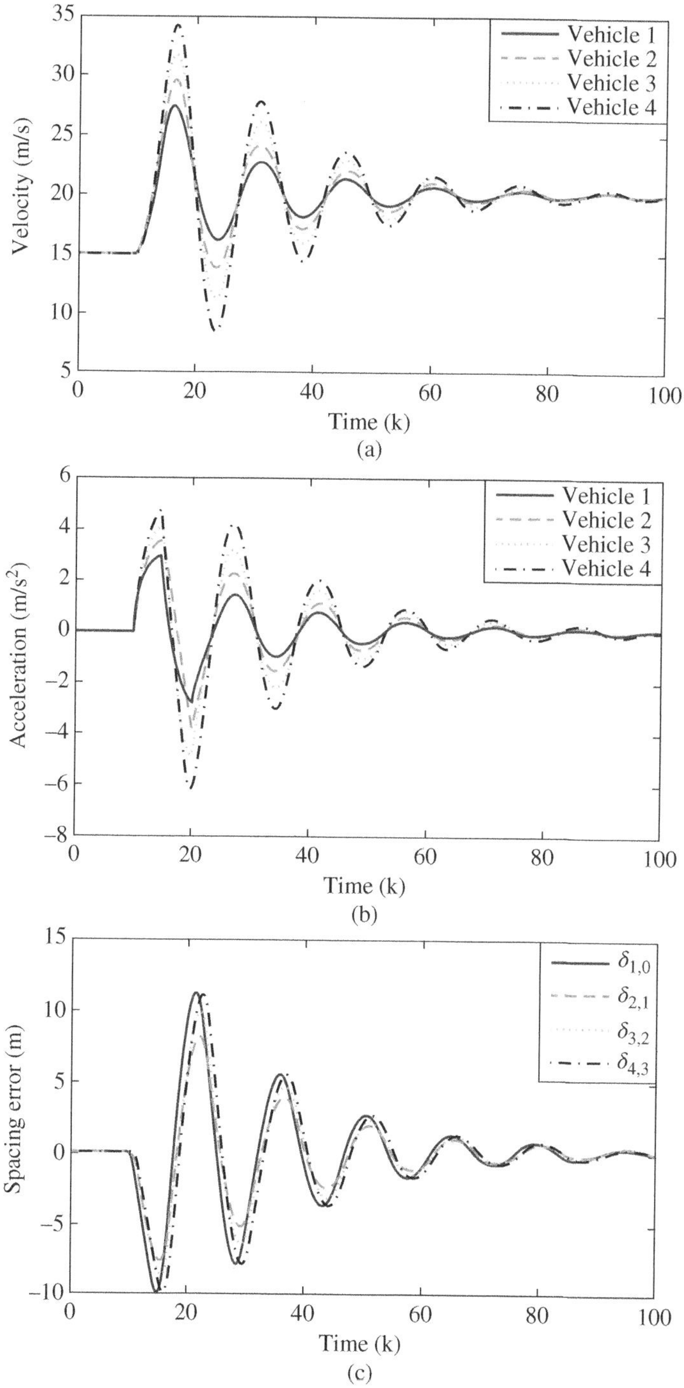

Figure 3.12 Tracking performance of the platoon with periodic switching topologies from set $_{\|x\|_p}$ using a controller in [6]: (a) velocity, (b) acceleration, and (c) spacing error. *Source:* Adapted from Firooznia et al. [6].

proposed controller in [2] subject to periodic switching IVCT from set $\vec{\mathcal{H}}_T^6$, packet dropouts and disturbances. It is observed that the control performance is degraded with a larger cost function, where the cost function of Figure 3.12 is $J(k) = 82\,513$ dB.

3.5 Conclusions and Future Work

In this study, we have investigated the IVCT assignment and control co-design for vehicle platoons in the LTE-V2V network. A sampled-data platoon control model is established considering the effect of IVCT, packet dropouts, and disturbances. Based on the optimal control and dynamic programming techniques, a unified control framework is established to jointly determine the optimal IVCT and the associated feedback controller. During the co-design procedure, both the fixed and periodic switching IVCTs are considered. The established co-design framework is implemented using the proposed algorithm, which not only guarantees individual vehicle stability and stochastic string stability but also minimizes the defined cost function. We also present numerical examples to demonstrate the effectiveness of the proposed co-design algorithm.

References

1 Willke, T.L., Tientrakool, P., and Maxemchuk, N.F. (2009). A survey of inter vehicle communication protocols and their applications. *IEEE Commun. Surveys Tuts.* 11 (2): 3–20, 2nd Quart.

2 Zheng, Y., Li, S.E., Wang, J. et al. (2016). Stability and scalability of homogeneous vehicular platoon: study on the influence of information flow topologies. *IEEE Trans. Intell. Transp. Syst.* 17 (1): 14–26.

3 Zheng, Y., Li, S., Li, K., and Wang, L.-Y. (2016). Stability margin improvement of vehicular platoon considering undirected topology and asymmetric control. *IEEE Trans. Control Syst. Technol.* 24 (4): 1253–1265.

4 Zheng, Y., Li, S.E., Li, K., and Ren, W. (2018). Platooning of connected vehicles with undirected topologies: robustness analysis and distributed H-infinity controller synthesis. *IEEE Trans. Intell. Transp. Syst.* 19 (5): 1353–1364.

5 Li, S.E., Qin, X., Li, K. et al. (2017). Robustness analysis and controller synthesis of homogeneous vehicular platoons with bounded parameter uncertainty. *IEEE/ASME Trans. Mechatron.* 22 (2): 1014–1025.

6 Firooznia, A., Ploeg, J., Wouw, N., and Zwart, H. (2017). Co-design of controller and communication topology for vehicular platooning. *IEEE Trans. Intell. Transp. Syst.* 12 (4): 1184–1194.

7 Wen, S.X. and Guo, G. (2016). Sampled-data control for connected vehicles with markovian switching topologies and communication delay. *IEEE Trans. Intell. Transp. Syst.* 21 (2): 2930–2942.

8 Li, K., Bian, Y., Li, S.E. et al. (2020). Distributed model predictive control of multi-vehicle systems with switching communication topologies. *Part C: Emerg. Technol.* 118: 1–18.

9 Study on LTE-based V2X Services (Release 14), 3GPP TR 36.885 Std., Mar. 2016.

10 Li, Z., Hu, B., Li, M., and Luo, G. (2019). String stability analysis for vehicle platooning under unreliable communication links with event-triggered strategy. *IEEE Trans. Veh. Technol.* 68 (3): 2152–2164.

11 Darbha, S., Konduri, S., and Pagilla, P.R. (2019). Benefits of V2V communication for autonomous and connected vehicles. *IEEE Trans. Intell. Transp. Syst.* 20 (5): 1954–1963.

12 Mei, J. et al. (2018). Joint radio resource allocation and control for vehicle platooning in LTE-V2V network. *IEEE Trans. Veh. Technol.* 67 (12): 12218–12230.

13 Bertsekas, D.P. (2005). *Dynamics Programming and Optimal Control (Athena Scientific Optimization and Computation)*, vol. 1. Belmont, MA: Althena Scientific.

14 Bittanti, S. and Colaneri, P. (2009). *Periodic Systems-Filtering and Control.* London: Spring-Verlag.

15 Arfken, G.B. and Weber, H.J. (1970). *Mathematical Methods for Physicists.* New York: Academic Press.

16 Steele, J.M. (2004). *The Cauchy-Schwarz Master Class: An Introduction to the Art of Mathematical Inequalities.* Cambridge University Press.

17 Quevedo, D.E., Ahlen, A., and Johansson, K.H. (2013). State estimation over sensor networks with correlated wireless fading channels. *IEEE Trans. Autom. Control* 58 (3): 581–593.

18 Liu, Y., Pan, C., Gao, H., and Guo, G. (2017). Cooperative spacing control for interconnected vehicle systems with input delays. *IEEE Trans. Veh. Technol.* 66 (12): 10692–10704.

4

Control with Communication Delay and Switching Topologies

4.1 Introduction

It is worth mentioning that a fixed network topology or a regularly switching intervehicle communication topology as studied in Chapters 1–3 is hardly the case in practice since the intervehicle connections in VANET may be varying or switching depending on circumstances and time. Owing to high mobility [1], the condition of the wireless channel varies randomly. Therefore, steady connectivity of the moving vehicles cannot be guaranteed and frequent handoffs are inevitable. How would a random switching topology and communication delays affect platoon control still remains an open and challenging problem? This chapter investigates sampled-data control of vehicular platoons with random switching topologies, communication delays, and external disturbance. We aim to establish a new framework for vehicular platoon control such that each vehicle can maintain a desired speed and keep a prespecified constant intervehicle spacing. Here, each follower is assumed to have access to its own states and those of its neighbors at each sampling instant and all sampled-data is transmitted via VANET with communication delay. To describe the varying information flow, we introduce a Markovian chain to represent the dynamical switching of communication topologies in the platoon. Then, using the theory of the Markovian jumping system with time delay, a set of sufficient conditions is derived to guarantee that the platooning tracking errors (i.e. errors in spacing, velocity, and acceleration) are mean-square exponentially stabilized. These conditions also serve as the design conditions for the controller gains of the connected vehicles. The results are extended to the case with partially unknown transition rates of the Markov chains. The contributions of this chapter are summarized as follows:

1) A Markovian chain-based sampled-data control model is established for the platoon with communication time delay, where the external disturbance is allowed to be nonlinear. The model in [2] with undirected topologies and disturbance can be seen as a special case of ours;
2) An improved stability analysis framework is proposed for the connected vehicular control system. Unlike the frequency domain analysis method in [2], here the initial conditions are not limited to zero, which makes our method more practical;
3) The controller synthesis of connected vehicles is implemented with both fixed-gain and variable-gain control methods, which can make the tracking errors mean-square exponential stabilized. The tracking errors can meet a given level of disturbance attenuation under zero initial conditions. Moreover, the results are extended to the case with partially unknown transition rates.

Connected Vehicular Systems: Communication, Control, and Optimization, First Edition. Ge Guo and Shixi Wen.

The rest of this chapter is organized as follows. Section 4.2 describes the control model for the vehicular platoon. Section 4.3 gives the internal stability analysis procedure for the connected vehicular control systems. Section 4.4 gives the design procedure for the controller. Section 4.5 illustrates the effectiveness of the proposed control method by the numerical simulations. Section 4.6 concludes the paper.

Notation $\mathbb{R}^n$ denotes the n-dimensional Euclidean space. diag{·} stands for a block-diagonal matrix. $I_3 \in \mathbb{R}^{3\times 3}$ is the expectation operator with respect to some probability measure. The notation $P \succ 0 (P \succeq 0)$ means that matrix P is real symmetric and positive definite (semi-positive-definite). $\|\cdot\|$ denotes the Euclidean norm, X^T the transpose of a matrix X. $\otimes$ the matrix Kronecker product, $*$ the symmetric elements of a matrix, $\lambda_{\min}(\cdot)$ are the minimal eigenvalues of a symmetric matrix.

4.2 Problem Formulation

Consider a platoon of $N + 1$ connected vehicles driving on a horizontal road (see Figure 4.1) in a VANET environment. Each vehicle can communicate with its neighbors to exchange information (e.g. position z_i, velocity v_i, and acceleration a_i) via VANET. Each follower has a sampled-data controller, which updates its output using the received information.

The vehicle dynamics depends on the engine, drive line, braking system, aerodynamic drag, tire friction, rolling resistance, gravitational force, etc. In order to derive a concise model for platoon control, the following reasonable assumptions are made [3–6]:

1) Tire longitudinal slip is negligible and the powertrain dynamics is integrated into a first-order inertial transfer function.
2) The vehicle body is considered to be rigid and symmetric.
3) The influences of pitch and yaw motions are neglected.
4) The driving and braking torques are controllable inputs.

Based on these assumptions, a simplified nonlinear vehicle dynamical model is given as follows:

$$\begin{cases} \dot{z}_i(t) = v_i(t) \\ \dot{v}_i(t) = \dfrac{1}{m_i}\left(\varpi_{T,i}\dfrac{T_i(t)}{R_i} - C_{A,i}v_i^2(t) - m_i g f\right) \\ \varsigma \dot{T}_i(t) + T_i(t) = T_{i,des}(t) \end{cases} \tag{4.1}$$

where m_i is the mass of the vehicle, $C_{A,i}$ is the combined aero dynamic drag coefficient, g is the acceleration due to gravity, f is the coefficient of rolling resistance, T_i denotes the actual driving/breaking torque, $T_{i,des}$ is the desired driving/breaking torque, ς is the inertial delay of vehicle dynamics, R_i denotes the tire radius, and $\varpi_{T,i}$ is the mechanical efficiency of the driveline.

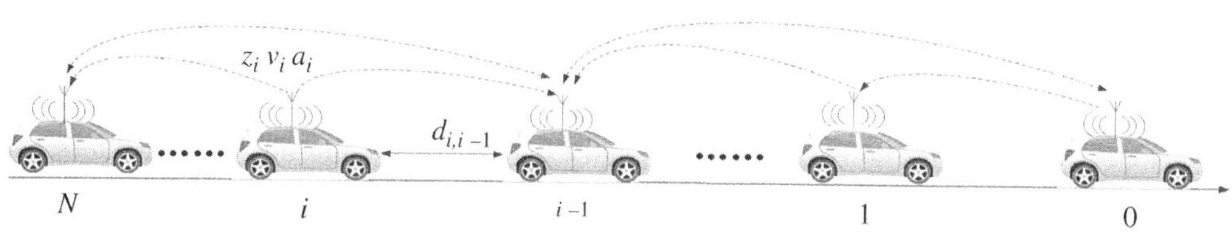

Figure 4.1 Vehicular platoon.

By introducing the following feedback linearization law [3, 4],

$$T_{i,des}(t) = \frac{1}{\varpi_{T,i}}(C_{A,i}v_i(t)(2\varsigma\dot{v}_i + v_i) + m_i g f + m_i u_i)R_i \tag{4.2}$$

where u_i is the new input to be designed, the nonlinear vehicle model (4.1) can be written as follows:

$$\begin{cases} \dot{z}_i(t) = v_i(t) \\ \dot{v}_i(t) = a_i(t) \\ \dot{a}_i(t) = -a_i(t)/\zeta + u_i(t)/\zeta \end{cases} \tag{4.3}$$

which will be used for control analysis and design. Let $x_i(t) = [z_i \ v_i \ a_i]^T$, we can then write the vehicle dynamics in a more concise form:

$$\dot{x}_i(t) = Ax_i(t) + B_1 u_i(t) \tag{4.4}$$

where

$$A = \begin{bmatrix} 0 & 1 & 0 \\ 0 & 0 & 1 \\ 0 & 0 & -\frac{1}{\varsigma} \end{bmatrix}, B_1 = \begin{bmatrix} 0 \\ 0 \\ -\frac{1}{\varsigma} \end{bmatrix}$$

In practice, the vehicle may have external disturbances from wind gust, air resistance, road slope, etc. Denoting by $\omega(x_i,t)$, the integrated external disturbance in the vehicle dynamics, we can have that

$$\dot{x}_i(t) = Ax_i(t) + B_1 u_i(t) + \omega(x_i,t) \tag{4.5}$$

It is assumed that $\omega(x_i, t)$ is uniformly Lipschitz continuous with respect to t, namely, there exists a positive constant ϱ such that the following inequality holds for all x_i, $x_0 \in \mathbb{R}^n$:

$$\| \omega(x_i, t) - \omega(x_0, t) \| \leq \varrho \, \| x_i - x_0 \| \tag{4.6}$$

The Lipschitz condition in (4.6) gives an upper bound on the change rate of the external disturbance. Due to the disturbance term $\omega(x_i, t)$, the stability analysis and controller synthesis method in [2] is not feasible.

The communication topology of the platoon of vehicles is represented by a directional and connected graph $\mathcal{G}_N = (\mathcal{V}_N, \mathcal{E}_N)$, where $\mathcal{V}_N = \{1, 2, ..., N\}$ is the set of vehicles and $\mathcal{E}_N \subseteq \mathcal{V}_N \times \mathcal{V}_N$ is the set of edges. An edge defined as $\varepsilon_{ij} \triangleq (i, j) \in \mathcal{E}_N$ implies that there exists a directed communication link from vehicle j to vehicle i, i.e. $(i, j) \in \mathcal{E}_N$. Let $N_i \triangleq \{j \mid \varepsilon_{ij} \in \mathcal{E}_N\}$ be the set of neighbors of vehicle i. $\mathcal{A} = [a_{ij}]_{n \times n}$ represents the adjacency matrix of graph $\mathcal{G}_N$, where $a_{ii} = 0$ for all i, and $a_{ij} = 1$ if $\varepsilon_{ij} \in \mathcal{E}_N$, otherwise $a_{ij} = 0$. The degree matrix of graph $\mathcal{G}_N$ is $\mathcal{D} = \text{diag}\{d_1, d_2, ..., d_n\}$, where the diagonal elements are represented as $d_i = \sum_{j \in N_i} a_{ij}$. Correspondingly, the Laplacian matrix of the directed graph $\mathcal{G}$ is defined as $\mathcal{L} = \mathcal{D} - \mathcal{A}$. A path is a sequence of connected edges in a graph. The graph $\mathcal{G}_N$ is said to be connected if there exists a path between any nodes. A neighbor set of vehicle i in the followers is defined as

$$\mathbf{N}_i = \{j \in \mathcal{V}_N \mid a_{ij} = 1\}$$

To model the communication from the leader to the followers, we define an augmented directed graph $\mathcal{G}_{N+1} = (\mathcal{V}_{N+1}, \mathcal{E}_{N+1})$, with a set of node $N+1$ vertices $\mathcal{V}_{N+1} = \{1, 2, ..., N+1\}$, which includes both the leader and the followers in a platoon. As in [2], a pining matrix $\mathcal{P} = \text{diag}\{p_1, p_2, ..., p_n\}$ is used to denote how each follower connects to the leader: $p_i = 1$ if $(0, i)$

$\in \mathcal{E}_{N+1}$, otherwise $p_i = 0$. Note that $(0, i) \in \mathcal{E}_{N+1}$ means that node i is pinned to the leader. The leader-accessible set of node i is defined as:

$$\mathbf{P}_i = \begin{cases} \{0\} & \text{if } p_i = 1 \\ \{i\} & \text{if } p_i = 0 \end{cases} \tag{4.7}$$

In this chapter, we consider a group of directed graph $\mathcal{G}_{\sigma(t)} \in \{\mathcal{G}_1, \mathcal{G}_2, ..., \mathcal{G}_q\}$, where $\sigma(t)$ is a continuous-time Markov process with values in a finite set $\mathcal{S} = \{1, 2, ..., q\}$. Such switching communication topologies may be due to varying information flow in VANET [7]. The transition rate of the Markov process is defined as

$$\text{prob}\{\sigma(t+\Delta t) = s|\sigma(t) = r\} = \begin{cases} \pi_{rs}\Delta t + O(\Delta t), r \neq s \\ 1 + \pi_{rr}\Delta t + O(\Delta t), r = s \end{cases}$$

where $\Delta(t) > 0$, $O(\Delta t) \to 0$, as $\Delta(t) \to 0$, and π_{rs} is the transition rate from mode r to model s, which satisfies $\pi_{rs} > 0$ for $r \neq s$, and $\pi_{rr} = -\sum_{s=1, s\neq r}^{m} \pi_{rs}$ for $r \in \mathcal{S}$. Furthermore, the corresponding transition rate matrix is defined by $\pi = [\pi_{rs}]$. Denoting the communication topology as $H_{\sigma(t)} = \mathcal{L}_{\sigma(t)} + \mathcal{P}_{\sigma(t)}$, we have that $H_{\sigma(t)} \in \{H_1, H_2, ..., H_q\}$. In this chapter, we make the following assumption [2, 6]:

Assumption 4.1 The augmented graph $\mathcal{G}_{N+1}$ contains at least one spanning tree rooting at the leader.

This assumption implies the leader is globally reachable in $\mathcal{G}_{N+1}$. Namely, each following vehicle can obtain the leader information directly or indirectly, which is a prerequisite to guarantee the internal stability of a platoon.

Since intervehicle communication is not perfect, we consider sampled-data transmission with delays. Let τ_{ij} denote the transmission delay from vehicle j in $\mathbf{I}_i = \mathbf{N}_i \cup \mathbf{P}_i$ to vehicle i. As a result, the maximum time delay when the k-th sampled data of all the vehicles in set I_i arrive at vehicle i can be defined as $\tau_k = \max_j\{\tau_{ij} \,|\, j \in \mathbf{I}_i\}$. Then, for $t \in [kh + \tau_k, (k+1)h + \tau_{k+1})$, $k \in \mathbb{N}$ with sampling period h, the local controller of the ith following vehicle is given as:

$$u_i(t) = -\sum_{j \in \mathbf{I}_i} k_p\left(z_i(kh) - z_j(kh) - d_{i,j}\right) + k_v\left(v_i(kh) - v_j(kh)\right) + k_a\left(a_i(kh) - a_j(kh)\right) \tag{4.8}$$

where d_{ij} denotes the desired distance between vehicle i and vehicle j. Here, we use the constant spacing policy, i.e. $d_{i,i-1} = d_0$ with constant $d_0 > 0$. k_p, k_v and k_a are feedback controller gains to be designed.

In order to have a compact form of (4.8), we define the following tracking errors for vehicle i

$$\begin{cases} \hat{z}_i(t) = z_i(t) - z_0(t) - d_{i,0} \\ \hat{v}_i(t) = v_i(t) - v_0(t) \\ \hat{a}_i(t) = a_i(t) - a_0(t) \end{cases} \tag{4.9}$$

Then, the controller in (4.8) can be rewritten as

$$u_i(t) = -K\sum_{j \in \mathbf{I}_i} \left(\hat{x}_i(kh) - \hat{x}_j(kh)\right) \tag{4.10}$$

where $i = 1, ..., N$, $t \in [kh + \tau_k, (k+1)h + \tau_{k+1})$, $K = \begin{bmatrix} k_p & k_v & k_a \end{bmatrix}$ and $\hat{x}_i(t) = \begin{bmatrix} \hat{z}_i(t) & \hat{v}_i(t) & \hat{a}_i(t) \end{bmatrix}^T$ are the vector of feedback gains and state, respectively. The controller in (4.10) requires full state measurement of the vehicles. However, some states may be unmeasurable or unreliable when the sensors are subject to failures or limited measurement capacity. Under such a circumstance, one can introduce an observer to estimate the unmeasurable states and then achieve platoon control using the observer-based control strategy [8].

Note that $\hat{x}_i(kh) = \hat{x}_i(t-(t-kh))$. Let $\tau_k(t) = t - kh$, $t \in [kh+\tau_k, (k+1)h+\tau_{k+1})$, $k \in \mathbb{N}$. It is clear that $\tau_k \le \tau_k(t) \le h + \tau_{k+1}$ for $t \in [kh+\tau_k, (k+1)h+\tau_{k+1})$, $k \in \mathbb{N}$. Let $\alpha = \min_k\{\tau_k \mid k \in \mathbb{N}\}$ and $\beta = h + \max_k\{\tau_k \mid k \in \mathbb{N}\}$. Next, defining $x(t) = \left[\hat{x}_1^T(t), \hat{x}_2^T(t), \dots \hat{x}_N^T(t)\right]^T$, and $u(t) = [u_1(t), u_2(t), \dots, u_N(t)]^T$ for $t \in [kh+\tau_k, (k+1)h+\tau_{k+1})$, we have

$$u(t) = H_{\sigma(kh)} \otimes Kx(t-\tau_k(t)) \tag{4.11}$$

For $t \in [-\beta, 0]$, the initial condition is assumed to be $x(t) = \varphi(t) = \varphi(t_0)$. Then, the closed-loop connected vehicular control systems are given as

$$\begin{cases} \dot{x}(t) = (I_N \otimes A)x(t) - H_{\sigma(t)} \otimes BKx(t-\tau_k(t)) + W(x,t), & t \in [kh+\tau_k, (k+1)h+\tau_{k+1}) \\ x(t) = \phi(t) = \phi(t_0), & t \in [-\beta, 0] \end{cases} \tag{4.12}$$

where $B = I_N \otimes B_1$, and external disturbances vector $W(x,t) = \left[W^T(x_1,t),\ W^T(x_2,t),\ \dots,\ W^T(x_N,t)\right]^T$ with $W(x_i, t) = \omega(x_i, t) - \omega(x_0, t)$. For the case with fixed intervehicle communication topology $H_{\sigma(t)} = H = \mathcal{L} + \mathcal{P}$ and zero communication delay, $\tau_k(t) = 0$ Equation (4.12) becomes the model of connected vehicles in [2].

Define the tracking error of positions as the output vector $y(t)$ of the connected vehicles. Then we have

$$y(t) = [\hat{z}_1(t),\ \hat{z}_2(t),\ \dots,\ \hat{z}_N(t)]^T = Cx(t) \tag{4.13}$$

where $C = I_N \otimes C_1$ and $C_1 = [1\ \ 0\ \ 0]$.

The objective of this chapter is to establish a connected vehicle control synthesis method such that

1) each following vehicle can maintain the same speed with the leader and keep the desired inter-vehicle spacing for any initial tracking errors, namely

$$\begin{cases} \lim\limits_{t\to\infty} \|v_i(t) - v_0(t)\| = 0 \\ \lim\limits_{t\to\infty} \|z_i(t) - z_{i-1}(t) - d_{i,i-1}\| = 0 \end{cases} \tag{4.14}$$

2) under zero-initial condition, the platoon system output $y(t)$ satisfies the following disturbance attenuation level with $\gamma > 0$ for all nonzero $W(x, t)$.

$$\int_0^\infty \mathbb{E}\{\|y(t)\|\}dt \le \int_0^\infty \mathbb{E}\{\gamma^2 \|W(x,t)\|\}dt \tag{4.15}$$

Inequality (4.15) is equivalent to

$$\mathbb{E}\left\{\left\|\frac{y(t)}{W(x,t)}\right\|_\infty\right\} \le \gamma^2$$

which is a typical notion of system robustness with respect to external disturbances and is used here to address disturbance propagation. Usually, the concept of uniform string stability (see [9, 10]) requires that the relative distance errors $\Delta_i(t)$ satisfy $\|(\Delta_i(t))/(\Delta_{i-1}(t))\| \le 1$. Such a strong string stability highly depends on the communication topology [2, 11] and, hence, is rather difficult to meet when switching topologies and communication delays exist. So, as in [2, 11], we use (4.15) to avoid disturbance propagation by attenuating its effect to a certain small level.

Definition 4.1 The connected vehicular control systems (4.12) are said to be exponentially mean-square stable if there exists constant $\lambda > 0$ and $\rho > 0$ such that

$$\mathbb{E}\{\|x(t)\|^2\} \le \lambda e^{-\rho t}\mathbb{E}\{\|\phi(t_0)\|^2\}.$$

4.3 Stability Analysis

In this section, we will derive a stability criterion for the connected vehicular control systems (4.12) with the completely/partially known transition rates of the Markovian chain. In order to accommodate the control objective presented in Section 4.1, we first establish the following Lyapunov-Krasovskii functional (LKF) as

$$V(t,x_t,\dot{x}_t) = \sum_{i=1}^{3} V_i(t,x_t,\dot{x}_t) \tag{4.16}$$

where

$$V_1(t,x_t,\dot{x}_t) = x^T(t)P_r x(t)$$

$$V_2(t,x_t,\dot{x}_t) = \int_{t-\alpha}^{t} x^T(t)Q_1 x(t)ds + \int_{t-\beta}^{t-\alpha} x^T(t)Q_2 x(t)ds$$

$$V_3(t,x_t,\dot{x}_t) = \int_{-\alpha}^{0}\int_{t+\theta}^{t} \dot{x}^T(s)R_1\dot{x}(s)dsd\theta + \int_{-\beta}^{-\alpha}\int_{t+\theta}^{t} \dot{x}^T(s)R_2\dot{x}(s)dsd\theta$$

with $x_t(\theta) = x(t+\theta)$, $\theta \in [-\beta, 0]$, $P_r \succ 0 (r = 1, 2, \ldots, q)$, $Q_1 \succ 0$, $Q_2 \succ 0$, $R_1 \succ 0$, $R_2 \succ 0$. Let $\psi(t) = \left[x^T(t),\ \ W^T(x,t),\ \ x^T(t-\tau_k(t)),\ \ x^T(t-\alpha),\ \ x^T(t-\beta)\right]^T$ and ℓ_i a block entry matrix with $\ell_{i,\ j} = \ell_i - \ell_j$. To make it clear, we give example: $\ell_1 = [I_{3N}\ \ 0\ \ 0\ \ 0\ \ 0]$ and $\ell_{12} = [I_{3N}\ \ -I_{3N}\ \ 0\ \ 0\ \ 0]$.

We first consider the case that the transition rates of the Markovian switching topologies are completely known. We have the following results:

Theorem 4.1 Consider a platoon of connected vehicles in VANET subject to Markovian switching topologies, communication delay, and external disturbance. Under Assumption 4.1, for given scalars $\beta > \alpha > 0$, $\varrho > 0$ and $\gamma > 0$, the tracking error of the connected vehicular system is exponentially mean-square stable with completely known transition rates if there exists real symmetric and positive definite matrices $P_r \in \mathbb{R}^{3N\times 3N}(r = 1, 2, \ldots, q)$, $Q_1 \in \mathbb{R}^{3N\times 3N}$, $Q_2 \in \mathbb{R}^{3N\times 3N}$, $R_1 \in \mathbb{R}^{3N\times 3N}$, $R_2 \in \mathbb{R}^{3N\times 3N}$ and some matrices $Y_m \in \mathbb{R}^{3N\times 15N}$, $(m = 1, 2)$ of appropriate dimensions such that the following linear matrix inequality (LMI) is held

$$\Omega_{rm} = \begin{bmatrix} \Upsilon_r & \Gamma_r^T(\alpha^2R_1 + \delta R_2) & \delta Y_m^T & \ell_2^T \\ * & -(\alpha^2R_1 + \delta R_2) & 0 & 0 \\ * & * & -\delta R_2 & 0 \\ * & * & * & -\gamma^{-2}I_{3N} \end{bmatrix} \prec 0 \tag{4.17}$$

for $r = 1, 2, \ldots, q$ and $m = 1, 2$, where $\delta = \beta - \alpha$

$$\Upsilon_r = \ell_1^T\left[\sum_{s=1}^{q}\pi_{rs}P_s + Q_1 + \varrho^2 I_{3N} - C^T C\right]\ell_1 + \ell_1^T P_r\Gamma_r + \Gamma_r P_r \ell_1 + -\ell_2^T\ell_2$$
$$+ \ell_4^T(Q_2 - Q_1)\ell_4 - \ell_5^T Q_2 \ell_5 - \ell_{14}^T R_1 \ell_{14} - \ell_{35}^T Y_1 - Y_1^T \ell_{35} - \ell_{43}^T Y_2 - Y_2^T \ell_{43}$$
$$\Gamma_r = (I_N \otimes A)\ell_1 + \ell_2 - (H_r \otimes BK)\ell_3$$

Moreover, the tracking errors satisfy the defined disturbance attenuation level under zero initial conditions.

Proof: Define the weak infinitesimal operator $\Im$ of $V(t, x_t, \dot{x}_t)$ with respect to $t \in [kh + \tau_k, (k+1)h + \tau_{k+1})$ as:

$$\Im V(t, x_t, \dot{x}_t) = \lim_{\Delta \to 0^+}\frac{1}{\Delta}\{\mathbb{E}\{V(t+\Delta, x_{t+\Delta}, \dot{x}_{t+\Delta})|x_t, \sigma(t) = r\} - V(t, x_t, \dot{x}_t)\} \tag{4.18}$$

Then, along with (4.18), we have

$$\Im V(t, x_t, \dot{x}_t) = \sum_{i=1}^{3}\Im V_i(t, x_t, \dot{x}_t) \tag{4.19}$$

where

$$\Im V_1(t, x_t, \dot{x}_t) = 2\psi(t)\ell_1^T\left[P_r\Gamma_r + \frac{1}{2}\sum_{s=1}^{q}\pi_{rs}P_s\ell_1\right]\psi(t)$$
$$\Im V_2(t, x_t, \dot{x}_t) = \psi^T(t)\{\ell_1^T Q_1 \ell_1 + \ell_4^T(Q_2 - Q_1)\ell_4 - \ell_5^T Q_2 \ell_5\}\psi(t)$$
$$\Im V_3(t, x_t, \dot{x}_t) = \psi^T(t)\{\Gamma_r^T[(\alpha^2 R_1 + \delta R_2)]\Gamma_r\}\psi(t) + \eta_1 + \eta_2$$

with

$$\eta_1 = -\alpha\int_{t-\alpha}^{t}\dot{x}^T(s)R_1\dot{x}(s)ds$$

$$\eta_2 = -\int_{t-\alpha}^{t-\beta}\dot{x}^T(s)R_2\dot{x}(s)ds$$

In order to deal with η_1 and η_2, the following two lemmas are needed:

Lemma 4.1 [12] For any constant matrix $R > 0 \in \mathbb{R}^{n\times n}$, scalar $\alpha > 0$, and vector function $\dot{x}: [-\alpha, 0] \to \mathbb{R}^n$, the following inequality holds:

$$-\alpha\int_{t-\alpha}^{t}\dot{x}^T(s)R\dot{x}(s)ds$$
$$\le -[x(t) - x(t-\alpha)]^T R[x(t) - x(t-\alpha)] \tag{4.20}$$

Lemma 4.2 [13] Let $\int_{t-\alpha(t)}^{t}\dot{x}^T(s)ds = E\varphi(t)$, where $E \in \mathbb{R}^{n\times k}$ and $\varphi(t) \in \mathbb{R}^k$. For any constant matrix $R > 0 \in \mathbb{R}^{n\times n}$, $Y \in \mathbb{R}^{n\times k}$, time-varying function $\tau(t)$ satisfying $0 < \tau(t) \le \beta$, and vector function $\dot{x}: [-\alpha, 0] \to \mathbb{R}^n$, the following inequality holds:

$$-\int_{t-\alpha(t)}^{t}\dot{x}^T(s)R\dot{x}(s)ds$$
$$\le \varphi^T(t)(\alpha(t)Y^T R^{-1} Y - E^T Y - Y^T E)\varphi(t) \tag{4.21}$$

Applying Lemma 4.1, we have that

$$\eta_1 \leq \psi(t)^T \ell_{14}^T R_1 \ell_{14} \psi(t) \tag{4.22}$$

Notice that

$$\eta_2 = -\int_{t-\beta}^{t-\tau_k(t)} \dot{x}^T(s) R_2 \dot{x}(s) ds - \int_{t-\tau_k(t)}^{t-\alpha} \dot{x}^T(s) R_2 \dot{x}(s) ds \tag{4.23}$$

By Lemma 4.2, we obtain

$$\eta_2 = -\int_{t-\beta}^{t-\tau_k(t)} \dot{x}^T(s) R_2 \dot{x}(s) ds - \int_{t-\tau_k(t)}^{t-\alpha} \dot{x}^T(s) R_2 \dot{x}(s) ds \tag{4.24}$$

According to the property of the Lipschitz function given in (4.6), one can obtain the following inequality

$$\| W(x_i, t) - W(x_0, t) \| \leq \varrho \| x_i(t) - x_0(t) \| \tag{4.25}$$

which is equivalent to

$$W^T(x, t) W(x, t) \leq \varrho^2 x(t)^T x(t) \tag{4.26}$$

Taking the mathematical expectation on both sides of Equation (4.19) and from Equations (4.22) to (4.26) for $t \in [kh + \tau_k, (k+1)h + \tau_{k+1})$, one can obtain

$$\mathbb{E}\{\Im V(t, x_t, \dot{x}_t)\} \leq \mathbb{E}\{\psi^T(t) \Sigma_r \psi(t)\} \tag{4.27}$$

where $\Sigma_r = \Upsilon_r + \Gamma_r^T(\alpha^2 R_1 + \delta R_2)\Gamma_r + (\beta - \tau_k(t)) Y_1^T R_2^{-1} Y_1 + (\alpha - \tau_k(t)) Y_2^T R_2^{-1} Y_2$. It is clear that if $\Sigma_r < 0$, then there exists a small enough $c_1 > 0$ such that

$$\mathbb{E}\{\Im V(t, x_t, \dot{x}_t)\} \leq -c_1 \mathbb{E}\{\|x(t)\|^2\} \tag{4.28}$$

which means that the connected vehicular system (4.12) is mean-square stable. Namely, condition (4.14) of the control objective is true.

Adding the term $-y^T(t)y(t) + \gamma^2 W^T(x, t)W(x, t)$ on the right side of (4.27), if the following LMI is satisfied

$$\Sigma_r' = \Sigma_r - \ell_1^T C^T C \ell_1 + \gamma^2 \ell_2^T \ell_2 \psi(t) 0 \leq 0 \tag{4.29}$$

then we can obtain that

$$\mathbb{E}\{\Im V(t, x_t, \dot{x}_t) + y^T(t)y(t) - \gamma^2 W^T(x, t) W(x, t)\} \leq 0 \tag{4.30}$$

Integrating both sides of (4.30) from $t = 0$ to $t = \infty$, under the zero initial condition, it is straightforward that

$$\int_0^\infty \mathbb{E}\{\| y(t) \|\} dt \leq \int_0^\infty \mathbb{E}\{\gamma^2 \| W(x, t) \|\} dt \tag{4.31}$$

namely, condition (4.15) of the control objectives is held.

In order to illustrate mean-square exponential stability, we define a new function $\mathcal{V} = e^{-\varepsilon t} V(t, x_t, \dot{x}_t)$, where $\varepsilon > 0$ is a scalar to be determined. Then, we can obtain that

$$\Im \mathcal{V} = e^{-\varepsilon t}[\varepsilon V(t, x_t, \dot{x}_t) + \Im V(t, x_t, \dot{x}_t)] \tag{4.32}$$

Integrating both sides of (4.32) from 0 to $T > 0$, and taking the expectation, one can obtain that

$$\begin{aligned}&\mathbb{E}\{e^{-\varepsilon T}V(t,x_t,\dot{x}_t)\} - \mathbb{E}\{V(0,x_0,\dot{x}_0)\}\\&= \int_0^T \varepsilon e^{-\varepsilon t}\mathbb{E}\{V(t,x_t,\dot{x}_t)\}dt + \int_0^T e^{-\varepsilon t}\mathbb{E}\{V(t,x_t,\dot{x}_t)\}dt\end{aligned} \tag{4.33}$$

In the same way as in [14], it is clear that there exists a scalar $\rho > 0$, such that for $T \geq 0$

$$\mathbb{E}\{V(T,x_T,\dot{x}_T)\} \leq \rho e^{-\varepsilon T}\mathbb{E}\{\|\phi(t_0)\|^2\} \tag{4.34}$$

Due to the fact that $V(T,x_T,\dot{x}_T) \geq \kappa x^T(T)x(T)$, where $\kappa = \min_{r=1,2,\ldots,q}\{\lambda_{\min}(P_r)\}$

$$\mathbb{E}\{\|x(T)\|^2\} \leq \frac{\rho}{\kappa}e^{-\varepsilon T}\mathbb{E}\{\|\phi(\theta)\|^2\} \tag{4.35}$$

It is clear that connected vehicular control systems (4.12) are exponentially mean-square stable. Notice that Σ_r is a convex combination of $Y_1^T R_2^{-1} Y_1$ and $Y_2^T R_2^{-1} Y_2$ on $\tau_k(t) \in [\alpha, \beta]$. Therefore, $\Sigma_r' \preceq 0$ if the following inequalities hold:

$$\Upsilon_r + \Gamma_r^T(\alpha^2 R_1 + \delta R_2)\Gamma_r + \delta Y_m^T R_2^{-1} Y_m + \gamma^2 \ell_2^T \ell_2 \preceq 0 \tag{4.36}$$

for $r = 1, 2, \ldots, q$ and $m = 1, 2$. Applying Schur's complement to Equation (4.36), we can obtain inequality (4.17) to guarantee that $\Sigma_r' \preceq 0$. The proof is completed.

Theorem 4.1 deals with the case where the Markovian chain has completely known transition rates. In practice, however, some of the transition rates are not known (e.g. costly or time-consuming to determine them all). For such circumstances, we give in the next subsection a feasible solution.

In this section, we extend these obtained results to cover Markovian switching topologies with partially unknown transition rates, e.g.

$$\begin{bmatrix} \pi_{11} & ? & \pi_{13} & \cdots & ? \\ \pi_{21} & ? & ? & \cdots & \pi_{2q} \\ \vdots & \vdots & \vdots & \ddots & \vdots \\ ? & \pi_{q2} & ? & \cdots & \pi_{qq} \end{bmatrix} \tag{4.37}$$

where "?" represents the unknown transition rate. For notational clarity, $\forall r \in \mathcal{S}$, the set Π^r denotes

$$\Pi^r = \Pi_k^r \cup \Pi_{uk}^r$$

with $\Pi_k^r = \{s : \pi_{rs} \text{ is known for } s \in \mathcal{S}\}$. $\Pi_{uk}^r = \{s : \pi_{rs} \text{ is unknown for } s \in \mathcal{S}\}$

Moreover, if $\Pi_k^r \neq \emptyset$, it is further described as

$$\Pi_k^r = \{k_1^r, k_2^r, \ldots, k_m^r\} \tag{4.38}$$

where m is a non-negation integer with $1 \leq m \leq q$, $k_r^s \in Z^+$, $1 \leq k_r^s \leq q$, $s = 1, 2, \ldots, m$ represents the sth known element of the set Π_k^r in the rth row of the transition rate matrix π.

Then, the stability condition for the Markovian switching time delay connected vehicular control system (4.12) with incomplete transition rates is given as follows.

Theorem 4.2 Consider a platoon of connected vehicles in VANET subject to Markovian switching topologies, communication delay, and external disturbance, where the transition rates of the Markovian chain is partially known. Under Assumption 4.1, for given scalars $\beta > \alpha > 0$, $\varrho > 0$, and $\gamma > 0$, connected vehicular systems (4.12) are exponentially mean-square stable with partially known transition rates if there exist real symmetric and positive definite matrices $P_r \in \mathbb{R}^{3N\times 3N}(r = 1, 2, \ldots, q)$, $Q_1 \in \mathbb{R}^{3N\times 3N}$, $Q_2 \in \mathbb{R}^{3N\times 3N}$, $R_1 \in \mathbb{R}^{3N\times 3N}$, $R_2 \in \mathbb{R}^{3N\times 3N}$, $Z_r \in \mathbb{R}^{3N\times 3N}$ and some matrices $Y_m \in \mathbb{R}^{3N\times 3N}(m = 1, 2)$ of appropriate dimensions such that the following LMIs are satisfy

$$\hat{\Omega}_{rm} = \begin{bmatrix} \hat{\Upsilon}_r & \Gamma_r^T(\alpha^2 R_1 + \delta R_2) & \delta Y_m^T & \ell_2^T \\ * & -(\alpha^2 R_1 + \delta R_2) & 0 & 0 \\ * & * & -\delta R_2 & 0 \\ * & * & * & -\gamma^{-2} I_{3N} \end{bmatrix} < 0 \tag{4.39}$$

$$P_s - Z_r \leq 0, P_s - Z_r \leq 0, r \in \Pi_{uk}^r, s \neq r, s \neq r \tag{4.40}$$

$$P_s - Z_r \leq 0, P_s - Z_r \leq 0, r \in \Pi_{uk}^r, s = r \tag{4.41}$$

where

$$\hat{\Upsilon}_r = \overline{\Upsilon}_r + \ell_1^T \textstyle\sum_{s\in\Pi_k^r} \pi_{rs}(P_s - Z_r)\ell_1$$

$$\overline{\Upsilon}_r = \ell_1^T\left[Q_1 + \varrho^2 I_{3N} - C^T C\right]\ell_1 + \ell_1^T P_r \Gamma_r + \Gamma_r P_r \ell_1 + \gamma^2 - \ell_2^T \ell_2 + \ell_4^T(Q_2 - Q_1)\ell_4$$
$$- \ell_5^T Q_2 \ell_5 - \ell_{14}^T R_1 \ell_{14} - \ell_{35}^T Y_1 - Y_1^T \ell_{35} - \ell_{43}^T Y_2 - Y_2^T \ell_{43}$$

$$\Gamma_r = (I_N \otimes A)\ell_1 + \ell_2 - (H_r \otimes BK)\ell_3$$

Moreover, the tracking errors satisfy the defined disturbance attenuation level under zero initial conditions.

Proof: Taking into account the case that the transition rate is not accessible completely, the following zero equations hold for arbitrary matrices $Z_r = Z_r^T$ due to $\sum_{s=1}^{q} \pi_{rs} = 0$.

$$-\psi^T(t)\ell_1^T \sum_{s=1}^{q} \pi_{rs} Z_r \ell_1 \psi(t) = 0 \tag{4.42}$$

Choose the same LKF defined as in (4.16) and add the left side of Equation (4.42) into the weak infinitesimal operator $\mathfrak{I}$ ∘

$$\mathbb{E}\{\mathfrak{I}V(t, x_t, \dot{x}_t)\} \leq \mathbb{E}\left\{\psi^T(t)\left(\Sigma_r + \ell_1^T \sum_{s=1}^{q} \pi_{rs} Z_r \ell_1\right)\psi(t)\right\} \tag{4.43}$$

According to the fact that

$$\sum_{s=1}^{q} \pi_{rs}(P_s - Z_r) = \sum_{s\in\Pi_k^r} \pi_{rs}(P_s - Z_r) + \sum_{s\in\Pi_{uk}^r} \pi_{rs}(P_s - Z_r)$$

Then (4.43) is rewritten as

$$\mathbb{E}\{\mathfrak{I}V(t, x_t, \dot{x}_t)\} \leq \mathbb{E}\left\{\psi^T(t)\left[\hat{\Sigma}_r + \ell_1^T \sum_{s\in_{uk}^r} \pi_{rs}(P_s - Z_r)\ell_1\right]\psi(t)\right\} \tag{4.44}$$

where

$$\hat{\Sigma}_r = \overline{\Sigma}_r + \ell_1^T \sum_{s\in\Pi_k^r} \pi_{rs}(P_s - Z_r)\ell_1$$

$$\overline{\Sigma}_r = \overline{\Upsilon}_r + \Gamma_r^T(\alpha^2 R_1 + \delta R_2)\Gamma_r + (\beta - \tau_k(t))Y_1^T R_2^{-1} Y_1 + (\alpha - \tau_k(t))Y_2^T R_2^{-1} Y_2$$

Note that $\pi_{rr} = -\sum_{s=1, s\neq r}^{m} \pi_{rs}$ and $\pi_{rs} \geq 0$ for all $r \neq s$, namely, $\pi_{rr} < 0$ for all $r \in S$, therefore, if $r \in \Pi_k^r$, LMI (4.40) and $\hat{\Sigma}_r \leq 0$ imply that there exists a small enough constant $c_1 > 0$ such that inequality (4.28) is held, which means that platoon control system (4.12) is mean-square stable. On the other hand, if $r \in \Pi_{uk}^r$, LMIs (4.40)–(4.41) and $\hat{\Sigma}_r \preceq 0$ also imply that there exists a small enough constant $c_1 > 0$ such that inequality (4.28) is held.

Note that $\hat{\Sigma}_r \preceq 0$ is also a convex combination of $Y_1^T R_2^{-1} Y_1$ and $Y_2^T R_2^{-1} Y_2$ on $\tau_k(t) \in [\alpha, \beta]$. Therefore, $\hat{\Sigma}_r \leq 0$, if the following LMIs hold:

$$\tilde{\Upsilon}_r + \Gamma_r^T(\alpha^2 R_1 + \delta R_2)\Gamma_r + \delta Y_m^T R_2^{-1} Y_m + \gamma^2 \ell_2^T \ell_2 + \ell_1^T \sum_{s \in \Pi_k^r} \pi_{rs}(P_s - Z_r)\ell_1 \preceq 0 \tag{4.45}$$

for $r = 1, 2, ..., q$ and $m = 1, 2$. Applying Schur's complement to LMI (4.39), we can obtain inequality (4.45). By using the same guidelines as in the proof of Theorem 4.1, it is easily known that the connected vehicular control system is exponentially mean-square stable and satisfies the disturbance attenuation level with γ. The proof is completed.

Remark 4.1 The existing results on connected vehicles ignore either the communication delay or switching topologies or both. One example is the work in [2, 4], which give sufficient and necessary conditions for controller design for platoons with undirected fixed topologies. These results cannot be extended to the case studied in this chapter since the effects of communication delay and external disturbance are investigated under directed switching topologies. The results in Theorems 4.1 and 4.2 provide sufficient conditions for platoon controller design. The feasibility of Theorems 4.1 and 4.2 are tested by the centralized computation of LMI (4.17) and (4.39). As a result, the dimension of LMI (4.17) and (4.39) is dependent on the platoon size N. For a larger N, there might be a considerable computational cost of the variable controller. However, this is not a big problem because most of the calculations can be done offline in a computer system.

Remark 4.2 Theorems 4.1 and 4.2 establish the relationship among communication topology, sampling period, communication delay, and controller gain for the platoon control system. For instance, given a group of communication topology H_r, sampling period h, and controller gain K, one can use Theorem 4.1 to obtain the allowable maximum transmission time delay $\tau_k(t)$ by solving a set of LMIs in Theorems 4.1 and 4.2, respectively.

4.4 Controller Synthesis

Based on the stability criterion in Theorems 4.1 and 4.2, we are in a position to give a sufficient condition on the existence of a fixed-gain controller and variable-gain controller for connected vehicles, respectively.

Without considering the effect of switching topologies and communication delay, we rewrite Equation (4.12) as

$$\dot{x}(t) = (I_N \otimes A)x(t) - H_r \otimes BKx(t) \tag{4.46}$$

For connected vehicular control system (4.46), the following lemma provides an approach to finding a stabilizing controller.

Lemma 4.3 [11] Let $\lambda_{ri}, \lambda_{ri}, i = 1, 2, ..., l$ be the real eigenvalues of H_r including multiplicity, $\lambda_{ri}^*, i = (l+1), (l+2), ..., (l+m)$ be the conjugate complex eigenvalues including multiplicity. For a homogeneous platoon (4.46), if there exists a positive scalar υ and symmetric matrix $Q \succ 0$, such that

$$AQ + QA^T - \upsilon B_1 B_1^T \prec 0 \tag{4.47}$$

$$\upsilon \leq \min_{1 \leq r \leq q} \left(\frac{\lambda_{ri}(H_r) + \lambda_{ri}^*(H_r)}{2} \right) \tag{4.48}$$

then the stability of the connected vehicular control system (4.46) can be guaranteed, and the feasible controller gain K can be constructed as

$$K = \frac{1}{2} B_1^T Q \tag{4.49}$$

Combining (4.49) with Theorems 4.1 and 4.2, we can easily obtain corresponding controller gains for connected vehicular control system (4.12) with completely and partially known transition rates.

A fixed-gain controller cannot always guarantee connected vehicle control performance under switching topologies and large transmission delays. Therefore, the following variable-gain controller is considered

$$u(t) = H_{\sigma(kh)} \otimes K_{\sigma(kh)} x(t - \tau_k(t)), t \in [kh + \tau_k, (k+1)h + \tau_{k+1})$$

which is motivated by switching control idea. In this subsection, we give results on the design of variable gain K_r of the platoon controller which is dependent on the communication topologies.

1) Feedback Gains Solution With Completely Known Transition Rates:

Theorem 4.3 Consider connected vehicle control is subject to Markovian switching topologies, communication delay, and external disturbance, where the transition rates of the Markovian chain are completely known. Under Assumption 4.1, for given scalars $\beta > \alpha > 0$, $\mu > 0$, $\gamma > 0$, and $\varrho > 0$, connected vehicular control system (4.12) is exponentially mean-square stable and satisfies the defined disturbance attenuation level under zero initial conditions, if there exist real symmetric and positive definite matrices $\tilde{P}_r \in \mathbb{R}^{3 \times 3} (r = 1, 2, ..., q)$, $\tilde{Q}_{1r} \in \mathbb{R}^{3N \times 3N}$, $\tilde{Q}_{2r} \in \mathbb{R}^{3N \times 3N}$, $\tilde{R}_{1r} \in \mathbb{R}^{3N \times 3N}$, $\tilde{R}_{2r} \in \mathbb{R}^{3N \times 3N}$, and some matrices $\tilde{K}_r \in \mathbb{R}^{1 \times 3}$ and $\tilde{Y}_m \in \mathbb{R}^{3N \times 15N} (m = 1, 2)$ of appropriate dimensions such that

$$\tilde{\Omega}_{rm} = \begin{bmatrix} \tilde{\Upsilon}_r & \tilde{\Gamma}_r^T & \tilde{\Xi}_{2r}^T & \tilde{\Xi}_{3r}^T & \delta Y_{rm}^T & \ell_2^T \\ * & \tilde{\Xi}_{1r} & 0 & 0 & 0 & 0 \\ * & * & -\tilde{\Lambda}_r & 0 & 0 & 0 \\ * & * & * & -I_{3N} & 0 & 0 \\ * & * & * & 0 & -\sigma\tilde{R}_{2r} & 0 \\ * & * & * & * & * & -\gamma^{-2} I_{3N} \end{bmatrix} \prec 0 \tag{4.50}$$

for $r = 1, 2, ..., q$, and $m = 1, 2$, where $\delta = \beta - \alpha$,

$$\tilde{\Upsilon}_r = \ell_1^T [\tilde{Q}_{1r} + \varrho^2 I_{3N} + C^T C] \ell_1 + \ell_1^T \tilde{\Gamma}_r + \tilde{\Gamma}_r \ell_1 + (\gamma^2 - 1) \ell_2^T \ell_2 + \ell_4^T (\tilde{Q}_{2r} - \tilde{Q}_{1r}) \ell_4 - \ell_5^T \tilde{Q}_{2r} \ell_5$$
$$- \ell_{14}^T \tilde{R}_{1r} \ell_{14} - \ell_{35}^T Y_{1r} - \tilde{Y}_{1r}^T \ell_{35} - \ell_{43}^T \tilde{Y}_{2r} - \tilde{Y}_{2r}^T \ell_{43}$$

$$\tilde{\Gamma}_r = (I_N \otimes A\tilde{P}_r) \ell_1 + (I_N \otimes \tilde{P}_r) \ell_2 - (H_r \otimes B\tilde{K}_r \ell_3)$$

$$\tilde{\Xi}_{1r} = \mu^2 (\alpha^2 \tilde{R}_{r1} + \delta \tilde{R}_{2r}) - 2\mu I_N \otimes \ddot{P}r$$

$$\tilde{\Xi}_{2r} = \left[\sqrt{\pi_{r1}} (I_N \otimes \tilde{P}r), ..., \sqrt{\pi_{rr-1}} (I_N \otimes \tilde{P}r), \sqrt{\pi_{rr+1}} (I_N \otimes \tilde{P}r), ..., \sqrt{\pi_{rq}} (I_N \otimes \tilde{P}r) \right]^T \ell_1$$

$$\tilde{\Xi}_{3r} = \varrho (I_N \otimes \tilde{P}_r)$$

$$\tilde{\Lambda}_r = \text{diag}\{I_N \otimes \tilde{P}_1, ..., I_N \otimes \tilde{P}_{N-1}, ..., I_N \otimes \tilde{P}_q\}$$

Moreover, the controller gains are given by

$$K_r = \tilde{K}_r\tilde{P}_r^{-1} \tag{4.51}$$

Proof: Let $P_r = I_N \otimes U_r$ for the conditions (4.17) in Theorem 4.1, and define $\tilde{P}_r = U_r^{-1}$, $\tilde{Q}_{1r} = P_r^{-1}Q_1P_r^{-1}$, $\tilde{Q}_{2r} = P_r^{-1}Q_2P_r^{-1}$, $\tilde{R}_{1r} = P_r^{-1}R_1P_r^{-1}$, $\tilde{R}_{2r} = P_r^{-1}R_2P_r^{-1}$, $J_r = \text{diag}\{P_r^{-1}, P_r^{-1}, P_r^{-1}, P_r^{-1}, P_r^{-1}\}$, $\tilde{Y}_{mr} = J_rY_{mr}J_r (m = 1, 2)$, and $\tilde{K}_r = K_rU_r^{-1}$, Pre- and post-multiplying both sides of (4.17) by $\text{diag}\{J_r, (\alpha^2R_{r1} + \delta R_{2r}), P_r^{-1}, I_{3N}\}$, and using Schur's complement, one can obtain (4.50), where the inequality $-P_r^{-1}(\alpha^2R_{r1} + \delta R_{2r})P_r^{-1}$ is replaced with $\mu^2(\alpha^2R_{r1} + \delta R_{2r}) - 2\mu P_r^{-1}$ due to $-P_r^{-1}(\alpha^2R_{r1} + \delta R_{2r})P_r^{-1} \preceq \mu^2(\alpha^2R_{r1} + \delta R_{2r}) - 2\mu P_r^{-1}$. The proof of Theorem 4.3 is completed.

2) Feedback Gains Solution With Partially Known Transition Rates:

Theorem 4.4 Consider platoon control is subject to Markovian switching topologies, communication delay, and external disturbance, where the transition rates of the Markovian chain are partially known. Under Assumption 4.1, for given scalars $\beta > \alpha > 0$, $\mu > 0$, $\gamma > 0$, $\varrho > 0$ Under Assumption 4.1, for given scalars $\tilde{P}_r \in \mathbb{R}^{3\times 3} (r = 1, 2, ..., q)$, $\tilde{Q}_{1r} \in \mathbb{R}^{3N\times 3N}$, $\tilde{Q}_{2r} \in \mathbb{R}^{3N\times 3N}$, $\tilde{R}_{1r} \in \mathbb{R}^{3N\times 3N}$, $\tilde{R}_{2r} \in \mathbb{R}^{3N\times 3N}$, $\tilde{Z}_r \in \mathbb{R}^{3\times 3}$ and some matrices $\tilde{Y}_m \in \mathbb{R}^{3N\times 15N} (m = 1, 2)$ and $\tilde{K}_r \in \mathbb{R}^{1\times 3}$ of appropriate dimensions such that

$$\begin{bmatrix} \tilde{\Upsilon}_r^1 & \tilde{\Gamma}_r^T & \left(\tilde{\Xi}_{2r}^1\right)^T & \tilde{\Xi}_{3r}^T & \delta Y_{rm}^T & \ell_2^T \\ * & \tilde{\Xi}_{1r} & 0 & 0 & 0 & 0 \\ * & * & -\left(\tilde{\Lambda}_r^1\right)^T & 0 & 0 & 0 \\ * & * & * & -I_{3N} & 0 & 0 \\ * & * & * & 0 & -\sigma\tilde{R}_{2r} & 0 \\ * & * & * & * & * & -\gamma^{-2}I_{3N} \end{bmatrix} \prec 0 \tag{4.52}$$

$r \in \Pi_k^r$

$$\begin{bmatrix} \tilde{\Upsilon}_r^2 & \tilde{\Gamma}_r^T & \left(\tilde{\Xi}_{2r}^2\right)^T & \tilde{\Xi}_{3r}^T & \delta Y_{rm}^T & \ell_2^T \\ * & \tilde{\Xi}_{1r} & 0 & 0 & 0 & 0 \\ * & * & -\left(\tilde{\Lambda}_r^2\right)^T & 0 & 0 & 0 \\ * & * & * & -I_{3N} & 0 & 0 \\ * & * & * & 0 & -\sigma\tilde{R}_{2r} & 0 \\ * & * & * & * & * & -\gamma^{-2}I_{3N} \end{bmatrix} \prec 0 \tag{4.53}$$

$r \in \Pi_{uk}^r$

$$\begin{bmatrix} -\tilde{Z}_r & \tilde{P}_r \\ * & \tilde{P}_s \end{bmatrix} \prec 0, \quad r \in \Pi_{uk}^r, s \neq r \tag{4.54}$$

$$\tilde{P}_s - \tilde{Z}_r \succeq 0, \quad r \in \Pi_{uk}^r, s = r \tag{4.55}$$

with $k_1^r, k_2^r, \cdots, k_m^r$ defined in (4.38) and $k_m^r = r$ for $r = 1, 2, ..., q$ and $m = 1, 2$, where $\delta = \beta - \alpha$

$$\tilde{\Upsilon}_r^1 = \tilde{\Upsilon}_r + \pi_{rr}\ell_1^T(I_N \otimes \tilde{P}_r)\ell_1 - \sum_{s \in \Pi_k^r} \pi_{rs}\ell_1^T(I_N \otimes \tilde{Z}_r)\ell_1$$

$$\tilde{\Upsilon}_r^2 = \tilde{\Upsilon}_r - \sum_{s \in \Pi_k^r} \pi_{rs}\ell_1^T(I_N \otimes \tilde{Z}_r)\ell_1$$

$$\tilde{\Xi}_{2r}^1 = \left[\sqrt{\pi_{rk_1^r}}(I_N \otimes \tilde{P}r), \sqrt{\pi_{rk_2^r}}(I_N \otimes \tilde{P}r), ..., \sqrt{\pi_{rk_{l-1}^r}}(I_N \otimes \tilde{P}r), \sqrt{\pi_{rk_{l+1}^r}}(I_N \otimes \tilde{P}r), ..., \sqrt{\pi_{rk_m^r}}(I_N \otimes \tilde{P}r)\right]\ell_1$$

$$\tilde{\Xi}_{2r}^2 = \left[\sqrt{\pi_{rk_1^r}}(I_N \otimes \tilde{P}r), \sqrt{\pi_{rk_2^r}}(I_N \otimes \tilde{P}r), ..., \sqrt{\pi_{rk_m^r}}(I_N \otimes \tilde{P}r)\right]\ell_1$$

$$\tilde{\Lambda}_r^1 = \text{diag}\left\{I_N \otimes \tilde{P}_{k_1^r}, I_N \otimes \tilde{P}_{k_2^r}, ..., I_N \otimes \tilde{P}_{k_{l-1}^r}, I_N \otimes \tilde{P}_{k_{l+1}^r}, ..., I_N \otimes \tilde{P}_{k_m^r}\right\}$$

$$\tilde{\Lambda}_r^2 = \text{diag}\{I_N \otimes \tilde{P}_{k_1^r}, I_N \otimes \tilde{P}_{k_2^r}, ..., I_N \otimes \tilde{P}_{k_m^r}\}$$

Moreover, the controller gains are given by

$$K_r = \tilde{K}_r\tilde{P}_r^{-1} \tag{4.56}$$

Proof: Matrix variables P_r, $\tilde{P}_r$, $\tilde{Q}_{r1}$, $\tilde{Q}_{r2}$, $\tilde{R}_{r1}$, $\tilde{R}_{r2}$, $\tilde{Y}_{mr}$, and $\tilde{K}_r$ has the same form as in the proof of Theorems 4.1 and 4.3. Let $I_N \otimes \tilde{Z}_r = P_r^{-1}Z_rP_r^{-1}$. Pre- and postmultiplying both sides of (4.39) by $\text{diag}\{J_r, (\alpha^2 R_{r1} + \delta R_{2r}), P_r^{-1}, I_{3N}\}$, and using Schur complement, one can obtain that

$$\begin{aligned}&\tilde{\Upsilon}_r + \tilde{\Gamma}_r^T(\alpha^2\tilde{R}_{1r} + \delta\tilde{R}_{2r})\tilde{\Gamma}_r + \delta\tilde{Y}_m^T\tilde{R}_{2r}^{-1}\tilde{Y}_m + \gamma^2\ell_2^T\ell_2 - \sum_{s \in \Pi_k^r}\pi_{rs}\ell_1^T(I_N \otimes \tilde{Z}_r)\ell_1 \\ &+ \sum_{s \in \Pi_k^r}\pi_{rs}\ell_1^T\left(I_N \otimes \tilde{P}_r\tilde{P}_s^{-1}\tilde{P}_r\right)\ell_1 \preceq 0\end{aligned} \tag{4.57}$$

Pre- and postmultiplying both sides of (4.40) and (4.41) by P_r^{-1}, respectively, one can derive that

$$\tilde{P}_r\tilde{P}_s^{-1}\tilde{P}_r - \tilde{Z}_r \preceq 0, r \in \Pi_{uk}^r, s \neq r \tag{4.58}$$

$$\tilde{P}_s - \tilde{Z}_r \succeq 0, r \in \Pi_{uk}^r, s = r \tag{4.59}$$

As $\pi_{rr} \leq 0$, $r \in \mathcal{S}$, if $r \in \Pi_k^r$, LMI (4.57) is dealt as

$$\begin{aligned}&\tilde{\Upsilon}_r + \tilde{\Gamma}_r^T(\alpha^2\tilde{R}_{1r} + \delta\tilde{R}_{2r})\tilde{\Gamma}_r + \delta\tilde{Y}_m^T\tilde{R}_{2r}^{-1}\tilde{Y}_m + \gamma^2\ell_2^T\ell_2 - \sum_{s \in \Pi_k^r}\pi_{rs}\ell_1^T(I_N \otimes \tilde{Z}_r)\ell_1 \\ &+ \sum_{s \in \Pi_k^r, s \neq r}\pi_{rs}\ell_1^T\left(I_N \otimes \tilde{P}_r\tilde{P}_s^{-1}\tilde{P}_r\right)\ell_1 + \pi_{rr}(I_N \otimes \tilde{P}_r) \preceq 0\end{aligned} \tag{4.60}$$

which is equivalent to LMI (4.50) by applying a Schur's complement and a similar manipulation in the proof of Theorem 4.1. If $r \in \Pi_{uk}^r$, LMI (4.57) is equivalent to LMI (4.53) by applying a Schur complement and the similar manipulation in the proof of Theorem 4.1. In addition, LMI (4.58) is equivalent to LMI (4.54) by applying a Schur's complement. Therefore, if LMI (4.52)–(4.55) holds, the platoon control system (4.12) is exponentially mean-square stable with a partially known transition rate according to Theorem 4.2. Then, the control objectives (4.14) and (4.15) can be done by the feedback controller (4.10), and the desired controller gain are given by (4.56). The proof is completed.

4.5 Simulation Studies

In this section, numerical simulations are given to verify the effectiveness of the proposed control method. The platoon control system uses a hierarchical architecture (see Figure 4.2), where the upper layer determines the desired acceleration a_{ides} as in (4.8), while the lower layer generates the throttle torque a_{thrdes} or brake torque b_{brkdes} based on a_{ides} and (4.2). The parameters in (4.1) are given in Table 4.1. The acceleration of the leader $a_0(t)$ is set as follows:

$$a_0(t) = \begin{cases} 1\,\text{m/s}^2, t \leq 10\,\text{s} \\ -0.15t + 2.5\,\text{m/s}^2, 10\,\text{s} < t \leq 20\,\text{s} \\ 0.15t - 3.5\,\text{m/s}^2, 20s < t \leq 30\,\text{s} \\ 1\,\text{m/s}^2, 30\,\text{s} < t \leq 40\,\text{s} \\ -0.1t + 5\,\text{m/s}^2, 40\,\text{s} < t \leq 50\,\text{s} \\ 0\,\text{m/s}^2, t > 50\,\text{s} \end{cases}.$$

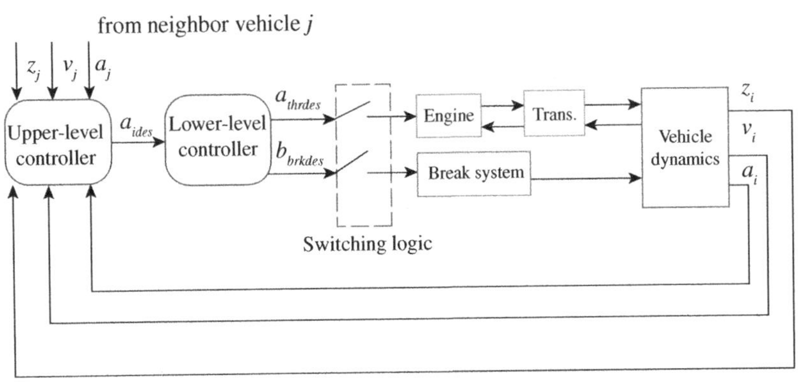

Figure 4.2 Controller structure for vehicle platoons.

Table 4.1 Vehicle parameters used in simulation.

Parameter	Notation	Value
Aerodynamic drag coefficient	$C_{A,i}$	0.492
Acceleration of gravity	g	$0.98\ \text{m/s}^2$
Coefficient of rolling resistance	f_i	0.01
Mechanical efficiency of the driveline	$\varpi_{T,i}$	0.9
Mass of the vehicle	m_i	2.9×10^3 kg
Inertial lag	ς	0.25 s
Sampling period	h	0.1 s
Length of vehicle	L	7 m
Communication delay	$\tau(t)$	0.25 s
Desired spacing	$d_{i,i-1}$	25 m
H_∞ disturbance attention level	γ	1.25
Tire radius	R_i	0.3

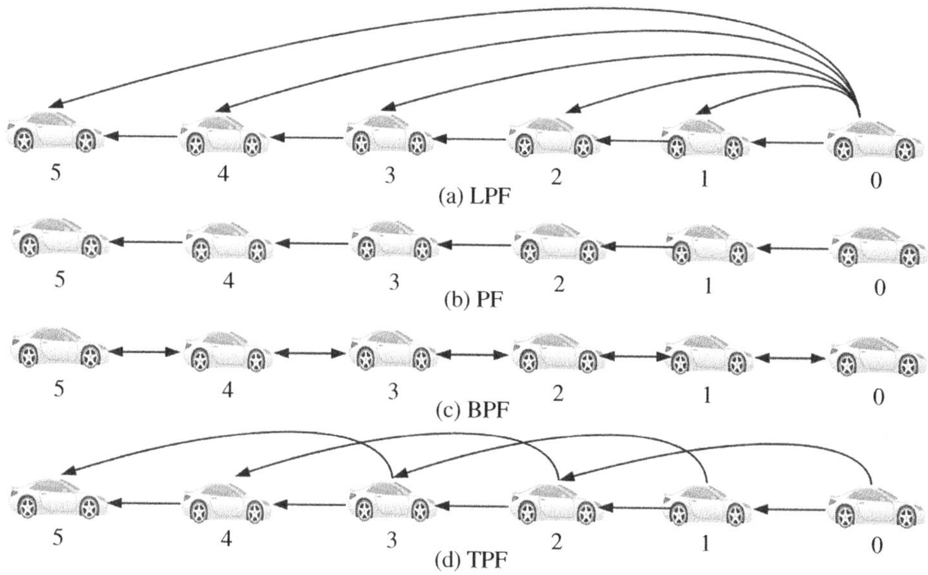

Figure 4.3 Possible communication topologies of vehicle platoons in simulation. (a) LPF, (b) PF, (c) BPF, (d) TPF.

Figure 4.3 shows the communication topologies of LPF, PF, BPF, and TPF used for platooning. It is obviously seen from Figure 4.3 that the leader vehicle is globally reachable in the union graph of $\mathcal{G}_r$, $r = 1, 2, 3, 4$, where $\mathcal{G}_1$ = '*LBF*', $\mathcal{G}_2$ = '*PF*', $\mathcal{G}_3$ = '*BPF*', and $\mathcal{G}_4$ = '*TPF*'. As the eigenvalues of communication topology H_r affect the stability and convergence speed of the platoon control system, we show the eigenvalues of different topologies under different platoon sizes in Figure 4.4, where the y-axis is the largest real part of the eigenvalues.

For Figure 4.4, we can see that PF mode has only forward communication links and hence the eigenvalue is always "1." TPF and LPF modes have the same eigenvalues with the largest one equal to "2." Since the BPF mode has backward communication links, a larger platoon size leads to bigger eigenvalues of the communication topology H_3.

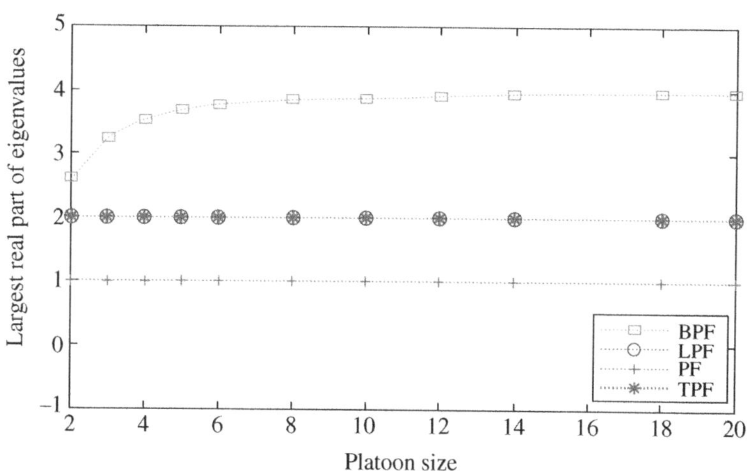

Figure 4.4 Comparison of the largest real part of the eigenvalues with respect to platoon size and communication topologies.

Table 4.2 Feasibility of LMIs in Theorems 4.1 and 4.2 with increasing communication delay τ_{ij}.

$\mathcal{G}_r$ \ τ_{ij}	**0.05 s**	**0.10 s**	**0.15 s**	**0.20 s**
$r = 1$	Feasible	Feasible	Feasible	Unfeasible
$r = 2$	Feasible	Feasible	Feasible	Feasible
$r = 3$	Feasible	Feasible	Unfeasible	Feasible
$r = 4$	Feasible	Feasible	Feasible	Feasible

Table 4.3 Feasibility of LMIs in Theorems 4.1 and 4.2 with decreasing disturbance attenuation level γ.

$\mathcal{G}_r$ \ γ	**1.5**	**1.25**	**1.00**	**0.75**
$r = 1$	Feasible	Feasible	Feasible	Feasible
$r = 2$	Feasible	Feasible	Unfeasible	Unfeasible
$r = 3$	Feasible	Feasible	Feasible	Unfeasible
$r = 4$	Feasible	Feasible	Feasible	Unfeasible

By Matlab LMI toolbox, the results on the feasibility of LMIs in Theorems 4.1 and 4.2 are shown in Tables 4.2 and 4.3. According to Lemma 4.3, the feedback controller gain is obtained as $K = [4.24 \;\; 20.91 \;\; 28.77]$, where $\upsilon = 0.0603$ and the partially known transition rate matrix of the Markovian switching communication topologies is given as

$$\pi = \left[\pi_{ij}\right] = \begin{bmatrix} -1.1 & ? & ? & 0.15 \\ ? & -1 & ? & 0.3 \\ 0.4 & ? & -1 & ? \\ ? & 0.5 & ? & -0.95 \end{bmatrix} \tag{4.61}$$

The switching states are shown in Figure 4.5, where the values 1, 2, 3, and 4 of the y-axis denote the value of the switching signal $\sigma(t)$. For example, $\sigma(t) = 1$ when $\mathcal{G}_{\sigma(t)} = \mathcal{G}_1$. With parameter $\varrho = 0.01$, communication delay $\tau_{ij} = 0.1$ s, and disturbance term $\omega(x_i, t) = 0.01 \sin(x_i)$, the LMIs of Theorem 4.2 is feasible by using the MATLAB LMI toolbox. The spacing, velocity, and acceleration of each follower are given in Figure 4.6, respectively. Different from [2], the initial tracking errors of each following vehicle in Figure 4.6 are not required to be zero. From those figures, we know that the connected vehicular control system is stable with the effect of Markovian switching topologies, communication delay, and external disturbance.

This fixed-gain controller may not be able to stabilize platoon control system when the communication delay is large enough, i.e. $\tau_{ij} = 0.15$ s in this example. The reason is that, according to the results on fixed-gain control in Theorem 4.2, the allowable maximum delay $\tau_k(t) = 0.225$ s. So, we used the variable-gain controllers as given in Theorem 4.4 for the case $\tau_{ij} = 0.15$ s, with the set of variable gains obtained as $K_1 = [4 \;\; 5 \;\; 8]$, $K_2 = [3 \;\; 2.5 \;\; 6]$, $K_3 = [2 \;\; 4.5 \;\; 7]$, $K_4 = [0.9 \;\; 2 \;\; 10]$. Under the same set of parameters in the previous simulation, the allowable maximum delay $\tau_k(t)$ is 0.37 s. And the spacing, velocity, and acceleration of the vehicle platoon under the variable-gain controller are given in Figure 4.7. It is clear that the proposed variable-gain controller can effectively stabilize vehicular platoons.

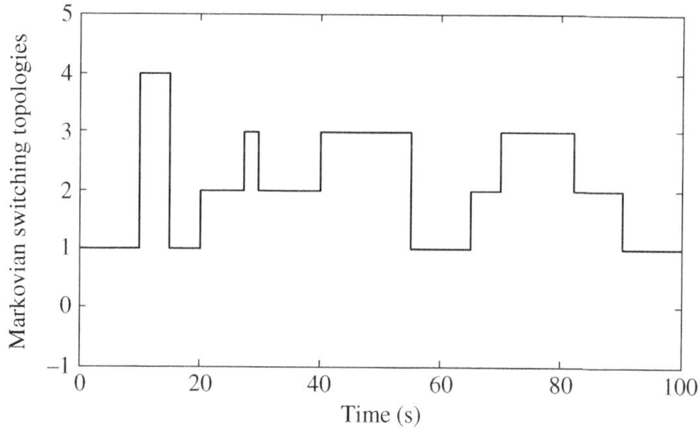

Figure 4.5 Markovian switching communication topologies for the platoon.

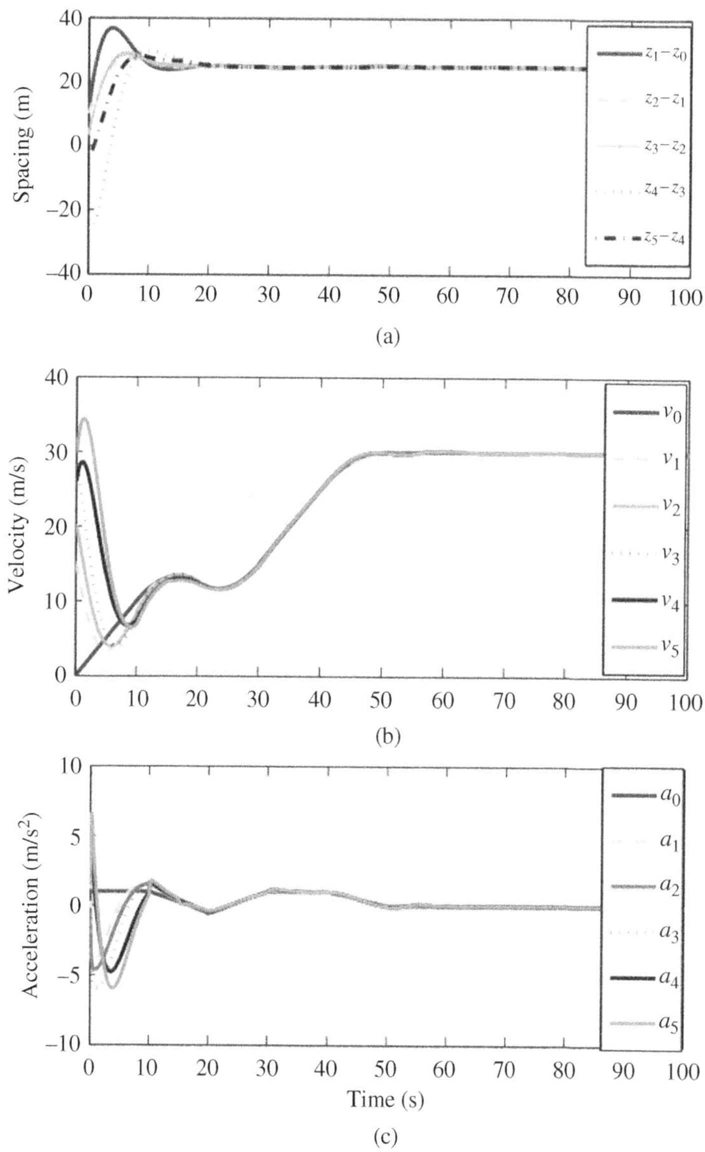

Figure 4.6 Spacing error (a), velocity (b), and acceleration (c) for the vehicle platoons using a designed fixed-gain controller with communication delay $\tau_{ij} = 0.1$ s.

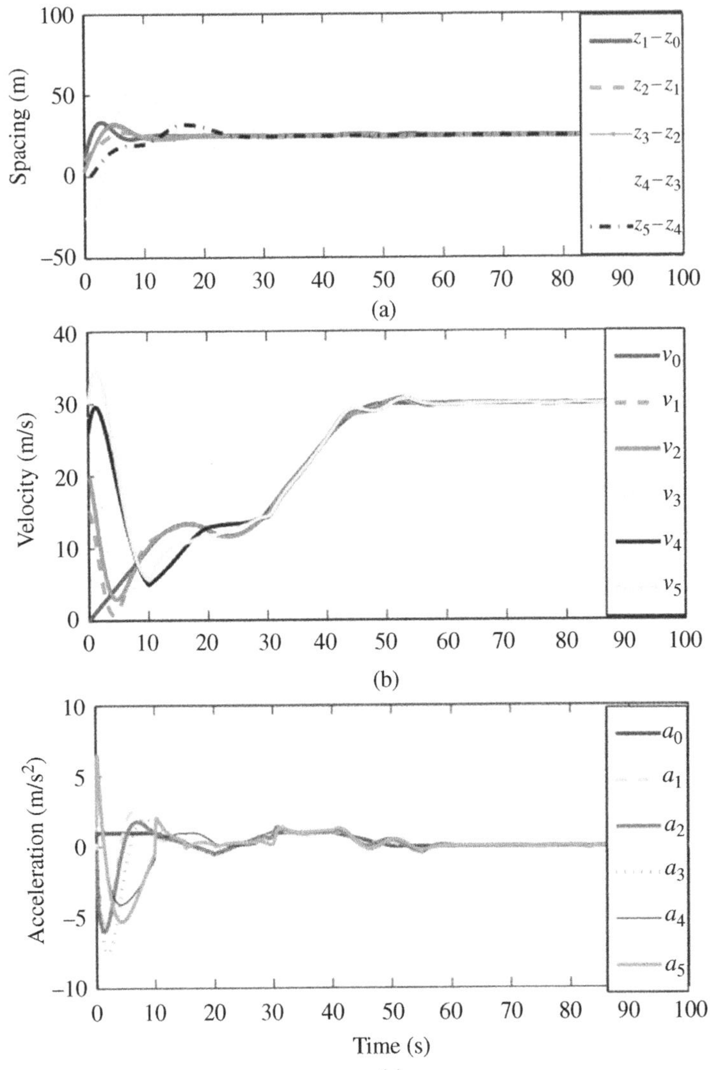

Figure 4.7 Spacing error (a), velocity (b), and acceleration (c) for the vehicle platoons using variable-gain controller with communication delay τ_{ij} = 0.15 s.

In this section, we give some comparative results with the platoon control method proposed in [2], where the controller gain (4.37) in [2] is used. Considering the same parameters used in the simulation of Section 4.4, as shown in Figure 4.8, the platoon is unstable under the effect of Markovian switching topologies and communication delay. Due to the fact that controller gain (4.37) in [2] is obtained with perfect intervesicle communication and undirected fixed communication topologies. In addition, we cannot find the feasible solution for LMIs (4.39)–(4.41) of Theorem 4.2 with controller gain (4.37) in [2]. Therefore, the stability requirement of the platoon cannot be guaranteed.

In the simulation, we found that the property of the Markovian switching topologies has a relation with the stability performance of the platoon. Here, the fixed-gain controller is involved with the same controller gain as obtained in Figures 4.6 and 4.7. By using the transition rate matrix (4.46), we obtain another group of communication topologies, which is shown in Figures 4.9 and 4.11. For comparison, we define some notations. Define $G_r(t)$ as the total time that topology

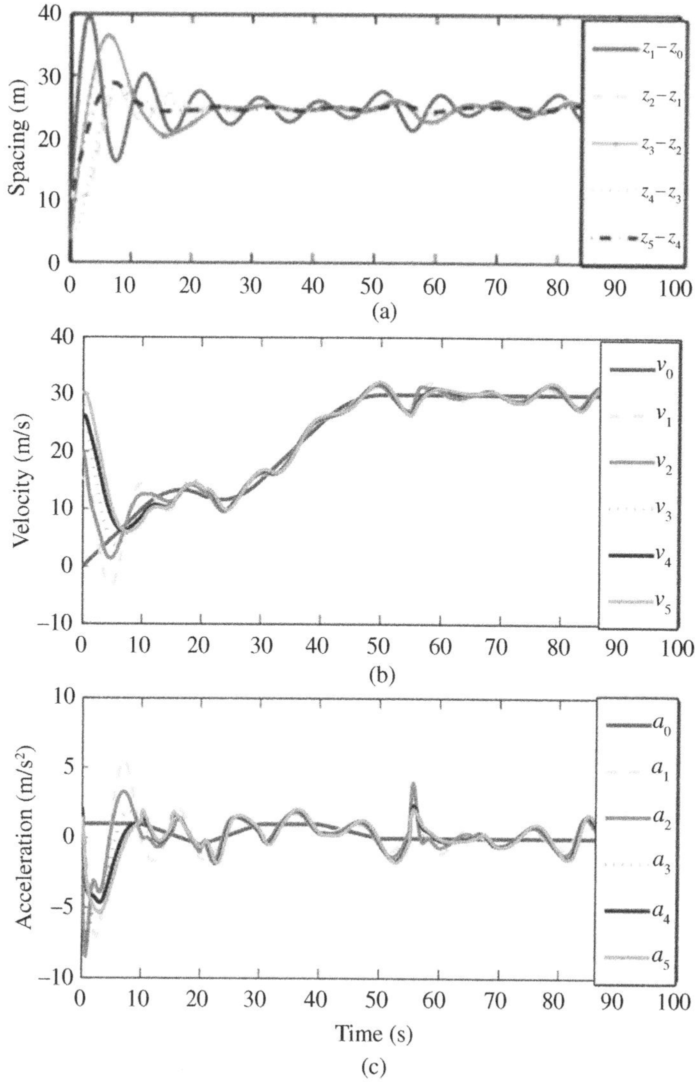

Figure 4.8 Spacing error (a), velocity (b), and acceleration (c) for vehicle platoons using a fixed-gain controller designed in [2] with communication delay τ_{ij} = 0.15 s. *Source:* Adapted from Zheng et al. [2].

$\mathcal{G}_r$ appears during the time interval [0, t), then $g_r = G_r(t)/t$ is the "mode rate" for topology $\mathcal{G}_r$. Let $N(t)$ denote the number of switching of different topologies during the time interval [0, t]. Then, $t/(N(t) + 1)$ is called the average dwell time. In order to measure the tracking performance during the transient process, as suggested in [4], the following performance index is used:

$$J = \sum_{k=0}^{T}\sum_{i=1}^{N}\frac{1}{2}(\hat{x}_i(kh) - \hat{x}_{i-1}(kh))^T\Lambda(\hat{x}_i(kh) - \hat{x}_{i-1}(kh)) \tag{4.62}$$

where $\Lambda = \text{diag}\{K^p, K^v, K^a\}$ is the weighting factor, T is the number of sampling times during the time interval [0, t]. With this performance index, we give some comparative simulation results and discussions in terms of mode rate and the average dwell time.

1) The platoon with a bigger mode rate g_1 (i.e. $\mathcal{G}_1$ = 'LPF') has better stability performance. This can be derived by comparing the defined performance index J (as listed in Table 4.5) in Figures 4.6 and 4.10,

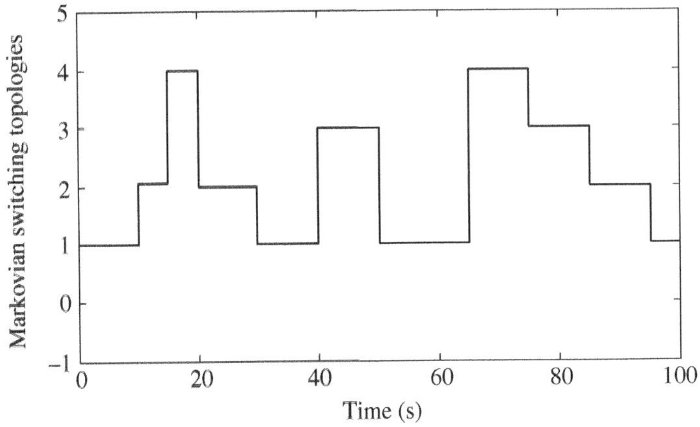

Figure 4.9 Markovian switching communication topologies for the platoon with mode rate g_1 = 40%.

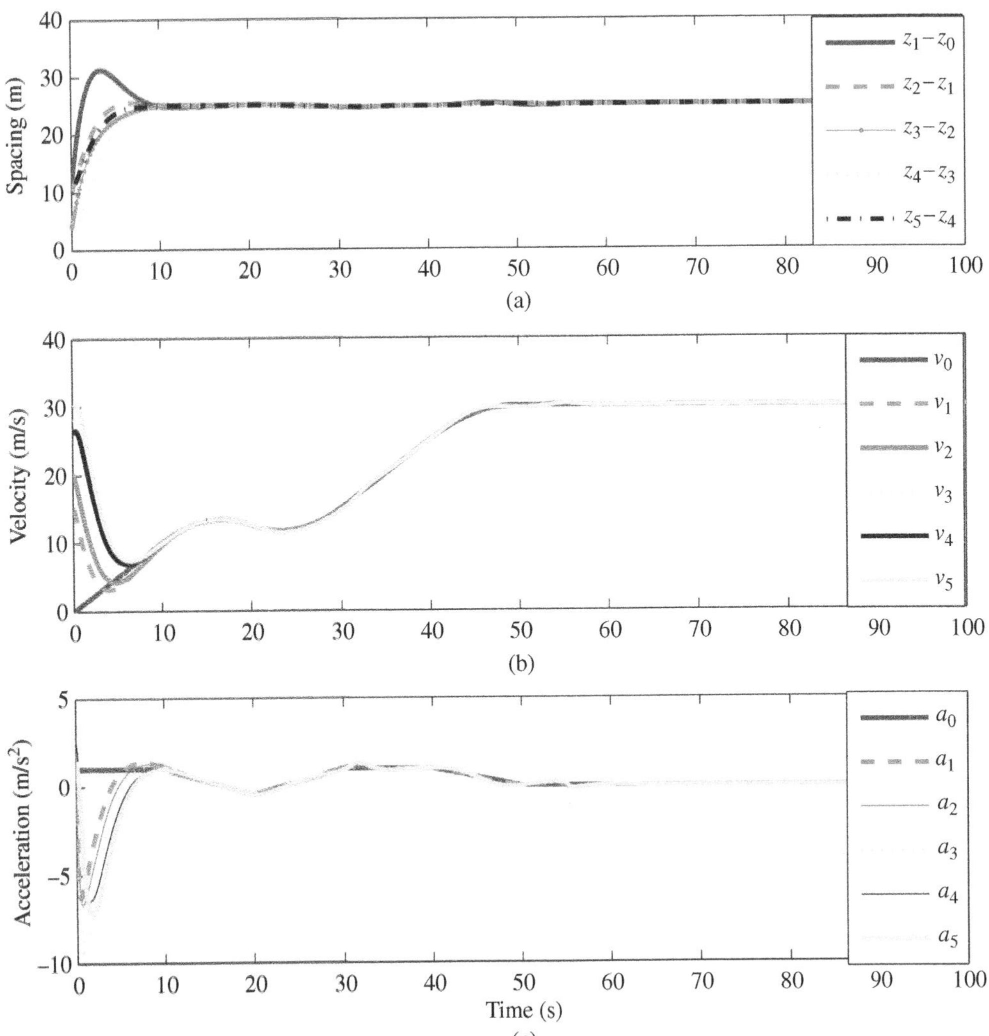

Figure 4.10 Spacing (a), velocity (b), and acceleration (c) for the vehicle platoons using designed fixed-gain controller with mode rate g_1 = 40%.

respectively. The mode rate g_r is listed in Table 4.4 for the Markovian switching topologies in Figures 4.5 and 4.9. This conclusion agrees with the analysis results in [2, 4, 9].

2) For the same mode rate g_1, we found that a platoon with a longer average dwell time has better stability performance. Compared with the value of the performance index J (as listed in Table 4.5) in Figures 4.10 and 4.12, it is easy to obtain. The average dwell time of the Markovian switching topologies in Figures 4.9 and 4.11 is 5 and 2.5 s, respectively.

Note that the results of this chapter are based on a linear platoon model. In practice, uncertain nonlinear dynamics (e.g. unmodeled dynamics and disturbances) are inevitable. In the simulations, we investigated the influence of nonlinear dynamics on the designed feedback controller using the

Table 4.4 Mode rate of communication topology.

g_r	g_1	g_2	g_3	g_4
$r=1$	0.35	0.30	0.30	0.05
$r=2$	0.40	0.25	0.20	0.15

Table 4.5 Value J computed by using the state response of the platoon in different simulation results.

	Figure 4.6	**Figure 4.10**	**Figure 4.12**
$J=2$	573	432	751

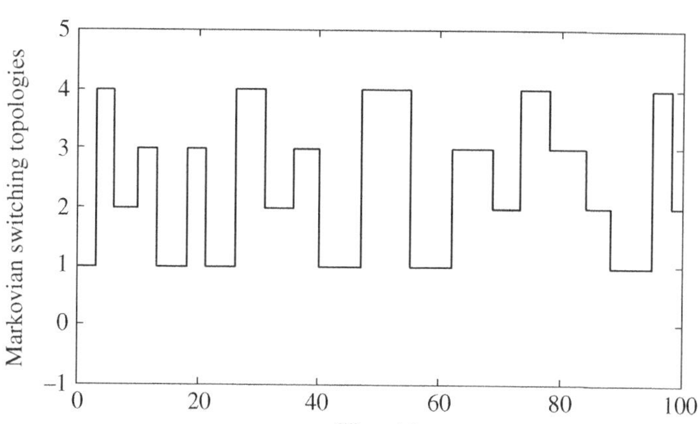

Figure 4.11 Markovian switching communication topologies for the platoon with average dwell time 2.5 s.

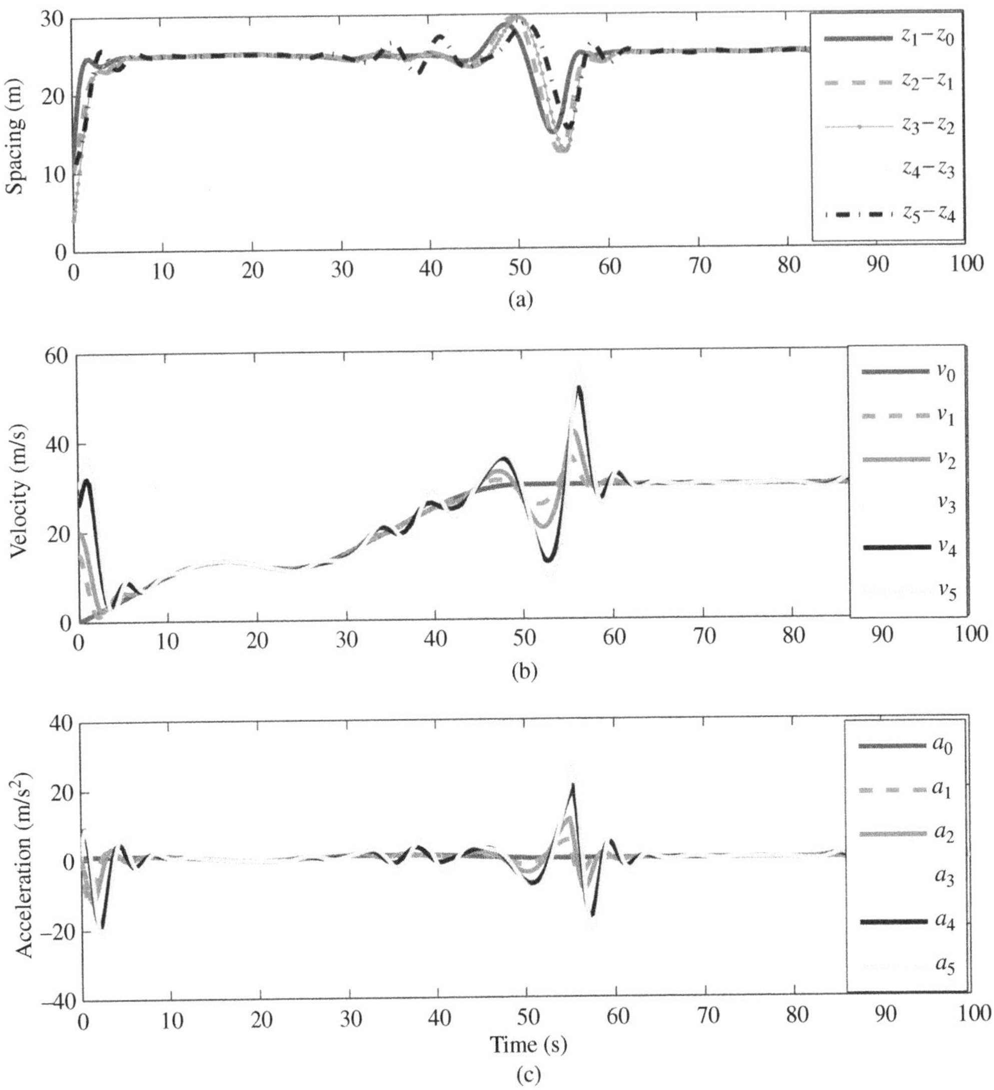

Figure 4.12 Spacing (a), velocity (b), and acceleration (c) for the vehicle platoons using designed fixed-gain controller with an average dwell time equal to 2.5 s.

platoon model with parameters in (4.1) and Table 4.3 of [15]. The spacing, velocity, and acceleration of the platoon with our proposed controller are shown in Figure 4.13. It is obvious that the control performance is degraded when uncertain nonlinear dynamics, communication limitations, and external disturbance are simultaneously considered.

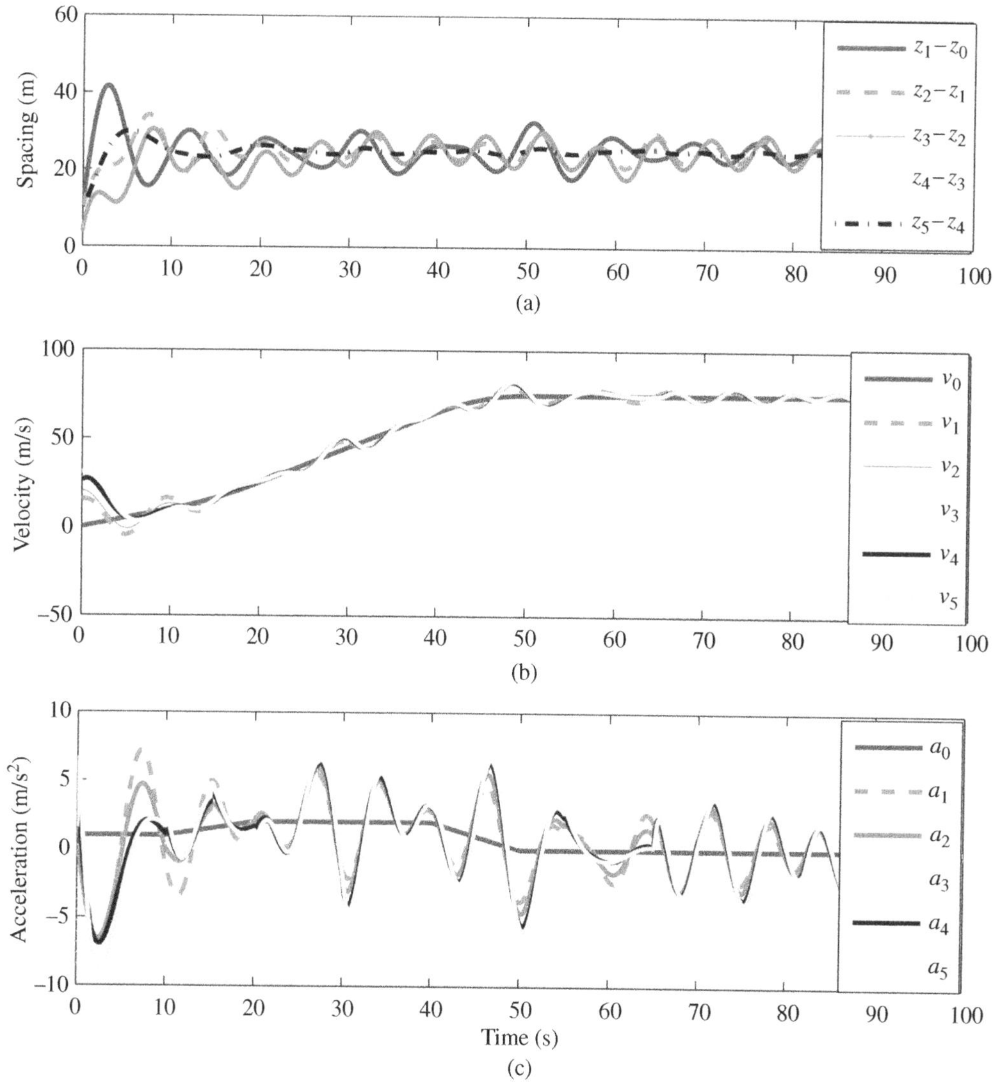

Figure 4.13 Spacing (a), velocity (b), and acceleration (c) for vehicle platoons of [15] using the designed fixed-gain controller. *Source:* Adapted from Ma et al. [15].

4.6 Conclusions and Future Work

We have developed a sampled-data control method to simultaneously deal with the issues of switching topologies, communication delay, and external disturbance for a platoon of vehicles. The method is derived from the framework of the Markovian jumping system with time delays. We show that both fixed-gain and variable-gain control schemes are feasible for platooning, but a variable-gain controller has the advantage of improved robustness and design flexibility.

References

1 Fu, F. and van der Schaar, M. (2012). Structure-aware stochastic control for transmission scheduling. *IEEE Trans. Veh. Technol.* 61 (9): 3931–3945.

2 Zheng, Y., Li, S.E., Li, K., and Ren, W. (2018). Platooning of connected vehicles with undirected topologies: robustness analysis and distributed H-infinity controller synthesis. *IEEE Trans. Intell. Transp. Syst.* 19 (5): 1353–1364.

3 Zheng, Y., Li, S.E., Wang, J. et al. (2016). Stability and scalability of homogeneous vehicular platoon: study on the influence of information flow topologies. *IEEE Trans. Intell. Transp. Syst.* 17 (1): 14–26.

4 Zheng, Y., Li, S., Li, K., and Wang, L.-Y. (2016). Stability margin improvement of vehicular platoon considering undirected topology and asym metric control. *IEEE Trans. Control Syst. Technol.* 24 (4): 1253–1265.

5 Li, S.E., Qin, X., Li, K. et al. (2017). Robustness analysis and controller synthesis of homogeneous vehicular platoons with bounded parameter uncertainty. *IEEE/ASME Trans. Mechatron.* 22 (2): 1014–1025.

6 Ren, W. and Beard, R.W. (2005). Consensus seeking in multiagent systems under dynamically changing interaction topologies. *IEEE Trans. Autom. Control* 50 (5): 655–661.

7 Toor, Y., Muhlethaler, P., and Laouiti, A. (2008). Vehicle ad hoc networks applications and related technical issues. *IEEE Commun. Surveys Tuts.* 10 (3): 74–88, 3rd Quart.

8 Wen, S. and Guo, G. (2019). Observer-based control of vehicle platoons with random network access. *Robot. Auton. Syst.* 115: 28–39.

9 Seiler, P., Pant, A., and Hedrick, K. (2004). Disturbance propagation in vehicle strings. *IEEE Trans. Autom. Control* 49 (10): 1835–1842.

10 Middleton, R.H. and Braslavsky, J.H. (2010). String instability in classes of linear time invariant formation control with limited communication range. *IEEE Trans. Autom. Control* 55 (7): 1519–1530.

11 Li, S. et al. (2019). Distributed platoon control under topologies with complex eigenvalues: stability analysis and controller synthesis. *IEEE Trans. Control Syst. Technol.* 27 (1): 206–220.

12 Han, Q. (2005). Absolute stability of time-delay systems with sector-bounded nonlinearity. *Automatica* 41 (12): 2171–2176.

13 Zhang, X.-M. and Han, Q.-L. (2013). Novel delay-derivative-dependent stability criteria using new bounding techniques. *Int. J. Robust Nonlin. Control* 23 (13): 1419–1432.

14 Mao, X. (1996). Robustness of exponential stability of stochastic differential delay equations. *IEEE Trans. Autom. Control* 41 (3): 442–447.

15 Ma, Y., Li, Z., Malekian, R. et al. (2019). Hierarchical fuzzy logic-based variable structure control for vehicles platooning. *IEEE Trans. Intell. Transp. Syst.* 20 (4): 1329–1340.

5

Control with Event-Triggered Communication

5.1 Introduction

It should be noted that the obtained results in Chapters 1–4 including the references therein are based on the periodic time-triggered communication mechanism (TCM), meaning the data transmission between successive vehicles is performed continually at every time step or periodically at equidistant sampling points. Such a communication mechanism places a heavy burden on intervehicle communication systems for the following reasons: (i) limited wireless network bandwidth or limited coverage of the roadside network, (ii) the intercommunication of moving vehicles cannot be guaranteed at all times. The TCM follows a periodic schedule to transmit sampled data and update the control input for the CACC system, even for two successive sampled data points with very little fluctuation. This is a waste of the limited communication resources available in the IEEE 802.11p communication protocol based on vehicular ad hoc networks (VANETs) and dedicated short-range communications [1]. Based on the digital nature of intervehicle communication networks, the rate at which data packages can be transmitted is limited. As reported in [2], high communication rates degrade the reliability of intervehicle communication channels and increase transmission delays. There is no doubt that intervehicle communication imperfections can have a significant impact on the performance of CACC systems [3, 4]. In order to secure the reliability and quality of intervehicle communication, it is important that only the information that is actually required to establish a stable CACC system is transmitted and that the transmission of unnecessary information is avoided. Hence, how to efficiently use scarce communication resources has become a major concern for CACC systems. To the best of our knowledge, the problem of a resource-efficient communication mechanism for CACC systems is still open and few results have been published on this topic in the literature. Difficulties may arise from handling intermittent data transmission and determining when intervehicle communication must be performed to guarantee the desired performance for the vehicle platoon, which gives rise to the first motivation for this study.

Recently, as an alternative to the TCM, the event-triggered communication mechanism (ECM) has become a hot research topic in the field of networked systems [5–8]. The core idea behind the ECM is that the transmission of the sampled data is only executed after the occurrence of an event, rather than after a certain amount of time has passed. Generally, an event is defined by an event-triggering condition, which is related to the system states and a threshold parameter. The sampled data is transmitted by the communication network only when a defined event is satisfied. In this manner, the transmission frequency over the communication network can be significantly reduced to save limited communication resources. Generally, there are two types of ECM: static ECM

Connected Vehicular Systems: Communication, Control, and Optimization, First Edition. Ge Guo and Shixi Wen.

(SECM) and dynamic ECM (DECM). Unlike SECM, which has a prescribed constant threshold parameter, the threshold parameter of DECM can be determined real time by adjusting rules [7, 8]. The threshold parameter determines when or how frequently data transmission occurs, which is closely related to data transmission rates over intervehicle communication network. Since the IEEE 802.11p standard can provide various data rates at 3, 4.5, 6, 9, 12, 18, 24, and 27 Mbps, the threshold parameter should dynamically change over time to reflect these time-varying data transmission rates. However, how the DECM can be extended to CACC systems is unclear. Therefore, the second motivation of this chapter is to design a DECM for CACC systems and apply it to facilitate velocity tracking and maintain intervehicle spacing.

In this chapter, we aim to design a resource-efficient communication mechanism for CACC systems considering the effects of actuator delay (e.g. fueling and braking delay) and disturbances (caused by acceleration of preceding vehicles). For maintaining distance and tracking velocity, a tracking-error-based sampled-data feedback controller for CACC systems is designed by considering heterogeneous information feedback. Here we assume that the controller gain of the sampled-data feedback controller has a prescribed value that can guarantee string stability requirement conditions. Specifically, spacing error can be computed by onboard sensors directly, whereas the velocity and acceleration error of the i-th following vehicle must be calculated from the velocities and accelerations of preceding vehicle, which are transmitted via the intervehicle communication. To save the limited communication resources of intervehicle communication networks, the DECM is used as a resource-efficient communication mechanism for a CACC system, where the threshold parameter can be determined in real time by predefined adjusting rules. Under the DECM, the sampled velocity and acceleration of the i-th following vehicle will be authorized to be transmitted to the next vehicle only when a defined event is trigged. This significantly reduces the transmission frequency of sampled data. Next, a series of sufficient conditions for the stability of the platoon tracking error system under the DECM-based sampled-data feedback controller are obtained. Additionally, design criteria for determining the parameters of the DECM are established to guarantee the performance of the CACC system. A simulation study demonstrates that the CACC system with the DECM not only reduces the transmission frequency of intervehicle communication but also guarantees a tracking error level below the H_∞ disturbance attenuation level. The main contributions of this chapter are summarized as follows:

1) A resource-efficient control architecture is proposed for CACC systems to save the limited communication resources of the intervehicle communication network. In contrast to the traditional sampled-data vehicle control system, the transmission of sampled velocity and acceleration value is determined by an event-triggered condition.
2) A series of sufficient conditions are obtained to guarantee the stability of a DECM-based sampled-data CACC system subject to actuator delay and disturbance. Those conditions manage the tradeoff between communication resource consumption and control performance.
3) A resource-efficient H_∞ robust communication control design framework is established for the CACC system. In this framework, the parameters of the DECM are determined based on the communication resource consumption and stability requirements of the CACC system.

The remainder of this chapter is organized as follows. Section 5.2 discusses the modeling of a single vehicle and the CACC system. It also contains a description of the DECM, model transformation, and control objective of the CACC system. Section 5.3 presents stability analysis results and the parameter design of the DECM for the CACC system. Numerical simulations and comparative results are presented in Section 5.4 to illustrate the effectiveness of the proposed algorithm. Section 5.4 gives the summary of this chapter.

Notations $\arg\min_p\{f(p)\}$ is the value of p for which $f(p)$ attains its minimum value. The symbol $\otimes$ denotes a matrix Kronecker product. diag$\{a_i\}$ is a diagonal matrix containing diagonal entries of a_i. I is an identity matrix of the appropriate dimensions. $P > 0$ indicates that P is a positive definite matrix. The symbol $*$ denotes the symmetric term in a symmetric matrix.

5.2 Problem Formulation

Consider a CACC system containing $N+1$ vehicles driving on a horizontal road (see Figure 5.1) in a VANET environment. z_i, v_i, and a_i denote the i-th ($i = 0, 1, \ldots, N$) vehicle's position, velocity, and acceleration, respectively, where $i = 0$ represents the leading vehicle. Each following vehicle is equipped with an onboard sensor to measure the distance between it and the preceding vehicle. Each vehicle transmits its state information (i.e. velocity and acceleration) to its following vehicle over the VANET. Considering the limited communication resources in the VANET, the distributed DECM will be applied to reduce the transmission frequency of the sampled velocity and acceleration values.

For each following vehicle, its dynamics include the engine, drive line, braking system, aerodynamic drag, tire friction, rolling resistance, and gravitational force, etc. In order to derive a concise model for platoon control, the following reasonable assumptions are added [9]:

1) Tire longitudinal slip in negligible and the power train dynamics are combined into a first-order inertial transfer function.
2) The vehicle body is considered to be rigid and symmetric.
3) The influence of pitch and yaw motions is neglected.
4) The driving and braking torques are controllable inputs.

According to these assumptions, a simplified nonlinear vehicle dynamic model can be written as follows:

$$\begin{cases} \dot{z}_i(t) = v_i(t) \\ \dot{v}_i(t) = \dfrac{1}{m_i}\left(\varpi_{T,i}\dfrac{T_i(t)}{R_i} - C_{A,i}v_i^2(t) - m_i g f\right) \\ \varsigma_i \dot{T}_i(t) + T_i(t) = T_{i,des}(t) \end{cases} \tag{5.1}$$

where m_i is the mass of the vehicle, $C_{A,i}$ is the combined aerodynamic drag coefficient, g is the acceleration due to gravity, f is the coefficient of rolling resistance, $T_i(t)$ denotes the actual driving/breaking torque, $T_{i,des}(t)$ is the desired driving/breaking torque, ς_i is the inertial delay of vehicle dynamics, R_i denotes the tire radius, and $\varpi_{T,i}$ is the mechanical efficiency of the driveline.

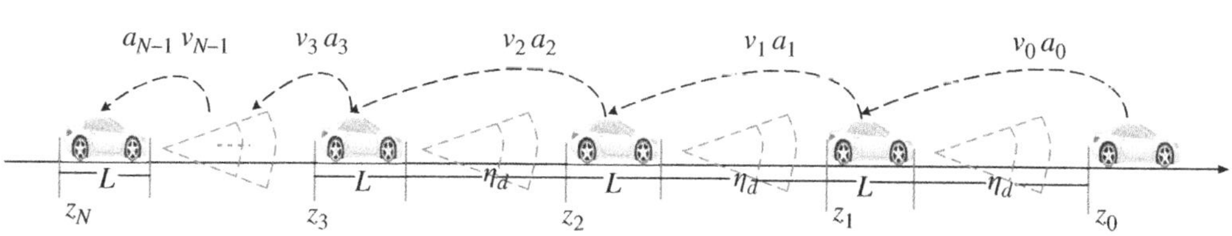

Figure 5.1 Vehicle platoon control system.

Based on the widely used exact feedback linearization technique [9–11], the nonlinear vehicle dynamic model (5.1) can be converted into a linear model for controller design. The position output combined with a relative degree tree is used to construct a feedback linearization law, which is written as follows:

$$T_{i,des}(t) = \frac{1}{\varpi_{T,i}}(C_{A,i}v_i(t)(2\tau_i\dot{v}_i + v_i) + m_igf + m_iu_i)R_i \tag{5.2}$$

where u_i is the new input signal after linearization. Then, we obtain the following linear model for vehicle dynamics:

$$\begin{cases} \dot{z}_i(t) = v_i(t) \\ \dot{v}_i(t) = a_i(t) \\ \dot{a}_i(t) = -a_i(t)/\varsigma_i + u_i(t)/\varsigma_i \end{cases} \tag{5.3}$$

It is assumed that vehicles are homogeneous (i.e. $\varsigma_i = \varsigma$), and their controllers $u_i(t)$ are designed to be identical. However, a more realistic dynamic model should consider actuating delay in the vehicles (e.g. fueling delay and braking delay). Here, for simplicity, we utilize a single lumped delay τ to represent the actuator time delay. Therefore, the third equation in (5.3) can be rewritten as:

$$\dot{a}_i(t) = -a_i(t)/\varsigma + u_i(t-\tau)/\varsigma \tag{5.4}$$

We define the spacing error, velocity error, and acceleration error, respectively, as

$$\begin{cases} e_i^\delta(t) = \delta_i(t) - e_i^d(t) \\ e_i^v(t) = v_{i-1}(t) - v_i(t) \\ e_i^a(t) = a_{i-1}(t) - a_i(t) \end{cases} \tag{5.5}$$

where $\delta_i(t) = z_{i-1}(t) - z_i(t) - L$ and $\delta_i^d(t) = h_v v_i(t) + z_0$ are the distance between successive vehicles and the desired distance, respectively. h_v is the time gap, z_0 is a given minimum distance, and L is the length of a vehicle. Here, a CTHS policy is adopted for the CACC system to regulate the spacing between vehicles.

The tracking error (e.g. spacing error, velocity error, and acceleration error)-based controller for the i-th following vehicle is designed as follows:

$$u_i(t) = k_p e_i^\delta(t) + k_v e_i^v(t) + k_a e_i^a(t) \tag{5.6}$$

where k_p, k_v, and k_a are the prescribed feedback controller gains.

Assumption 5.1 The gain for the tracking-error-based sampled-data feedback controller (5.6) was designed by utilizing the method in [11] such that the magnitude of the string stability transfer function satisfies $\|G_{\Delta_i}(s)\| \leq 1$, where $G_{\Delta_i}(s) = (\Delta_i(s))/(\Delta_{i-1}(s))$ and Δ_i is the signal of interest (e.g. relative distance error, velocity, or acceleration).

Note that the feedback controller (5.6) refers to heterogeneous information feedback. Specifically, the spacing error $e_i^\delta(t)$ can be computed from its own velocity $v_i(t)$ and the distance $\delta_i(t)$ measured by onboard sensors. However, the velocity error $e_i^v(t)$ and acceleration error $e_i^a(t)$ must be calculated based on the velocity and acceleration of the preceding vehicle, which are transmitted via the inter-communication network. For this reason, the overall measurement output state $\left[e_i^\delta \;\; v_i(t) \;\; a_i(t)\right]^T$ of the i-th following vehicle is split into two parts, $x_i^o(t)$ and $x_i^c(t)$, which respectively denote the portion of measurement supplied by the onboard sensors and that transmitted via the VANETs with

$x_i^o(t) = \left[e_i^\delta(t)\ 0\ 0\right]^T$ and $x_i^c(t) = [0\ v_i(t)\ a_i(t)]^T$. Correspondingly, the control input is also divided into two parts such as:

$$u_i(t) = u_i^o(t) + u_i^c(t) = K_i^o x_i^o(t) + K_i^c\left[x_i^c(t) - x_{i-1}^c(t)\right] \tag{5.7}$$

where $K_i^o = \left[k_p\ 0\ 0\right]$ and $K_i^c = [0\ k_v\ k_a]$.

For conventional CACC systems, each vehicle's velocity and acceleration are sampled periodically by onboard sensors at sampling times kh, $k \in \mathcal{N}$, where $h > 0$ is the sampling period. The i-th vehicle then transmits the sampled data $v_i(kh)$ and $a_i(kh)$ to the next vehicle at each sampling time kh via the VANETs, regardless of the state of the CACC system and communication resource occupation. This periodic communication TCM has advantages in terms of simplicity of modeling and implementation. However, it inevitably leads to excessive consumption of intervehicle communication resources, such as limited bandwidth and data transmission rates. From a resource conservation perspective, an efficient communication mechanism should be applied to CACC systems to avoid unnecessary data exchange between the vehicles to achieve better communication resource efficiency.

Instead of utilizing the TCM, in this chapter, the DECM is adopted (as shown in Figure 5.2). The core idea behind the DECM is that whether or not the sampled data $v_i(kh)$ and $a_i(kh)$ should be transmitted depends on the following defined event:

$$\alpha_i^T(kh)\Phi\alpha_i(kh) \geq \sigma_i(kh)y_i^T(kh)\Phi y_i(kh), k > t_m^i, m \in Z^+ \tag{5.8}$$

where $t_m^i h$ denotes the m-th transmitted instance of sampled data from the i-th vehicle. $\Phi > 0$ is a weighting matrix and $\alpha_i(kh) = x_i^c(kh) - x_i^c\left(t_m^i h\right)$ and $y_i(kh) = x_i^c(kh) - x_{i-1}^c\left(t_{m'_{i-1}}^{i-1} h\right)$, where $x_{i-1}^c\left(t_{m'_{i-1}}^{i-1} h\right)$ is the latest local measurement from its preceding vehicle with $m'_{i-1} = \arg\min_p\left\{t_m^i + q_i - t_p^{i-1} \mid t_m^i + q_i > t_p^{i-1},\ p \in Z^+\right\}$ and $\sigma_i\left(t_m^i h + q_i h\right)$ represents a dynamic threshold parameter. Therefore, the velocity $v_i(kh)$ and acceleration $a_i(kh)$ are transmitted to the next vehicle only if the defined event (5.8) is satisfied (i.e. the event is triggered). One can see from the defined event condition (5.8) that at the k-th sampling instance, the defined event-triggered condition for vehicle i is closely related to the sampled data error $\alpha_i(kh)$ and the sampled data $y_i(kh)$ including the latest transmitted sampled data $x_i^c\left(t_m^i h\right)$ from vehicle i and the latest transmitted sampled data from the preceding vehicle. Therefore, the defined event condition is related to the vehicle's state.

Remark 5.1 It can be seen from (5.8) that the next broadcasting instance of vehicle i is not only related to its own sampled data but also to the preceding vehicles' sampled data. Because the CACC system is a cascaded control system, varying velocity and acceleration values (e.g. harsh brakes or harsh accelerations) can be transmitted to the next vehicles under condition (5.8) in a timely manner. Therefore, the defined event condition (5.8) can help the CACC system maintain its desired formation model.

According to the defined event condition (5.8), the next transmission instance $t_{m+1}^i h$ of the i-th following vehicle is determined by

$$\begin{aligned} t_{m+1}^i h = t_m^i h + \min_{q_i \geq 1}\{q_i h \mid \alpha_i^T\left(t_m^i h + q_i h\right)\Phi\alpha_i\left(t_m^i h + q_i h\right) \\ \geq \sigma_i\left(t_m^i h + q_i h\right)y_i^T\left(t_m^i h + q_i h\right)\Phi y_i\left(t_m^i h + q_i h\right)\} \end{aligned} \tag{5.9}$$

where $q_i \in \mathcal{N}$.

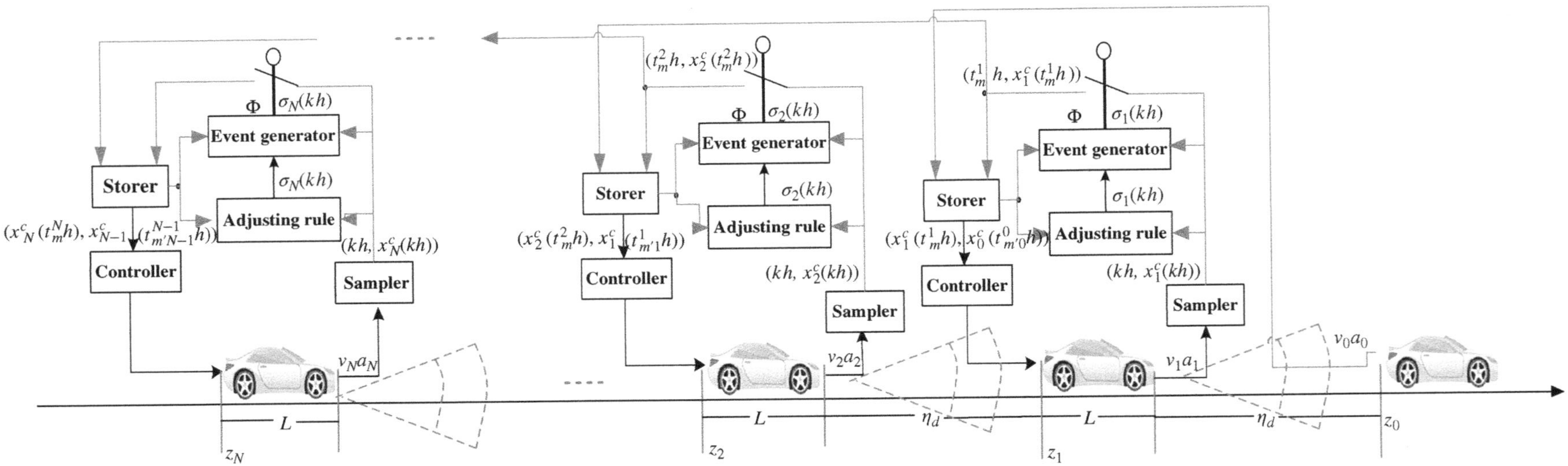

Figure 5.2 DECM.

The dynamic rule for adjusting the threshold parameter $\sigma\left(t_m^i h + q_i h\right)$ is chosen as

$$\begin{aligned}\sigma\left(t_m^i h + q_i h\right) &= \sigma\left(t_m^i h + q_i h - h\right)\\ &\quad - \theta\sigma\left(t_m^i h + q_i h\right)\sigma\left(t_m^i h + q_i h - h\right)y_i^T\left(t_m^i h + q_i h - h\right)\Phi y_i\left(t_m^i h + q_i h - h\right)\end{aligned} \tag{5.10}$$

where $\theta > 0$ is a predefined constant and $\sigma(0) = \sigma_0 \in [0, 1)$ is the given initial condition.

Remark 5.2 It should be noted that the "storer" in Figure 5.2 is employed to gather the sampled velocity and acceleration of the i-th vehicle and its preceding vehicle. When the new sampled data from the i-th vehicle and its preceding vehicle are detected, the storer will immediately update its information. Then, the controller of the i-th following vehicle updates its input based on the received sampled data from the storer, which implies that the controller update of the i-th vehicle is driven by both its own and its preceding vehicle's event instances.

Remark 5.3 For the given initial condition $\sigma_0 \in [0, 1)$, the defined event-triggered condition (5.9) with the dynamic adjusting law (5.10) has the following feature [7, 8]:

$$\alpha_i^T(kh)\Phi\alpha_i(kh) \leq \sigma_i(kh)y_i^T(kh)\Phi y_i(kh) \tag{5.11}$$

where the threshold parameter $\sigma(kh)$ satisfies the following inequality:

$$0 \leq \sigma(kh) \leq \sigma_0 < 1 \tag{5.12}$$

Furthermore, $\{\sigma(kh)\}$ is a monotone nonincreasing sequence for all $k \in \mathcal{N}$.

Remark 5.4 The DECM can efficiently save intervehicle communication resources. For example, if

$$\alpha_i^T\left(t_m^i h + q_i h\right)\Phi\alpha_i\left(t_m^i h + q_i h\right) < \sigma\left(t_m^i h + q_i h\right)y_i^T\left(t_m^i h + q_i h\right)\Phi y_i\left(t_m^i h + q_i h\right)$$

then no triggering signal is created by the event generator. Therefore, the current sampled data is not transmitted to the next vehicle because the fluctuation between the two successive sets of sampled data is very small. In this manner, the frequency of sampled data transmission via the intervehicle communication network can be significantly reduced.

Remark 5.5 The DECM with the threshold-adjusting rule (5.10) includes both the static ECM and TCM. Specifically, if $\theta = 0$ in (5.10) and the threshold parameter of (5.9) is set to $\sigma_i\left(t_m^i h + q_i h\right) = \sigma_i$ with a constant σ_i, then the DECM becomes an SECM with the following form:

$$\begin{aligned}t_{m+1}^i h = t_m^i h + \min_{q_i \geq 1}\{q_i h \mid \alpha_i^T\left(t_m^i h + q_i h\right)\Phi\alpha_i\left(t_m^i h + q_i h\right)\\ \geq \sigma_i y_i^T\left(t_m^i h + q_i h\right)\Phi y_i\left(t_m^i h + q_i h\right)\}\end{aligned}$$

Furthermore, if we set $\theta = 0$ in (5.10) and $\sigma(\cdot) \to 0$ in (5.9), then we can derive that $\alpha_i^T\left(t_m^i h + q_i h\right)\Phi\alpha_i\left(t_m^i h + q_i h\right) \geq 0$, which implies that the sampled velocity and acceleration of the i-th vehicle are transmitted at each sampling instance. Therefore, the ECM is converted into the TCM.

Under the DECM using (5.9), $u_i^c(t)$ from the sampled data feedback controller can be represented as:

$$u_i^c(t) = K_i^c\left[x_i^c\left(t_m^i h\right) - x_{i-1}^c\left(t_{m'_{i-1}(t)}^{i-1} h\right)\right], t \in \left[t_m^i h,\ t_{m+1}^i h\right) \tag{5.13}$$

where $m'_{i-1}(t) = \arg\ \min_p\left\{t - t_p^{i-1} \mid t > t_p^{i-1},\ p \in Z^+\right\}$.

We define the measurement error for sampled data at the k-th sampling time for each vehicle as

$$e_i^c(kh) = x_i^c(kh) - x_i^c\left(t_m^i h\right), t_m^i < k < t_{m+1}^i \tag{5.14}$$

Because the release time intervals $\left[t_m^i h, t_{m+1}^i h\right) = \cup_{k=t_m^i}^{t_{m+1}^i - 1}[kh, (k+1)h)$ and $\left[t_m^i h, t_{m+1}^i h\right)$ can be divided into $t_{m+1}^i - t_m^i$ sampling intervals, $u_i^c(t)$ in (5.13) for $[kh, (k+1)h)$ can be represented as

$$u_i^c(t) = K_i^c\left[x_i^c(kh) - x_{i-1}^c(kh) - e_i^c(kh) + e_{i-1}^c(kh)\right] \tag{5.15}$$

By combining (5.7) and (5.15), the sampled data feedback controller $u_i(t)$, $t \in [kh, (k+1)h)$ for each vehicle can be represented as follows:

$$\begin{aligned} u_i(t) &= K_i^o x_i^o(kh) + K_i^c\left[x_i^c(kh) - x_{i-1}^c(kh) - e_i^c(kh) + e_{i-1}^c(kh)\right] \\ &= K_i x_i(th) + K_i\left[e_{i-1}^c(kh) - e_i^c(kh)\right] \end{aligned} \tag{5.16}$$

where $x_i(t) = x_i^o(t) + x_i^c(t) - x_{i-1}^c(t)$ denotes the tracking error vector and $K_i = K_i^o + K_i^c$ is the feedback controller gain.

We now define an "artificial delay" $\tau'(t) = t - kh$, where $\tau'(t) = t - kh$. It is clear that $\tau'(t)$ is piecewise linear and discontinuous at $t = kh$, meaning it satisfies $0 \le \tau'(t) < h$ and $\dot{\tau}'(t) = 1$ at $t \ne kh$. Therefore, the controller $u_i(t)$ in (5.16) can be rewritten with the following form

$$u_i(t) = K_i x_i(t - \tau'(t)) + K_i\left[e_{i-1}^c(t - \tau'(t)) - e_i^c(t - \tau'(t))\right] \tag{5.17}$$

We define $x(t) = \mathrm{Col}[x_i(t)]_{i=1}^N$, $u(t) = \mathrm{Col}[u_i(t)]_{i=1}^N$ and $w(t) = \mathrm{Col}[a_i(t)]_{i=0}^{N-1}$ as the tracking error, control input vector, and disturbance vector, respectively, where "Col" denotes a column vector. Based on (5.4), (5.5), and (5.17), the state-space equation for the platoon can be written as

$$\dot{x}(t) = Ax(t) + Bu(t - \tau) + Dw(t), t \in [kh,\ (k+1)h) \tag{5.18}$$

where

$$A = \begin{bmatrix} A_1 & 0 & \cdots & 0 \\ 0 & A_2 & \cdots & 0 \\ \cdots & \ddots & \ddots & \cdots \\ 0 & \cdots & 0 & A_N \end{bmatrix}, B = \begin{bmatrix} B_1 & 0 & \cdots & 0 \\ -B_1 & B_2 & \cdots & 0 \\ \cdots & \ddots & \ddots & \cdots \\ 0 & \cdots & -B_{N-1} & B_N \end{bmatrix},$$

$$D = \begin{bmatrix} d_1 & & \\ & \ddots & \\ & & d_N \end{bmatrix}, A_i = \begin{bmatrix} 0 & 1 & h_v \\ 0 & 0 & 1 \\ 0 & 0 & -\dfrac{1}{\varsigma} \end{bmatrix}, B_i = \begin{bmatrix} 0 \\ 0 \\ -\dfrac{1}{\varsigma} \end{bmatrix}, d_i = \begin{bmatrix} h_v \\ 0 \\ 0 \end{bmatrix}.$$

Then, the tracking-error-based sampled-data feedback controller for the CACC system can be defined as

$$u(t - \tau) = K[x(t - \tau(t)) + \varepsilon(t - \tau(t))], t \in [kh, (k+1)h) \tag{5.19}$$

where $\tau(t) = \tau'(t) + \tau$, $0 \le \tau(t) < h'$ with $h' = h + \tau$, $\dot{\tau}(t) = 1$ at $t \neq kh$. Additionally, $K = \text{diag}\{K_1, ..., K_N\}$ and $\varepsilon(t) = (L \otimes I_3)\, e(t)$ with $e(t) = \text{Col}\left[e_i^c(t)\right]_{i=1}^{N}$ and

$$L = \begin{bmatrix} 1 & 0 & \cdots & 0 \\ -1 & 1 & \cdots & 0 \\ \cdots & \ddots & \ddots & \cdots \\ 0 & \cdots & -1 & 1 \end{bmatrix}.$$

By substituting (5.19) into (5.18), the state space equation for the CACC system in $t \in [kh, (k+1)h)$ can be rewritten as

$$\dot{x}(t) = Ax(t) + BKx(t - \tau(t)) + BK\varepsilon(t - \tau(t)) + Dw(t) \tag{5.20}$$

where the initial condition of the state is assumed to be $x(t) = \varphi(t) = \varphi(t_0)$ for $t \in [-h', 0]$.

The control objective for this chapter is to design a DECM (5.9) utilizing the threshold parameter adjusting rule (5.10) for the CACC systems under the sampled-data feedback controller (5.16) such that each vehicle can track the speed of the leading vehicle while maintaining the desired distance between any two consecutive vehicles. Specifically, the following criteria must be met:

1) Steady-state performance: the tracking error $x_i(t)$ is asymptotically stable.
2) Disturbance rejection performance: with an initial value of zero, the tracking error $x(t)$ of the CACC system satisfies the following H_∞ disturbance attenuation level with $\gamma > 0$:

$$\|x(t)\|_2 \le \gamma \|w(t)\|_2 \tag{5.21}$$

for all nonzero $w(t) \in l_2[0, \infty)$.

5.3 Event-Triggered Communication and Platoon Control

We are now in a position to derive sufficient conditions for the existence of the proposed DECM (5.9) and sampled-data feedback controller (5.16) for the CACC system. The main results of this derivation are contained in the following theorem.

Theorem 5.1 Consider a CACC system with the DECM (5.9) with the threshold parameter-adjusting rule (5.10). For given scalars $h > 0$, $\tau > 0$, $\sigma_0 \in [0, 1)$, and $\gamma > 0$, and the feedback gain matrix K, if there exist real matrices $P = P^T > 0$, $Q = Q^T > 0$, and $R = R^T > 0$, and appropriate dimensions such that the following inequalities hold:

$$\begin{bmatrix} R & S \\ * & R \end{bmatrix} \ge 0 \tag{5.22}$$

$$\begin{bmatrix} \Gamma & h'\Omega & \gamma^{-1} I_{3N} \\ * & -R & 0 \\ * & * & -I_{3N} \end{bmatrix} < 0 \tag{5.23}$$

where $\Gamma = [\Gamma_{pq}]_{5\times 5}$ is a symmetric block matrix with its entries given by $\Gamma_{11} = PA + A^TP + Q - R$, $\Gamma_{12} = PBK + R + S$, $\Gamma_{13} = -S$, $\Gamma_{14} = PB$, $\Gamma_{15} = PD$, $\Gamma_{22} = -S - S^T - 2R + (L \otimes I_3)^T(\sigma_0 I_N \otimes \Phi)(L \otimes I_3)$, $\Gamma_{23} = R + S$, $\Gamma_{24} = (L \otimes I_3)^T(\sigma_0 I_N \otimes \Phi)$, $\Gamma_{33} = -R - Q$, and $\Gamma_{44} = (L \otimes I_3)^T(\sigma_0 I_N \otimes \Phi)(L \otimes I_3) - I_N \otimes \Phi$, and the other entries are zero, then the tracking error of the CACC system is asymptotically stable with an H_∞ disturbance attenuation level of γ.

Proof: We choose the following Lyapunov-Krasovskii function for the CACC systems:

$$V(t) = \sum_{i=1}^{3} V_i(t), t \in [kh, (k+1)h) \tag{5.24}$$

where $V_1(t) = x^T(t)Px(t)$, $V_2(t) = \int_{t-h'}^{t} x^T(s)Qx(s)ds$ and $V_3(t) = \int_{-h'}^{0}\int_{t+\theta}^{t} \dot{x}^T(s)R\dot{x}(s)dsd\theta$ for the positive definite matrices $\boldsymbol{P}$, $\boldsymbol{Q}$, and $\boldsymbol{R}$.

Taking the time derivative of $V(t)$ along the trajectories of (5.17), one can obtain that

$$\begin{aligned}\dot{V}_1(t) &= x^T(t)\left[PA + A^TP\right]x(t) + 2x^T(t)PBKx(t-\tau(t)) \\ &\quad + 2x^T(t)PBK\varepsilon(t-\tau(t)) + 2x^T(t)PDw(t)\end{aligned} \tag{5.25}$$

$$\dot{V}_2(t) = x^T(t)Qx(t) - x^T(t-h)Qx(t-h) \tag{5.26}$$

$$\dot{V}_3(t) = h^2\dot{x}^T(s)R\dot{x}(s) - h\int_{t-\tau(t)}^{t} \dot{x}^T(s)R\dot{x}(s)ds \tag{5.27}$$

According to the reciprocally convex inequality given in lemma 5.1 of [12], we find that

$$-h\int_{t-\tau(t)}^{t} \dot{x}^T(s)R\dot{x}(s)ds \le \begin{bmatrix}\varphi_1^T & \varphi_2^T\end{bmatrix}\begin{bmatrix}-R & S \\ * & -R\end{bmatrix}\begin{bmatrix}\varphi_1 \\ \varphi_2\end{bmatrix} \tag{5.28}$$

where $\varphi_1 = x(t-\tau(t)) - x(t)$, $\varphi_2 = x(t-h) - x(t-\tau(t))$ and $\boldsymbol{S}$ is a real matrix satisfying (5.22).

Based on the DECM from (5.9), (5.10), and (5.12), we can derive that

$$\begin{aligned}\varepsilon^T(kh)(I_N \otimes \Phi)\varepsilon(kh) &= e^T(kh)\left[(L\otimes I_3)^T(I_N \otimes \Phi)(L\otimes I_3)\right]e(kh) \\ &\le x^T(kh)\left[(L\otimes I_3)^T(\sigma_0 I_N \otimes \Phi)(L\otimes I_3)\right]x(kh)\end{aligned} \tag{5.29}$$

which means that

$$\begin{bmatrix}x(t-\tau(t)) \\ \varepsilon(t-\tau(t))\end{bmatrix}^T \begin{bmatrix}\Pi_1 & \Pi_2 \\ * & \Pi_3\end{bmatrix}\begin{bmatrix}x(t-\tau(t)) \\ \varepsilon(t-\tau(t))\end{bmatrix} \ge 0 \tag{5.30}$$

where $\Pi_1 = (L\otimes I_3)^T(\sigma_0 I_N \otimes \Phi)(L\otimes I_3)$, $\Pi_2 = \Gamma_{24}$ and $\Pi_3 = \Gamma_{44}$.

By combining (5.25)–(5.28) and (5.30), one can obtain that

$$\dot{V}(t) \le \Psi^T(t)\left(\Gamma + h'^2\Omega^T R\Omega\right)\Psi(t) \tag{5.31}$$

where $\Psi(t) = \mathrm{Col}\{x(t), x(t-\tau(t)), x(t-h), \varepsilon(t-\tau(t)), w(t)\}$ and $\Omega = [A \;\; BK \;\; 0 \;\; BK \;\; D]$.

By Schur's complement to (5.23), we can derive that

$$\begin{bmatrix}\Gamma & h'\Omega \\ * & -R\end{bmatrix} < 0 \tag{5.32}$$

Based on Schur's complement of (5.31), one can find that $\Gamma + h'^2\Omega^T R\Omega < 0$, meaning $\dot{V}(t) \le 0$. Therefore, the error-tracking system (5.18) is asymptotically stable.

In the following analysis, we consider the H_∞ disturbance attenuation for the CACC system. For this purpose, we add the term $x^T(t)x(t) - \gamma^2 w^T(t)w(t)$ to both sides of (5.31). According to condition (5.23) in Theorem 5.1 and the Schur complement, we know that

$$\dot{V}(t) + x^T(t)x(t) - \gamma^2 w^T(t)w(t) \le \Psi^T(t)\left(\Gamma + h'^2\Omega^T R\Omega\right)\Psi(t) + x^T(t)x(t) - \gamma^2 w^T(t)w(t) < 0$$

which means that $\int_0^{h'} \dot{V}(t) + x^T(t)x(t) - \gamma^2 w^T(t)w(t) < 0$.

22/8/2023 7:

Therefor e, we have $\int_0^{h'} x^T(t)x(t) \leq \gamma^2 \int_0^{\infty} w^T(t)w(t)$ under the condition of zero initial error, namely $\|x(t)\|_2 \leq \gamma \|w(t)\|_2$ for all nonzero $w(t) \in l_2[0, \infty)$. The proof is completed.

Remark 5.6 The proposed DECM-based sampled-data controller can significantly reduce the transmission frequency of sampled data but does not affect the control performance of the CACC system. We will now elaborate on this concept. According to the properties of the event-triggered condition, one can obtain condition (5.29), where the measurement error $e_i^c(kh)$ is defined in (5.14). The conditions in Theorem 5.1 guarantee that the state x(kh) of the CACC system can be stabilized. Therefore, the measurement error $e_i^c(kh)$ is stable, which means that the current sampled data $x_i^c(kh)$ has very little fluctuation with respect to the latest transmitted data $x_i^c(t_m^i h)$. Therefore, it is not necessary to transmit the current data via the VANET for controller input updating. Comparative results are presented in the simulation study section to illustrate that the control performance of the DECM-based sampled-data controller is similar to that of the TCM-based sampled-data controller.

Theorem 5.2 Consider a CACC system with the DECM (5.9) with the threshold parameter-adjusting rule (5.10). For given scalars $h > 0, \tau > 0, \sigma_0 \in [0, 1), \gamma > 0$, and $\kappa > 0$ and the feedback gain matrix $\boldsymbol{K}$, if there exist real matrices $\tilde{P} = \tilde{P}^T > 0$, $\tilde{Q} = \tilde{Q}^T > 0$, and $\tilde{R} = \tilde{R}^T > 0$, and appropriate dimensions $\tilde{S}$ such that the following inequalities hold:

$$\begin{bmatrix} \tilde{R} & \tilde{S} \\ * & \tilde{R} \end{bmatrix} \geq 0 \tag{5.33}$$

$$\begin{bmatrix} \Gamma & h'\Omega & \gamma^{-1} I_{3N} \\ * & -R & 0 \\ * & * & -I_{3N} \end{bmatrix} < 0 \tag{5.34}$$

where $\Gamma = [\Gamma_{pq}]_{5\times 5}$ is a symmetric block matrix with its entries given by $\Gamma_{11} = \tilde{P}A + A^T\tilde{P} + \tilde{Q} - \tilde{R}$, $\Gamma_{12} = B\tilde{K} + \tilde{R} + \tilde{S}$, $\Gamma_{13} = -\tilde{S}$, $\Gamma_{14} = B\tilde{K}$, $\Gamma_{15} = \tilde{P}D$, $\Gamma_{22} = -S - S^T - 2R + (L \otimes I_3)^T \sigma_0 \tilde{\Phi} (L \otimes I_3)$, $\Gamma_{23} = \tilde{R} + \tilde{S}$, $\Gamma_{24} = (L \otimes I_3)^T \sigma_0 \tilde{\Phi}$, $\Gamma_{33} = -\tilde{R} - \tilde{Q}$, and $\Gamma_{44} = (L \otimes I_3)^T \sigma_0 \tilde{\Phi} (L \otimes I_3) - I_N \otimes \tilde{\Phi}$, and the other entries are zero, then the tracking error of the CACC system is asymptotically stable with an H_∞ disturbance attenuation level of γ. Furthermore, the event-triggering weight matrix in (5.9) is defined as

$$\Phi = \tilde{P}^{-1} \tilde{\Phi} \tilde{P}^{-1} \tag{5.35}$$

5.4 Simulation Study

In this section, numerical examples are presented to illustrate how the proposed DECM can be applied to a six-vehicle CACC system with a sampled-data feedback controller. For the simulations, the necessary parameters for the DECM are listed in Table 5.1. The velocity and acceleration of the leading vehicle are presented in Figure 5.3.

By using the controller design method in [11] without sensor failures, the sampled-data feedback controller gain for the CACC systems can be obtained as $K_i = [10 \quad 11 \quad 12]$, which satisfies the stability conditions in Theorem 5.1 and guarantees the string stability requirement. By utilizing

Table 5.1 Parameters for the simulation.

Parameters	Notation	Value
Inertial lag	ς	0.25s
sampling period	h	0.1s
Length of vehicle	L	6 m
Actuator time delay	τ	0.1 s
Desired spacing	δ_d	5m
H_∞ disturbance attenuation level	γ	1.25

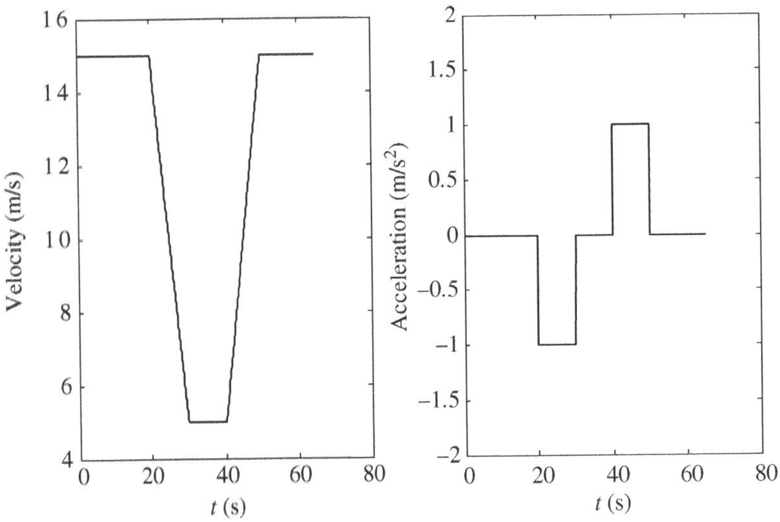

Figure 5.3 Velocity and acceleration of the leading vehicle.

Theorem 5.2 with the obtained controller gain and simulation parameters listed in Table 5.1, the design problem for the DECM for the CACC system is solvable. The designed parameter for the DECM was set to $\Phi = \begin{bmatrix} 0.061 & 0.007 & 0.005 \\ * & 0.053 & 0.006 \\ * & * & * \end{bmatrix}$ with $\theta = 8$, $\kappa = 1$ and $\sigma_0 = 0.6$.

In the DECM-based sampled-data feedback controller, the spacing error, velocity and, acceleration of each vehicle are presented in Figure 5.4 and the dynamic threshold parameters $\sigma(k)$ are presented in Figure 5.5. The corresponding transmission time instance and release time intervals are presented in Figure 5.6. From these figures, it is clear that each vehicle can accurately and rapidly trace the velocity of the leading vehicle and maintain the desired space between successive vehicles. In contrast to the spacing error, the velocity and acceleration of the following vehicle are presented in Figure 5.7 for the TCM-based sampled-data controller of [11]. It can be observed that the control performance of the DECM-based sampled-data controller appears very similar to that of the TCM-based sampled-data controller. The sampled-data transmission ratios (STR) and average release time intervals (ARTI) of each vehicle under the DECM are listed in Table 5.2, where the maximum

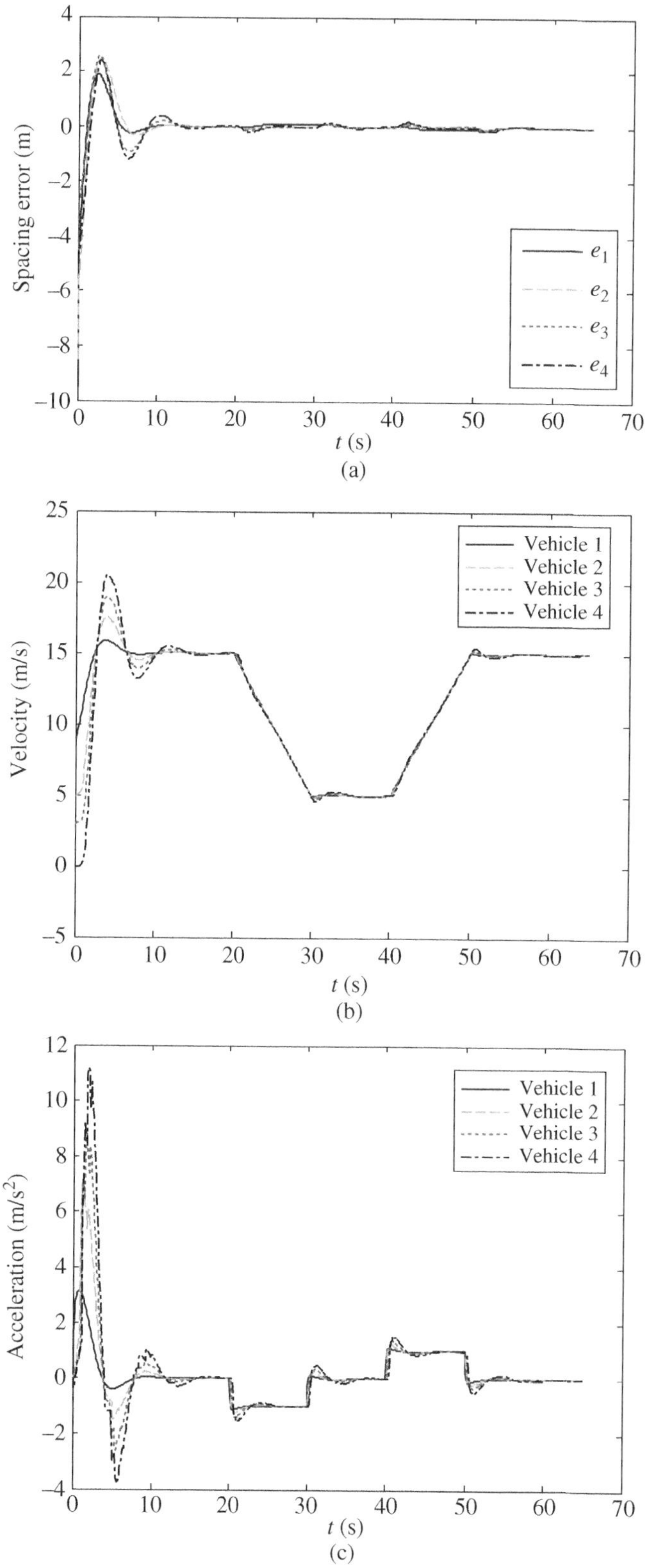

Figure 5.4 The spacing error (a), velocity (b), and acceleration (c) of each vehicle in the DECM-based sampled-data controller.

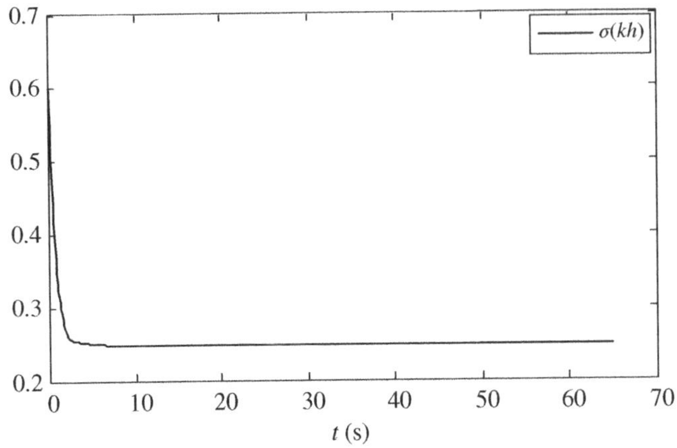

Figure 5.5 Dynamic threshold parameters $\sigma(k)$ with σ_0 = 0.6 and θ = 8.

value of the release time interval is approximately 1.2 s. The mean STR and ARTI values for the CACC system are 45.75% and 0.215 s, respectively. Therefore, only 298 data samples needed to be transmitted over the VANET with an average time interval of 0.22 s in [0, 65 s) when utilizing the DECM. However, all 650 sampled-data packets must be transmitted in [0, 65 s) under the TCM-based sampled-data controller from [11]. This is a waste of the limited communication resources in the CACC system. In comparison to the TCM, the DECM can significantly reduce the data transmission frequency over the VANET to save limited communication resources.

In the following analysis, we present the comparative results by utilizing the SECM. In this case, the parameters in (5.10) are set to $\sigma(kh) = \sigma_0 = 0.6$ and $\theta = 8$. Under the SECM, the spacing error, velocity, and acceleration of each vehicle are presented in Figure 5.8. The corresponding transmission time instance and release time intervals are presented in Figure 5.9. The STR and the ARTI of each vehicle can be found in Table 5.2. Based on the simulation results discussed here, we have presented the following conclusions:

1) The SECM generates fewer transmissions of sampled data for the CACC system (see from Table 5.2). The main reason for this is that the threshold parameter $\sigma(kh)$ is always fixed as σ_0. In contrast, under the DECM, $\sigma(kh)$ is monotonically nonincreasing, as shown in Figure 5.5. The larger the value of the threshold parameter, the fewer data packets are transmitted. The maximum release time interval of the CACC system under the SECM is 2s (see Figure 5.9b), which is larger than that under the DECM.
2) The CACC system with the DECM achieves better overall performance. Although the SECM facilitates fewer data packet transmissions, the untransmitted sampled data may contain important information for the CACC systems and its nontransmission may lead to performance degradation. This point is illustrated by Figures 5.4a–c and 5.8a–c. Compared to the DECM, the spacing error, velocity, and acceleration of each vehicle under the SECM present larger fluctuations in the time interval [0, 20 s).

1) The proposed DECM-based sampled-data feedback controller can facilitate a larger upper bound for the time-varying delay and has less computational complexity involved in the obtained stability criterion than those proposed in [12]. For the same Lyapunov-Krasovskii function (5.24), the reciprocally convex inequality (RCI) [12] is used in our paper to find the stability criteria for the time-delayed CACC system. Compared to the approach used in [11], the RCI can provide a less

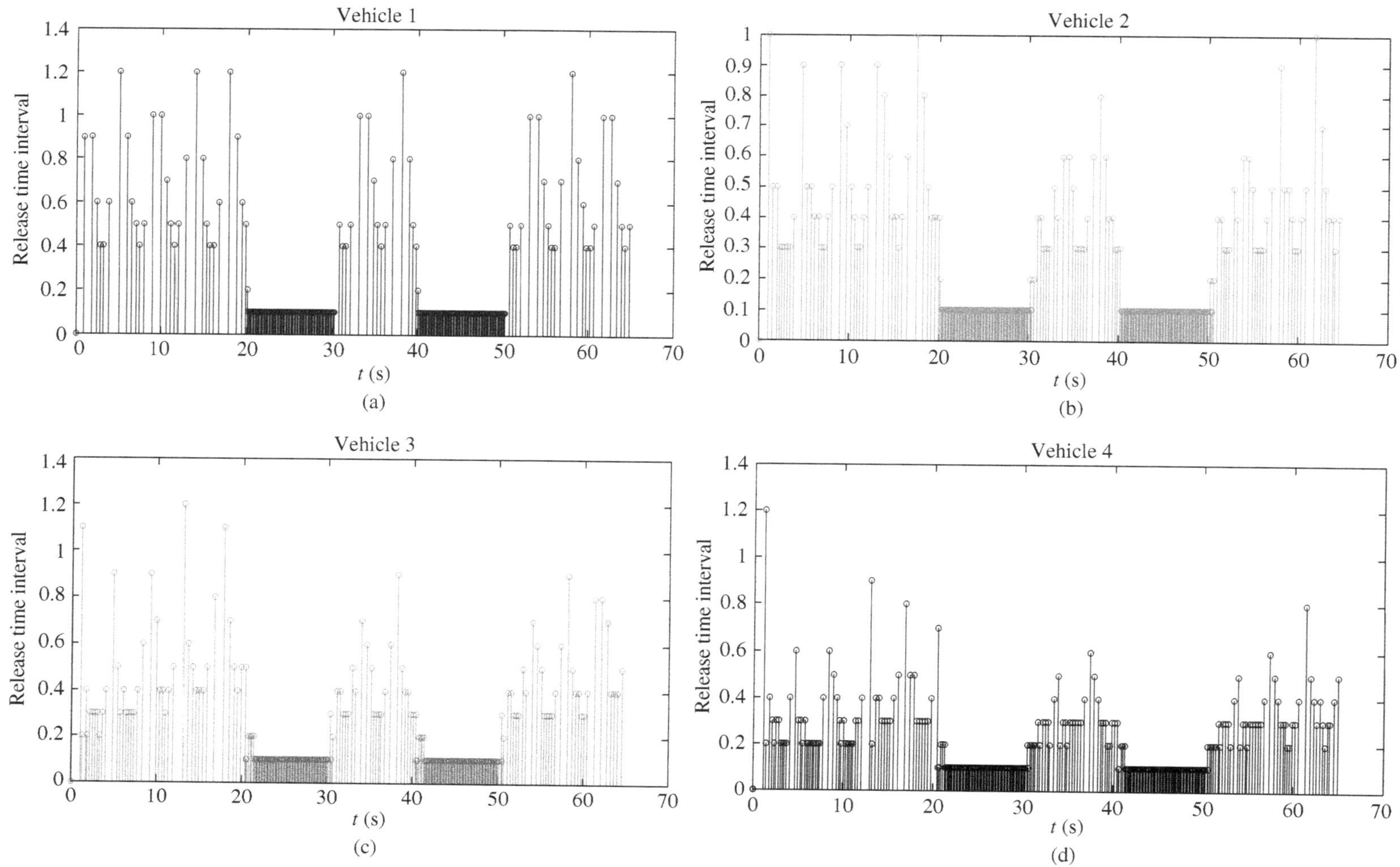

Figure 5.6 Transmission instance and release time intervals for intervehicle communication for each vehicle under the DECM: (a) Vehicle 1; (b) Vehicle 2; (c) Vehicle 3; (d) Vehicle 4.

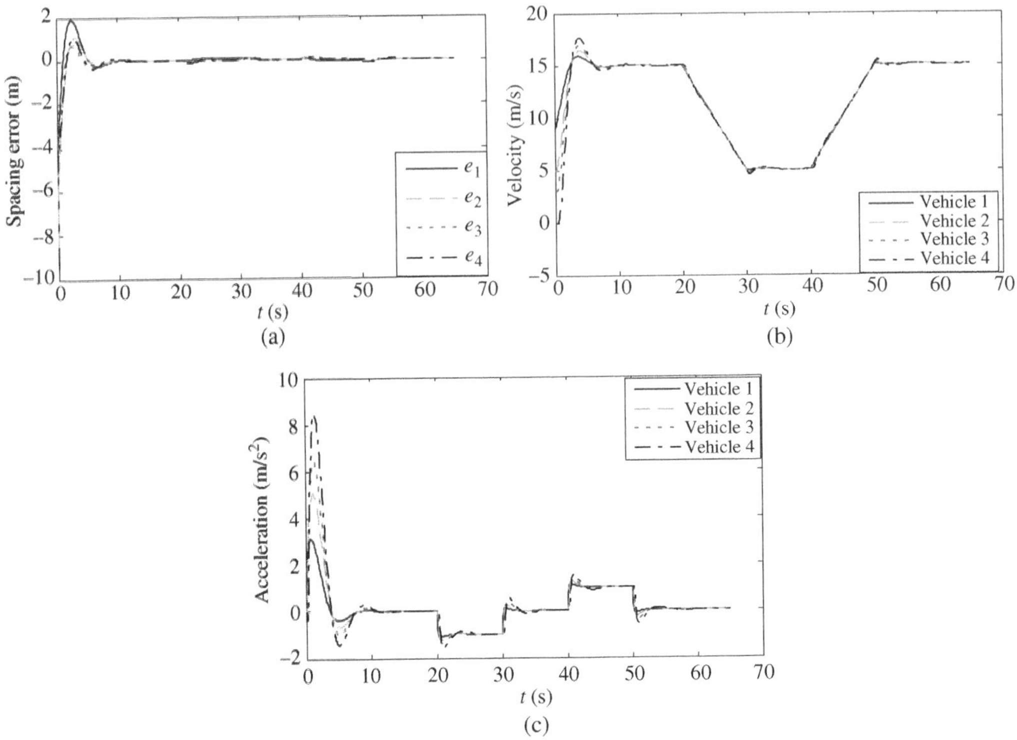

Figure 5.7 The spacing error (a), velocity (b), and acceleration (c) of each vehicle in the TCM-based sampled-data controller. *Source:* Adapted from Guo and Yue [11].

Table 5.2 STR and ARTI of the DECM and SECM.

	DECM		SECM	
	STR (%)	ARTI (s)	STR (%)	ARTI (s)
$i = 1$	42	0.24	36	0.29
$i = 2$	46	0.20	36	0.25
$i = 3$	44	0.22	39	0.20
$i = 4$	51	0.20	41	0.20

conservative stability condition with fewer slack matrix variables. We will discuss an example to illustrate this point. In our paper, the proposed stability criterion for Theorem 5.1 only has one slack matrix variable. However, the stability criterion in [11] requires three slack matrix variables, which adds to the computational complexity of the CACC algorithm. By utilizing the LMI toolbox to find a feasible solution for the stability criteria proposed in our paper and in [11], the obtained upper bounds for the time-varying delay are respectively 0.31 and 0.18 s. We set the lumped delay for the time-delay CACC system to 0.29 s. By using the sampled-data feedback controller with a gain of $K_i = \begin{bmatrix} 10 & 11 & 12 \end{bmatrix}$, the spacing error, velocity, and acceleration of each vehicle under the DECM-based sampled-data feedback controller and TCM-based sampled-data feedback controller

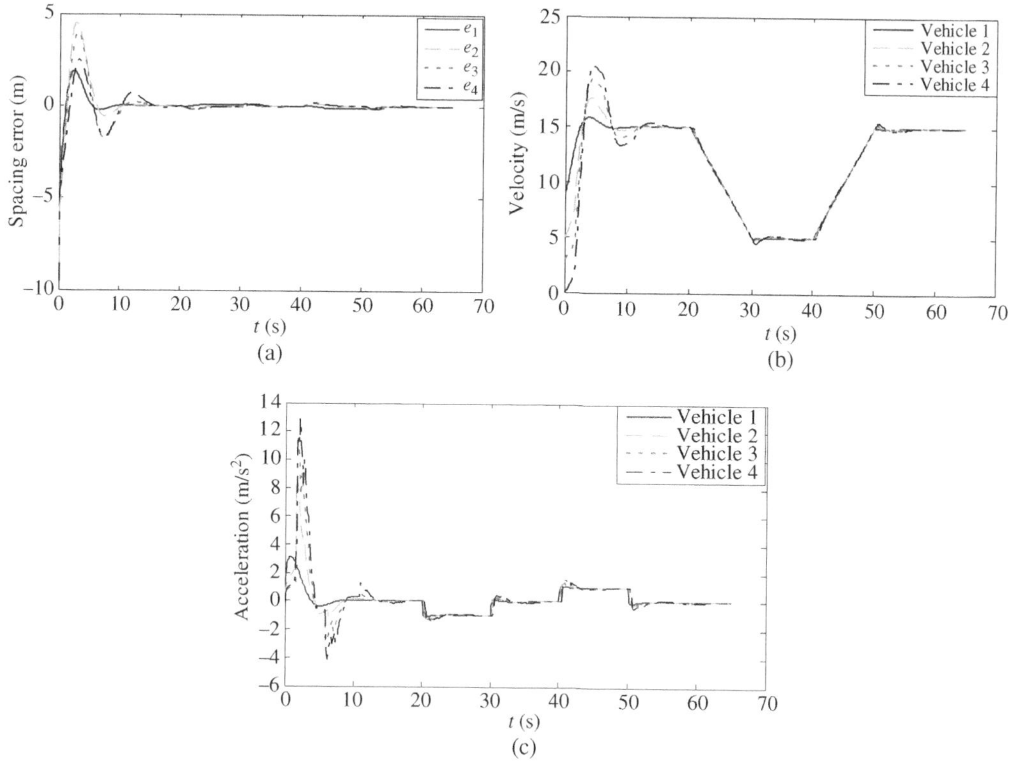

Figure 5.8 The spacing error (a), velocity (b), acceleration (c), and of each following vehicle under the SECM-based sampled-data controller.

are presented in Figures 5.10 and 5.11, respectively. The response to spacing errors, velocity, and acceleration of each following vehicle in [11] is not as stable as that in the proposed system, which is oscillating. Therefore, the proposed DECM-based sampled-data feedback controller can facilitate a larger time delay for the sampled-data CACC system.

2) In terms of the disturbances yielded by the varying acceleration of the preceding vehicles, the DECM-based sampled-data feedback controller has a more stable performance. We will discuss an example to illustrate this point. Here, it is assumed that the acceleration of the leading vehicle varies frequently, as shown in Figure 5.12. We aim to design the DECM-based sampled-data feedback controller such that the desired H_∞ disturbance attenuation level defined in (5.21) satisfies $\|x(t)\|_2{}^*\|w(t)\|_2{}^{-1} \leq \gamma = 8$. Under such circumstance, we found that the controller gain $K_i = [10 \quad 11 \quad 12]$ used in the previous simulation study cannot guarantee that the conditions in Theorem 5.1 have a feasible solution. So, we find a new controller gain $K_i = [13 \quad 15 \quad 11]$ by using the algorithm 5.1 from [11], which satisfies the condition of Theorem 5.1 in our paper and guarantees the string stability requirement for the CACC system. The spacing error, velocity and acceleration of each vehicle under the DECM-based sampled data feedback controller (controller gain is $K_i = [13 \quad 15 \quad 11]$) and TCM-based sampled-data feedback controller (controller gain is $K_i = [10\ 11\ 12]$) are presented in Figures 5.13a–d and 5.14a–d, respectively, where the value of $\|x(t)\|_2{}^*\|w(t)\|_2{}^{-1}$ is given in Figures 5.13d and 5.14d. For the proposed DECM-based sampled-data controller, the tracking error of the CACC system can satisfy the given H_∞ disturbance attenuation

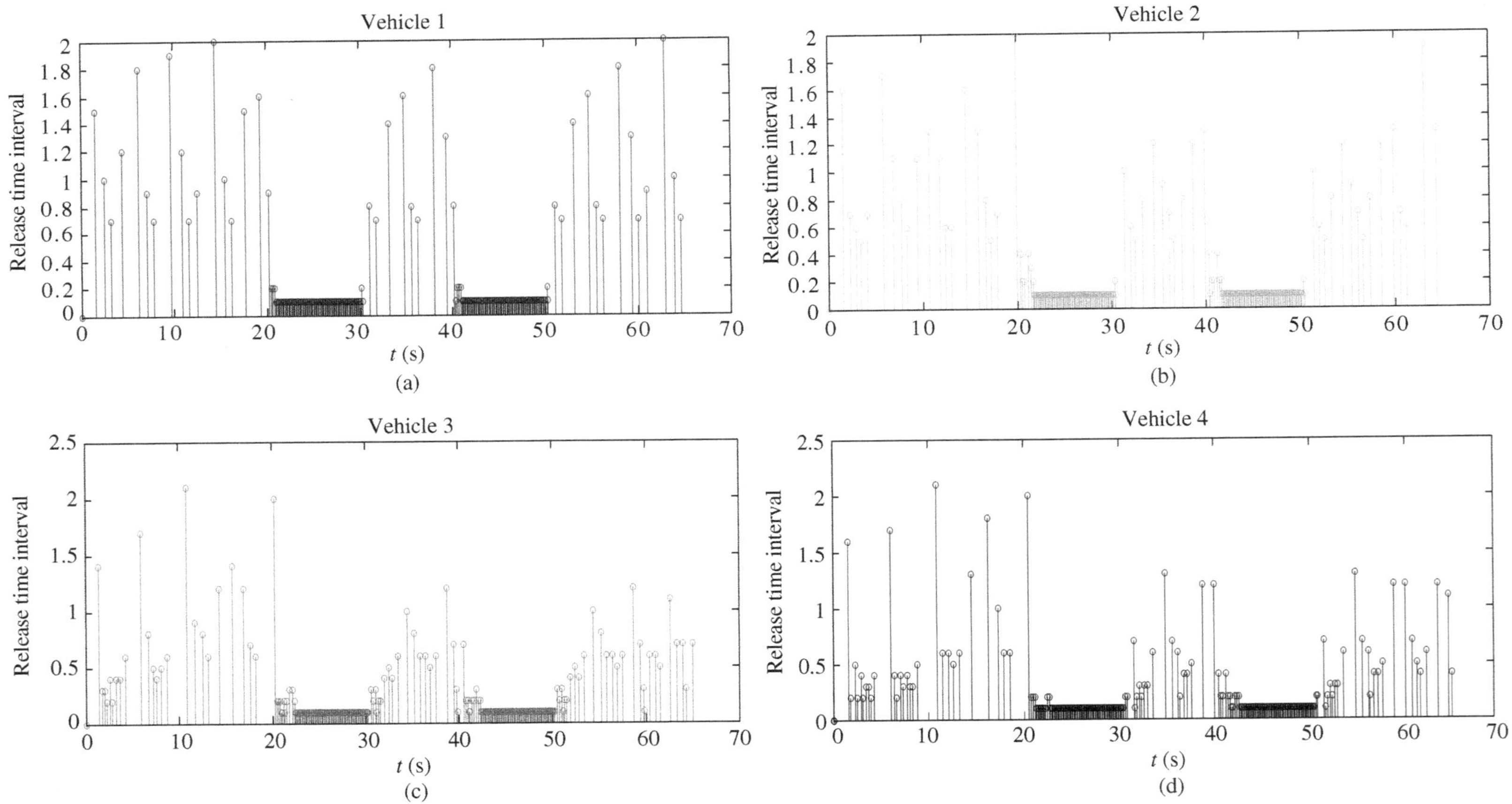

Figure 5.9 Transmission instances and release time intervals of intervehicle communication for each vehicle under the SECM: (a) Vehicle 1; (b) Vehicle 2; (c) Vehicle 3; (d) Vehicle 4.

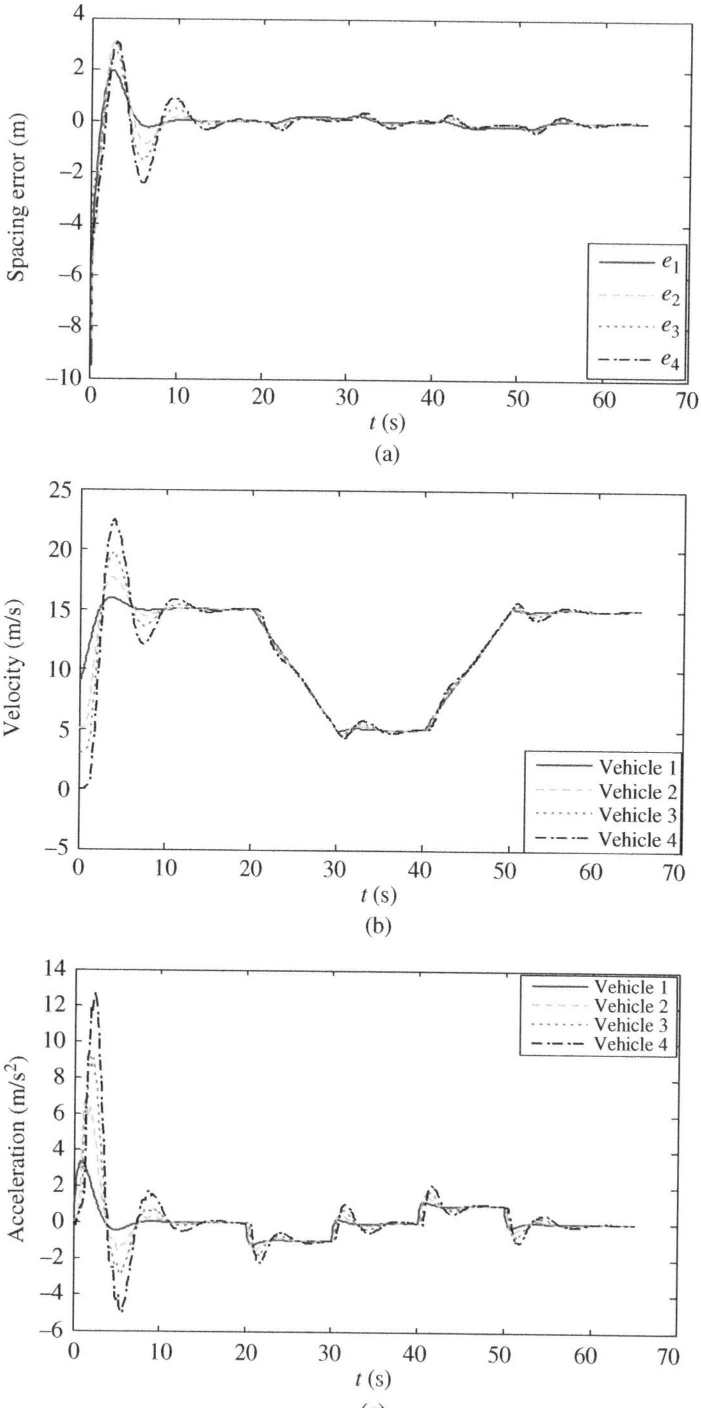

Figure 5.10 The spacing error (a), velocity (b), and acceleration (c) of each vehicle in the DECM-based sampled-data controller with a lumped time delay of 0.29 s.

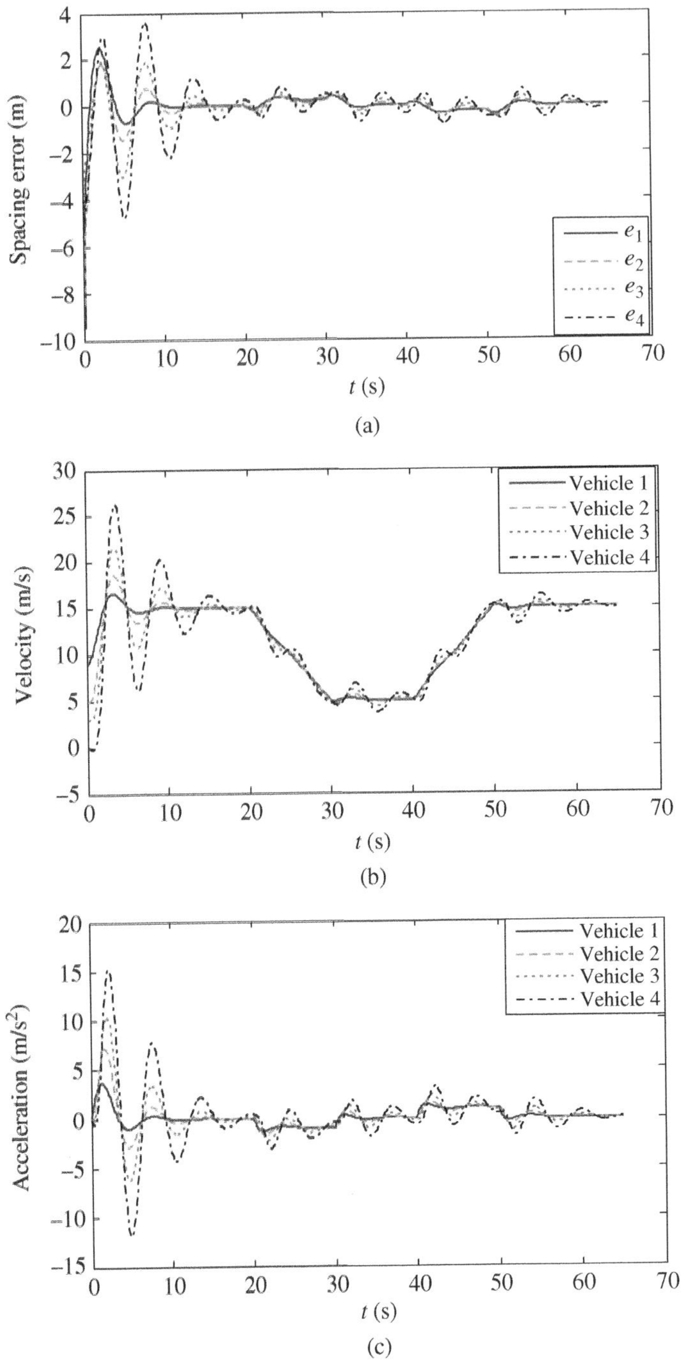

Figure 5.11 The spacing error (a), velocity (b), and acceleration (c) of each vehicle in the TCM-based sampled-data controller from [11] with a lumped time delay of 0.29 s. *Source:* Adapted from Guo and Yue [11].

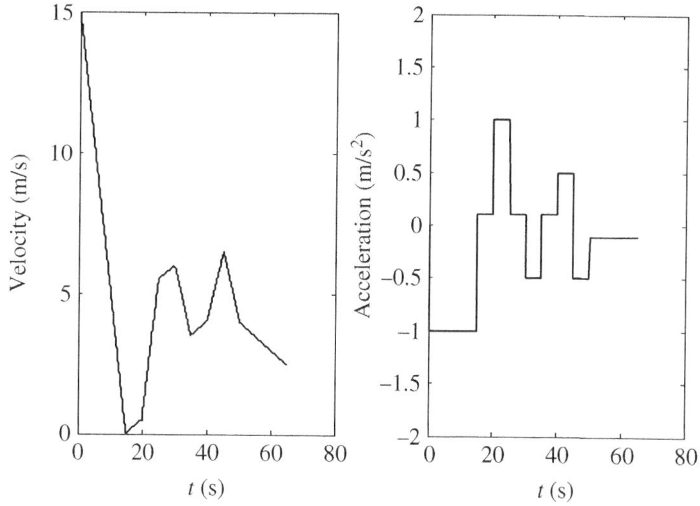

Figure 5.12 Varying acceleration of the leading vehicle.

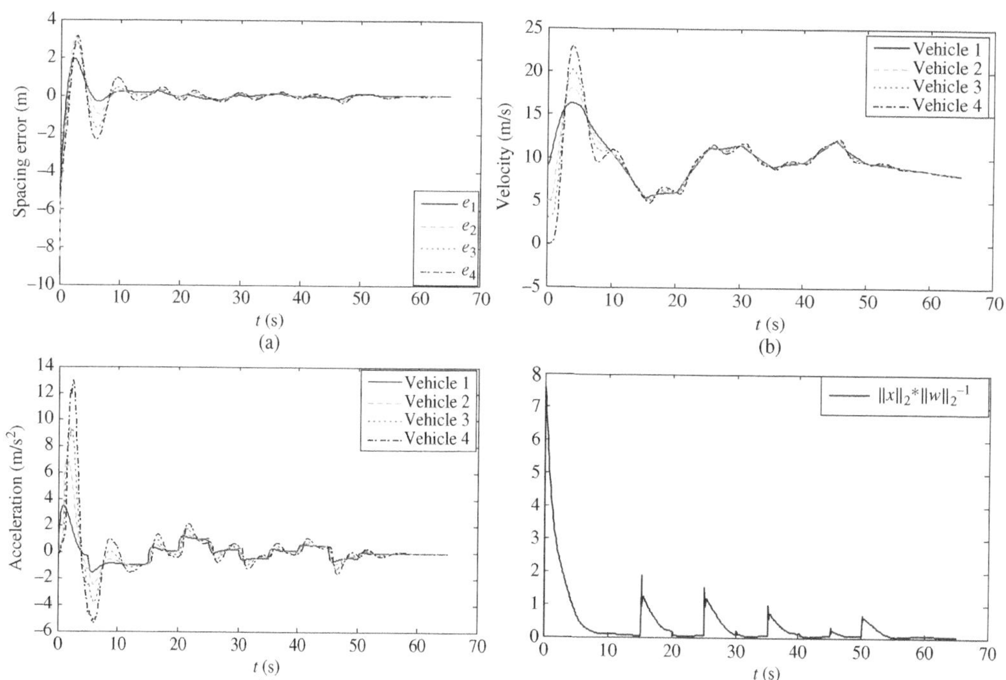

Figure 5.13 The spacing error (a), velocity (b), acceleration (c), and (d) $\|x(t)\|_2{}^*\|w(t)\|_2^{-1}$ of each vehicle in the DECM-based sampled-data controller with controller gain $K_i = [13 \quad 15 \quad 11]$.

level γ. This is because the H_∞ robust control method is used in the CACC algorithm in our system (e.g. the proposed stability conditions in Theorem 5.1 include the H_∞ disturbance attenuation level γ). However, the H_∞ disturbance attenuation level of the TCM-based sampled-data controller in [11] appears to reach values of $\gamma > 8$ (see Figure 5.14d). Furthermore, $\|x(t)\|_2{}^*\|w(t)\|_2^{-1}$ in our paper is smaller than

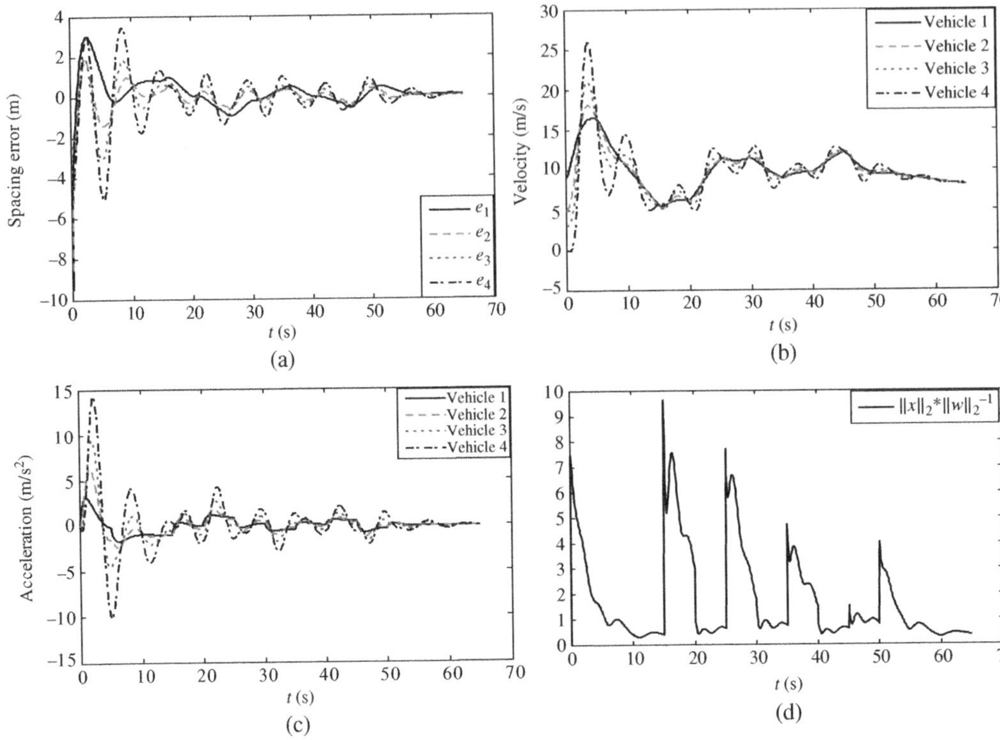

Figure 5.14 The spacing error (a), velocity (b), acceleration (c), and (d) $\|x(t)\|_2{}^{*}\|w(t)\|_2{}^{-1}$ of each vehicle in the TCM-based sampled-data controller from [11] with controller gain $K_i = [10 \quad 11 \quad 12]$. *Source:* Adapted from Guo and Yue [11].

it is in [11]. From Figures 5.13 and 5.14, it is clear that the DECM-based sampled-data controller has better stability than the TCM-based sampled-data controller proposed in [11] when subjected to acceleration disturbances.

5.5 Conclusions and Future Work

A DECM has been designed for CACC systems to save limited intervehicle communication resources. By utilizing the DECM, the number of sampled velocity and acceleration that needs to be transmitted to successive vehicles is significantly reduced. The DECM-based sampled-data controller for CACC systems was transformed into a stabilization problem for an error-tracking system with time-varying delays. According to Lyapunov stability theory, sufficient conditions have been obtained under which all tracking errors of each vehicle can be asymptotically stabilized by the DECM and sampled-data feedback controller. By utilizing the obtained conditions, we established a unified framework to design the parameters for the DECM. The effectiveness and advantages of the proposed DECM-based sampled-data feedback controller over the existing SECM and TCM were demonstrated through numerical experiments.

In future work, we will focus on the optimization of CACC systems in terms of fuel consumption and communication resources. An optimal event-triggered sampled-data controller is required for CACC systems to optimize these performance indexes. In contrast to the generalized optimal problem, the performance indexes should include the transmission statuses of sampled velocity and acceleration values. Another interesting problem is to extend the proposed resource-efficient communication mechanism to nonlinear CACC systems with uncertainty, such as uncertain vehicle mass and engine time constants. The difficulty in this problem is how to obtain the parameters for the DECM by utilizing the adaptive control approach.

References

1 Yasser, T., Paul, M., and Anis, L. (2008). Vehicle ad hoc networks: applications and related technical issues. *IEEE Commun. Surv. Tut.* 10 (3): 74–88.

2 Llatser, I., Festag, A., and Fettweis, G. (2016). Vehicular communication performance in convoys of automated vehicles. *Proc. IEEE Int. Conf. Commun. (ICC)*, Kuala Lumpur, Malaysia (22–27 May 2016), pp. 1–6.

3 Gao, F., Li, S., Zheng, Y., and Kum, D. (2016). Robust control of heterogeneous vehicular platoon with uncertain dynamics and communication delay. *IET Intell. Transp. Syst.* 10 (7): 503–513.

4 Harfouch, Y.A., Yuan, S., and Baldi, S. (2017). An adaptive switched control approach to heterogeneous platooning with inter-vehicle communication losses. *IEEE Trans. Control Netw. Syst.*, DOI:https://doi.org/10.1109/TCNS.2017.2718359, 2017.

5 Ding, L., Han, Q.L., Ge, X., and Zhang, X.M. (2018). An overview of recent advances in event-Triggered consensus of multi-agent systems. *IEEE Trans. Cybern.* 48 (4): 1110–1123.

6 Zhang, X.M., Han, Q.L., and Zhang, B.L. (2017). An overview and deep investigation on sampled-data-based event-triggered control and filtering for networked systems. *IEEE Trans. Ind. Informat.* 48 (4): 4–16.

7 Ge, X. and Han, Q.L. (2017). Distributed formation control of networked multi-agent systems using a dynamic event-triggered communication mechanism. *IEEE Trans. Ind. Electron.* 64 (10): 8118–8127.

8 Yin, X., Yue, D., and Hu, S. (2016). Adaptive periodic event-triggered consensus for multi-agent systems subject to input saturation. *Int. J. Control.* 89 (4): 653–667.

9 Zheng, Y., Li, S.E., and Wang, J. (2016). Stability and scalability of homogeneous vehicular platoon: study on the influence of information flow topologies. *IEEE Trans. Intell. Transp. Syst.* 17 (1): 14–26.

10 Guo, G. and Yue, W. (2012). Autonomous platoon control allowing range-limited sensors. *IEEE Trans. Veh. Technol.* 61 (7): 2901–2912.

11 Guo, G. and Yue, W. (2014). Sampled-data cooperative adaptive cruise control of vehicles with sensor failures. *IEEE Trans. Intell. Transp. Syst.* 15 (6): 2404–2418.

12 Park, P., Ko, J.W., and Jeong, C. (2011). Reciprocally convex approach to stability of systems with time-varying delays. *Automatica* 47 (1): 235–238.

19

Part II

Performance Guarantee Under Actuator Limitation

This part discusses control of connected vehicles with performance guarantees, which means that the prescribed tracking performance can be guaranteed in the presence of actuator saturation/fault, input quantization and dead-zone nonlinearity, and uncertainty and unknown disturbances. **Chapter 6** addresses a fault-tolerant control problem for connected vehicles with nonlinear dynamics and subject to actuator faults and saturation. A quadratic spacing policy is introduced to remove the assumption of zero initial spacing errors. The nonlinear vehicle dynamics is approximated by a radial basis function neural network (RBFNN). The PID-type sliding mode fault-tolerant control method developed can guarantee individual vehicle stability, string stability, and traffic flow stability. **Chapter 7** revisits the fault-tolerant control problem considering actuator faults, input quantization, and dead-zone nonlinearity. An adaptive fault-tolerant control scheme is developed by employing RBFNN and PID-type sliding mode control method. **Chapter 8** investigates prescribed performance concurrent control of connected vehicles with unknown parameters, disturbances, and actuator saturation. A closer spacing policy is introduced to achieve string stability with a virtual leader-bidirectional communication topology. By introducing an auxiliary system introduced to deal with actuator saturation, a distributed adaptive tracking controller is derived that can achieve individual vehicle stability and uniform ultimate bounded string stability. **Chapter 9** revisits the prescribed performance platoon control problem subject to actuator saturation, uncertain parameters, and unknown disturbances. The actuator saturation nonlinearity is approximated with a smooth hyperbolic tangent function. The effects of uncertain parameters and exogenous disturbances are dealt with by introducing some adaptation laws. Two adaptive sliding mode control schemes based on leader-predecessor and leader-bidirectional communication topologies, respectively, are given with ensured string stability, strong string stability, and prescribed tracking performance.

Connected Vehicular Systems: Communication, Control, and Optimization, First Edition. Ge Guo and Shixi Wen.

6

Adaptive Fault-Tolerant Control with Actuator Saturation

6.1 Introduction

It is worth mentioning that the works of Chapters 1–6 for connected vehicles generally assume fault-free environments. However, contingent failures might happen to an actuator in practical vehicles for reasons such as wearouts, aging or external stress damage. For example, partial loss of effectiveness and bias faults are two common actuator faults [1]. On the other hand, the adopted constant time headway (CTH) policy may result in traffic flow instability [2] or even jams [3]. Therefore, various spacing strategies have been developed to guarantee string stability and traffic flow stability [4–6]. However, these results ignore the issue of actuator faults, and might not work in the presence of actuator faults. What is clear is that platooning with guaranteed string stability and traffic flow stability is rather challenging in the occurrence of actuator faults.

Actuator saturation (due to physical limitations) is another challenging issue for platooning [7]. Many platoon control methods, including adaptive control [8] and function approximation [9, 10], were proposed to compensate for actuator saturation. To be specific, the authors of [8] provide an adaptive control algorithm for coping with actuator saturation and unknown system parameters. A smooth hyperbolic tangent function is used to approximate input constraints in [9, 10]. However, these works all assume fault-free environments. As mentioned earlier, the occurrence of actuator faults may lead to large compensating control signals [11], which can potentially result in actuator saturation and cause further performance degradation or even instability [12, 13]. Therefore, it is necessary to study actuator faults and saturation in vehicular platoon systems simultaneously. How to develop an effective control scheme under such a circumstance is a demanding task of practical interest as well as theoretical significance. This is another motivation of this article.

This chapter aims to address the fault-tolerant control problem for heterogeneous vehicular platoons with actuator faults and saturation. By employing radial basis function neural networks (RBFNN) and the PID-type sliding-mode control method, a platoon control scheme is derived based on the Lyapunov stability theory, which is capable of guaranteeing individual vehicle stability and string stability for the whole vehicular platoon. In addition, traffic flow stability is also ensured thanks to the new quadratic spacing policy. The main contributions are summarized as follows:

1) An adaptive fault-tolerant control method is proposed to compensate for the effects of actuator faults and saturation, where an RBFNN is involved to estimate the nonlinear vehicle dynamics via parameter adaptation.
2) To tackle the problem of large transient engine and brake torques due to initial condition uncertainties, a new quadratic spacing policy is given which removes the assumption of zero initial spacing errors.

Connected Vehicular Systems: Communication, Control, and Optimization, First Edition. Ge Guo and Shixi Wen.

3) A minimal learning parameter mechanism is employed for the RBFNN, where the number of parameters depends only on the platoon size, and hence avoids explosion of parameters and reduces the computational burden significantly.

The remaining aspects of this chapter are arranged as follows. Section 6.2 introduces the system modeling with actuator faults and saturation. Section 6.3 proposes an improved quadratic spacing policy. The controller design and stability analysis are given in Section 6.4. Simulation results are illustrated in Section 6.5 and conclusions are presented in Section 6.6 – Fault-Tolerant Platooning With Actuator Saturation.

Notations Throughout this article, $|\cdot|$ denotes the absolute value of real numbers, $\|\cdot\|$ denotes the Euclidean norm of a vector, R^n represents the n-dimensional Euclidean space, and $\Omega_N = \{1, 2, ..., N\}$ denotes the integer set from 1 to N.

6.2 System Modeling and Problem Formulation

Consider a vehicular platoon with a leader and N following vehicles (see Figure 6.1). The leader dynamics is described by

$$\dot{x}_0(t) = v_0(t), \quad \dot{v}_0(t) = a_0(t) \tag{6.1}$$

where $x_0(t)$, $v_0(t)$, and $a_0(t)$ are the position, velocity, and acceleration of the leader, respectively, with $a_0(t)$ being a given function.

The dynamics of vehicle i can be described by the following nonlinear third-order model [1, 2]:

$$\begin{aligned}
\dot{x}_i(t) &= v_i(t) \\
\dot{v}_i(t) &= a_i(t) \\
\dot{a}_i(t) &= -\frac{1}{\tau_i}\left(a_i(t) + \frac{vA_iC_{di}}{2m_i}v_i(t)^2 + \frac{d_{mi}}{m_i}\right) \\
&\quad - \frac{vA_iC_{di}v_i(t)a_i(t)}{m_i} + \frac{c_i(t)}{\tau_i m_i} + w_i(t)
\end{aligned} \tag{6.2}$$

where $c_i(t)$ is the engine/brake input, τ_i is the engine time constant, v is the specific mass of the air, m_i, A_i, C_{di}, and d_{mi} denote the mass, cross-sectional area, drag coefficient, and mechanical drag of vehicle i, respectively, $(vA_iC_{di})/(2m_i)$ denotes the air resistance, and $\omega_i(t)$ is the unknown external disturbances caused by a wind gust and/or rough road condition.

Let

$$u_i(t) = \frac{c_i(t)}{m_i}$$

$$f_i(v_i, a_i, t) = -\frac{1}{\tau_i}\left(a_i(t) + \frac{vA_iC_{di}}{2m_i}v_i(t)^2 + \frac{d_{mi}}{m_i}\right) - \frac{vA_iC_{di}v_i(t)a_i(t)}{m_i}$$

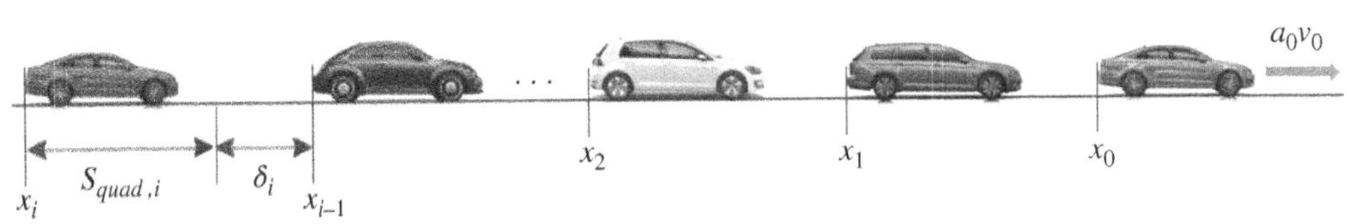

Figure 6.1 Heterogeneous intelligent platoon configuration.

Then, the longitudinal dynamics of vehicle i becomes

$$\begin{aligned}\dot{x}_i(t) &= v_i(t)\\ \dot{v}_i(t) &= a_i(t)\\ \dot{a}_i(t) &= f_i(v_i, a_i, t) + \frac{u_i(t)}{\tau_i} + \omega_i(t)\end{aligned} \tag{6.3}$$

where $u_i(t)$ is used here instead of $c_i(t)$ so that the input does not depend on the vehicle's mass. The mechanical drag d_{mi} and the air resistance $(vA_iC_{di})/(2m_i)$ cannot be obtained accurately in practice, and hence, the nonlinear function $f_i(v_i, a_i, t)$ is unknown, which is to be approximated by an RBFNN.

In practical systems, partial loss of effectiveness and bias faults are two common actuator faults [1]. Partial loss of effectiveness refers to faults due to wear-outs and bias faults represent those caused by aging or external stress damage. As in [1], we adopt the following fault model

$$u_{ai}(t) = \rho_i(t)u_i(t) + r_i(t) \tag{6.4}$$

where $\rho_i(t)$ indicates the actuation effectiveness and $r_i(t)$ denotes the bias fault. Considering actuator faults, the longitudinal dynamics of vehicle i becomes

$$\begin{aligned}\dot{x}_i(t) &= v_i(t)\\ \dot{v}_i(t) &= a_i(t)\\ \dot{a}_i(t) &= f_i(v_i, a_i, t) + \frac{\rho_i(t)u_i(t) + r_i(t)}{\tau_i} + \omega_i(t)\end{aligned} \tag{6.5}$$

In addition, taking into account the saturation effect, the third equation in (6.5) becomes

$$\dot{a}_i(t) = f_i(v_i, a_i, t) + \frac{\rho_i(t)u_i(t) + r_i(t)}{\tau_i} + \omega_i(t) \tag{6.6}$$

where $\text{sat}(u_i(t))$ denotes the asymmetric nonlinear saturation characteristics and is defined as

$$\text{sat}(u_i(t)) = \begin{cases} u_{r,i\max} & \text{if} \quad u_i(t) > b_{r,i}\\ g_{r,i}(u_i(t)) & \text{if} \quad 0 \le u_i(t) \le b_{r,i}\\ g_{l,i}(u_i(t)) & \text{if} \quad b_{l,i} \le u_i(t) < 0\\ u_{l,i\min} & \text{if} \quad u_i(t) < b_{l,i}\end{cases} \tag{6.7}$$

with constants $u_{r,\max} > 0$ and $u_{l,\min} < 0$ being the unknown bounds of the control force, $b_{r,i} > 0$ and $b_{l,i} < 0$ begin the saturation amplitudes, $g_{r,i}(u_i(t))$ and $g_{l,i}(u_i(t))$ being unknown nonlinear functions. Here we consider the following saturating function:

$$\text{sat}(u_i(t)) = \chi_{ui}(t)u_i(t) \tag{6.8}$$

where $\chi_{ui}(t) \in (0, 1]$ is an indicator of saturation degree. As mentioned in [9], there exists a constant χ_l satisfying

$$0 < \chi_l \le \chi_{ui}(t) \le 1 \tag{6.9}$$

When $\chi_{ui}(t)$ approaches zero, the input $u_i(t)$ is almost completely saturated, whereas $\chi_{ui}(t) = 1$ implies no saturation. The longitudinal dynamics of vehicle i could further be rewritten as

$$\begin{aligned}\dot{x}_i(t) &= v_i(t)\\ \dot{v}_i(t) &= a_i(t)\\ \dot{a}_i(t) &= \frac{\chi_{ui}(t)\rho_i(t)u_i(t)}{\tau_i} + f_i(v_i, a_i, t) + \omega_i(t) + \frac{r_i(t)}{\tau_i}\end{aligned} \tag{6.10}$$

Assumption 6.1 The actuation effectiveness $\rho_i(t)$ and bias fault $r_i(t)$ are unknown bounded time-varying functions, and there exist unknown positive constants ρ_{i0} and r_i^* satisfying $0 < \rho_{i0} \leq \rho_i(t) \leq 1$ and $|r_i(t)| \leq r_i^* < \infty$.

Assumption 6.2 $\omega_i(t)$ is an unknown bounded disturbance with $|\omega_i(t)| \leq \omega_i^*$, where ω_i^* is a positive constant.

Remark 6.1 The problem of platooning with actuator faults and saturation has not been addressed up to now. A similar fault-tolerant control problem was studied for high-speed trains in [1]. However, since trains are scheduled to be separated with a large time headway, the method cannot be used to deal with the closely spaced vehicle platooning problem. In addition, aiming at a steady traffic flow, here, we introduce a quadratic spacing policy (instead of CTH) to address this issue.

The objective of this article is to develop a platoon control scheme, in situations with actuator faults and saturation, to meet the following criteria:

1) *Individual vehicle stability*: All the follower's velocities and accelerations track the trajectory of the leader with bounded spacing and velocity errors.
2) *String stability*: The platoon system in (6.10) is string-stable if

$$|\delta_N(s)| \leq |\delta_{N-1}(s)| \leq \cdots \leq |\delta_1(s)| \tag{6.11}$$

i.e. the error transfer function $H_i(s), = \delta_{i+1}(s)/\delta_i(s)$ satisfies $|H_i(s) \leq 1|$ for all $i = \Omega_N$, where $\delta_i(s)$ denotes the Laplace transform $\delta_i(t)$.
3) *Traffic flow stability [3]*: It requires that the gradient $\partial Q/\partial p$ is positive, that is

$$\partial Q/\partial p > 0 \tag{6.12}$$

where Q and p denote the traffic flow rate and traffic density, respectively.

Before proceeding further, some useful lemmas are introduced and will be used later.

Lemma 6.1 [14] RBFNNs are usually employed to deal with unknown nonlinear functions through their inherent approximation capabilities. Using the RBFNNs technology, any continuous function $f(Z)$ can be represented as

$$f(Z) = \omega^{*T}\xi(Z) + \varsigma$$

where Z is the input vector, $\mathbf{W}^* = \left[\omega_1^*, \omega_2^*, \ldots, \omega_M^*\right]^T$ is the ideal weight vector with M being the number of neurons, ς represents the reconstruction error and $\boldsymbol{\xi}(Z) = [\xi_1(Z), \xi_2(Z), \ldots, \xi_M(Z)]^T$ is the Gaussian basis function vector with $\xi_k(Z)$ being described as

$$\xi_i^k(Z_i) = \exp\left[-\frac{\left(Z_i - c_i^k\right)^T\left(Z_i - c_i^k\right)}{2\left(d_i^k\right)^2}\right], k = 1, 2, 3, \ldots, M$$

where b_k is the width of the Gaussian function and ϕ_k is the center vector.

According to Lemma 6.1, the unknown nonlinear function $f_i(v_i, a_i, t)$ can be approximated as

$$f_i(v_i, a_i, t) = W_i^{*T}\xi_i(Z_i) + \varsigma_i, i \in \Omega_N \tag{6.13}$$

where W_i^* is the ideal weight vector and ς_i represents the reconstruction error satisfying $|\varsigma_i| \le \varsigma_i^*$ with ς_i^* being an unknown positive constant.

Lemma 6.2 [15] For $\forall \epsilon > 0$ and $\chi \in R$, the following inequality holds:

$$0 \le |\chi| - \chi \tanh\left(\frac{\chi}{\epsilon}\right) \le \kappa\epsilon$$

where κ is a constant satisfying $\kappa = e^{(-k-1)}$, that is $\kappa = 0.2785$.

Lemma 6.3 (Young's inequality [16]) For $\forall (a, b) \in R^2$, the following inequality holds:

$$ab \le \frac{\zeta^p}{p}|a|^p + \frac{1}{q\zeta^q}|b|^q$$

where $\zeta > 0,\ p > 1,\ q > 1$, and $(p-1)(q-1) = 1$.

6.3 Quadratic Spacing Policy and Adaptive PID-Type Sliding Surface

This section gives the improved quadratic spacing policy and an adaptive PID-type sliding surface, which will be used for controller design due to its excellent performance in eliminating steady-state spacing errors.

The assumption of zero initial spacing errors has been made in numerous existing literature, such as [17, 18]. However, in practical engineering, the initial spacing errors are not zero in most cases, which makes the assumption restrictive. To remove this assumption, an improved quadratic spacing policy (inspired by [6]) is introduced, which is based on the vehicle speed, safety coefficient, and braking capacity, as shown here:

$$\begin{aligned}
\delta_i(t) &= \tilde{\delta}_i(t) - \Gamma_i(t) \\
\tilde{\delta}_i(t) &= x_{i-1}(t) - x_i(t) - L_i - \Delta_{i-1,t} - hv_i(t) - \frac{\sigma v_i^2(t)}{2A_m} \\
\Gamma_i(t) &= \left\{\tilde{\delta}_i(0) + \left[\pi_i\tilde{\delta}_i(0) + \dot{\tilde{\delta}}(0)\right]t + \frac{1}{2}\left[\pi_i^2\tilde{\delta}_i(0) + 2\pi_i\dot{\tilde{\delta}}(0) + \ddot{\tilde{\delta}}(0)\right]t^2\right\}e^{-\pi_i t}
\end{aligned} \tag{6.14}$$

where π_i is a positive constant, L_i denotes the length of vehicle i and $\Delta_{i-1,\,i}$ is a given minimum distance, h is the delay in the platoon system, which compensates for the delay in braking or acceleration (about 70–160 ms, see [5]), σ represents safety coefficient depending on the road or weather and A_m is the absolute value of the maximum possible deceleration. Further, we have

$$\delta_i(0) = 0,\ \dot{\delta}_i(0) = 0,\ \ddot{\delta}_i(0) = 0 \tag{6.15}$$

which indicates that the initial values of the improved spacing errors are always zero, removing the assumption of zero initial spacing errors in the existing results.

The desired intervehicle distance including vehicle length can then be defined as

$$s_{quad,i} = L_i + \Delta_{i-1,i} + hv_i(t) + \frac{\sigma v_i^2(t)}{2A_m} + \Gamma_i(t) \tag{6.16}$$

Remark 6.2 A quadratic spacing policy [6] was proposed to guarantee string stability and traffic flow stability, which can also ensure a higher traffic capacity compared with the commonly used CTH spacing policy. However, the quadratic spacing policy [6] does not consider the problem of nonzero initial errors, which may cause large transient engine and brake torques or lead to string instability [9, 19]. Thus, an improved quadratic spacing policy is proposed to transform the nonzero initial spacing error problem into a zero initial spacing error problem.

Sliding-mode control, in particular, the PID-type sliding-mode control, has been widely adopted to deal with external disturbances and model uncertainties, owing to excellent performance in eliminating steady-state spacing errors. We consider an adaptive PID-type sliding surface:

$$s_i(t) = K_p\delta_t(t) + K_i\int_0^t \delta_i(\tau)d\tau + K_d\dot{\delta}_i(t) \tag{6.17}$$

where positive scalars K_p, K_ι, and K_d are proportional, integral, and derivative coefficients, respectively. As in [20], to deal with the relation of $\delta_i(t)$ and $\delta_{i+1}(t)$ better, we introduce the following coupled sliding surface:

$$S_i = \begin{cases} \lambda s_i - s_{i+1} & i = 1, 2, \ldots, N-1 \\ \lambda s_i & i = N \end{cases} \tag{6.18}$$

where λ is a positive parameter coupling the sliding surfaces $s_i(t)$ and $s_{i+1}(t)$. When $S_i(t)$ reaches the sliding surface, $S_i(t)$ does the same thing at the same time.

6.4 Controller Design and Stability and Analysis

In this section, by combining the adaptive RBFNN approximation technique with the PID-type sliding-mode control method, an adaptive fault-tolerant platoon control scheme is developed to handle actuator faults and saturation.

As mentioned earlier, the RBFNN technique is used to approximate the unknown nonlinear function $f_i(v_i, a_i, t)$ in the vehicle dynamics. Let the estimation errors be

$$\begin{aligned} \tilde{\theta}_i(t) &= \theta_i^* - \rho_{i0}\hat{\theta}_i(t) \\ \tilde{\eta}_i(t) &= \eta_i^* - \rho_{i0}\hat{\eta}_i(t) \end{aligned} \tag{6.19}$$

where ρ_{i0} is given in Assumption 6.1, $\hat{\theta}_i(t)$ and $\hat{\eta}_i(t)$ are the estimates of θ_i^* and η_i^* defined here:

$$\begin{aligned} \eta_i^* &\geq \varsigma_i^* + \omega_i^* + \frac{r_i^*}{\tau_i} \\ \theta_i^* &= \|W_i^*\|^2 = W_i^{*T}W_i^* \end{aligned} \tag{6.20}$$

Then, we introduce the following fault-tolerant controller

$$\begin{aligned} u_i(t) = \frac{\tau_i}{\chi_l}\Bigg\{ &\frac{K_i S_i(t)}{\lambda K_d(h + ((\sigma v_i(t))/(A_m)))} + \hat{\eta}_i(t)\tanh\left(\frac{S_i(t)}{\mu_i}\right) \\ &+ \frac{\Lambda_i}{\lambda K_d(h + ((\sigma v_i(t))/A_m))}\frac{Z_i(t)^2 S_i(t)}{|Z_i(t)S_i(t)| + \vartheta_i} + \frac{b_i^2}{2}\hat{\theta}_i(t)\xi_i^T(Z_i)\xi_i(Z_i)S_i(t)\Bigg\} \end{aligned} \tag{6.21}$$

where $1 \le 1/\rho_{i0} \le \Lambda_i$, ϑ_i, μ_i, and b_i are positive constants, k_i is the feedback gain, $Z_i(t)$ is given by

$$\begin{cases} Z_i(t) = \lambda K_p \dot{\delta}_i(t) + \lambda K_i \delta_i(t) + \lambda K_d \big(a_{i-1}(t) - a_i(t) - ((\sigma a_i^2(t))/A_m) - \ddot{\Gamma}_i(t)\big) \\ \qquad - K_p \dot{\delta}_{i+1}(t) - K_i \delta_{i+1}(t) - K_d \ddot{\delta}_{i+1}(t), \quad \text{for } i = 1, 2, \ldots, N-1 \\ Z_i(t) = \lambda K_p \dot{\delta}_i(t) + \lambda K_i \delta_i(t) + \lambda K_d \big(a_{i-1}(t) - a_i(t) - ((\sigma a_i^2(t))/(A_m)) - \ddot{\Gamma}_i(t)\big), \quad \text{for } i = N. \end{cases} \tag{6.22}$$

and the following adaptation laws

$$\begin{aligned} \dot{\hat{\theta}}(t) &= \alpha_i \lambda K_d \left(h + \frac{\sigma v_i(t)}{A_m} \right) \frac{b_i^2}{2} \xi_i^T(Z_i) \xi_i(Z_i) S_i^2(t) \\ &= \Xi_{1i}(t) \hat{\theta}_i(t) \\ \dot{\hat{\eta}}_i(t) &= \beta_i \lambda K_d \left(h + \frac{\sigma v_i(t)}{A_m} \right) S_i(t) \tanh\left(\frac{S_i(t)}{\mu_i} \right) \\ &\quad - \Xi_{2i}(t) \hat{\eta}_i(t) \end{aligned} \tag{6.23}$$

Here $\hat{\theta}_i(0) \ge 0$ and $\hat{\eta}_i(0) \ge 0$, k_i, λ, b_i, K_p, K_i, K_d, α_i, and β_i are any positive constants, while ϑ_i and μ_i are small constants, $\Xi_{1i}(t)$ and $\Xi_{2i}(t)$ are any continuous and bounded functions satisfying

$$\Xi_{ji}(t) > 0, \quad \lim_{t \to \infty} \int_{t_0}^{t} \Xi_{ji}(\tau) d\tau \le \overline{\Xi}_{ji} < +\infty, \quad j = 1, 2 \tag{6.24}$$

with Ξ_{ji} being any finite positive constants.

It should be mentioned that this adaptive fault-tolerant control law can compensate for the effects of actuator faults and saturation by adjusting parameters Λ_i and χ_l.

Remark 6.3 One shortcoming of the RBFNN control method is the computational burden might be excessively large since the number of parameters will increase with the increase of NN nodes [14]. Inspired by [19], by adopting a minimal learning parameter mechanism, the number of parameters depends only on the number of vehicles, which can alleviate the explosion of learning parameters.

Remark 6.4 For practical systems, the method in [20] is subject to chattering problems due to the use of a discontinuous sign function. Here, the problem is avoided by using a continuous function $\tanh((S_i(t))/\mu)$, where μ is a small positive scalar. It should be mentioned that if μ is too small, there may also be chattering, whereas a large μ may degrade the control performance. Thus, selecting suitable controller parameter μ is crucial to maintain a desirable performance.

We now give the main result in the following theorem.

Theorem 6.1 Consider the heterogeneous vehicular platoon (6.3) with actuator faults (6.4) and saturation (6.7), under the improved quadratic spacing policy (6.14), the adaptive fault-tolerant controller (6.21) and the adaptation laws (6.23) can drive $\delta_i(t)$ into a small set around the origin $\Omega = \{\delta_i(t) : |\delta_i(t)| < \vartheta\}$ by choosing appropriate design parameters. Moreover, string stability (6.11) can be guaranteed when $0 < \lambda \le 1$ and traffic flow stability (6.12) is also guaranteed thanks to the new quadratic spacing policy (6.14).

Proof: The proof consists of the following three parts:

a) *Boundedness of* $S_i(t)$: Choose the following Lyapunov function candidate:

$$V_i = \frac{1}{2}S_i^2(t) + \frac{1}{2\rho_{i0}\alpha_i}\tilde{\theta}_i^2 + \frac{1}{2\rho_{i0}\beta_i}\tilde{\eta}_i^2$$

where $\tilde{\theta}_i(t)$ and $\tilde{\eta}_i(t)$ are defined in (6.19).
Taking the derivative of $V_i(t)$ yields

$$\dot{V}_i(t) = S_i(t)\dot{S}_i(t) - \frac{1}{\alpha_i}\tilde{\theta}_i(t)\dot{\hat{\theta}}_i(t) - \frac{1}{\beta_i}\tilde{\eta}_i(t)\dot{\hat{\eta}}_i(t) \tag{6.25}$$

Then, the derivative of $S_i(t)$ can be deduced using (6.14), (6.17), and (6.18) as

$$\begin{aligned}
\dot{S}_i(t) &= \lambda\left(K_p\dot{\delta}_i(t) + K_i\delta_i(t) + K_d\ddot{\delta}_i(t)\right) - K_p\dot{\delta}_{i+1}(t) - K_i\delta_{i+1}(t) + K_d\ddot{\delta}_{i+1}(t) \\
&= \lambda\left\{K_p\dot{\delta}_i(t) + K_i\delta_i(t) + K_d\left(a_{i-1}(t) - a_i(t) - h\dot{a}_i(t) - \frac{\sigma a_i^2(t)}{A_m} - \frac{\sigma v_i(t)}{A_m}\dot{a}_i(t) - \ddot{\Gamma}_i(t)\right)\right\} \\
&\quad - K_p\dot{\delta}_{i+1}(t) - K_i\delta_{i+1}(t) - K_d\ddot{\delta}_{i+1}(t) \\
&= -\lambda K_d\left(h + \frac{\sigma v_i(t)}{A_m}\right)\dot{a}_i(t) + Z_i(t)
\end{aligned}$$

for $i = 1, 2, \ldots, N-1$

$$\begin{aligned}
\dot{S}_i(t) &= \lambda\left(K_d\dot{\delta}_i(t) + K_i\ddot{\delta}_i(t)\right) \\
&= \lambda\{K_p\dot{\delta}_i(t) + K_i\delta_i(t) + K_d(a_{i-1}(t) - a_i(t) \\
&\quad - h\dot{a}_i(t) - \frac{\sigma a_i^2(t)}{A_m} - \frac{\sigma v_i(t)}{A_m}\dot{a}_i(t) - \ddot{\Gamma}_i(t)\Big)\Big\}, \quad \text{for } i = N \\
&= -\lambda K_d\left(h + \frac{\sigma v_i(t)}{A_m}\right)\dot{a}_i(t) + Z_i(t)
\end{aligned} \tag{6.26}$$

where $Z_i(t)$ is defined in (6.22). Then, substituting (6.10), (6.13), and (6.26) into $S_i(t)\dot{S}_i(t)$ yields

$$\begin{aligned}
S_i(t)\dot{S}_i(t) &= -\lambda K_d\left(h + \frac{\sigma v_i(t)}{A_m}\right)S_i(t)\left(\frac{\chi_{ui}(t)\rho_i(t)u_i(t)}{\tau_i} + f_i(v_i, a_i, t) + \omega_i(t) + \frac{r_i(t)}{\tau_i}\right) + Z_i(t)S_i(t) \\
&= -\lambda K_d\left(h + \frac{\sigma v_i(t)}{A_m}\right)S_i(t)\frac{\chi_{ui}(t)\rho_i(t)u_i(t)}{\tau_i} - \lambda K_d\left(h + \frac{\sigma v_i(t)}{A_m}\right)S_i(t)W_i^{*T}\xi_i(Z_i) \\
&\quad - \lambda K_d\left(h + \frac{\sigma v_i(t)}{A_m}\right)S_i(t)\left(\varsigma_i + \omega_i(t) + \frac{r_i(t)}{\tau_i}\right) + Z_i(t)S_i(t) \\
&\le -\lambda K_d\left(h + \frac{\sigma v_i(t)}{A_m}\right)S_i(t)\frac{\chi_{ui}(t)\rho_i(t)u_i(t)}{\tau_i} - \lambda K_d\left(h + \frac{\sigma v_i(t)}{A_m}\right)S_i(t)W_i^{*T}\xi_i(Z_i) \\
&\quad + \lambda K_d\left(h + \frac{\sigma v_i(t)}{A_m}\right)\eta_i^*|S_i(t)| + |Z_i(t)S_i(t)|
\end{aligned} \tag{6.27}$$

where $\varsigma_i^* + w_i^* + ((r_i^*)/(\tau_i)) \leq \eta_i^*$ is used. Then, according to Lemma 6.3, there exists a design parameter $b_i > 0$ such that

$$\begin{aligned}
&-\lambda K_d\left(h + \frac{\sigma v_i(t)}{A_m}\right) S_i(t) W_i^{*T}\xi_i(Z_i) \\
&\leq \lambda K_d\left(h + \frac{\sigma v_i(t)}{A_m}\right)\frac{b_i^2}{2} S_i^2(t)\theta_i^*\xi_i^T(Z_i)\xi_i(Z_i) \\
&+ \frac{1}{2b_i^2}\lambda K_d\left(h + \frac{\sigma v_i(t)}{A_m}\right)
\end{aligned} \tag{6.28}$$

with $\theta_i^* = \|W_i^*\|^2 = W_i^{*T} W_i^*$. Here, a minimal learning parameter mechanism is employed to only estimate the constants θ_i^* instead of estimating the ideal weights W_i^* directly. Thus, the number of parameters will reduce drastically from $\sum_{i=1}^{N} N_i$ to N, which avoids explosion of parameters and reduces the computational burden significantly. Then, substituting (6.28) into (6.27), one has

$$\begin{aligned}
S_i(t)\dot{S}_i(t) \leq &-\lambda K_d\left(h + \frac{\sigma v_i(t)}{A_m}\right) S_i(t)\frac{\chi_{ui}(t)\rho_i(t)u_i(t)}{\tau_i} + \lambda K_d\left(h + \frac{\sigma v_i(t)}{A_m}\right)\frac{b_i^2}{2}\theta_i^*\xi_i^T(Z_i)\xi_i(Z_i)S_i^2(t) \\
&+ \frac{1}{2b_i^2}\lambda K_d\left(h + \frac{\sigma v_i(t)}{A_m}\right) + |Z_i(t)S_i(t)| + \lambda K_d\left(h + \frac{\sigma v_i(t)}{A_m}\right)\eta_i^*|S_i(t)| \\
\leq &-\lambda K_d\left(h + \frac{\sigma v_i(t)}{A_m}\right) S_i(t)\frac{\chi_{ui}(t)\rho_i(t)u_i(t)}{\tau_i} + \lambda K_d\left(h + \frac{\sigma v_i(t)}{A_m}\right)\frac{b_i^2}{2}\theta_i^*\xi_i^T(Z_i)\xi_i(Z_i)S_i^2(t) \\
&+ \frac{1}{2b_i^2}\lambda K_d\left(h + \frac{\sigma v_i(t)}{A_m}\right) + \Lambda_i\rho_{i0}|Z_i(t)S_i(t)| + \lambda K_d\left(h + \frac{\sigma v_i(t)}{A_m}\right)\eta_i^*|S_i(t)|
\end{aligned} \tag{6.29}$$

where the fact $1 \leq 1/\rho_{i0} \leq \Lambda_i$ is used.

It follows from (6.23) that $\dot{\hat{\theta}}_i(t) + \Xi_{1i}(t)\hat{\theta}_i(t) \geq 0$, which implies that $\hat{\theta}_i(t) \geq e^{-\int_0^t \Xi_{1i}(\tau)d\tau}\hat{\theta}_i(0)$. Thus, the adaptive parameter $\hat{\theta}_i(t)$ will be non-negative if we select $\hat{\theta}_i(0) \geq 0$. Similarly, the adaptive parameter $\hat{\eta}_i(t)$ is also non-negative. Then, incorporating (6.8) with (6.21), we have

$$\begin{aligned}
&-\lambda K_d\left(h + \frac{\sigma v_i(t)}{A_m}\right) S_i(t)\frac{\chi_{ui}(t)\rho_i(t)u_i(t)}{\tau_i} \\
&\leq -k_i\rho_{i0}S_i^2(t) - \Lambda_i\rho_{i0}\frac{Z_i^2(t)S_i^2(t)}{|Z_i(t)S_i(t)| + \vartheta_i} \\
&-\lambda K_d\left(h + \frac{\sigma v_i(t)}{A_m}\right)\rho_{i0}\hat{\eta}_i(t)S_i(t)\tanh\left(\frac{S_i(t)}{\mu_i}\right) \\
&-\lambda K_d\left(h + \frac{\sigma v_i(t)}{A_m}\right)\frac{\rho_{i0}b_i^2}{2}\hat{\theta}_i(t)\xi_i^T(Z_i)\xi_i(Z_i)S_i^2(t) \\
&\leq -k_i\rho_{i0}S_i^2(t) - \Lambda_i\rho_{i0}|Z_i(t)S_i(t)| + \Lambda_i\rho_{i0}\vartheta_i \\
&-\lambda K_d\left(h + \frac{\sigma v_i(t)}{A_m}\right)\rho_{i0}\hat{\eta}_i(t)S_i(t)\tanh\left(\frac{S_i(t)}{\mu_i}\right) \\
&-\lambda K_d\left(h + \frac{\sigma v_i(t)}{A_m}\right)\frac{\rho_{i0}b_i^2}{2}\hat{\theta}_i(t)\xi_i^T(Z_i)\xi_i(Z_i)S_i^2(t)
\end{aligned} \tag{6.30}$$

where the facts that $-\dfrac{Z_i^2(t)S_i^2(t)}{|Z_i(t)S_i(t)|+\vartheta_i} \le -|Z_i(t)S_i(t)|+\vartheta_i$ and $0<\rho_{i0}\le\rho_i(t)\le 1$ have been used.
Then, using (6.23), we can have that

$$
\begin{aligned}
&-\frac{1}{\alpha_i}\tilde{\theta}_i(t)\dot{\hat{\theta}}_i(t)-\frac{1}{\beta_i}\tilde{\eta}_i(t)\dot{\hat{\eta}}_i(t)\\
&=-\lambda K_d\left(h+\frac{\sigma v_i(t)}{A_m}\right)\frac{b_i^2}{2}\tilde{\theta}_i(t)\xi_i^T(Z_i)\xi_i(Z_i)S_i^2(t)\\
&\quad-\lambda K_d\left(h+\frac{\sigma v_i(t)}{A_m}\right)\tilde{\eta}_i(t)S_i(t)\tanh\left(\frac{S_i(t)}{\mu_i}\right)\\
&\quad+\frac{\Xi_{1i}}{\alpha_i}\tilde{\theta}_i(t)\hat{\theta}_i(t)+\frac{\Xi_{2i}}{\beta_i}\tilde{\eta}_i(t)\hat{\eta}_i(t)
\end{aligned}
\tag{6.31}
$$

Substituting (6.29), (6.30), and (6.31) into (6.25) and using simple calculations yields

$$
\begin{aligned}
\dot{V}_i(t)\le&-k_i\rho_{i0}S_i^2(t)+\frac{\Xi_{1i}(t)}{\alpha_i}\tilde{\theta}_i(t)\dot{\hat{\theta}}_i(t)+\frac{\Xi_{2i}(t)}{\beta_i}\tilde{\eta}_i(t)\dot{\hat{\eta}}_i(t)\\
&+\lambda K_d\left(h+\frac{\sigma v_i(t)}{A_m}\right)\eta_i^*|S_i(t)|-\lambda K_d\left(h+\frac{\sigma v_i(t)}{A_m}\right)\eta_i^*S_i(t)\tanh\left(\frac{S_i(t)}{\mu_i}\right)\\
&+\frac{1}{2b_i^2}\lambda K_d\left(h+\frac{\sigma v_i(t)}{A_m}\right)+\Lambda_i\rho_{i0}\vartheta_i
\end{aligned}
\tag{6.32}
$$

Further, recalling (6.19) and applying Lemma 6.3, we have

$$
\begin{aligned}
\frac{\Xi_{1i}(t)}{\alpha_i}\tilde{\theta}_i(t)\hat{\theta}_i(t)&\le-\frac{\Xi_{1i}(t)}{2\rho_{i0}\alpha_i}\tilde{\theta}_i^2(t)+\frac{\Xi_{1i}(t)}{2\rho_{i0}\alpha_i}\theta_i^{*2}\\
\frac{\Xi_{2i}(t)}{\beta_i}\tilde{\eta}_i(t)\hat{\eta}_i(t)&\le-\frac{\Xi_{2i}(t)}{2\rho_{i0}\beta_i}\tilde{\eta}_i^2(t)+\frac{\Xi_{2i}(t)}{2\rho_{i0}\beta_i}\eta_i^{*2}
\end{aligned}
\tag{6.33}
$$

which, together with Lemma 6.2, (6.33), and (6.32), implies that

$$
\begin{aligned}
\dot{V}_i(t)\le&-k_i\rho_{i0}S_i^2(t)-\frac{\Xi_{1i}(t)}{2\rho_{i0}\alpha_i}\tilde{\theta}_i^2(t)\\
&-\frac{\Xi_{2i}(t)}{2\rho_{i0}\beta_i}\tilde{\eta}_i^2(t)+\frac{\Xi_{1i}(t)}{2\rho_{i0}\alpha_i}\theta_i^{*2}+\frac{\Xi_{2i}(t)}{2\rho_{i0}\beta_i}\eta_i^{*2}+0.2785\mu_i\lambda K_d\left(h+\frac{\sigma v_i(t)}{A_m}\right)\eta_i^*\\
&+\frac{1}{2b_i^2}\lambda K_d\left(h+\frac{\sigma v_i(t)}{A_m}\right)+\Lambda_i\rho_{i0}\vartheta_i
\end{aligned}
\tag{6.34}
$$

Let

$$
V(t)=\sum_{i=1}^{N}V_i(t)
$$

Then, according to (6.34), we can obtain that

$$\dot{V}(t) \leq -\zeta_1 V(t) + \zeta_2$$

where

$$\zeta_1 = \min\{2\rho_{i0}k_i, \min_{1 \leq i \leq N}\Xi_{1i}(t), \min_{1 \leq i \leq N}\Xi_{2i}(t)\}$$

$$\zeta_2 = \sum_{i=1}^{N}\left\{\frac{\Xi_{1i}(t)}{2\rho_{i0}\alpha_i}\theta_i^{*2} + \frac{\Xi_{2i}(t)}{2\rho_{i0}\beta_i}\eta_i^{*2}n + 0.2785\mu_i\lambda K_d\left(h + \frac{\sigma v_i(t)}{A_m}\right)\eta_i^{*}n\right.$$
$$\left. + \frac{1}{2b_i^2}\lambda K_d\left(h + \frac{\sigma v_i(t)}{A_m}\right) + \Lambda_i\rho_{i0}\vartheta_i\right\}$$

Then, based on corollary 1 in [17], we have

$$V(t) \leq \left(V(0) - \frac{\zeta_2}{\zeta_1}\right)e^{i\zeta_i t} + \frac{\zeta_2}{\zeta_1} \leq V(0) + \frac{\zeta_2}{\zeta_1} \tag{6.35}$$

where

$$V(0) = \sum_{i=1}^{N}\left\{\frac{1}{2}S_i^2(0) + \frac{1}{2\rho_{i0}\alpha_i}\tilde{\theta}_i^2(0) + \frac{1}{2\rho_{i0}\beta_i}\tilde{\eta}_i^2(0)\right\}$$

Then, we can conclude that all the states of the closed-loop system are uniformly ultimately bounded with $V(0) \geq 0$. In particular, we have

$$\lim_{t\to\infty}\sum_{i=1}^{N}\frac{1}{2}S_i^2(t) \leq V(0) + \frac{\zeta_2}{\zeta_1} \tag{6.36}$$

Hence, based on the definitions (6.17) and (6.18), the spacing errors $\delta_i(t)$ are uniformly ultimately bounded.

As shown in many existing results [9, 19], reducing ζ_2 or increasing ζ_1 will cause a smaller bound of $S_i(t)$, and $S_i(t)$ can converge to a small set around the origin by choosing appropriate design parameters.

b) *String stability*: Along similar lines as in the proof of [20], string stability can be proved. Since $S_i(t) = \lambda s_i(t) - s_{i+1}(t)$, when $S_i(t)$ converges to a small set around the origin by choosing appropriate design parameters, we have

$$\begin{aligned}&\lambda\left(K_p\delta_i(t) + K_i\int_0^t \delta_i(\tau)d\tau + K_d\dot{\delta}_i(t)\right)\\ &\approx K_p\delta_{i+1}(t) + K_i\int_0^t \delta_{i+1}(\tau)d\tau + K_d\dot{\delta}_{i+1}(t)\end{aligned} \tag{6.37}$$

the Laplace transform of which yields

$$\begin{aligned}&\lambda\left(K_p\delta_i(s) + sK_i\dot{\delta}_i(s) + \frac{K_d}{s}\dot{\delta}_i(s)\right)\\ &\approx K_p\delta_{i+1}(t) + K_i\int_0^t \delta_{i+1}(\tau)d\tau + K_d\dot{\delta}_{i+1}(t)\end{aligned} \tag{6.38}$$

Then, $H_i(s) = \delta_{i+1}(s)/\delta_i(s) \approx \lambda$. Thus, string stability can be guaranteed when $0 < \lambda \leq 1$.

c) *Traffic flow stability*: It is reasonable to assume that at a steady state, the separation and velocity of all vehicles are the same, i.e. $S_{quad,\ i}(t) = S_{quad}(t)$ and $v_i(t) = v(t)$, for all i. Thus, by definition and the fact that $\Gamma_i(t) = 0$ at a steady state, one can have

$$S_{quad} = L + \Delta + hv(t) + \frac{\sigma v^2(t)}{2A_\mathrm{m}} \tag{6.39}$$

Then, the steady traffic density is

$$p = \frac{1}{L + \Delta + hv + ((\sigma v^2(t))/(2A_\mathrm{m}))}$$

Since the flow rate $Q(p) = pv$, we have

$$Q(p) = p\left(\sqrt{\left(\frac{hA_m}{\sigma}\right)^2 - \frac{2A_m}{\sigma}\left(L + \Delta - \frac{1}{p}\right)} - \frac{hA_m}{\sigma}\right)$$

and

$$\frac{\partial Q}{\partial p} = \left(\sqrt{\left(\frac{hA_m}{\sigma}\right)^2 - \frac{2A_m}{\sigma}\left(L + \Delta - \frac{1}{p}\right)} - \frac{hA_m}{\sigma}\right) - \frac{A_m}{\sigma p\sqrt{\left(\frac{hA_m}{\sigma}\right)^2 - \frac{2A_m}{\sigma}\left(L + \Delta - \frac{1}{p}\right)}}$$

The critical traffic density p_{cr} can be obtained by $\partial Q/\partial p = 0$, which is given as follows:

$$p_{cr} = \frac{1}{2(L + \Delta) + h\sqrt{((2(L + \Delta)A_m)/(\sigma))}} \tag{6.40}$$

The proof of the theorem is completed.

Remark 6.5 It is clear from (6.40) that the parameters A_m and σ determine the critical traffic density. It is noted that if we select a smaller A_m or a bigger σ, the critical traffic density p_{cr} will be higher, but results in larger intervehicle spacings, thereby reducing traffic throughput. This disadvantage becomes more pronounced if parameter A_m is too small and/or σ is too big. Therefore, these parameters should be adjusted carefully according to the required control performance.

To show the comparison with the developed scheme, the following corollary can be obtained by similar techniques using Theorem 6.1. Noted that the controller (6.21) can be reduced to the standard controller when $\chi_l = 1$ and $\Lambda_i = 1$.

Corollary 6.1 Consider the heterogeneous vehicular platoon (6.3) without actuator faults and saturation, under the improved quadratic spacing policy (6.14), the standard controller and the adaptation laws (6.23) can drive $\delta_i(t)$ into a small set around the origin $\Omega = \{\delta_i(t) : |\delta_i(t)| < \vartheta\}$ by choosing appropriate design parameters. Moreover, string stability (6.11) can be guaranteed when $0 < \lambda \leq 1$ and traffic flow stability (6.12) is also guaranteed thanks to the new quadratic spacing policy (6.14).

Corollary 6.2 Consider the heterogeneous vehicular platoon (6.3) with actuator faults (6.4) and saturation (6.7), assuming zero initial spacing errors for the traditional quadratic spacing policy

[14], the adaptive fault-tolerant control laws (6.21)–(6.23) with $\Gamma_i(t) = 0$ can drive $\delta_i(t)$ into a small set around the origin $\Omega = \{\delta_i(t) : |\delta_i(t)| < \vartheta\}$ by choosing appropriate design parameters. Moreover, string stability (6.11) can be guaranteed when $0 < \lambda \leq 1$ and traffic flow stability (6.12) is also guaranteed thanks to the quadratic spacing policy [14].

6.5 Simulation Results

In this section, we show the effectiveness of the developed scheme via computer simulations, which runs in a virtual environment established using the System Build software package in MATLAB. In the simulations, the desired vehicle spacing is $\Delta_{i-1,i} = 7$ m, the delay time is $h = 0.12$ s, the safety coefficient is $\sigma = 0.2$, and the absolute value of the maximum possible deceleration is $A_m = 7\ \text{m/s}^2$. The other parameters in [21] are given in Tables 6.1 and 6.2. The RBF vectors $W_i^{*T}\xi_i(Z_i)$ contains $M = 20$ nodes. The centers ϕ_k for the input $Z_i(t) = v_i(t)$ are evenly spaced in $[-2, 5]$, and the widths are $b_k = 2(k = 1, 2, \ldots, 20)$. The external disturbances are $\omega_i = 0.01\sin(t),\ i \in \Omega_N$, and the desired acceleration of the leader is modeled as

$$a_0(t) = \begin{cases} 0.5t & \text{m/s}^2, \quad 1\,\text{s} \leq t < 4\,\text{s} \\ 2 & \text{m/s}^2, \quad 4\,\text{s} \leq t < 8\,\text{s} \\ -0.5t + 6.5 & \text{m/s}^2, \quad 8\,\text{s} \leq t < 12\,\text{s} \\ 0 & \text{m/s}^2, \quad \text{otherwise} \end{cases}$$

In addition, we chose the saturation input $\text{sat}(u_i(t))$ as

$$\text{sat}(u_i(t)) = \begin{cases} 15, & \text{if} \quad u_i(t) > 15 \\ u_i(t), & \text{if} \quad 0 \leq u_i(t) \leq 15 \\ u_i(t), & \text{if} \quad -20 \leq u_i(t) < 0 \\ -20, & \text{if} \quad u_i(t) < -20. \end{cases}$$

Table 6.1 The dynamics parameters for followers

Parameter name	Simulation values
Specific mass of the air	$v = 1.2\,\text{kg/m}^3$
Cross-sectional area	$A_i = 2.2\,\text{m}^2$
Drag coefficient	$C_{di} = 0.35$
Mechanical drag	$d_{mi} = 5N$

Table 6.2 The parameters m_i, τ_i, and L_i for followers

i	1	2	3	4	5	6
m_i(kg)	1500	1600	1550	1650	1500	1400
τ_i(s)	0.1	0.3	0.2	0.4	0.25	0.4
L_i(m)	4	4.5	5	5	4.5	3.5

Table 6.3 Initial states for all vehicles.

i	0	1	2	3	4	5	6
$x_i(0)$(m)	150	135	125.5	112.5	99.5	87	75.5
$v_i(0)$(m/s)	1	4	2	0	5	3	1
$a_i(0)$(m/s^2)	0	1	5	2	1	3	1

Table 6.4 Control parameters for followers.

Parameter name	Simulation values
Controller parameters	$k_i = 45,\ \lambda = 0.75,\ \pi_i = 3$
PID parameters	$K_p = 1,\ K_i = 2.8,\ K_d = 0.15$
Initial adaptive values	$\hat{\theta}_i(0) = 1,\ \hat{\eta}_i(0) = 0$
Adaptive parameters 1	$\alpha = 15,\ \beta = 10$
Adaptive parameters 2	$\Xi_{1i} = 0.01e^{-20t},\ \Xi_{2i} = 0.1e^{-10t}$
Tuning parameters	$b_i = 2,\ \mu_i = 0.3,\ \vartheta_i = 0.8$

Then, we assume $\chi_l = 0.5$. In the case of actuator faults, we choose the actuation effectiveness as $\rho_i = 0.75 + 0.25\sin(0.1it)$ and the bias fault as $r_i(t) = 0.01\sin(t)$. Then, we can obtain $\rho_{i0} = 0.5$. Using the fact $1 \leq 1/\rho_{i0} \leq \Lambda_i$, we assume $\Lambda_i = 2$. The initial states are listed in Table 6.3 and the control parameters are given in Table 6.4. Note that some general guidelines on design parameter selection were given in [19]. So, it is omitted in this article.

In this subsection, comparative simulations are made between Corollary 6.1 and Theorem 6.1. First of all, when there do not exist actuator faults and saturation, the simulation results under Corollary 6.1 are shown in Figure 6.2. The string stability is guaranteed in Figure 6.2a since the amplitude of the spacing error decreases along the platoon. Figure 6.2b shows that collisions are avoided since there are no crossed or overlapped positions. Figure 6.2c, d show that the velocities and accelerations of all followers eventually track those of the leader. The sliding surfaces and control input in Figure 6.2e, f can converge to the neighborhood of origin and the chattering is eventually eliminated. Figures 6.2g, h show that the estimations of the adaptive laws $\hat{\theta}_i(t)$ and $\hat{\eta}_i(t)$ are all bounded.

However, when actuator faults and saturation occur, the system performance deteriorates significantly. More specifically, Figure 6.3a shows that the string stability cannot be guaranteed in the initial transient and the spacing errors cannot converge to the neighborhood of origin. The sliding surfaces and control input in Figures 6.3e, f show that the chattering is serious. Figures 6.3g, h show that the estimations of the adaptive laws $\hat{\theta}_i(t)$ and $\hat{\eta}_i(t)$ cannot converge to constants. The results in Figure 6.3a–f demonstrate the need for effective control schemes to handle actuator faults and saturation.

Then, the simulation results under Theorem 6.1 are shown in Figure 6.4. In the presence of actuator faults and saturation, string stability is still guaranteed and the performance is very similar to those in Figure 6.2, which shows that the proposed scheme by adjusting fault and saturation parameters is effective to handle actuator faults and saturation.

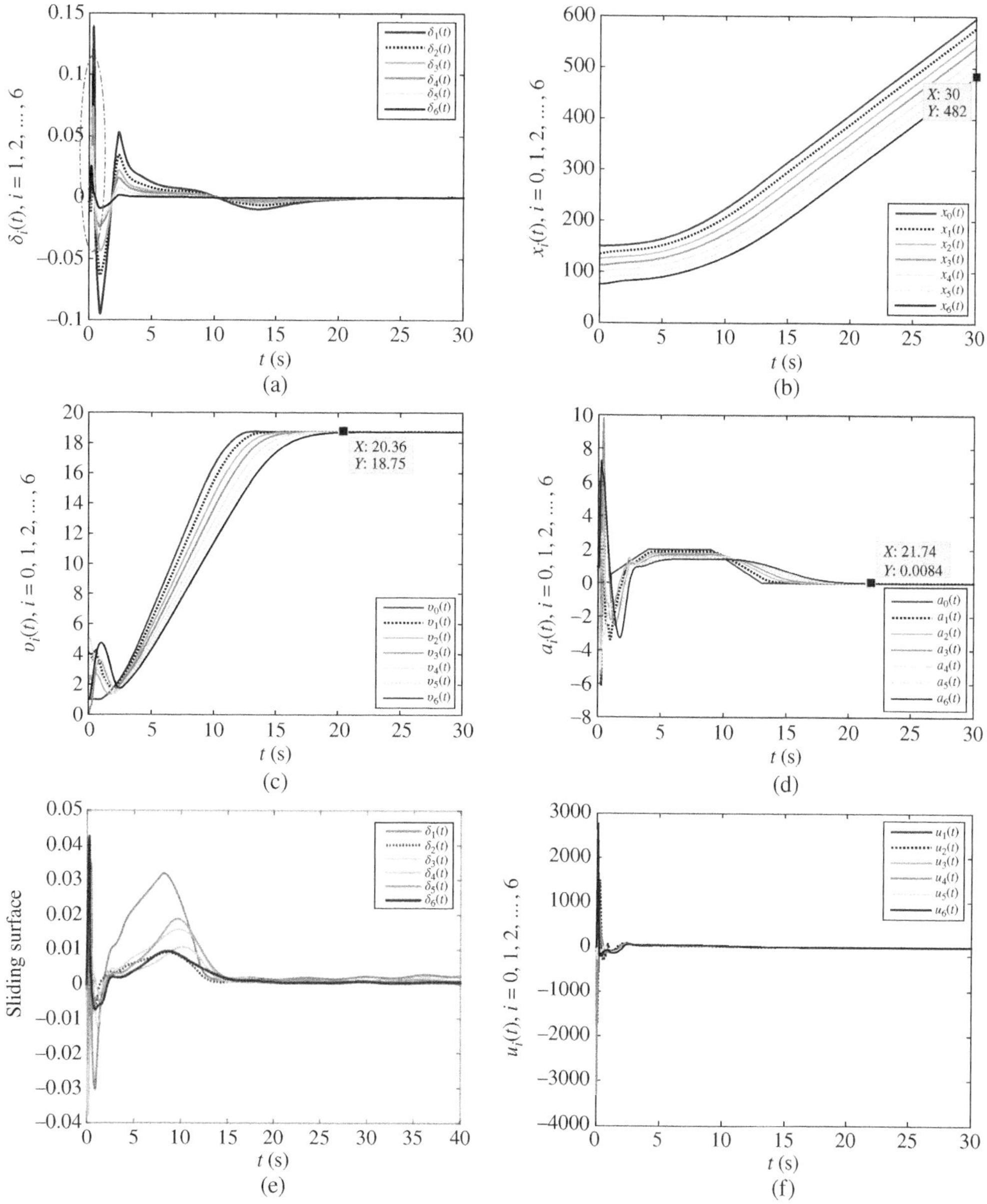

Figure 6.2 Simulation results under Corollary 6.1 without actuator faults and saturation. (a) Spacing error $\delta_i(t)$. (b) Position $x_i(t)$. (c) Velocity $v_i(t)$. (d) Acceleration $A_i(t)$. (e) Sliding surface $S_i(t)$. (f) Control input of the actuator. (g) Estimation $\widehat{\theta}_i(t)$. (h) Estimation $\widehat{\eta}_i(t)$.

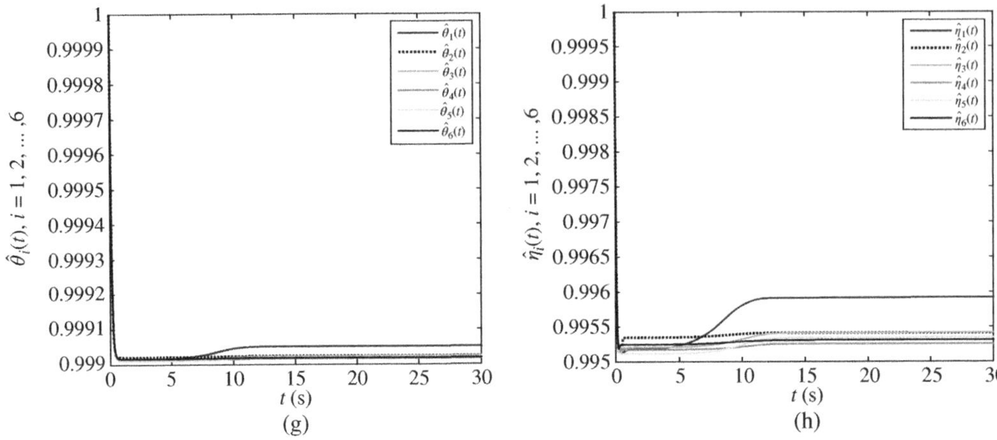

Figure 6.2 (Continued)

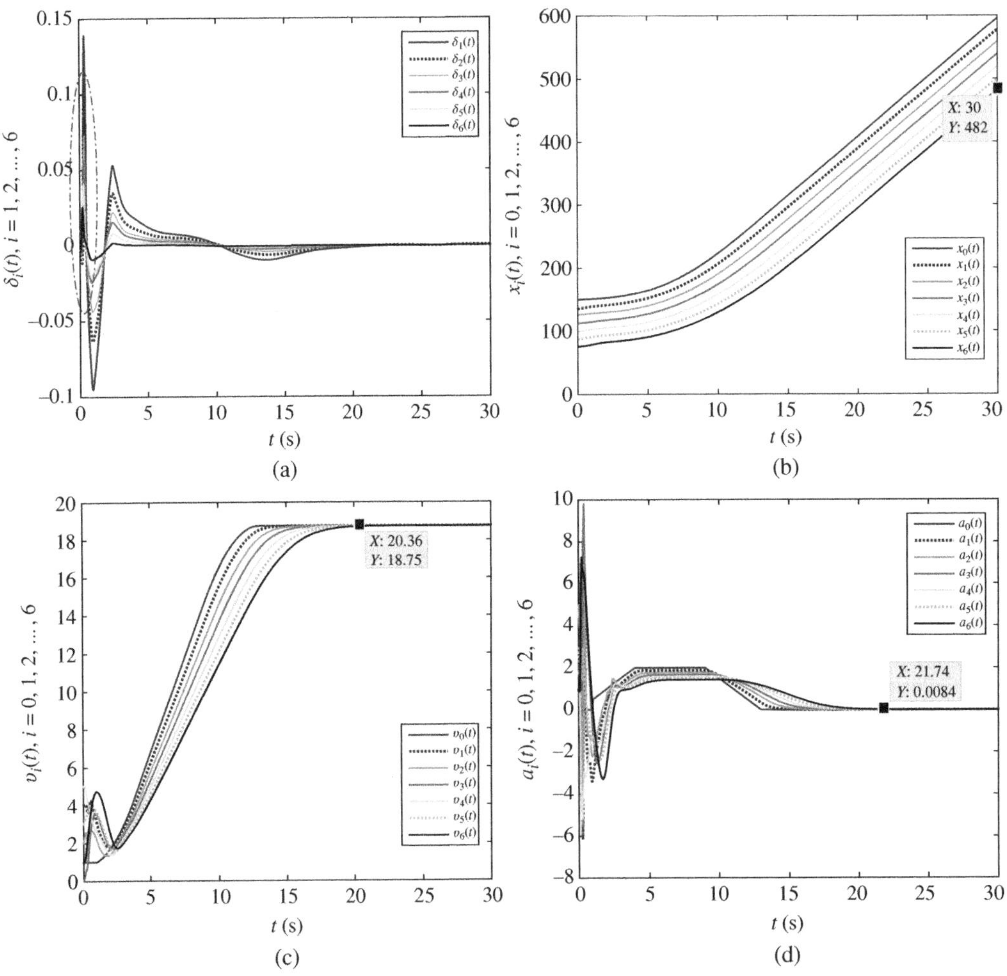

Figure 6.3 Simulation results under Corollary 6.1 with actuator faults and saturation. (a) Spacing error $\delta_i(t)$. (b) Position $x_i(t)$. (c) Velocity $v_i(t)$. (d) Acceleration $A_i(t)$. (e) Sliding surface $S_i(t)$. (f) Control input of the actuator. (g) Estimation $\widehat{\theta}_i(t)$. (h) Estimation $\widehat{\eta}_i(t)$.

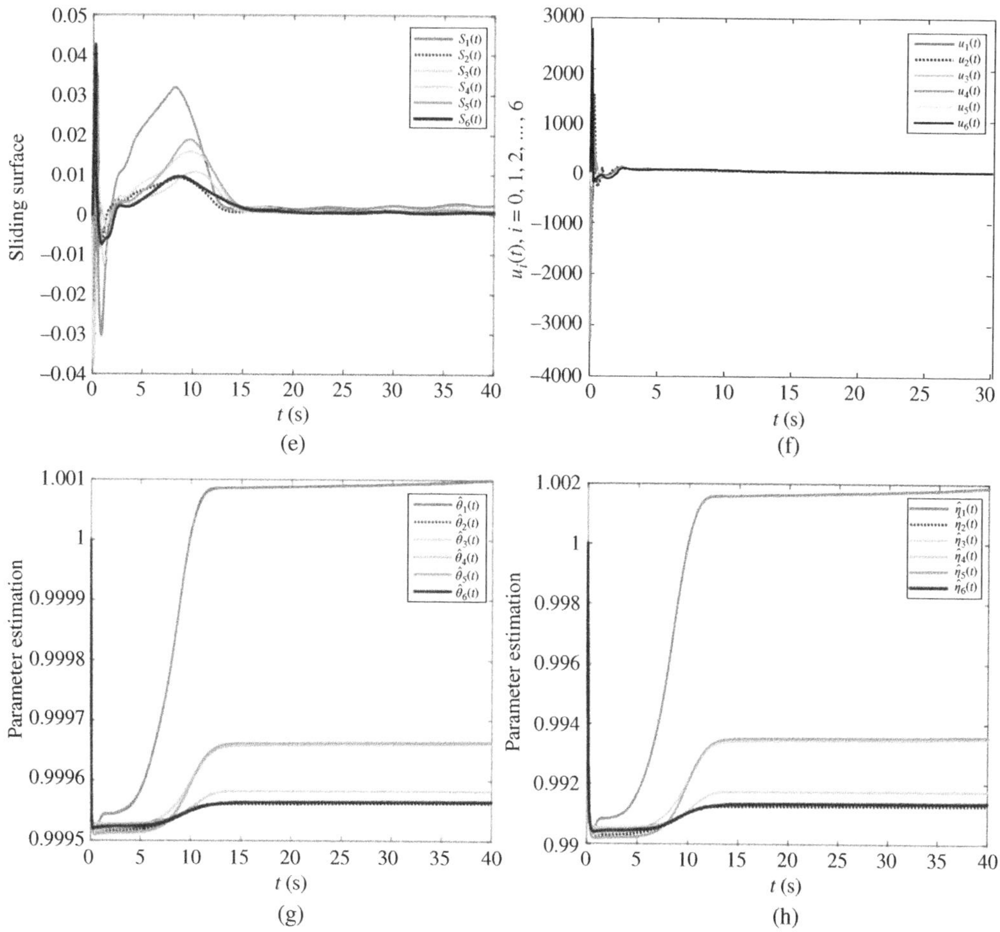

Figure 6.3 (Continued)

To indicate the effect of nonzero initial spacing errors, this subsection shows the simulation results under Corollary 6.2 in Figure 6.5. The sliding surface, input control, and adaptive update laws are similar to those in Figure 6.4. Thus, they are omitted here.

From Figure 6.5, we can conclude that the nonzero initial spacing errors may cause string instability. Therefore, it is important to consider the effect of nonzero initial spacing errors.

6.6 Conclusions and Future Work

An adaptive fault-tolerant control scheme is proposed based on an improved quadratic spacing policy for vehicular platoons with actuator faults and saturation. The presented scheme can guarantee individual vehicle stability, string stability, and traffic flow stability while dealing with the effects of actuator faults and saturation and removing the requirement of zero initial spacing errors at the same time.

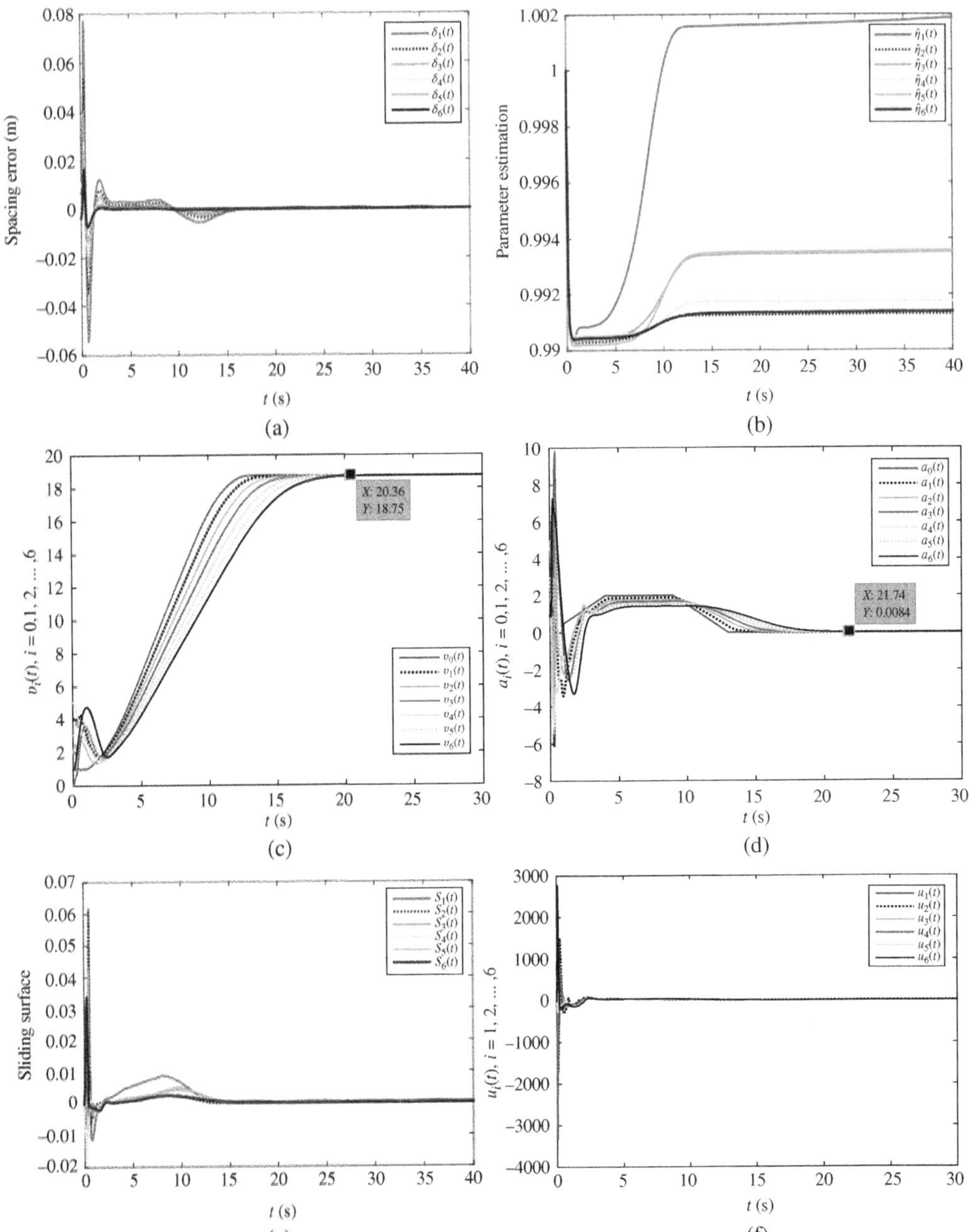

Figure 6.4 Simulation results under Theorem 6.1 with actuator faults and saturation. (a) Spacing error $\delta_i(t)$. (b) Position $x_i(t)$. (c) Velocity $v_i(t)$. (d) Acceleration $A_i(t)$. (e) Sliding surface $S_i(t)$. (f) Control input of the actuator. (g) Estimation $\widehat{\theta}_i(t)$. (h) Estimation $\widehat{\eta}_i(t)$.

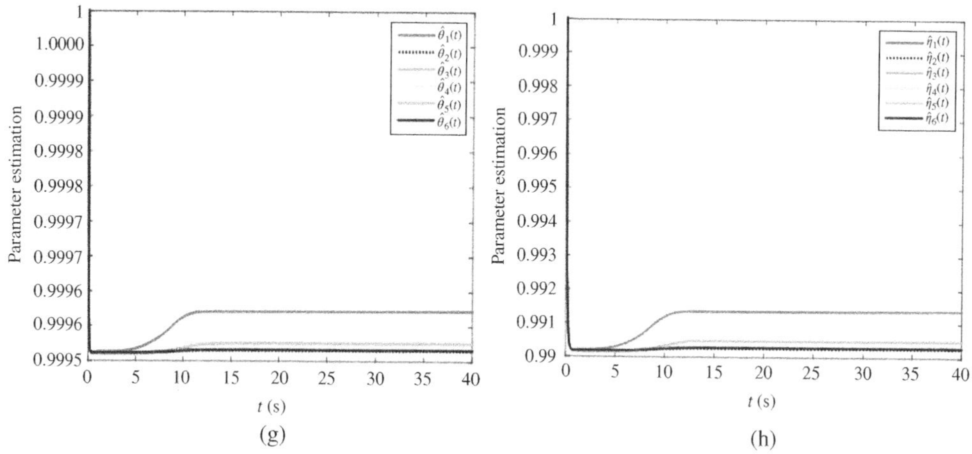

Figure 6.4 (Continued)

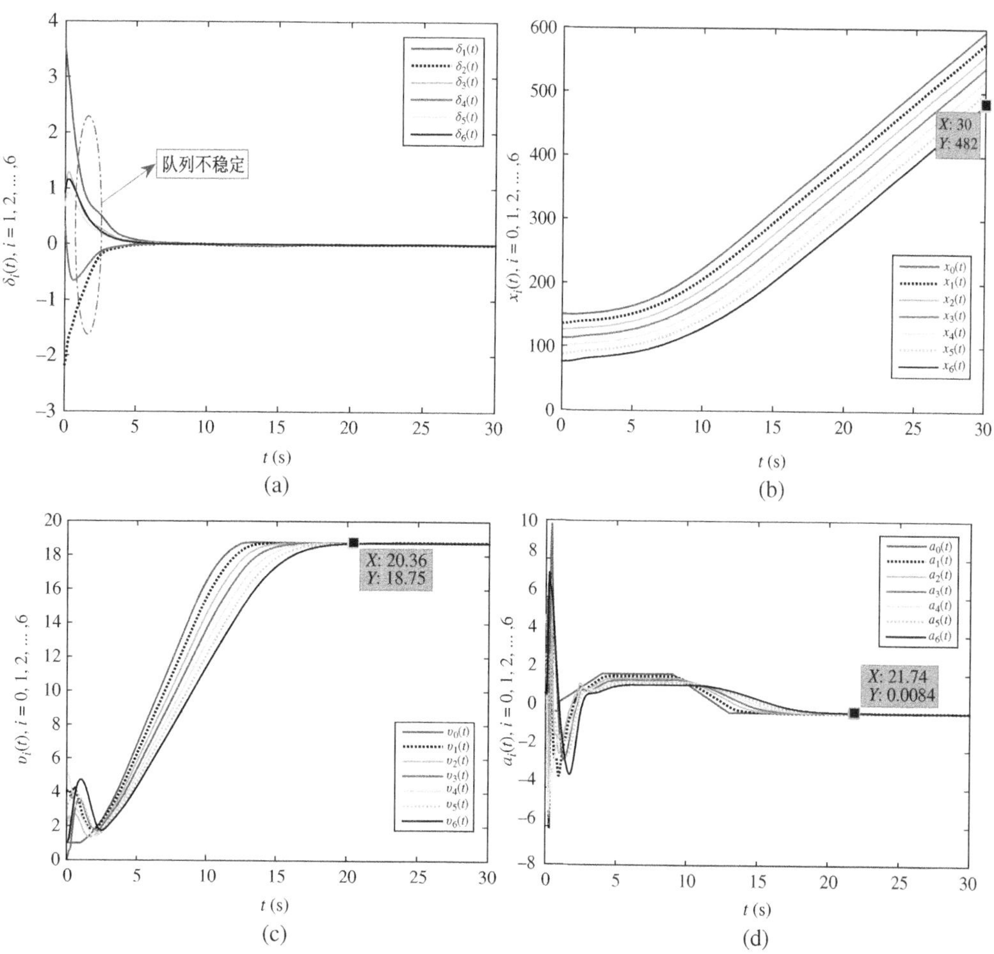

Figure 6.5 Simulation results under Corollary 6.2 with actuator faults and saturation. (a) Spacing error $\delta_i(t)$. (b) Position $x_i(t)$. (c) Velocity $v_i(t)$. (d) Acceleration $A_i(t)$.

References

1 Guo, X.G., Wang, J.L., and Liao, F. (2017). Adaptive fuzzy fault-tolerant control for multiple high-speed trains with proportional and integral-based sliding mode. *IET Control Theory Appl.* 11 (8): 1234–1244.

2 Santhanakrishnan, K. and Rajamani, R. (2003). On spacing policies for highway vehicle automation. *IEEE Trans. Intell. Transp. Syst.* 4 (4): 198–204.

3 Darbha, S. and Rajagopal, K.R. (1999). Intelligent cruise control systems and traffic flow stability. *Transp. Res. C Emerg. Technol.* 7 (6): 329–352.

4 Zhou, J. and Peng, H. (2005). Range policy of adaptive cruise control vehicle for improved flow stability and string stability. *IEEE Trans. Intell. Transp. Syst.* 6 (2): 229–237.

5 Zhao, J., Oya, M., and El Kamel, A. (2009). A safety spacing policy and its impact on highway traffic flow. *Proc. IEEE Intell. Vehicles Symp.* Xi'an, China (June 2009), pp. 960–965.

6 Sungu, H.E., Inoue, M., and Imura, J.-I. (2015). Nonlinear spacing policy based vehicle platoon control for local string stability and global traffic flow stability. *Proc. Eur. Control Conf. (ECC).* Linz, Austria (July 2015), pp. 3396–3401.

7 Warnick, S.C. and Rodriguez, A.A. (2000). A systematic antiwindup strategy and the longitudinal control of a platoon of vehicles with control saturations. *IEEE Trans. Veh. Technol.* 49 (3): 1006–1016.

8 Guo, X., Wang, J., Liao, F. et al. (2018). A low-complexity control for nonlinear vehicular platoon with asymmetric actuator saturation. *Proc. IEEE 14th Int. Conf. Control Automat. (ICCA).* Anchorage, AK (June 2018), pp. 387–392.

9 Guo, X.G., Wang, J.L., Liao, F., and Teo, R.S.H. (2018). CNN-based distributed adaptive control for vehicle-following platoon with input saturation. *IEEE Trans. Intell. Transp. Syst.* 19 (10): 3121–3232.

10 Guo, G. and Li, D. (2019). Adaptive sliding mode control of vehicular platoons with prescribed tracking performance. *IEEE Trans. Veh. Technol.* 68 (8): 7511–7520.

11 Hao, L.Y., Zhang, H., Guo, G., and Li, H. (2019). Quantized sliding mode control of unmanned marine vehicles: various thruster faults tolerated with a unified model. *IEEE Trans. Syst. Man Cybern. Syst.*, 2019, doi:https://doi.org/10.1109/TSMC.2019.2912812.

12 Xiao, B., Hu, Q., and Shi, P. (2013). Attitude stabilization of spacecrafts under actuator saturation and partial loss of control effectiveness. *IEEE Trans. Control Syst. Technol.* 21 (6): 2251–2263.

13 Shen, Q., Wang, D., Zhu, S., and Poh, K. (2015). Finite-time fault-tolerant attitude stabilization for spacecraft with actuator saturation. *IEEE Trans. Aerosp. Electron. Syst.* 51 (3): 2390–2405.

14 Chen, M., Tao, G., and Jiang, B. (2015). Dynamic surface control using neural networks for a class of uncertain nonlinear systems with input saturation. *IEEE Trans. Neural Netw. Learn. Syst.* 26 (9): 2086–2097.

15 Polycarpou, M.M. (1996). Stable adaptive neural control scheme for nonlinear systems. *IEEE Trans. Autom. Control* 41 (3): 447–451.

16 Deng, H. and Krstic, M. (1997). Stochastic nonlinear stabilization – I: a backstepping design. *Syst. Control Lett.* 32 (3): 143–150.

17 Ali, A., Garcia, G., and Martinet, P. (2015). The flatbed platoon towing model for safe and dense platooning on highways. *IEEE Intell. Transp. Syst. Mag.* 7 (1): 58–68.

18 Xiao, L. and Gao, F. (2011). Practical string stability of platoon of adaptive cruise control vehicles. *IEEE Trans. Intell. Transp. Syst.* 12 (4): 1184–1194.

19 Guo, X., Wang, J., Liao, F., and Teo, R.S.H. (2019). Neuroadaptive quantized PID sliding-mode control for heterogeneous vehicular platoon with unknown actuator dead zone. *Int. J. Robust Nonlinear Control* 29 (1): 188–208.

20 Kwon, J.-W. and Chwa, D. (2014). Adaptive bidirectional platoon control using a coupled sliding mode control method. *IEEE Trans. Intell. Transp. Syst.* 15 (5): 2040–2048.

21 Ghasemi, A., Kazemi, R., and Azadi, S. (2013). Stable decentralized control of a platoon of vehicles with heterogeneous information feedback. *IEEE Trans. Veh. Technol.* 62 (9): 4299–4308.

7

Fault-Tolerant Control with Input Quantization and Dead Zone

7.1 Introduction

Different from the discussions in Chapter 6, here we consider the case that the actuators may have imperfect input due to controller quantization [1], or may suffer from dead-zone nonlinearity [2]. Quantized control problem has received particular research attention [3–5]. Dead-zone nonlinearity of throttle/brake is quite common and may cause performance degradation or even instability [6, 7]. Recently, the combined effect of controller quantization and dead-zone nonlinearity is investigated in [8], which proposes a neuroadaptive quantized control method for heterogeneous vehicular platoons. However, the work in [8] is based on constant time-headway spacing, which may result in traffic flow instability and, hence, lower traffic flow or even complete jam [9, 10]. To be specific, the desired separation S is defined as $S = L + \Delta + hv$, with L being the length of the vehicle, Δ being a given minimum distance, and h being the constant time headway. Such a spacing scheme yields the gradient $\partial Q/\partial p = -((L + \Delta)/v) \leq 0$, where Q and p, respectively, are the traffic flow rate and traffic density, indicating that traffic flow is not stable [11, 12]. Furthermore, the work in [8] does not consider the possibility of actuator faults. In practice, however, the controller may lose its efficiency or even completely collapse during operation due to various kinds of reasons, such as overvoltage in the traction transformer, overcurrent in the traction converter, overheating in the asynchronous motor, and so on [13]. The occurrence of actuator faults may lead to system performance deterioration and instability [14, 15] and, hence, degradation of safety and comfort.

The purpose of this chapter is, therefore, to fill the aforementioned gaps in two aspects. Namely, for heterogeneous vehicular platoons, we aim to develop a fault-tolerant control scheme to deal with actuator faults and find out a new spacing policy to fix the traffic flow instability issue raising from the constant time headway spacing policy. To reduce the conservativeness of traditional quadratic spacing policy, an improved quadratic spacing policy with the lower bound of fault factor is proposed, which can also remove the assumption of zero initial spacing errors. By using a radial basis function neural network (RBFNN) to approximate nonlinear vehicle dynamics and an adaptive controller to compensate for the approximation errors and model uncertainties, an adaptive sliding-mode fault-tolerant platoon controller is derived based on Lyapunov stability theory, which is capable of guaranteeing individual vehicle stability and string stability. In addition, traffic flow stability is ensured thanks to the new spacing policy. The main contributions are summarized as follows:

1) An improved quadratic spacing policy is suggested, which takes into account the lower bound of the fault factor.

Connected Vehicular Systems: Communication, Control, and Optimization, First Edition. Ge Guo and Shixi Wen.

2) To tackle the problem of large transient engine and brake torques due to initial condition uncertainties, the new spacing policy removes the assumption of zero initial spacing errors.
3) A minimal learning parameter mechanism is employed, where the number of parameters depends only on the platoon size, and hence avoids the explosion of parameters and reduces the computational burden significantly.

The remainder of this chapter is arranged as follows. Section 7.2 introduces the system modeling with actuator faults, input quantization and dead-zone nonlinearity. Section 7.3 proposes an improved quadratic spacing policy with the lower bound of the fault factor. The controller design and stability analysis are given in Section 7.4. Simulation results are illustrated in Section 7.4, and conclusions are stated in Section 7.5.

Notations Throughout this chapter, $|\cdot|$ denotes the absolute value of real numbers, $\|\cdot\|$ denotes the Euclidean norm of a vector, and R^n denotes the n-dimensional Euclidean space. In this chapter, $\Omega_N = \{1, 2, ...N\}$ represents the integer set from 1 to N, and $\Omega_N \backslash \{N\}$ represents the relative complement of $\{N\}$ in Ω_N.

7.2 System Modeling and Problem Formulation

Consider a vehicular platoon, as shown in Figure 7.1, containing a lead vehicle and N following vehicles. The leader's dynamics is described as

$$\dot{x}_0 = v_0, \dot{v} = a_0 \tag{7.1}$$

where x_0, v_0, and a_0 are the position, velocity, and acceleration of the leader, respectively, with a_0 being a given function.

The dynamics of vehicle i is described by the following nonlinear third-order model [16]

$$\begin{aligned} \dot{x}_i &= v_i \\ \dot{v}_i &= a_i \\ \dot{a}_i &= f_i(v_i, a_i) + \frac{p_i}{\tau_i m_i} + \omega_i \end{aligned} \tag{7.2}$$

where p_i is the engine/brake input, ω_i is the unknown external disturbances caused by a wind gust and/or rough road condition, $f_i(v_i, a_i)$ is given by

$$f_i(v_i, a_i) = -\frac{1}{\tau_i}\left(a_i + \frac{\nu A_i C_{di}}{2m_i} v_i^2 + \frac{d_{mi}}{m_i}\right) - \frac{\nu A_i C_{di} v_i a_i}{m_i}$$

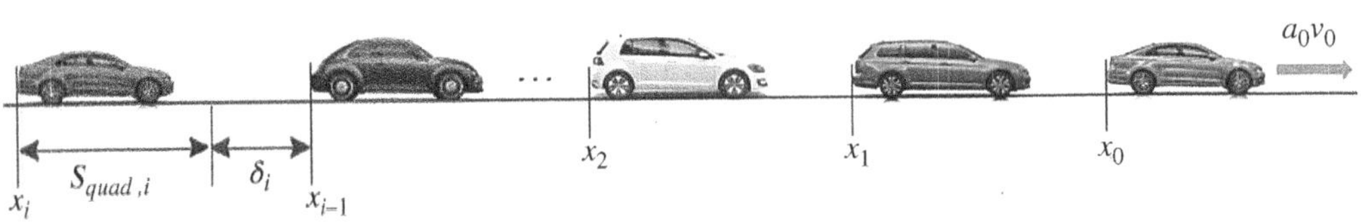

Figure 7.1 Heterogeneous intelligent platoon configuration.

where τ_i is the engine time constant; v is the specific mass of the air; m_i, A_i, and C_{di} denote the mass, cross-sectional area, and drag coefficient of vehicle i, respectively; d_{mi} is the mechanical drag; and $(vA_iC_{di})/(2m_i)$ denotes the air resistance

Let

$$u_i = \frac{p_i}{m_i}$$

Then, the longitudinal dynamics of vehicle i becomes

$$\begin{aligned} \dot{x}_i &= v_i \\ \dot{v}_i &= a_i \\ \dot{a}_i &= f_i(v_i, a_i) + \frac{u_i}{\tau_i} + \omega_i \end{aligned} \tag{7.3}$$

where u_i is used here instead of p_i so that the input does not depend on the vehicle's mass. The mechanical drag d_{mi} and the air resistance $(vA_iC_{di})/(2m_i)$ cannot be obtained accurately in practice, and hence, the nonlinear function $f_i(v_i, a_i)$ is unknown. In this chapter, it will be approximated by RBFNN method [6] as

$$f_i(v_i, a_i) = w_i^{*T}\xi_i(Z_i) + \varsigma_i \tag{7.4}$$

where Z_i is the input vector, $\boldsymbol{W}_i^* = \left[\omega_1^*, \omega_2^*, \ldots, \omega_M^*\right]^T$ denotes the ideal weight vector with M being the number of neurons, ς_i is the reconstruction error satisfying $|\varsigma_i| \leq \varsigma_i^*$ with $\varsigma_i^* > 0$, and $\xi_i(Z_i) = \left[\xi_i^1(Z_i), \xi_i^2(Z_i), \ldots, \xi_i^M(Z_i)\right]^T$ denotes the Gaussian basis function vector given by

$$\xi_i^k(Z_i) = \exp\left[-\frac{\left(Z_i - c_i^k\right)^T\left(Z_i - c_i^k\right)}{2\left(d_i^k\right)^2}\right], \quad k = 1, 2, 3, \ldots, M$$

with d_i^k being the width of the Gaussian function and c_i^k the center of the receptive field.

In practical systems, partial loss of effectiveness and bias faults are two common actuator faults [13]. Partial loss of effectiveness refers to faults due to wear-outs and bias faults represent those caused by aging or external stress damage. As in [13], we adopt the following fault model:

$$u_i^f = \rho_i(t)u_i^d + r_i(t) \tag{7.5}$$

where $\rho_i(t)$ indicates the actuation effectiveness and $r_i(t)$ denotes the bias fault.

For input quantization and dead-zone nonlinearity, we use the same modeling of [8]. To be specific, the unknown dead-zone input is expressed as

$$\begin{aligned} u_i^d &= D_i(u_i^q) \\ &= \begin{cases} n_i^r\left(u_i^q - b_i^r\right), & \text{if} \quad u_i^q \geq b_i^r \\ 0, & \text{if} \quad -b_i^l < u_i^q < b_i^r \\ n_i^l\left(u_i^q + b_i^l\right), & \text{if} \quad u_i^q \leq -b_i^l \end{cases} \end{aligned} \tag{7.6}$$

where n_i^r and n_i^l stand for the right and the left slopes of the dead-zone characteristic, respectively, b_i^r and b_i^l represent the breakpoints of the input nonlinearity, u_i^q represents the quantized input, which is described as

$$\begin{aligned} u_i^q &= Q_i(u_i) \\ Q_i(u_i) &\triangleq \phi_i^d q\left(u_i/\phi_i^c\right) \end{aligned} \tag{7.7}$$

where the function $q(\cdot) \triangleq round(\cdot)$ rounds toward the nearest integer, $q(u_i/\phi_i^c)$ denotes the finite precision implementation of the control input, and $\phi_i^c > 0$ and $\phi_i^d > 0$ are quantization adjustable parameters at the coder and decoder sides, respectively. Normally, $\phi_i^c = \phi_i^d$, but here we consider mismatched coder-decoder, which is due to hardware imperfections [5]. Let $\phi_i = \phi_i^c/\phi_i^d$ be the mismatch ratio, which may be time-varying, then the quantization operator $Q_i(\cdot)$ becomes

$$\begin{aligned} Q_i(u_i) &= \phi_i q_{\phi_i^c}(u_i) \\ q_{\phi_i^c}(u_i) &= \phi_i^c q_{\phi_i^c}(u_i/\phi_i^c) \end{aligned} \tag{7.8}$$

Assumption 7.1 ω_i is an unknown bounded disturbance with $|\omega_i| \le \omega_i^*$, where ω_i^* is a positive constant.

Assumption 7.2 The actuation effectiveness $\rho_i(t)$ and bias fault $r_i(t)$ are unknown bounded time-varying functions, and there exist unknown positive constants ρ_{i0} and r_i^* satisfying $0 < \rho_{i0} \le \rho_i(t) \le 1$ and $|r_i(t)| \le r_i^* < \infty$.

Assumption 7.3 The dead-zone parameters n_i^r, n_i^l, b_i^r and b_i^l are positive but unknown. There exists a sufficiently small parameter $\varsigma_i > 0$ such that $0 < \varsigma_i \le n_i^r$ and $0 < \varsigma_i \le n_i^l$ for all $i \in \Omega_N$.

Assumption 7.4 The mismatch ratio ϕ_i in (7.8) is bounded by $\phi_i^m \le \phi_i \le \phi_i^M$ for all $t > 0$, where ϕ_i^m and ϕ_i^M are unknown positive constants.

To overcome the difficulty arising from the quantizer $Q_i(u_i)$ in (7.7), we decompose the quantizer $q_{\phi_i^c}(u_i)$ in (7.8) into a linear part and a nonlinear part as follows:

$$q_{\phi_i^c}(u_i) = u_i + \Delta u_i \tag{7.9}$$

where the quantization error is given by

$$\Delta u_i \triangleq q_{\phi_i^c}(u_i) - u_i \tag{7.10}$$

Then, based on the quantizer characteristics (7.8), each component of Δu_i at time t is bounded by half of the quantization level ϕ_i^c, i.e.

$$\|\Delta u_i\|_\infty \le 0.5 \max{}_{t \ge 0} \phi_i^c \tag{7.11}$$

In addition, as in [8], the dead-zone input can be transformed into the following form:

$$D_i(u_i^q) = n_i u_i^q + \Delta n_i \tag{7.12}$$

where

$$n_i = \begin{cases} n_i^l, & \text{if } u_i^q \le 0 \\ n_i^r, & \text{if } u_i^q > 0 \end{cases}$$

and

$$\Delta n_i = \begin{cases} -n_i^r b_i^r, & \text{if } u_i^q \ge b_i^r \\ -n_i u_i^q, & \text{if } -b_i^l < u_i^q < b_i^r \\ n_i^l b_i^l, & \text{if } u_i^q \le -b_i^l, \end{cases}$$

with Δn_i being bounded as $|\Delta n_i| \le \Delta n_i^*$ and $\Delta n_i^* > 0$ an unknown constant.

Then, by substituting (7.8), (7.9), and (7.12) into (7.5), the following fault model u_i^f is finally decomposed into the following form:

$$\begin{aligned} u_i^f &= \rho_i(t)n_i\phi_i u_i + \Delta N_i \\ \Delta N_i &= \rho_i(t)n_i\phi_i\Delta u_i + \rho_i(t)\Delta n_i + r_i(t) \end{aligned} \tag{7.13}$$

where $0 < \zeta_i\phi_i^m \leq n_i\phi_i \leq \phi_i^M \max\{n_i^l, n_i^r\}$, $|\Delta N_i| \leq \frac{1}{2}\phi_i^M \max\{n_i^l, n_i^r\}$, $\max_{t\geq 0}\phi_i^c + \Delta n_i^* + r_i^* \triangleq \Delta N_i^*$. It should be pointed out that the value of the control gain $n_i\phi_i$ is unknown since ζ_i, ϕ_i^m, ϕ_i^M, and $\max\{n_i^l, n_i^r\}$ are unknown. Taking this into consideration, an online adaptive updating law ψ_i is used to estimate the reciprocal $\psi_i^* = 1/\zeta_i\phi_i^m$ of the lower bound of the control gain in (7.13).

Remark 7.1 Note that the work in [8] ignores the issue of actuator faults, which may cause system performance deterioration or even instability. We take actuator faults into consideration in vehicular platoon systems. In addition, the spacing policy in their work is simply a constant time headway scheme, which is adverse for traffic flow stability. Here we propose a quadratic spacing policy to address this issue.

Definition 7.1 (Weak string stability [17]) The spacing errors of following vehicles in a platoon are uniformly bounded if, for some $\varrho > 0$, $\exists\vartheta > 0$ such that

$$\begin{aligned} &\sup_i \max\{|\delta_i(0), \dot{\delta}_i(0)|\} < \varrho \\ &\Rightarrow \sup_i \sup_{t\geq 0} \max\{|\delta_i(t), \dot{\delta}_i(t)|\} < \vartheta \end{aligned} \tag{7.14}$$

Definition 7.2 (String stability [18, 19]) The platoon system is string-stable if

$$|\delta_N(t)| \leq |\delta_{N-1}(t)| \leq \cdots \leq |\delta_1(t)| \tag{7.15}$$

i.e. the error transfer function $G_i(s) := \delta_{i+1}(s)/\delta_i(s)$ satisfies $|G_i(s)| \leq 1$ for all $i = \Omega_N$, where $\delta_i(s)$ denotes the Laplace transform of $\delta_i(t)$.

The objective of this chapter is to develop a platoon control scheme, in situations with actuator faults, input quantization, and dead-zone nonlinearity, to meet the following criteria:

1) *Individual vehicle stability [18]*: All the follower's velocities and accelerations track the trajectory of the leader with bounded spacing and velocity errors.
2) Weak string stability is guaranteed for transient state and strict string stability is guaranteed for steady state.
3) *Traffic flow stability [9]*: It requires that the gradient $\partial Q/\partial p$ is positive, that is

$$\partial Q/\partial p > 0 \tag{7.16}$$

where Q and p denote the traffic flow rate and traffic density, respectively.

Before proceeding further, some useful lemmas are introduced and will be used later.

Lemma 7.1 ([20]) For $\forall \epsilon > 0$ and $\chi \in R$, the following inequality holds:

$$0 \leq |\chi| - \chi \tanh\left(\frac{\chi}{\epsilon}\right) \leq \kappa\epsilon$$

where κ is a constant satisfying $\kappa = e^{-(k+1)}$, that is $\kappa = 0.2785$.

Lemma 7.2 (Young's inequality [21] $\forall(a, b) \in R^2$, the following inequality holds) For

$$ab \leq \frac{\varphi^p}{p}|a|^p + \frac{1}{q\varphi^q}|b|^q$$

where $\varphi > 0$, $p > 1$, $q > 1$ and $(p-1)(q-1) = 1$.

7.3 Improved Quadratic Spacing Policy and Adaptive PID-Type Sliding Surface

This section gives an improved quadratic spacing policy and an adaptive PID-type sliding surface which will be used for controller design due to its excellent performance in eliminating steady-state spacing errors.

To reduce the conservativeness of traditional quadratic spacing policy [12], an improved quadratic spacing policy is introduced, which is given here based on the vehicle speed, braking capacity, and the lower bound of fault factor:

$$\begin{aligned} \delta_i &= \tilde{\delta}_i - \Gamma_i \\ \tilde{\delta}_i &= x_{i-1} - x_i - L_i - \Delta_{i-1,i} - hv_i - \frac{\sigma v_i^2}{2A_m} \\ \Gamma_i &= \left\{\tilde{\delta}_i(0) + \left[\pi_i\tilde{\delta}_i(0)\dot{\tilde{\delta}}(0)\right]t + \frac{1}{2}\left[\pi_i^2\tilde{\delta}_i(0) + 2\pi_i\dot{\tilde{\delta}}(0) + \ddot{\tilde{\delta}}(0)\right]t^2\right\}e^{-\pi_i t} \end{aligned} \tag{7.17}$$

where π_i is a positive constant, L_i denotes the length of vehicle i and $\Delta_{i-1,\,i}$ is a given minimum distance, h is the delay in the platoon system, which compensates for the delay in braking or acceleration (about 10 ms to 80 ms, see [12]), σ represents safety coefficient depending on the road or weather and A_m is the absolute value of the maximum possible deceleration and ρ_{i0} is the lower bound of fault factor. Further, we have

$$\delta_i(0) = 0, \dot{\delta}_i(0) = 0, \ddot{\delta}_i(0) = 0 \tag{7.18}$$

which indicates that the initial values of the improved spacing errors are always zero, removing the assumption of zero initial spacing errors in the existing results.

The desired intervehicle distance including vehicle length can then be defined as

$$s_{quad,i} = L_i + \Delta_{i-1,i} + hv_i + \frac{\sigma v_i^2}{2A_m\rho_{i0}} + \Gamma_i \tag{7.19}$$

Remark 7.2 A nonlinear quadratic spacing policy [12] was proposed to guarantee string stability and traffic flow stability, which can also ensure a higher traffic capacity compared with the commonly used constant time-headway spacing policy. However, when actuator faults occur, the maximum acceleration changes, which may invalidate the quadratic spacing policy [12]. In addition, the quadratic spacing policy [12] does not consider the problem of nonzero initial errors, which may cause string instability.

Remark 7.3 To reduce the conservativeness of the traditional quadratic spacing policy [12], an improved quadratic spacing policy is suggested, which takes into account the lower bound of the fault factor. In addition, the new spacing policy removes the assumption of zero initial spacing errors.

We consider an adaptive PID-type sliding surface to eliminate steady-state spacing errors under external disturbances and uncertainties:

$$s_i = K_p\delta_i + K_i\int_0^t \delta_i(\tau)d\tau + K_d\dot{\delta}_i \tag{7.20}$$

where positive scalars K_p, K_i, and K_d are proportional, integral, and derivative coefficients, respectively. As in [19], to deal with the relation of δ_i and δ_{i+1} better, we introduce the following coupled sliding surface:

$$S_i = \begin{cases} \lambda s_i - s_{i+1} & i = 1, 2, \ldots, N-1 \\ \lambda s_i & i = N \end{cases} \tag{7.21}$$

where λ is a positive parameter coupling the sliding surfaces s_i and s_{i+1}. When S_i reaches the sliding surface, s_i does the same thing at the same time.

7.4 Controller Design and Stability Analysis

In this section, by combining the adaptive RBFNN approximation technique with the PID-type sliding mode control method, an adaptive fault-tolerant platoon control scheme is developed for heterogeneous vehicular platoons with actuator faults, input quantization, and dead-zone nonlinearity.

As mentioned earlier, the RBFNN technology is used to approximate the unknown nonlinear function $f_i(v_i, a_i)$ in vehicle dynamics. Let the estimation errors be

$$\begin{aligned} \tilde{\theta}_i &= \theta_i^* - \rho_{i0}\hat{\theta}_i \\ \tilde{\eta}_i &= \eta_i^* - \rho_{i0}\hat{\eta}_i \\ \tilde{\psi}_i &= \psi_i^* - \hat{\psi}_i \end{aligned} \tag{7.22}$$

where ρ_{i0} is given in Assumption 7.2, and

$$\begin{aligned} \psi_i^* &= \frac{1}{\zeta_i\phi_i^m} \\ \eta_i^* &\geq \varsigma_i^* + \omega_i^* + \frac{\Delta N_i^*}{\tau_i} \\ \theta_i^* &= \|W_i^*\|^2 = W_i^{*T}W_i^* \end{aligned} \tag{7.23}$$

Then, we introduce the following fault-tolerant control law

$$
\begin{aligned}
u_i &= \hat{\psi}_i \overline{u}_i \\
\overline{u}_i &= \tau_i \left\{ \frac{k_i S_i}{\lambda K_d (h + ((\sigma v_i)/(A_m \rho_{i0})))} + \frac{b_i^2}{2} \hat{\theta}_i \xi_i^T(Z_i) \xi_i(Z_i) S_i \right. \\
&\quad \left. + \hat{\eta}_i \tanh\left(\frac{S_i}{\mu_i}\right) + \frac{\Lambda_i}{\lambda K_d (h + ((\sigma v_i)/(A_m \rho_{i0})))^{|R_i S_i| + \vartheta_i}} \right\}
\end{aligned} \tag{7.24}
$$

where $1 \leq 1/\rho_{i0} \leq \Lambda_i$, ϑ_i, μ_i, and b_i are positive constants, k_i is the feedback gain, R_i is given by

$$
\begin{cases}
R_i = \lambda K_p \dot{\delta}_i + \lambda K_i \delta_i + \lambda K_d \left(a_{i-1} - a_i - \dfrac{\sigma a_i^2}{A_m \rho_{i0}} - \ddot{\Gamma}_i \right) \\
\qquad - K_p \dot{\delta}_{i+1} - K_i \delta_{i+1} - K_d \ddot{\delta}_{i+1}, \quad \text{for } i \in \Omega_N \backslash \{N\} \\
R_i = \lambda K_p \dot{\delta}_i + \lambda K_i \delta_i + \lambda K_d \left(a_{i-1} - a_i - \dfrac{\sigma a_i^2}{A_m \rho_{i0}} - \ddot{\Gamma}_i \right), \quad \text{for } i = N
\end{cases} \tag{7.25}
$$

and the following adaptation laws

$$
\begin{aligned}
\dot{\hat{\theta}}_i &= \alpha_i \lambda K_d \left(h + \frac{\sigma v_i}{A_m \rho_{i0}} \right) \frac{b_i^2}{2} \xi_i^T(Z_i) \xi_i(Z_i) S_i^2 - \Xi_{1i} \hat{\theta}_i \\
\dot{\hat{\eta}}_i &= \beta_i \lambda K_d \left(h + \frac{\sigma v_i}{A_m \rho_{i0}} \right) S_i \tanh\left(\frac{S_i}{\mu_i}\right) - \Xi_{2i} \hat{\eta}_i \\
\dot{\hat{\psi}}_i &= \gamma_i \lambda K_d \left(h + \frac{\sigma v_i}{A_m \rho_{i0}} \right) \frac{1}{\tau_i} \rho_{i0} \overline{u}_i S_i - \Xi_{3i} \hat{\psi}_i
\end{aligned} \tag{7.26}
$$

Here $\hat{\theta}_i(0) \geq 0$, $\hat{\eta}(0) \geq 0$, and $\hat{\Psi}_i(0) \geq 0$, λ, k_i, b_i, K_p, K_i, K_d, α_i, β_i, and γ_i are any positive constants, while ϑ_i and μ_i are small constants Ξ_{1i}, Ξ_{2i}, and Ξ_{3i} are any continuous and bounded functions satisfying

$$
\Xi_{ji}(t) > 0, \ \lim_{t \to \infty} \int_{t_0}^{t} \Xi_{ji}(\tau) d\tau \leq \overline{\Xi}_{ji} < +\infty, \quad j = 1, 2, 3 \tag{7.27}
$$

with $\overline{\Xi}_{ji}$ being any finite positive constants.

Remark 7.4 One shortcoming of the RBFNN's control method is that the computational burden might be excessively large since the number of parameters to be estimated will increase dramatically with the increase of NN nodes [6]. Inspired by [8], by adopting a minimal learning parameter mechanism, the number of parameters depends only on the number of vehicles, which can alleviate the explosion of learning parameters.

We now give the main result in the following theorem.

Theorem 7.1 For a vehicular platoon (7.3) with actuator faults (7.5), input quantization (7.8), and unknown dead zone (7.6), under the improved quadratic spacing policy (7.17), the adaptive fault-tolerant control law (7.24) and the adaptation laws (7.26) can drive the spacing errors into a small set around the origin $\Omega = \{\delta_i(t) : |\delta_i(t)| < \vartheta\}$, i.e. weak string stability (7.14) is guaranteed. Moreover, strict string stability (7.15) for the steady state can be guaranteed when $0 < \lambda \leq 1$ and traffic flow stability (7.16) is also guaranteed thanks to the new quadratic spacing policy.

Proof: The proof consists of the following three parts.

(a) *Boundedness of* S_i: Choose the following Lyapunov function candidate:

$$V_i = \frac{1}{2}S_i^2 + \frac{1}{2\rho_{i0}\alpha_i}\tilde{\theta}_i^2 + \frac{1}{2\rho_{i0}\beta_i}\tilde{\eta}_i^2 + \frac{1}{2\gamma_i}\zeta_i\phi_i^m\tilde{\psi}_i^2 \tag{7.28}$$

where $\tilde{\theta}_i$, $\tilde{\eta}_i$, and $\tilde{\psi}_i$ are the estimation errors defined in (7.22).

Taking the derivative of V_i yields

$$\dot{V}_i = S_i\dot{S}_i - \frac{1}{\alpha_i}\tilde{\theta}_i\dot{\hat{\theta}}_i - \frac{1}{\beta_i}\tilde{\eta}_i\dot{\hat{\eta}}_i - \frac{1}{\gamma_i}\zeta_i\phi_i^m\tilde{\psi}_i\dot{\hat{\psi}}_i \tag{7.29}$$

Then, by utilizing (7.17), (7.20), and (7.21), the derivative of S_i can be obtained

$$\begin{aligned}
\dot{S}_i &= \lambda\left(K_p\dot{\delta}_i + K_i\delta_i + K_d\ddot{\delta}_i\right) - K_p\dot{\delta}_{i+1} - K_i\delta_{i+1} + K_d\ddot{\delta}_{i+1}\\
&= \lambda\left\{K_p\dot{\delta}_i + K_i\delta_i + K_d\left(a_{i-1} - a_i - ha_i - \frac{\sigma a_i^2}{A_m\rho_{i0}} - \frac{\sigma v_i\dot{a}_i}{A_m\rho_{i0}} - \ddot{\Gamma}_i\right)\right\}\\
&\quad - K_p\dot{\delta}_{i+1} - K_i\delta_{i+1} - K_d\ddot{\delta}_{i+1}\\
&= -\lambda K_d\left(h + \frac{\sigma v_i}{A_m\rho_{i0}}\right)\dot{a}_i + R_i, \quad \text{for } i \in \Omega_N\backslash\{N\}\\
\dot{S}_i &= \lambda\left\{K_p\dot{\delta}_i + K_i\delta_i + K_d\left(a_{i-1} - a_i - h\dot{a}_i - \frac{\sigma a_i^2}{A_m\rho_{i0}} - \frac{\sigma v_i\dot{a}_i}{A_m\rho_{i0}} - \ddot{\Gamma}_i\right)\right\}\\
&= -\lambda K_d\left(h + \frac{\sigma v_i}{A_m\rho_{i0}}\right)\dot{a}_i + R_i, \quad \text{for } i = N
\end{aligned} \tag{7.30}$$

where R_i is defined in (7.25). Then, substituting (7.3), (7.4), (7.13) and (7.30) into $S_i\dot{S}_i$ yields

$$\begin{aligned}
S_i\dot{S}_i &= -\lambda K_d\left(h + \frac{\sigma v_i}{A_m\rho_{i0}}\right)\left(\frac{1}{\tau_i}\rho_i(t)n_i\phi_i u_i + f_i(v_i, a_i) + \frac{\Delta N_i}{\tau_i} + \omega_i\right)S_i + R_iS_i\\
&= -\lambda K_d\left(h + \frac{\sigma v_i}{A_m\rho_{i0}}\right)\frac{1}{\tau_i}\rho_i(t)n_i\phi_i u_iS_i - \lambda K_d\left(h + \frac{\sigma v_i}{A_m\rho_{i0}}\right)W_i^{*T}\xi_i(Z_i)S_i\\
&\quad - \lambda K_d\left(h + \frac{\sigma v_i}{A_m\rho_{i0}}\right)\left(\varsigma_i + \omega_i + \frac{\Delta N_i}{\tau_i}\right)S_i + R_iS_i\\
&\leq -\lambda K_d\left(h + \frac{\sigma v_i}{A_m\rho_{i0}}\right)\frac{1}{\tau_i}\rho_i(t)\zeta_i\phi_i^m u_iS_i - \lambda K_d\left(h + \frac{\sigma v_i}{A_m\rho_{i0}}\right)W_i^{*T}\xi_i(Z_i)S_i\\
&\quad + \lambda K_d\left(h + \frac{\sigma v_i}{A_m\rho_{i0}}\right)\eta_i^*|S_i| + |R_iS_i|
\end{aligned} \tag{7.31}$$

where $\varsigma_i^* + \omega_i^* + \frac{\Delta N_i^*}{\tau_i} \leq \eta_i^*$ and $0 < \zeta_i\phi_i^m \leq n_i\phi_i$ are used. Then, according to Lemma 7.2, there exists a design parameter $b_i > 0$ such that

$$
\begin{aligned}
& -\lambda K_d\left(h+\frac{\sigma v_i}{A_m\rho_{i0}}\right)W_i^{*T}\xi_i(Z_i)S_i \\
& \leq \lambda K_d\left(h+\frac{\sigma v_i}{A_m\rho_{i0}}\right)\frac{b_i^2}{2}\theta_i^*\xi_i^T(Z_i)\xi_i(Z_i)S_i^2+\frac{1}{2b_i^2}\lambda K_d\left(h+\frac{\sigma v_i}{A_m\rho_{i0}}\right)
\end{aligned} \tag{7.32}
$$

with $\theta_i^* = \|W_i^*\|^2 = W_i^{*T}W_i^*$. Here a minimal learning parameter mechanism is employed to just estimate constants θ_i^* but not ideal weights W_i^* directly. Thus, the number of parameters reduces drastically from $\sum_{i=1}^N N_i$ to N, which avoids explosion-of-parameters and reduces the computational burden significantly. Then, substituting (7.32) into (7.31), one has

$$
\begin{aligned}
S_i\dot{S}_i \leq & -\lambda K_d\left(h+\frac{\sigma v_i}{A_m\rho_{i0}}\right)\frac{1}{\tau_i}\rho_i(t)\zeta_i\phi_i^m u_iS_i+\lambda K_d\left(h+\frac{\sigma v_i}{A_m\rho_{i0}}\right)\frac{b_i^2}{2}\theta_i^*\xi_i^T(Z_i)\xi_i(Z_i)S_i^2 \\
& +\lambda K_d\left(h+\frac{\sigma v_i}{A_m\rho_{i0}}\right)\eta_i^*|S_i|+|R_iS_i|+\frac{1}{2b_i^2}\lambda K_d\left(h+\frac{\sigma v_i}{A_m\rho_{i0}}\right) \\
\leq & -\lambda K_d\left(h+\frac{\sigma v_i}{A_m\rho_{i0}}\right)\frac{1}{\tau_i}\rho_{i0}\zeta_i\phi_i^m u_iS_i+\lambda K_d\left(h+\frac{\sigma v_i}{A_m\rho_{i0}}\right)\frac{b_i^2}{2}\theta_i^*\xi_i^T(Z_i)\xi_i(Z_i)S_i^2 \\
& +\lambda K_d\left(h+\frac{\sigma v_i}{A_m\rho_{i0}}\right)\eta_i^*|S_i(t)|+\rho_{i0}\Lambda_i|R_iS_i|+\frac{1}{2b_i^2}\lambda K_d\left(h+\frac{\sigma v_i}{A_m\rho_{i0}}\right)
\end{aligned} \tag{7.33}
$$

where $0<\rho_{i0}\leq\rho_i(t)\leq 1$ and $1\leq 1/\rho_{i0}\leq\Lambda_i$ are used.

It follows from (7.26) that $\dot{\hat{\theta}}+\Xi_{1i}\hat{\theta}_i\geq 0$, which implies that $\hat{\theta}_i\geq e^{-\Xi_{1i}t}\hat{\theta}_i(0)$. Thus, the adaptive parameter $\hat{\theta}_i$ will be non-negative if we select $\hat{\theta}_i(0)>0$. Similarly, the adaptive parameter $\hat{\eta}_i$ and $\hat{\psi}_i$ are also non-negative. Then, using the control law (7.24), we have

$$
\begin{aligned}
& -\lambda K_d\left(h+\frac{\sigma v_i}{A_m\rho_{i0}}\right)\frac{1}{\tau_i}\rho_{i0}\zeta_i\phi_i^m u_iS_i \\
& = -\lambda K_d\left(h+\frac{\sigma v_i}{A_m\rho_{i0}}\right)\frac{1}{\tau_i}\rho_{i0}\bar{u}_iS_i+\lambda K_d\left(h+\frac{\sigma v_i}{A_m\rho_{i0}}\right)\frac{1}{\tau_i}\rho_{i0}\zeta_i\phi_i^m\tilde{\psi}_i\bar{u}_iS_i \\
& \leq -\rho_{i0}k_iS_i^2-\rho_{i0}\Lambda_i|R_iS_i|+\rho_{i0}\Lambda_i\vartheta_i-\lambda K_d\left(h+\frac{\sigma v_i}{A_m\rho_{i0}}\right)\frac{b_i^2}{2}\rho_{i0}\hat{\theta}_i\xi_i^T(Z_i)\xi_i(Z_i)S_i^2 \\
& -\lambda K_d\left(h+\frac{\sigma v_i}{A_m\rho_{i0}}\right)\rho_{i0}\hat{\eta}_iS_i\tanh\left(\frac{S_i}{\mu_i}\right)+\lambda K_d\left(h+\frac{\sigma v_i}{A_m\rho_{i0}}\right)\frac{1}{\tau_i}\rho_{i0}\zeta_i\phi_i^m\tilde{\psi}_i\bar{u}_iS_i
\end{aligned} \tag{7.34}
$$

where the facts $u_i=\tilde{\psi}_i\bar{u}_i$ and $\left(-\left(R_i^2S_i^2\right)/\left(|R_iS_i|+\vartheta_i\right)\right)\leq -|R_iS_i|+\vartheta_i$ have been used. Then, using (7.26), we can have that

$$
\begin{aligned}
& -\frac{1}{\alpha_i}\tilde{\theta}_i\dot{\hat{\theta}}_i-\frac{1}{\beta_i}\tilde{\eta}_i\dot{\hat{\eta}}_i-\frac{1}{\gamma_i}\zeta_i\phi_i^m\tilde{\psi}_i\dot{\hat{\psi}}_i \\
& = -\lambda K_d\left(h+\frac{\sigma v_i}{A_m\rho_{i0}}\right)\frac{b_i^2}{2}\tilde{\theta}_i\xi_i^T(Z_i)\xi_i(Z_i)S_i^2 \\
& -\lambda K_d\left(h+\frac{\sigma v_i}{A_m\rho_{i0}}\right)\tilde{\eta}_iS_i\tanh\left(\frac{S_i}{\mu_i}\right) \\
& +\frac{\Xi_{1i}}{\alpha_i}\tilde{\theta}_i\hat{\theta}_i+\frac{\Xi_{2i}}{\beta_i}\tilde{\eta}_i\hat{\eta}_i+\frac{\Xi_{3i}}{\gamma_i}\zeta_i\phi_i^m\tilde{\psi}_i\hat{\psi}_i
\end{aligned} \tag{7.35}
$$

Substituting (7.33), (7.34), and (7.35) into (7.29) and using simple calculations yields

$$\begin{aligned}\dot{V}_i(t) \leq & -\rho_{i0}k_iS_i^2 + \frac{\Xi_{li}}{\alpha_i}\tilde{\theta}_i\hat{\theta}_i + \frac{\Xi_{2i}}{\beta_i}\tilde{\eta}_i\hat{\eta}_i + \frac{\Xi_{3i}}{\gamma_i}\zeta_i\phi_i^m\tilde{\psi}_i\hat{\psi}_i \\ & + \lambda K_d\left(h + \frac{\sigma v_i}{A_m\rho_{i0}}\right)\eta_i^*|S_i| - \lambda K_d\left(h + \frac{\sigma v_i}{A_m\rho_{i0}}\right)\eta_i^* S_i \tanh\left(\frac{S_i}{\mu_i}\right) \\ & + \frac{1}{2\mathrm{b}_i^2}\lambda K_d\left(h + \frac{\sigma v_i}{A_m\rho_{i0}}\right) + \rho_{i0}\Lambda_i\vartheta_i\end{aligned} \tag{7.36}$$

Further, recalling (7.22) and applying Lemma 7.2, we have

$$\begin{aligned}\frac{\Xi_{1i}}{\alpha_i}\tilde{\theta}_i\hat{\theta}_i &\leq -\frac{\Xi_{1i}}{2\rho_{i0}\alpha_i}\tilde{\theta}_i^2 + \frac{\Xi_{1i}}{2\rho_{i0}\alpha_i}\theta_i^{*2} \\ \frac{\Xi_{2i}}{\beta_i}\tilde{\eta}_i\hat{\eta}_i &\leq -\frac{\Xi_{2i}}{2\rho_{i0}\beta_i}\tilde{\eta}_i^2 + \frac{\Xi_{2i}}{2\rho_{i0}\beta_i}\eta_i^{*2} \\ \frac{\Xi_{3i}}{\gamma_i}\tilde{\psi}_i\hat{\psi}_i &\leq -\frac{\Xi_{3i}}{2\gamma_i}\tilde{\psi}_i^2 + \frac{\Xi_{3i}}{2\gamma_{ii}}\psi_i^{*2}\end{aligned} \tag{7.37}$$

which, together with Lemma 7.1, (7.36), and (7.37), implies that

$$\begin{aligned}\dot{V}_i \leq & -\rho_{i0}k_iS_i^2 - \frac{\Xi_{1i}}{2\rho_{i0}\alpha_i}\tilde{\theta}_i^2 - \frac{\Xi_{2i}}{2\rho_{i0}\beta_i}\tilde{\eta}_i^2 - \frac{\Xi_{3i}}{2\gamma_i}\zeta_i\phi_i^m\tilde{\psi}_i^2 \\ & + \frac{\Xi_{1i}}{2\rho_{i0}\alpha_i}\theta_i^{*2} + \frac{\Xi_{2i}}{2\rho_{i0}\beta_i}\eta_i^{*2} + \frac{\Xi_{3i}}{2\gamma_i}\zeta_i\phi_i^m\psi_i^{*2} \\ & + 0.2785\mu_i\lambda K_d\left(h + \frac{\sigma v_i}{A_m\rho_{i0}}\right)\eta_i^* \\ & + \frac{1}{2b_i^2}\lambda K_d\left(h + \frac{\sigma v_i}{A_m\rho_{i0}}\right) + \rho_{i0}\Lambda_i\vartheta_i\end{aligned} \tag{7.38}$$

Let

$$V = \sum_{i=1}^{N} V_i$$

Then, according to (7.38), we can obtain that

$$\dot{V} \leq -\zeta_1 V + \zeta_2$$

where

$$\zeta_1 = \min\left\{2\rho_{i0}k_i, \min_{1\leq i\leq N}\Xi_{1i}, \min_{1\leq i\leq N}\Xi_{2i}, \min_{1\leq i\leq N}\Xi_{3i}\right\}$$

$$\begin{aligned}\zeta_2 = \sum_{i=1}^{N}\left\{\frac{\Xi_{1i}}{2\rho_{i0}\alpha_i}\theta_i^{*2} + \frac{\Xi_{2i}}{2\rho_{i0}\beta_i}\eta_i^{*2} + \frac{\Xi_{3i}}{2\gamma_i}\zeta_i\phi_i^m\psi_i^{*2} + 0.2785\mu_i\lambda K_d\left(h + \frac{\sigma v_i}{A_m\rho_{i0}}\right)\eta_i^*\right. \\ \left. + \frac{1}{2b_i^2}\lambda K_d\left(h + \frac{\sigma v_i}{A_m\rho_{i0}}\right) + \rho_{i0}\Lambda_i\vartheta_i\right\}\end{aligned}$$

Then, based on Corollary 1 in [22], we have

$$V \leq \left(V(0) - \frac{\zeta_2}{\zeta_1} \right) e^{i\zeta_i t} + \frac{\zeta_2}{\zeta_1} \leq V(0) + \frac{\zeta_2}{\zeta_1} \tag{7.39}$$

where

$$V(0) = \sum_{i=1}^{N} \left\{ \frac{1}{2} S_i^2(0) + \frac{1}{2\rho_{i0}\alpha_i} \tilde{\theta}_i^2(0) + \frac{1}{2\rho_{i0}\beta_i} \tilde{\eta}_i^2(0) \right\}$$

Then, we can conclude that all the states of the closed-loop system are uniformly ultimately bounded with $V(0) \geq 0$. In particular, we have

$$\lim_{t \to \infty} \sum_{i=1}^{N} \frac{1}{2} S_i^2 \leq V(0) + \frac{\zeta_2}{\zeta_1} \tag{7.40}$$

Hence, based on the definitions (7.20) and (7.21), the spacing errors δ_i are uniformly ultimately bounded.

As mentioned in [8, 13], reducing ζ_2 or increasing ζ_1 will cause a smaller bound of S_i, and the coupled sliding surface S_i can converge to a small set around the origin by choosing appropriate design parameters.

(b) *String stability*: The proof of string stability is divided into two parts: i) weak string stability for transient process, and ii) strict string stability for steady state. Before arriving at the sliding surfaces, according to the previous proof, the spacing errors are bounded, i.e. (7.14) is satisfied and weak string stability is guaranteed.

After arriving at the sliding surfaces, strict string stability of the vehicular platoon will be proved along similar lines as in [19].

Since $S_i = \lambda s_i - s_{i+1}$, when S_i converges to a small set around the origin by choosing appropriate design parameters, we have

$$\begin{aligned} &\lambda \left(K_p \delta_i + K_i \int_0^t \delta_i(\tau) d\tau + K_d \dot{\delta}_i \right) \\ &\approx K_p \delta_{i+1} + K_i \int_0^t \delta_{i+1}(\tau) d\tau + K_d \dot{\delta}_{i+1} \end{aligned} \tag{7.41}$$

the Laplace transform of which yields

$$\begin{aligned} &\lambda \left(K_p \delta_i(s) + s K_i \delta_i(s) + \frac{K_d}{s} \delta_i(s) \right) \\ &\approx K_p \delta_{i+1}(s) + s K_i \delta_{i+1}(s) + \frac{K_d}{s} \delta_{i+1}(s) \end{aligned} \tag{7.42}$$

Then, $G_i(s) = \delta_{i+1}(s)/\delta_i(s) \approx \lambda$. Thus, strict string stability can be guaranteed when $0 < \lambda \leq 1$.

(c) *Traffic flow stability*: It is reasonable to assume that, at a steady state, the separation and velocity of all vehicles are the same, i.e. $S_{quad,\, i} = S_{quad}$ and $v_i = v$, for all i. Thus, by definition and the fact that $\Gamma_i = 0$ at a steady state, one can have

$$S_{quad} = L + \Delta + hv(t) + \frac{\sigma v^2}{2A_{\mathrm{m}}\rho_{i0}} \tag{7.43}$$

Then, the steady traffic density is

$$p = \frac{1}{L + \Delta + hv + ((\sigma v^2)/(2A_{\mathrm{m}}\rho_{\mathrm{i}0}))}$$

Since the flow rate $Q(p) = pv$, we have

$$Q(p) = p\left(\sqrt{\frac{h^2 A_m^2 \rho_{i0}^2}{\sigma^2} - \frac{2(L+\Delta)A_m\rho_{i0}}{\sigma} + \frac{2A_m\rho_i}{\sigma p}} - \frac{hA_m\rho_{i0}}{\sigma}\right)$$

To analyze the traffic flow stability, the gradient $\partial Q/\partial p$ needs to be calculated. Then, we have

$$\frac{\partial Q}{\partial p} = \left(\sqrt{\frac{h^2 A_m^2 \rho_{i0}^2}{\sigma^2} - \frac{2(L+\Delta)A_m\rho_{i0}}{\sigma} - \frac{2A_m\rho_{i0}}{\sigma p}}\right) - \frac{A_m\rho_{i0}}{\sigma p\sqrt{((h^2A_m^2\rho_{i0}^2)/\sigma^2) - ((2(L+\Delta)A_m\rho_{i0})/\sigma) - ((2A_m\rho_{i0})/(\sigma p))}} - \frac{hA_m\rho_{i0}}{\sigma}$$

The critical traffic density p_{cr1} can be obtained by p_{cr1}, which is given here

$$p_{cr1} = \frac{1}{2(L+\Delta) + h\sqrt{(2(L+\Delta)A_m\rho_{i0})/(\sigma)}} \tag{7.44}$$

The proof of the theorem is completed.

Remark 7.5 It should be mentioned that if ζ_1 is too big or ζ_2 is too small, the control signal will be large and may potentially result in saturation. Therefore, selecting suitable controller parameters is crucial to maintain desirable performance.

Remark 7.6 It can be easily shown that the improved quadratic spacing policy can provide a higher traffic capacity than the traditional quadratic spacing [12]. In the latter case, the critical traffic density is

$$p_{cr2} = \frac{1}{2(L+\Delta) + h\sqrt{(2(L+\Delta)A_m)/\sigma}},$$

which is much smaller than p_{cr1}.

7.5 Simulation Results

In this section, the effectiveness of the developed scheme is illustrated via computer simulations, which runs in a virtual environment established using System Build software package in MATLAB. In the simulations, the desired vehicle spacing is $\Delta_{i-1,i} = 7$ m, the delay time is $h = 0.08$ s, the safety factor is $\sigma = 0.2$, and the absolute value of the maximum possible deceleration is $A_m = 7$ m/s^2. The other parameters in [23] are given in Tables 7.1 and 7.2. The RBF vectors $W_i^{*T}\xi_i(Z_i)$ contains $M = 20$ nodes. The centers c_i^k for the input $Z_i(t) = v_i(t)$ are evenly spaced in $[-2, 5]$, and the widths and $d_i^k = 2(k = 1, 2, ..., 20)$. The external disturbances are $\omega_i(t) = 0.01\sin(t)$, $i \in \Omega_N$ and the desired acceleration of the leader is modeled as

Table 7.1 The dynamics parameters for followers.

Parameter name	Simulation values
Specific mass of the air	$v = 1.2\,\text{kg/m}^3$
Cross-sectional area	$A_i = 2.2\,\text{m}^2$
Drag coefficient	$C_{di} = 0.35$
Mechanical drag	$d_{mi} = 5N$

Table 7.2 The parameters m_i, τ_i, and L_i for followers.

i	1	2	3	4	5	6
m_i(kg)	1500	1600	1550	1650	1500	1400
τ_i(s)	0.1	0.3	0.2	0.4	0.25	0.4
L_i(m)	4	4.5	5	5	4.5	3.5

$$a_0(t) = \begin{cases} 0.5t & \text{m/s}^2 \quad 1\,\text{s} \le t < 4\,\text{s} \\ 2 & \text{m/s}^2 \quad 4\,\text{s} \le t < 9\,\text{s} \\ -0.5t + 6.5 & \text{m/s}^2 \quad 9\,\text{s} \le t < 12\,\text{s} \\ 0 & \text{m/s}^2 \quad \text{otherwise} \end{cases}$$

In addition, we chose the actuation effectiveness as $\rho_i(t) = 0.75 + 0.25\sin(0.1it)$ and the bias fault as $r_i(t) = 0.01\sin(t)$, then, we can obtain $\rho_{i0} = 0.5$. Based on the relationship $1 \le 1/\rho_{i0} \le \Lambda_i$, we assume $\Lambda_i = 2$. The initial states are listed in Table 7.3 and the control parameters are given in Table 7.4.

To illustrate the performance of the proposed method, this subsection compares the simulation results with the existing result [8], where the combined effect of input quantization and dead-zone nonlinearity is investigated, but ignores the issue of actuator faults. It should be noticed that when there do not exist actuator faults, the simulation results in [8] are satisfactory. However, when actuator faults occur, the simulation results using the approach in [8] show that the system performance deteriorates significantly. More specifically, Figure 7.2a show that string stability cannot be

Table 7.3 The initial states for all vehicles.

i	0	1	2	3	4	5	6
$x_i(0)$(m)	150	135	125.5	112.5	99.5	87	75.5
$v_i(0)$(m/s)	1	4	2	0	5	3	1
$a_i(0)$(m/s^2)	0	1	5	2	1	3	1

Table 7.4 Control parameters of followers.

Parameter name	Simulation values
Quantizer parameters	$\phi_i^c = 0.05(1 + 5\,\mathrm{e}^{-0.25t})$, $\phi_i^d = 0.01$
Deadzone parameters	$n_i^r = 0.7$, $n_i^l = b_i^r = 1$, $b_i^l = 2$
Controller parameters	$k_i = 45$, $\lambda = 0.72$, $\pi_i = 3$
PID parameters	$K_p = 8$, $K_i = 6$, $K_d = 0.6$
Initial adaptive values	$\hat{\theta}_i(0) = 1$, $\hat{\eta}_i(0) = 1$, $\hat{\psi}_i(0) = 1$
Adaptive parameters 1	$\alpha = 15$, $\beta = 0.05$, $\gamma = 80$
Adaptive parameters 2	$\Xi_{1i} = \Xi_{2i} = \Xi_{3i} = 0.01e^{-10t}$
Tuning parameters	$b_i = 2$, $\mu_i = 0.5$, $\vartheta_i = 0.5$

guaranteed in the initial transient, which may result in a collision between consecutive vehicles in Figure 7.2b. These simulation results in Figure 7.2a–f show that actuator faults may result in a loss of string stability and may cause collisions among the vehicles, which indicates the need for an effective controller to tolerate actuator faults.

Then, the simulation results under Theorem 7.1 are shown in Figure 7.3. In the case of actuator faults, input quantization, and unknown dead-zone, the string stability in Figure 7.3a is still guaranteed. Figure 7.3b shows that collisions are avoided since there are no crossed or overlapped positions. The velocity and acceleration curves in Figure 7.3c, d show that good tracking performance can be achieved. The sliding surfaces and control input in Figure 7.3e, f can converge to the neighborhood of origin and chattering is eliminated. The estimations of the adaptive laws $\hat{\theta}_i(t)$, $\hat{\eta}_i(t)$, and $\hat{\psi}_i(t)$ are all bounded and omitted for brevity. In summary, the developed scheme is effective to deal with actuator faults, input quantization and unknown dead-zone.

To indicate the effect of nonzero initial spacing errors, this subsection shows the simulation results under the traditional quadratic spacing [12] in Figure 7.4. The sliding surface and input control are similar to those in Figure 7.3. Thus, they are omitted here. From Figure 7.4, we can conclude that the nonzero initial spacing errors may cause string instability. Therefore, it is important to consider the effect of nonzero initial spacing errors.

7.6 Conclusions and Future Work

An adaptive fault-tolerant control scheme is proposed based on an improved quadratic spacing policy for vehicular platoons with the integrated effects of actuator faults, input quantization, and dead-zone nonlinearity. The presented scheme can guarantee individual vehicle stability, string stability, and traffic flow stability while reducing the effects of actuator faults, input quantization, and dead-zone nonlinearity and removing the requirement of zero initial spacing errors at the same time.

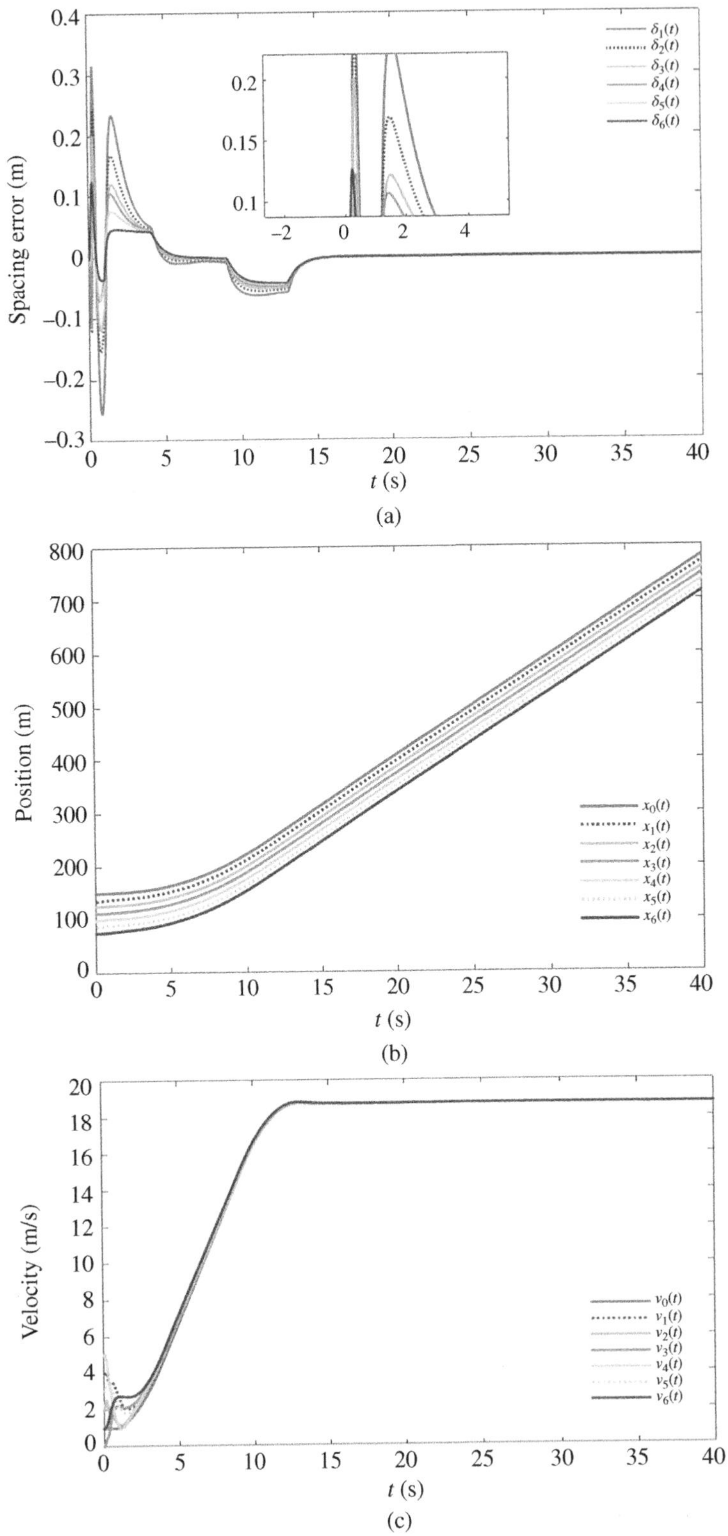

Figure 7.2 Simulation results by using the approach in [8] with actuator faults, input quantization and dead-zone nonlinearity. (a) Spacing error $\delta_i(t)$. (b) Position $x_i(t)$. (c) Velocity $v_i(t)$. (d) Acceleration $a_i(t)$. (e) Sliding surface $S_i(t)$. (f) Control input of the actuator. *Source:* Adapted from Guo et al. [8].

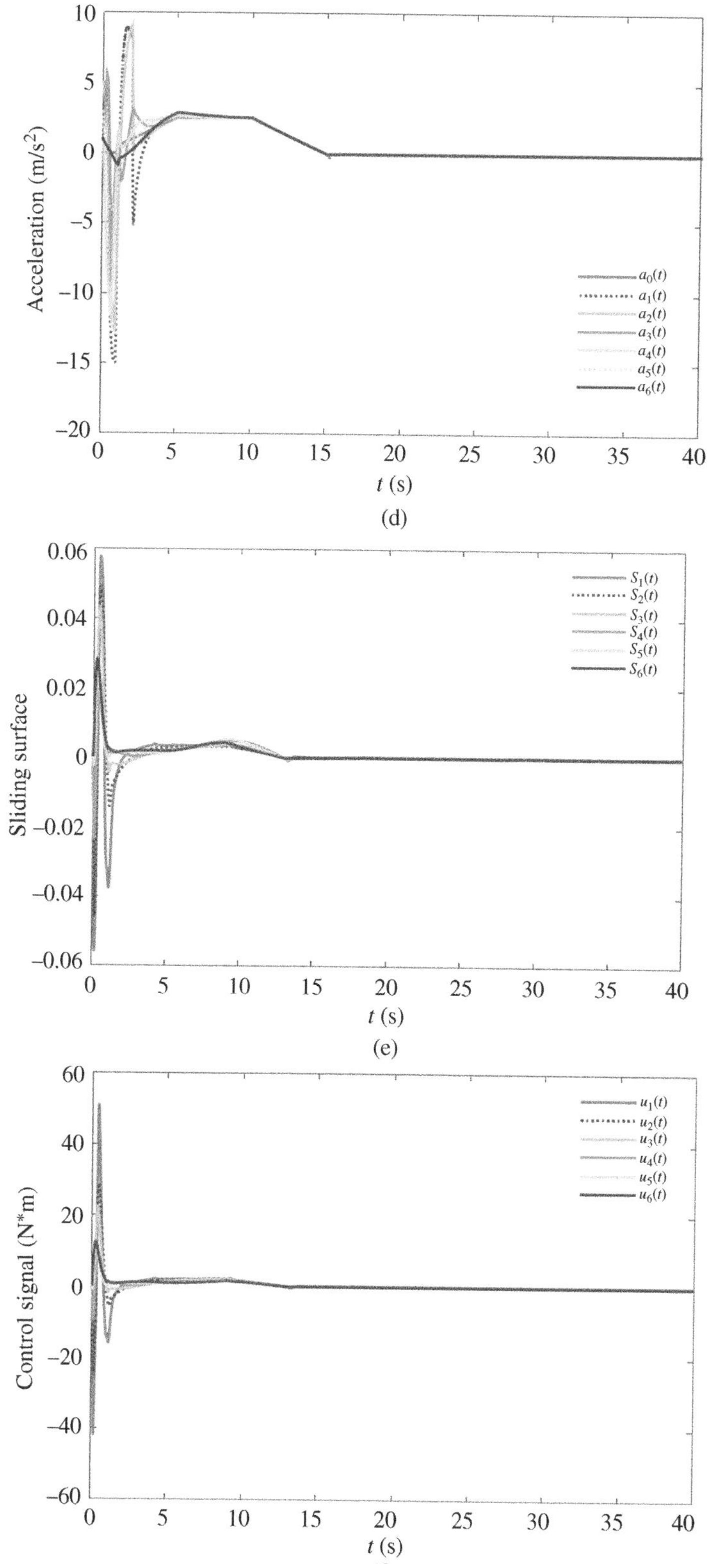

Figure 7.2 (Continued)

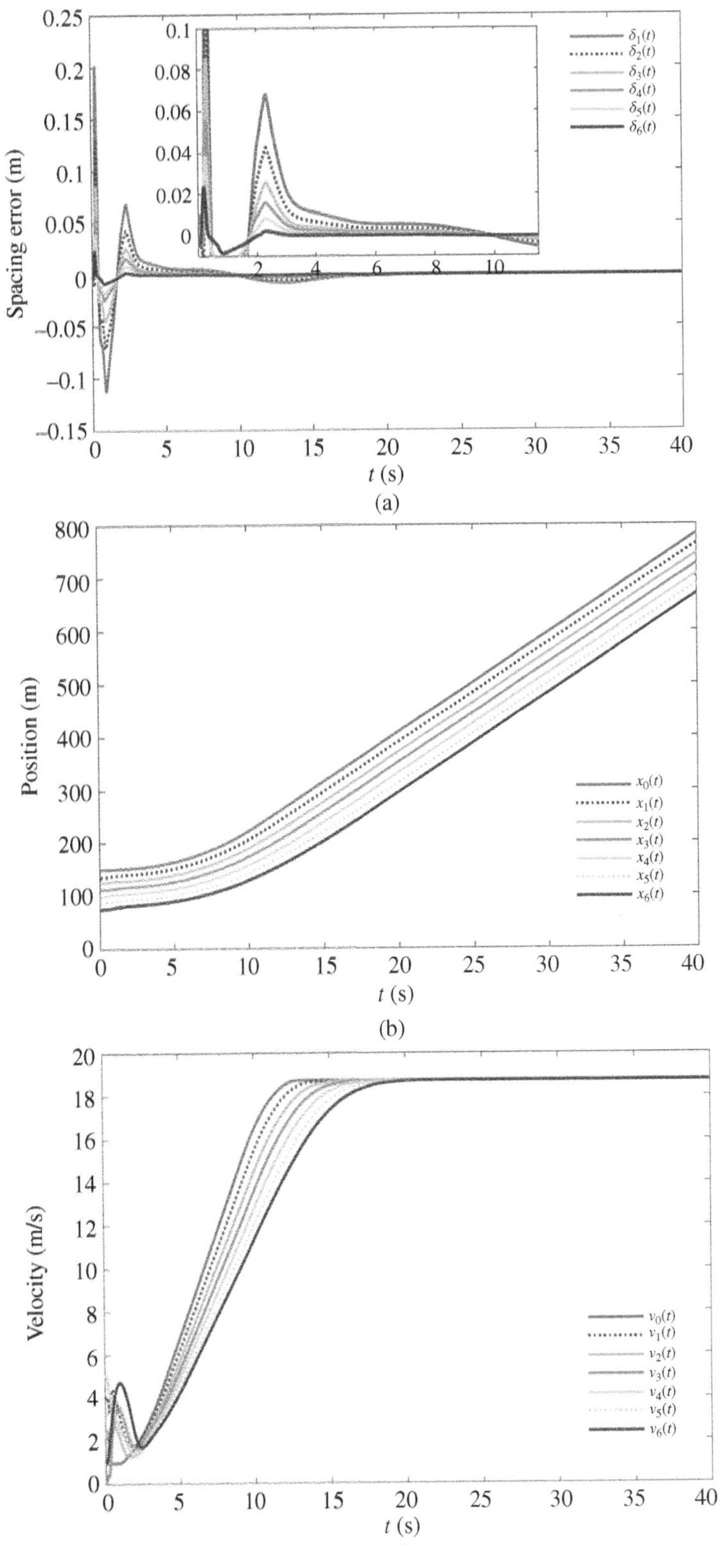

Figure 7.3 Simulation results under Theorem 7.1 with actuator faults, input quantization and dead-zone nonlinearity. (a) Spacing error $\delta_i(t)$. (b) Position $x_i(t)$. (c) Velocity $v_i(t)$. (d) Acceleration $a_i(t)$. (e) Sliding surface $S_i(t)$. (f) Control input of the actuator.

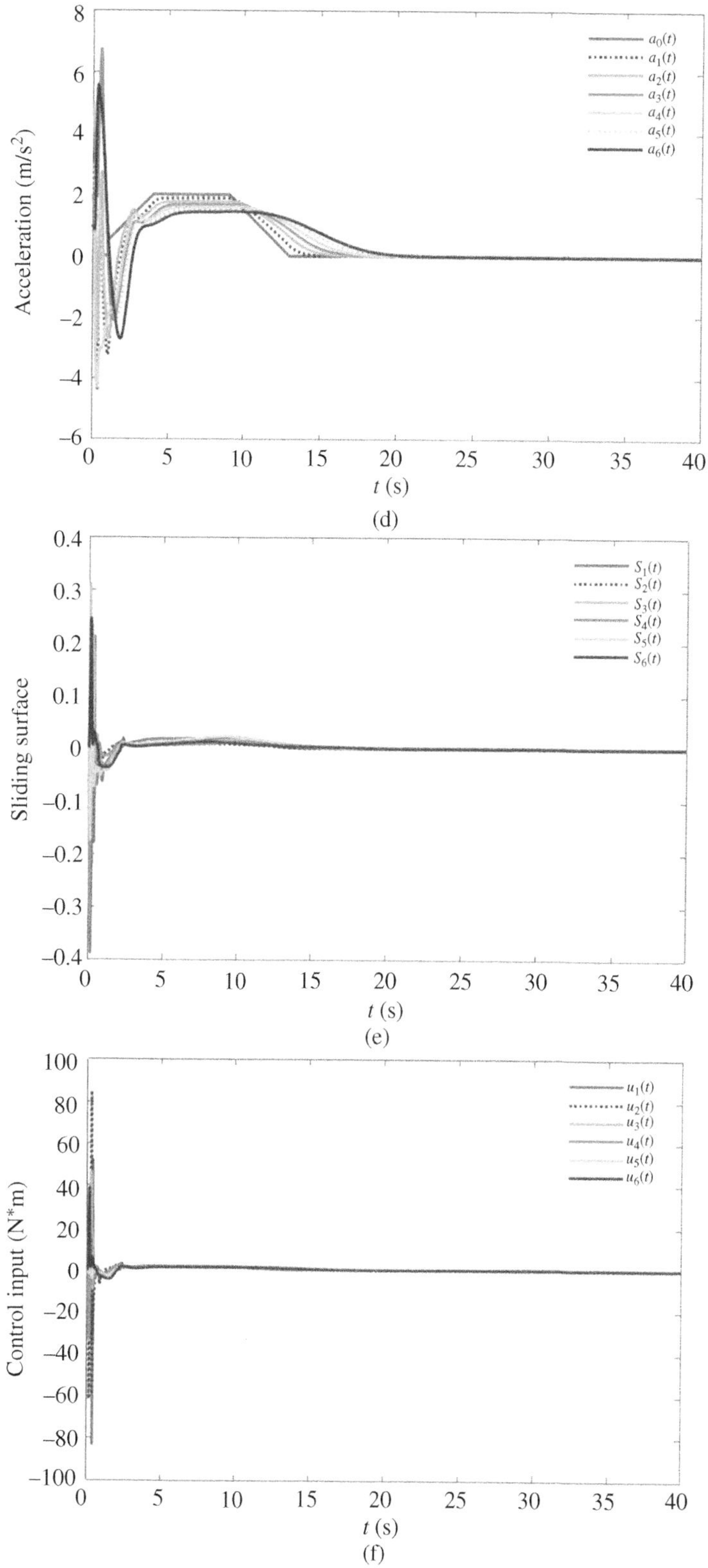

Figure 7.3 (Continued)

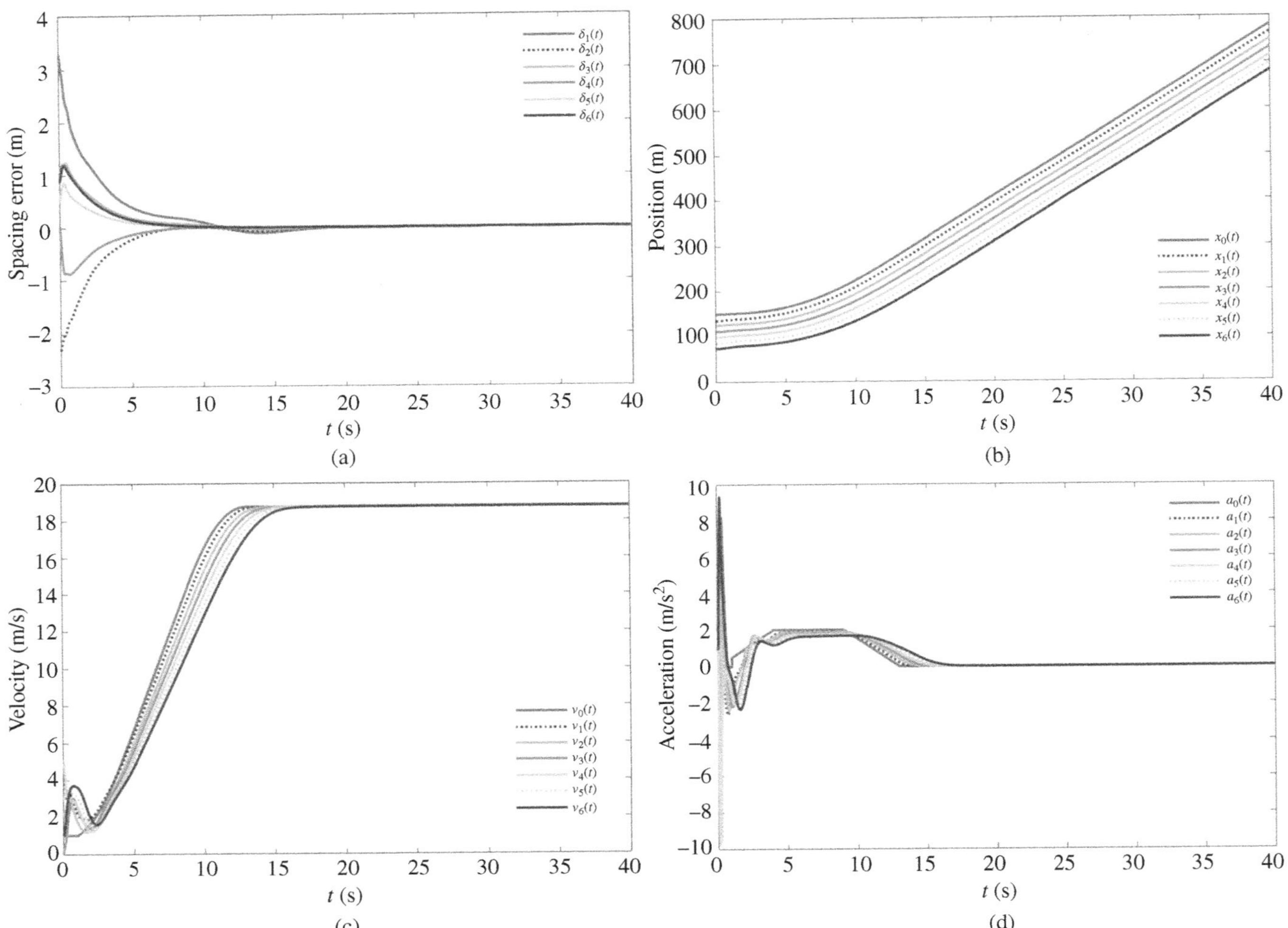

Figure 7.4 Simulation results under the traditional quadratic spacing with actuator faults, input quantization, and dead-zone nonlinearity. (a) Spacing error $\delta_i(t)$. (b) Position $x_i(t)$. (c) Velocity $v_i(t)$. (d) Acceleration $a_i(t)$. *Source:* Adapted from Sungu et al. [12].

References

1 Hao, L.Y., Zhang, H., Guo, G., and Li, H. (2021). Quantized sliding mode control of unmanned marine vehicles: various thruster faults tolerated with a unified model. *IEEE Trans. Syst. Man Cybern. Syst.* to be published, https://doi.org/10.1109/TSMC.2019.2912812.

2 Tong, S. and Li, Y. (2013). Adaptive fuzzy output feedback control of MIMO nonlinear systems with unknown dead-zone input. *IEEE Trans. Fuzzy Syst.* 21 (1): 134–146.

3 Liberzon, D. (2003). Hybrid feedback stabilization of systems with quantized signals. *Automatica* 39 (9): 1543–1554.

4 Yun, S.W., Choi, Y.J., and Park, P. (2009). H2 control of continuous-time uncertain linear systems with input quantization and matched disturbances. *Automatica* 45 (10): 2435–2439.

5 Kameneva, T. and Nešic, D. (2009). Robustness of quantized control systems with mismatch between coder/decoder initializations. *Automatica* 45 (3): 817–822.

6 Chen, M. and Tao, G. (2015). Adaptive fault-tolerant control of uncertain nonlinear large-scale systems with unknown dead zone. *IEEE Trans. Cybern.* 46 (8): 1851–1862.

7 Li, Y.X. and Yang, G.H. (2017). Adaptive fuzzy decentralized control for a class of large-scale nonlinear systems with actuator faults and unknown dead zones. *IEEE Trans. Syst., Man. Cybern. Syst.* 47 (5): 729–740.

8 Guo, X., Wang, J., Liao, F., and Teo, R.S.H. (2019). Neuroadaptive quantized PID sliding-mode control for heterogeneous vehicular platoon with unknown actuator deadzone. *Int. J. Robust Nonlinear Control* 29 (1): 188–208.

9 Darbha, S. and Rajagopal, K.R. (1999). Intelligent cruise control systems and traffic flow stability. *Transp. Res. Part C* 7 (6): 329–352.

10 Baskar, L.D., Schutter, B.D., and Hellendoorn, H. (2012). Traffic management for automated highway systems using model-based predictive control. *IEEE Trans. Intell. Transp. Syst.* 13 (2): 838–847.

11 Santhanakrishnan, K. and Rajamani, R. (2003). On spacing policies for highway vehicle automation. *IEEE Trans. Intell. Transp. Syst.* 4 (4): 198–204.

12 Sungu, H.E., Inoue, M., and Imura, J.I. (2015). Nonlinear spacing policy based vehicle platoon control for local string stability and global traffic flow stability. *Proc. Control Conf. Eur.* Linz, Austria (July 2015), pp. 3396–3401.

13 Guo, X. G., Wang, J. L., and Liao, F. (2017). Adaptive fuzzy fault-tolerant control for multiple high-speed trains with proportional and integral-based sliding mode. *IET Control Theory Appl.* 11 (8): 1234–1244.

14 Xiao, B., Hu, Q., and Shi, P. (2013). Attitude stabilization of spacecrafts under actuator saturation and partial loss of control effectiveness. *IEEE Trans. Control Syst. Technol.* 21 (6): 2251–2263.

15 Xu, S.S.D., Chen, C.C., and Wu, Z.L. (2015). Study of nonsingular fast terminal sliding-mode fault-tolerant control. *IEEE Trans. Ind. Electron.* 62 (6): 3906–3913.

16 Guo, G. and Yue, W. (2012). Autonomous platoon control allowing range-limited sensors. *IEEE Trans. V eh. Technol.* 61 (7): 2901–2912.

17 Swaroop, D., Hedrick, J.K., and Choi, S.B. (2001). Direct adaptive longitudinal control of vehicle platoons. *IEEE Trans. Veh. Technol.* 50 (1): 150–161.

18 Swaroop, D. (1994). String stability of interconnected systems: an application to platooning in automated highway systems. Ph.D. dissertation. Univ. California, Berkeley, CA, USA.

19 Kwon, J.W. and Chwa, D. (2014). Adaptive bidirectional platoon control using a coupled sliding mode control method. *IEEE Trans. Intell. Transp. Syst.* 15 (5): 2040–2048.

20 Polycarpou, M.M. (1996). Stable adaptive neural control scheme for nonlinear systems. *IEEE Trans. Autom. Control* 41 (3): 447–451.

21 Deng, H. and Krstic, M. (1997). Stochastic nonlinear stabilization – I: a backstepping design. *Syst. Control Lett.* 32 (3): 143–150.

22 Ge, S.S. and Wang, C. (2004). Adaptive neural control of uncertain MIMO nonlinear systems. *IEEE Trans. Neural Netw.* 15 (3): 674–692.

23 Ghasemi, A., Kazemi, R., and Azadi, S. (2013). Stable decentralized control of platoon of vehicles with heterogeneous information feedback. *IEEE Trans. Veh. Technol.* 62 (9): 4299–4308.

8

Prescribed Performance Concurrent Control

8.1 Introduction

Note that the obtained results of Chapters 1–7 generally focus on steady-state platoon performance. In practice, however, the transient performance (e.g. convergence rate and overshoot) is also of great significance for control systems [1]. To deal with this, a platooning scheme based on prescribed performance control (PPC) was proposed in [2], with which the tracking errors can be limited to a small residual set with a prescribed overshoot and a convergence rate no less than a given value. In doing so, a transformed error function (TEF) was introduced and the constrained tracking error was transformed into an equivalent unconstrained one. The result is interesting but limited in at least two aspects: (i) The TEF involved depends on the signs of the initial tracking errors (ITEs); (ii) The second-order derivatives of the transformed errors make the controller design complicated.

This chapter investigates a platoon control problem with nonlinear third-order vehicle dynamics and bidirectional information flow. A new spacing policy is introduced based on the difference between the vehicle velocity and the platoon reference velocity, which is given from the planning layer (considered to be a virtual leader). Unlike the serial scheme with a single performance function in [2], two performance functions (PFs) are introduced, which are, respectively, based on the tracking error and the sliding mode variable (see Figure 8.2), to make the controller design independent of the sign of the ITE. Then, a prescribed performance concurrent control (PPCC) method is presented, which makes the two variables meet the prescribed performance constraints concurrently, and hence, guarantees individual stability and string stability. The main contributions of this chapter are summarized as follows:

1) To improve string stability and increase road capacity, a new spacing policy is given, which is based on the relative velocity of the vehicle and the platoon reference velocity.
2) The PPCC framework involves two performance functions based, respectively, on the tracking error and the sliding mode variable that are related to individual stability and string stability. The TEFs are independent of the signs of ITEs, making the control algorithm convenient for practical use. Moreover, there is no need for the second-order derivatives of the transformed errors, unlike [2], which implies lower complexity of design and less computational cost.
3) A distributed adaptive sliding mode control (SMC) method with the prescribed performance is developed for nonlinear third-order vehicular platooning systems. Both transient and steady-state performances of the tracking errors are greatly improved with guaranteed string stability. The effect of nonzero ITE on the whole platoon is also removed.

Connected Vehicular Systems: Communication, Control, and Optimization, First Edition. Ge Guo and Shixi Wen.

The remainder of this chapter is organized as follows. The vehicle nonlinear model and preliminaries of PPC are given in Section 8.2. Section 8.3 contains the distributed PPC design and stability analysis. Simulations are given in Section 8.4 and the conclusions are in Section 8.5.

8.2 Problem Formulation

In this chapter, we consider a vehicular platoon consisting of one leading vehicle (indexed by 0) and N following vehicles (indexed by i accordingly) as illustrated in Figure 8.1, where p_i, v_i, and a_i are, respectively, the position, speed, and acceleration of vehicle i, $i \in \mathcal{V}_N$, $\mathcal{V}_N = \{0, 1, 2, \ldots N\}$. The vehicles in the platoon are equipped with a road map, a global positioning system (GPS), wireless communication devices, and onboard sensors to interact with each other. In addition, bidirectional information flow is considered in this chapter. Moreover, each vehicle including the leader can receive the platoon reference velocity v_p from the platoon planning layer, which can be considered as a virtual leader. Here it is also assumed that the communication channels are reliable.

Assumption 8.1 The trajectories of $v_p(t)$, $\dot{v}_p(t)$, and $\ddot{v}_p(t)$ are smooth and bounded, where $\dot{v}_p(t)$ and $\ddot{v}_p(t)$ are the first and second-time derivatives of $v_p(t)$, respectively.

In this chapter, the longitudinal dynamics of the ith vehicle is described as the following nonlinear third-order model [3]:

$$\dot{p}_i(t) = v_i(t) \tag{8.1a}$$

$$\dot{v}_i(t) = a_i(t) \tag{8.1b}$$

$$\dot{a}_i(t) = \frac{\text{sat}(u_i(t))}{m_i\tau_i} - \frac{a_i(t)}{\tau_i} - \frac{c_i\left(v_i^2 + 2\tau_i v_i a_i\right)}{m_i\tau_i} + D_i(t) \tag{8.1c}$$

where m_i denotes the ith vehicle's mass, c_i is the aerodynamic drag coefficient, τ_i is the engine time constant. $D_i(t)$ represents the lumped disturbances caused by unmodeled dynamics and external disturbances such as road surface and wind gusts. In this chapter, it is assumed that parameters c_i and m_i are unknown constants and that are bounded by $|m_i\tau_i D_i| \leq \overline{D}_i$ with $\overline{D}_i$ being an unknown constant. u_i represents the throttle/brake input, which is subject to asymmetric saturation nonlinearity due to physical limitations as described by:

$$\text{sat}(u_i(t)) = \begin{cases} u_{\max i} & u_i(t) \geq u_{\max i}/C_{ri} \\ C_{ri}u_i(t) & 0 < u_i(t) < u_{\max i}/C_{ri} \\ C_{li}u_i(t) & u_{\min i}/C_{li} < u_i(t) < 0 \\ u_{\min i} & u_i(t) \leq u_{\min i}/C_{li} \end{cases} \tag{8.2}$$

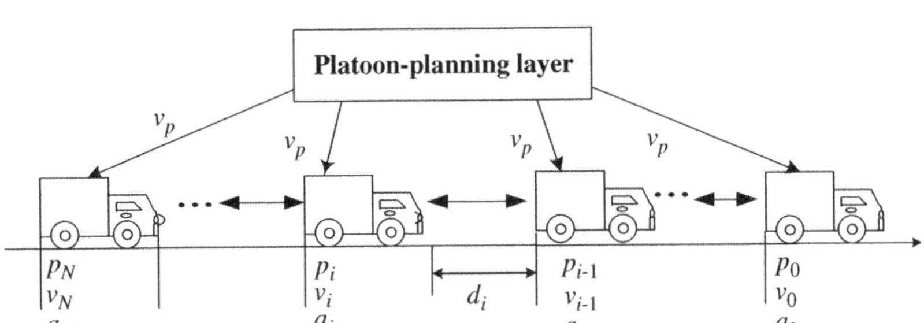

Figure 8.1 The interaction structure of a platoon of CAVs with an active leader.

where C_{li} and C_{ri} denote the left and right slopes of saturation characteristics, respectively. $u_{\max i}$ and $u_{\min i}$ are known upper and lower bounds of actuators.

Remark 8.1 In comparison with the second-order nonlinear models and third-order linear models in existing works, here we introduce a more accurate third-order nonlinear vehicle model considering the engine dynamics with asymmetric actuator saturations. Moreover, the dynamics of the leader vehicle here also has an input to track the platoon velocity rather than a given value.

To increase the traffic density at high speed and improve string stability, a new spacing policy is employed and the tracking error $e_i(t)$ is defined as:

$$e_i(t) = d_i(t) - d^* - h_i\left(v_i(t) - v_p(t)\right) \tag{8.3}$$

where $d_i(t)$ is the distance between two consecutive vehicles with $d_i(t) = p_{i-1}(t) - p_i(t) - l_i$, l_i be the length of the ith vehicle. d^* is the desired safe distance and h_i is the headway time.

From (8.3), the time derivatives of e_i are derived by

$$\dot{e}_i = v_{i-1} - v_i - h_i\left(\dot{v}_i - \dot{v}_p\right) \tag{8.4a}$$

$$\ddot{e}_i = \dot{v}_{i-1} - \dot{v}_i - h_i\left(\ddot{v}_i - \ddot{v}_p\right) \tag{8.4b}$$

Remark 8.2 It should be noted that the spacing policy in (8.3) is based on the relative velocity or the velocity difference between the current vehicle and the platoon reference velocity. This policy can avoid large intervehicle distances at high velocities.

$$-\underline{\xi}_i \rho_i(t) < e_i(t) < \overline{\xi}_i \rho_i(t) \tag{8.5}$$

where $\underline{\xi}_i, \overline{\xi}_i > 0$ are positive constants. The PF $\rho_i(t)$ is positive and strictly decreasing and chosen as: $\rho_i(t) = (\rho_{0i} - \rho_{\infty i})\exp(-\delta_i t) + \rho_{\infty i}$, where ρ_{0i}, $\rho_{\infty i}$, and δ_i are the designed strictly positive constants satisfying $\rho_{\infty i} < \rho_{0i}$, $\rho_{0i} = \rho_i(0)$, and $\lim_{t\to\infty} \rho_i(t) = \rho_{i\infty} > 0$. That is, the convergence rate of tracking errors $e_i(t)$ is no less than the constant δ_i, having a maximum overshoot of less than $\min\left(\underline{\xi}_i, \overline{\xi}_i\right)\rho_i(0)$. Furthermore, the ITEs $e_i(0)$ should satisfy $|e_i(0)| \le \min\left(\underline{\xi}_i, \overline{\xi}_i\right)\rho_i(0)$.

The tracking error constraint in (8.5) makes the controller design very difficult. To address this issue, the constrained tracking error is transformed into an unconstrained one by introducing the following error transformation:

$$\varepsilon_i(t) = \psi_i^{-1}(e_i(t)/\rho_i(t)) \tag{8.6}$$

where $\psi_i^{-1}(\bullet)$ is the inverse function of $\psi_i(\bullet)$ and $\psi_i(\bullet)$ is used-defined smooth, strictly increasing and has following properties: $\lim_{\varepsilon_i\to-\infty} \psi_i(\varepsilon_i) = -\underline{\xi}_i$, $\lim_{\varepsilon_i\to+\infty} \psi_i(\varepsilon_i) = \overline{\xi}_i$. In this chapter, we choose the function $\psi_i(\bullet)$ as:

$$\psi_i(\varepsilon_i) = \frac{\overline{\xi}_i \underline{\xi}_i\left(e^{\varepsilon_i} - e^{-\varepsilon_i}\right)}{\underline{\xi}_i e^{\varepsilon_i} + \overline{\xi}_i e^{-\varepsilon_i}} \tag{8.7}$$

and the transformed tracking error as:

$$\varepsilon_i(t) = \frac{1}{2}\ln\frac{\overline{\xi}_i\left(e_i(t) + \underline{\xi}_i \rho_i(t)\right)}{\underline{\xi}_i\left(\overline{\xi}_i \rho_i(t) - e_i(t)\right)} \tag{8.8}$$

Taking the time derivative of $\varepsilon_i(t)$, one can obtain

$$\dot{\varepsilon}_i = \partial \psi_i^{-1}/\partial(e_i/\rho_i)(1/\rho_i)(\dot{e}_i - e_i\dot{\rho}_i/\rho_i) = R_i(\dot{e}_i + e_i\eta_i) \tag{8.9}$$

where $R_i = \left(\partial \psi_i^{-1}/\partial(e_i/\rho_i)\right)(1/\rho_i)$, and $R_i > 0$, $\eta_i = -\dot{\rho}_i/\rho_i$.

Note that owing to the properties of the function $\psi_i(\varepsilon_i)$, if $-\underline{\xi}_i\rho_i(0) < e_i(0) < \overline{\xi}_i\rho_i(0)$ and the transformed error $\varepsilon_i(t)$ is guaranteed bounded by the designed controller, we will have:

$$-\underline{\xi}_i < \psi_i(\varepsilon_i) < \overline{\xi}_i \tag{8.10}$$

Remark 8.3 The TEF (8.8) is independent of the signs of ITEs, which guarantees that the steady-state tracking errors can converge to zero when the transformed errors approach zero.

The objective of this chapter can be described as follows:

1) Individual vehicle stability: Maintain a desired distance with respect to the preceding vehicle and track the platoon's velocity (i.e. $d_i \rightarrow d^*$, and $v_i(t) \rightarrow v_p(t)$);
2) String stability: The tracking errors must not increase along the platoon, i.e. $|e_N(t)| \leq |e_{N-1}(t)| \leq \cdots \leq |e_0(t)|$;
3) Prescribed tracking performance: Tracking errors $e_i(t)$ satisfy a prescribed transient and steady-state tracking performance, i.e. $-\underline{\xi}_i\rho_i(t) < e_i(t) < \overline{\xi}_i\rho_i(t)$.

8.3 Controller Design Guaranteed Prescribed Performance

To achieve the aforementioned objectives, a PPCC scheme based on the adaptive SMC method is developed in this section. The block diagram of the proposed platoon control structure is shown in Figure 8.2. Furthermore, the effect of nonzero ITE on the whole platoon is also considered.

To facilitate the proof of platoon stability analysis, the following lemma is given.

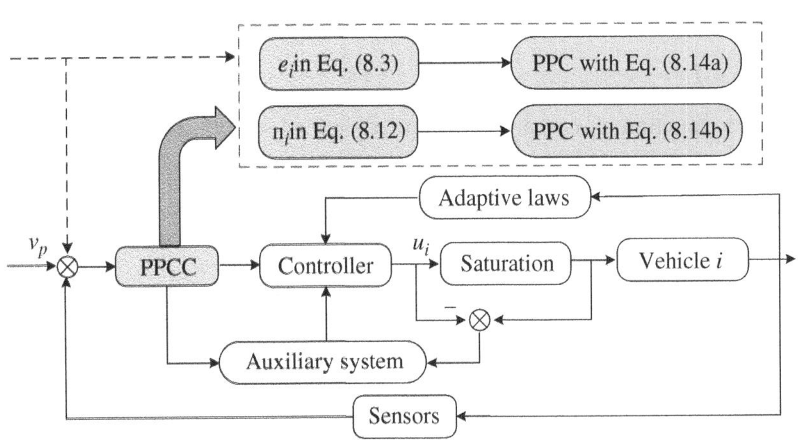

Figure 8.2 Block diagram of the proposed control scheme.

Lemma 8.1 [4] There is a continuous function $V(t) \geq 0$, and $V(0)$ is bounded. If the following inequality holds: $\dot{V}(t) \leq -q_1 V(t) + q_2$, then it is concluded that $V(t)$ is bounded with $q_1 > 0$, q_2 being a constant.

Define linear sliding mode variable $s_i(t)$ as:

$$s_i(t) = \dot{e}_i(t) + \lambda e_i(t) \tag{8.11}$$

where λ is a positive constant.

It is obvious that the convergence of $e_i(t)$ depends on the $s_i(t)$ and λ. However, string stability depends on the relationship between $e_i(t)$ and $e_{i+1}(t)$. For this reason, a coupled sliding surface $\Pi_i(t)$ is adopted for the platoon system as [5]:

$$\Pi_i(t) = qs_i(t) - s_{i+1} \tag{8.12}$$

where $i \in v_N$, $q > 0$ is a weighting factor. For the last vehicle, s_{N+1} does not exist, i.e. $s_{N+1} = 0$ and $e_{N+1} = \dot{e}_{N+1} = \ddot{e}_{N+1} = 0$. Then one can have

$$\boldsymbol{\Pi}(t) = \mathbf{Q}\mathbf{s}(t) \tag{8.13}$$

where

$$\boldsymbol{\Pi}(t) = [\Pi_1(t)\Pi_2(t)\cdots\Pi_N(t)]^T, \quad \mathbf{s}(t) = [s_1(t)s_2(t)\cdots s_N(t)]^T,$$

$$\boldsymbol{Q} = \begin{bmatrix} q & -1 & \cdots & 0 & 0 \\ 0 & q & -1 & \cdots & 0 \\ & & \vdots & & \\ 0 & 0 & \cdots & q & -1 \\ 0 & 0 & \cdots & 0 & q \end{bmatrix}.$$

As the convergence of $s_i(t)$ and $\Pi_i(t)$ is equivalent [5], for the PPCC strategy as shown in Figure 8.2, define the following constraints on the prescribed performance of tracking errors $e_i(t)$ and $\Pi_i(t)$, respectively:

$$-\underline{\xi}_{ei}\rho_{ei}(t) < e_i(t) < \bar{\xi}_{ei}\rho_{ei}(t) \tag{8.14a}$$

$$-\underline{\xi}_{\Pi i}\rho_{\Pi i}(t) < \Pi_i(t) < \bar{\xi}_{\Pi i}\rho_{\Pi i}(t) \tag{8.14b}$$

where $\rho_{ei}(t)$ and $\rho_{\Pi i}(t)$ are the PFs with respect to e_i and Π_i respectively, $\underline{\xi}_{ei}$, $\bar{\xi}_{ei}$, $\underline{\xi}_{\Pi i}$, and $\underline{\xi}_{\Pi i}$ are positive constants. Accordingly, the transformed errors are defined as ε_{ei} and $\varepsilon_{\Pi i}$, respectively, as:

$$\varepsilon_{ei}(t) = \psi_{ei}^{-1}(e_i(t)/\rho_{ei}(t)) \tag{8.15a}$$

$$\varepsilon_{\Pi i}(t) = \psi_{\Pi i}^{-1}(\Pi_i(t)/\rho_{\Pi i}(t)) \tag{8.15b}$$

From (8.9), one can have:

$$\begin{aligned} \dot{\varepsilon}_{ei} &= \partial\psi_{ei}^{-1}/\partial(e_i/\rho_{ei})(1/\rho_{ei})(\dot{e}_i - e_i\dot{\rho}_{ei}/\rho_{ei}) \\ &= R_{ei}(\dot{e}_i + e_i\eta_{ei}) \end{aligned} \tag{8.16a}$$

$$\begin{aligned} \dot{\varepsilon}_{\Pi i} &= \partial\psi_{\Pi i}^{-1}/\partial(\Pi_i/\rho_{\Pi i})(1/\rho_{\Pi i})(\dot{\Pi}_i - \Pi_i\dot{\rho}_{\Pi i}/\rho_{\Pi i}) \\ &= R_{\Pi i}(\dot{\Pi}_i + \Pi_i\eta_{\Pi i}) \end{aligned} \tag{8.16b}$$

where $R_{ei} = (\partial\psi_{ei}^{-1}/\partial(e_i/\rho_{ei}))(1/\rho_{ei}) > 0$, $\eta_{ei} = -\dot{\rho}_{ei}/\rho_{ei}$, and $R_{\Pi i} = (\partial\psi_{\Pi i}^{-1}/\partial(\Pi_i/\rho_{\Pi i}))(1/\rho_{\Pi i}) > 0$, $\eta_{\Pi i} = -\dot{\rho}_{\Pi i}/\rho_{\Pi i}$.

Furthermore, we can obtain the following property [6]:

$$\varepsilon_{ei}R_{ei}e_i \geq \ell\varepsilon_{ei}^2 \tag{8.17}$$

with ℓ is a positive constant.

Remark 8.4 The earlier idea is novel in at least two aspects. First, the prescribed performance in [2] is defined solely on the transformed tracking errors, while the performance constraints here are prescribed by two error variables, i.e. the tracking error and the coupled sliding surface. Second, the controller design in [2] requires the second time derivative of the transformed tracking errors, while this issue is avoided in our method. The proposed controller has quite simple structure, which largely reduces the computation complexity and thus is more flexible for practical applications [7, 8].

In this subsection, the design of the prescribed tracking performance controller is carried out under the assumption of zero ITEs.

The novel distributed adaptive prescribed performance tracking controller u_i is designed as:

$$\begin{aligned} u_i = \frac{1}{qR_{\pi i}h_i}\Big\{&k_s(\varepsilon_{\Pi i} - \phi_i) + qR_{\Pi i}h_i\hat{c}_i(v_i^2 + 2\tau_i v_i a_i) + qR_{\Pi i}h_i\hat{f}_i \\ &+ \tau_i R_{\Pi i}\hat{m}_i T_i + qR_{\Pi i}h_i\hat{\overline{D}}_i \tanh\left(\frac{\varepsilon_{\Pi i}}{\gamma_i}\right)\Big\} \end{aligned} \tag{8.18}$$

with the following adaptation laws:

$$\dot{\hat{c}}_i = \beta_{ci}\left[(v_i^2 + 2\tau_i v_i a_i)R_{\Pi i}\varepsilon_{\Pi i}qh_i - \sigma_{ci}\hat{c}_i\right] \tag{8.19a}$$

$$\dot{\hat{m}}_i = \beta_{mi}(\tau_i R_{\Pi i}T_i\varepsilon_{\Pi i} - \sigma_{mi}\hat{m}_i) \tag{8.19b}$$

$$\dot{\hat{\overline{D}}}_i = \beta_{\overline{D}i}\left[\varepsilon_{\Pi i}R_{\Pi i}qh_i \tanh\left(\frac{\varepsilon_{\Pi i}}{\gamma_i}\right) - \sigma_{\overline{D}i}\hat{\overline{D}}_i\right] \tag{8.19c}$$

And an auxiliary system [9] is given by:

$$\dot{\phi}_i = \begin{cases} -\dfrac{qR_{\Pi i}h_i|\varepsilon_{\Pi i}\Delta u_i| + 0.5(\Delta u_i)^2 + \cosh(|R_{ei}\varepsilon_{ei}s_i|)}{|\phi_i|^2}\phi_i \\ -k_\phi\phi_i + \Delta u_i, |\phi_i| \geq \mu_i \\ 0, |\phi_i| < \mu_i \end{cases} \tag{8.20}$$

where k_s, k_ϕ, and γ_i are the positive designed constants. $\hat{c}_i$, $\hat{m}_i$ and $\hat{\overline{D}}_i$ are the estimations of parameters c_i, m_i, and $\overline{D}_i$, respectively. β_{ci}, β_{mi}, and $\beta_{\overline{D}i}$, σ_{ci}, σ_{mi}, and $\sigma_{\overline{D}i}$, and μ_i are positive constants. ϕ_i is the state of the auxiliary design system and $\Delta u_i = \text{sat}(u_i) - u_i$. And T_i is denoted by:

$$\begin{aligned} T_i &= q\left[q_{i-1} - a_i + h_i\ddot{v}_p + h_i a_i/\tau_i + \lambda\dot{e}_i\right] - \dot{s}_{i+1} + \Pi_i\eta_{\Pi i} \quad i = 1, \ldots, \text{N} \\ T_0 &= q\left[h_i\ddot{v}_p + h_i a_i/\tau_i + \lambda\dot{e}_i\right] - \dot{s}_{i+1} + \Pi_i\eta_{\Pi i} \end{aligned} \tag{8.21}$$

Note that in (8.18)–(8.19), the adaptive control method [10] is employed to deal with uncertain parameters c_i, m_i and $\overline{D}_i$. The auxiliary system (8.20) is involved to compensate for actuator saturation and coupling of the two control layers. Due to the second term in (8.19), the proposed adaptation laws can prevent the estimated values of the adaptation parameters from drifting [2].

With the proposed controllers here, the following results are obtained in Theorem 8.1.

Theorem 8.1 For the platooning system in (8.1)–(8.2) with controller (8.18), adaptation law (8.19), and auxiliary system (8.20) under zero ITEs, the stability of vehicles can be guaranteed in the sense that the tracking errors ε_{ei} and $\varepsilon_{\Pi i}$ are uniformly ultimately bounded and approach zero. Moreover, the string stability of the platoon can be guaranteed if $0 < q \leq 1$.

Proof: First, consider the Lyapunov function as follows:

$$V(t) = \sum_{i=0}^{N} V_i(t) \tag{8.22a}$$

$$V_i = \frac{\varepsilon_{ei}^2}{2} + \frac{m_i \tau_i \varepsilon_{\Pi i}^2}{2} + \frac{\tilde{c}_i^2}{2\beta_{ci}} + \frac{\tilde{m}_i^2}{2\beta_{mi}} + \frac{\tilde{\overline{D}}_i^2}{2\beta_{\overline{D}i}} + \frac{\phi_i^2}{2} \tag{8.22b}$$

where $\tilde{(\cdot)} = \hat{(\cdot)} - (\cdot)$ is the estimation error. Differentiating (8.22b), we obtain:

$$\dot{V}_i = \varepsilon_{ei}\dot{\varepsilon}_{ei} + m_i \tau_i \varepsilon_{\Pi i}\dot{\varepsilon}_{\Pi i} + \frac{\tilde{c}_i \dot{\hat{c}}_i}{\beta_{ci}} + \frac{\tilde{m}_i \dot{\hat{m}}_i}{\beta_{mi}} + \frac{\tilde{\overline{D}}_i \dot{\hat{\overline{D}}}_i}{\beta_{\overline{D}i}} + \phi_i \dot{\phi}_i \tag{8.23}$$

From (8.4) and (8.12), the derivative $\dot{\Pi}_i$ is given by

$$\begin{aligned} \dot{\Pi}_i &= q\dot{s}_i - \dot{s}_{i+!} = q(\ddot{e}_i + \lambda \dot{e}_i) - \dot{s}_{i+1} \\ &= q\left[\dot{v}_{i-1} - \dot{v}_i - h_i\left(\ddot{v}_i - \ddot{v}_p\right) + \lambda \dot{e}_i\right] - \dot{s}_{i+1} \end{aligned} \tag{8.24}$$

Substituting (8.24) into (8.16b) yields:

$$\begin{aligned} \dot{\varepsilon}_{\Pi i} &= R_{\Pi i}\left(\dot{\Pi}_i + \Pi_i \eta_{\Pi i}\right) \\ &= R_{\Pi i}\left\{q\left[\dot{v}_{i-1} - \dot{v}_i - h_i\left(\ddot{v}_i - \ddot{v}_p\right) + \lambda \dot{e}_i\right] - \dot{s}_{i+1} + \Pi_i \eta_{\Pi i}\right\} \end{aligned} \tag{8.25}$$

Then according to (8.1) and (8.18), (8.25) can be written as

$$\begin{aligned} \dot{\varepsilon}_{\Pi i} &= R_{\Pi i}\left\{qh_i\left[\frac{-u_i}{m_i\tau_i} + \frac{c_i}{m_i\tau_i}\left(v_i^2 + 2\tau_i v_i a_i\right) - \frac{\Delta u_i}{m_i \tau_i} - D_i\right] + T_i\right\} \\ &= -R_{\Pi i} q h_i \frac{1}{m_i\tau_i}\left\{\frac{1}{qR_{\pi i}h_i}\left[k_s(\varepsilon_{\Pi i} - \phi_i) + \tau_i R_{\Pi i}\hat{m}_i T_i \right.\right. \\ &\quad \left.\left. + qR_{\Pi i}h_i\hat{c}_i\left(v_i^2 + 2\tau_i v_i a_i\right) + qR_{\Pi i}h_i\hat{\overline{D}}_i \tanh\left(\frac{\varepsilon_{\Pi i}}{\gamma_i}\right)\right]\right\} \\ &\quad + \frac{R_{\Pi i} q h_i c_i}{m_i \tau_i}\left(v_i^2 + 2\tau_i v_i a_i\right) - \frac{R_{\Pi i} q h_i \Delta u_i}{m_i \tau_i} - R_{\Pi i} q h D_i + R_{\Pi i} T_i \end{aligned} \tag{8.26}$$

Accordingly, one can have

$$\begin{aligned} m_i \tau_i \varepsilon_{\Pi i}\dot{\varepsilon}_{\Pi i} = &-\varepsilon_{\Pi i}\left\{\left[k_s(\varepsilon_{\Pi i} - \phi_i) + \tau_i R_{\Pi i}\hat{m}_i T_i\right.\right. \\ &\left.\left. + qR_{\Pi i}h_i\hat{c}_i\left(v_i^2 + 2\tau_i v_i a_i\right) + qR_{\Pi i}h_i\hat{\overline{D}}_i \tanh\left(\frac{\varepsilon_{\Pi i}}{\gamma_i}\right)\right]\right\} \\ &+ \varepsilon_{\Pi i}R_{\Pi i}qh_i c_i\left(v_i^2 + 2\tau_i v_i a_i\right) - \varepsilon_{\Pi i}R_{\Pi i}qh_i\Delta u_i \\ &- \varepsilon_{\Pi i}R_{\Pi i}qhm_i\tau_i D_i + \varepsilon_{\Pi i}R_{\Pi i}m_i\tau_i T_i \end{aligned} \tag{8.27}$$

Obviously, the following inequalities hold:

$$\begin{aligned} &-R_{\Pi i}\varepsilon_{\Pi i}qh_i m_i \tau_i D_i \leq R_{\Pi i}qh_i\overline{D}_i|\varepsilon_{\Pi i}|, \\ &-qR_i h_i \Delta u_i \varepsilon_{\Pi i} \leq qR_i h_i|\Delta u_i \varepsilon_{Ti}|, \end{aligned}$$

By using (8.19) and (8.20), (8.23) satisfies

$$\begin{aligned}\dot{V}_i \leq & -k_s\varepsilon_{\Pi i}^2 + k_s\varepsilon_{\Pi i}\phi_i - qR_{\Pi i}h_i\varepsilon_{\Pi i}\hat{\overline{D}}_i\tanh\left(\frac{\varepsilon_{\Pi i}}{\gamma_i}\right) \\ & + \varepsilon_{\Pi i}R_{\Pi i}qh_i\tilde{\overline{D}}_i\tanh(\varepsilon_{\Pi i}/\gamma_i) + R_{\Pi i}qh_i|\varepsilon_{\Pi i}\Delta u_i| \\ & + R_{\Pi i}qh\overline{D}_i|\varepsilon_{\Pi i}| + \varepsilon_{ei}R_{ei}(\dot{e}_i + e_i\eta_{ei}) - \sigma_{ci}\tilde{c}_i\hat{c}_i \\ & - \sigma_{mi}\tilde{m}_i\hat{m}_i - \sigma_{\overline{D}i}\tilde{\overline{D}}_i\hat{\overline{D}}_i - k_\phi\phi_i^2 - qR_{\Pi i}h_i|\varepsilon_{\Pi i}\Delta u_i| \\ & 0.5(\Delta u_i)^2 - \cosh(|R_{ei}\varepsilon_{ei}s_i|) + \Delta u_i\phi_i\end{aligned} \tag{8.28}$$

Then, based on the following inequality [11]:

$$0 \leq |\varepsilon_{\Pi i}| - \varepsilon_{\Pi i}\tanh\left(\frac{\varepsilon_{\Pi i}}{\gamma_i}\right) \leq k_\varepsilon\gamma_i, k_\varepsilon = 0.2785 \tag{8.29a}$$

and the following Young's inequalities:

$$\varepsilon_{\Pi i}\phi_i \leq 0.5\varepsilon_{\Pi i}^2 + 0.5\phi_i^2 \tag{8.29b}$$

$$\Delta u_i\phi_i \leq 0.5\Delta u_i^2 + 0.5\phi_i^2 \tag{8.29c}$$

$$-\sigma_{ici}\tilde{c}_i\hat{c}_i \leq \frac{\sigma_{ci}}{2}c_i^2 - \frac{\sigma_{ci}}{2}\tilde{c}_i^2 \tag{8.29d}$$

$$-\sigma_{mi}\tilde{m}_i\hat{m}_i \leq \frac{\sigma_{mi}}{2}m_i^2 - \frac{\sigma_{mi}}{2}\tilde{m}_i^2 \tag{8.29e}$$

$$-\sigma_{\overline{D}i}\tilde{\overline{D}}_i\hat{\overline{D}}_i \leq \frac{\sigma_{\overline{D}i}}{2}\overline{D}_i^2 - \frac{\sigma_{\overline{D}i}}{2}\tilde{\overline{D}}_i^2 \tag{8.29f}$$

one can obtain:

$$\begin{aligned}\dot{V}_i \leq & -\frac{1}{2}k_s\varepsilon_{\Pi i}^2 - \left(k_\phi - \frac{1}{2}k_s - \frac{1}{2}\right)\phi_i^2 - |R_{ei}\varepsilon_{ei}s_i| \\ & + \varepsilon_{ei}R_{ei}(e_i + e_i\eta_{ei}) - \frac{\sigma_{ci}}{2}\tilde{c}_i^2 - \frac{\sigma_{mi}}{2}\tilde{m}_i^2 - \frac{\sigma_{\overline{D}i}}{2}\tilde{\overline{D}}_i^2 \\ & + \frac{\sigma_{ci}}{2}c_i^2 + \frac{\sigma_{mi}}{2}m_i^2 + \frac{\sigma_{\overline{D}i}}{2}\overline{D}_i^2 + qR_{\Pi i}h_i\overline{D}_ik_\varepsilon\gamma_i\end{aligned} \tag{8.30}$$

Based on Equations (8.11) and (8.17), (8.30) becomes

$$\begin{aligned}\dot{V}_i \leq & -\frac{1}{2}k_s\varepsilon_{\Pi i}^2 - \left(k_\phi - \frac{1}{2}k_s - \frac{1}{2}\right)\phi_i^2 - |R_{ei}\varepsilon_{ei}s_i| \\ & + \varepsilon_{ei}R_{ei}s_i - \varepsilon_{ei}R_{ei}e_i(\lambda - \eta_{ei}) - \frac{\sigma_{ci}}{2}\tilde{c}_i^2 - \frac{\sigma_{mi}}{2}\tilde{m}_i^2 - \frac{\sigma_{\overline{D}i}}{2}\tilde{\overline{D}}_i^2 \\ & + \frac{\sigma_{ci}}{2}c_i^2 + \frac{\sigma_{mi}}{2}m_i^2 + \frac{\sigma_{\overline{D}i}}{2}\overline{D}_i^2 + \sigma_{\Pi i} + qR_{\Pi i}h_i\overline{D}_ik_\varepsilon\gamma_i \\ \leq & -\frac{1}{2}k_s\varepsilon_{\Pi i}^2 - \left(k_\phi - \frac{1}{2}k_s - \frac{1}{2}\right)\phi_i^2 - \ell(\lambda - \eta_{ei})\varepsilon_{ei}^2 \\ & - \frac{\sigma_{ci}}{2}\tilde{c}_i^2 - \frac{\sigma_{mi}}{2}\tilde{m}_i^2 - \frac{\sigma_{\overline{D}i}}{2}\tilde{\overline{D}}_i^2 + \frac{\sigma_{ci}}{2}c_i^2 \\ & + \frac{\sigma_{mi}}{2}m_i^2 + \frac{\sigma_{\overline{D}i}}{2}\overline{D}_i^2 + \sigma_{\Pi i} + qR_{\Pi i}h_i\overline{D}_ik_\varepsilon\gamma_i \\ \leq & -\zeta_iV_i + \chi_i\end{aligned} \tag{8.31}$$

where

$$\varsigma_i = \min\left\{\frac{k_s}{m_i\tau_i}, \sigma_{ci}\beta_{ci}, \sigma_{mi}\beta_{mi}, \sigma_{\overline{D}i}, \beta_{\overline{D}i}, 2k_\phi - k_s - 1, 2\ell(\lambda - \eta_{ei})\right\}$$

$$\chi_i = \frac{\sigma_{ci}}{2}c_i^2 + \frac{\sigma_{mi}}{2}m_i^2 + \frac{\sigma_{\overline{D}i}}{2}\overline{D}_i^2 + \sigma_{\Pi i} + qR_{\Pi_i}h_i\overline{D}_ik_\varepsilon\gamma_i$$

and with $\ell(\lambda - \eta_{ei}) > 0$, $2k_\phi - k_s - 1 > 0$.

According to Lemma 8.1, it is concluded that V_i is ultimately bounded by the following inequality:

$$\begin{aligned} V_i(t) &\leq [V_i(0) - \chi_i/\varsigma_i]e^{-\varsigma_i t} + \chi_i/\varsigma_i \\ &\leq V_i(0) + \chi_i/\varsigma_i \end{aligned} \tag{8.32}$$

where $V_i(0)$ is the initial value of V_i. Then one can also know that the transformed errors ε_{ei}, $\varepsilon_{\Pi i}$, the parameters estimation errors $\tilde{(\cdot)}$, and the auxiliary variable ϕ_i must remain bounded.

Last, for the global Lyapunov function, one can obtain

$$\dot{V} = -\varsigma\sum_{i=1}^{N} V_i \sum_{i=1}^{N}\chi_i = -\varsigma V + \chi \tag{8.33}$$

where $\varsigma = \min(\varsigma_i)$, $\chi = \sum_{i=1}^{N}\chi_i$.

In view of (8.32), we can have

$$0 \leq V \leq [V(0) - \chi/\varsigma]e^{-\varsigma t} + \chi/\varsigma \leq V(0) + \chi/\varsigma \tag{8.34}$$

The aforementioned analysis demonstrates that all the signals in the platoon systems are uniformly ultimately bounded. Hence, the prescribed tracking performance is guaranteed, i.e. the condition (8.14) is satisfied. Furthermore, the coupled sliding mode variable Π_i can converge to an arbitrarily small neighborhood of zero and satisfies:

$$|\Pi_i| \leq \min\left(\underline{\xi}_i, \overline{\xi}_i\right)\rho_\infty = \Pi_i^+ \tag{8.35}$$

Since $\Pi_i = qs_i - s_{i+1} = \Pi_i^+$, one can get

$$q(\dot{e}_i + \lambda e_i) = (\dot{e}_{i+1} + \lambda e_{i+1}) + \Pi_i^+ \tag{8.36}$$

Taking the Laplace transform of (8.36) under the condition of zero ITEs [12], it follows that

$$q[sE_i(s) + \lambda E_i(s)] = sE_{i+1}(s) + \lambda E_{i+1}(s) + \Pi_i^+ \tag{8.37}$$

Then one can obtain

$$G_i(s) = \frac{E_{i+1}(s)}{E_i(s)} \leq \frac{(s+\lambda)E_{i+1}(s) + \Pi_i^+}{(s+\lambda)E_i(s)} = q \tag{8.38}$$

Thus, the string stability can be guaranteed for $0 < q \leq 1$, i.e. $|e_{i+1}(t)| \leq |e_i(t)|$. The proof is completed.

To overcome the effects of the nonzero ITEs on string stability, a new tracking error $\bar{e}_i(t)$ is designed as

$$\bar{e}_i(t) = e_i(t) - \tilde{e}_i(t) \tag{8.39}$$

with

$$\tilde{e}_i = \left\{e_i(0) + \dot{e}_i(0)t + 0.5[2\alpha e_i(0) + \ddot{e}_i(0)]t^2\right\}e^{-\alpha t^2}$$

$$\dot{\tilde{e}}_i = \{\dot{e}_i(0) + [2ae_i(0) + \ddot{e}_i(0)]t\}e^{-at^2}$$
$$2ate^{-at^2}\{e_i0 + \dot{e}_i(0)t + 0.5[2ae_i(0) + \ddot{e}_i(0)]t^2\}$$
$$\ddot{\tilde{e}}_i = [2ae_i(0) + \ddot{e}_i(0)]e^{-at^2} - 4ate^{-at^2}\{\dot{e}_i(0) + [2ae_i(0) + \ddot{e}_i(0)]t\} - \left[2ae^{-at^2} - 4a^2t^2e^{-at^2}\right]\{e_i(0) + \dot{e}_i(0)t + 0.5[2ae_i(0) + \ddot{e}_i(0)]t^2\}$$

Then one can obtain that

$$\bar{e}_i(0) = \dot{\bar{e}}_i(0) = \ddot{\bar{e}}_i(0) = 0 \tag{8.40}$$

It is worth noting that the variable $\tilde{e}_i(t)$ can transform the nonzero ITEs problem into a zero ITEs case. That is to say, $\bar{e}_i(t)$ can converge to $e_i(t)$ when the $\tilde{e}_i(t)$ converges to zero due to the fast decay term e^{-at^2}. Then, similar to the case of zero ITEs, the new sliding mode variable $\bar{s}_i(t)$ is defined by

$$\bar{s}_i(t) = \dot{\bar{e}}_i(t) + \lambda\bar{e}_i(t) \tag{8.41}$$

Accordingly, the new coupled sliding variable $\overline{\Pi}_i(t)$ becomes

$$\overline{\Pi}_i(t) - q\bar{s}_i(t) - \bar{s}_{i+1} \tag{8.42}$$

After the error transformation, we can obtain the new transformed errors $\bar{\varepsilon}_{ei}$ and $\bar{\varepsilon}_{\overline{\Pi}i}$ with the related terms $\bar{R}_{\bar{e}i}$, $\bar{R}_{\overline{\Pi}i}$, $\bar{\eta}_{\bar{e}i}$, and $\bar{\eta}_{\overline{\Pi}i}$. Then the new distributed adaptive prescribed performance tracking controller in this case is expressed by:

$$u_i = \frac{1}{q\bar{R}_{\overline{\Pi}i}h_i}\{k_s(\bar{\varepsilon}_{\overline{\Pi}i} - \phi_i) + \tau_i\bar{R}_{\overline{\Pi}i}\hat{m}_i\overline{T}_i + q\bar{R}_{\overline{\Pi}i}h_i\hat{c}_i(v_i^2 + 2\tau_i v_i a_i) + q\bar{R}_{\overline{\Pi}i}h_i\hat{\bar{D}}_i\tanh\left(\frac{\bar{\varepsilon}_{\overline{\Pi}i}}{\gamma_i}\right)\right\} \tag{8.43}$$

where $\overline{T}_i = q\left[q_{i-1} - a_i + h_i\ddot{v}_p + h_i a_i/\tau_i + \lambda\dot{\bar{e}}_i\right] - \dot{\bar{s}}_{i+1} + \overline{\Pi}_i\bar{\eta}_{\overline{\Pi}i}$ for $i = 1, \dots, N$ and $\overline{T}_0 = q\left[h_i\ddot{v}_p + h_i a_i/\tau_i + \lambda\dot{\bar{e}}_i\right] - \dot{\bar{s}}_{i+1} + \overline{\Pi}_i\bar{\eta}_{\overline{\Pi}i}$. The adaptation laws are designed as follows:

$$\dot{\hat{c}}_i = \beta_{ci}\left[(v_i^2 + 2\tau_i v_i a_i)\bar{R}_{\overline{\Pi}i}\bar{\varepsilon}_{\overline{\Pi}i}qh_i - \sigma_{ci}\hat{c}_i\right] \tag{8.44a}$$

$$\dot{\hat{m}}_i = \beta_{mi}(\tau_i\bar{R}_{\overline{\Pi}i}\overline{T}_i\bar{\varepsilon}_{\overline{\Pi}i} - \sigma_{mi}\hat{m}_i) \tag{8.44b}$$

$$\dot{\hat{\bar{D}}}_i = \beta_{\bar{D}i}\left[\bar{\varepsilon}_{\overline{\Pi}i}\bar{R}_{\overline{\Pi}i}qh_i\tanh\left(\frac{\bar{\varepsilon}_{\overline{\Pi}i}}{\gamma_i}\right) - \sigma_{\bar{D}i}\hat{\bar{D}}_i\right] \tag{8.44c}$$

And the auxiliary system becomes:

$$\dot{\phi}_i = \begin{cases} -\dfrac{q\bar{R}_{\Pi i}h_i|\bar{\varepsilon}_{\Pi i}\Delta u_i| + 0.5(\Delta u_i)^2 + \cosh\left(\left|\bar{R}_{ei}\bar{\varepsilon}_{ei}\bar{s}_i\right|\right)}{|\phi_i|^2}\phi_i \\ + \Delta u_i - k_\phi\phi_i, |\phi_i| \geq \mu_i \\ 0, |\phi_i| < \mu_i \end{cases} \tag{8.45}$$

Theorem 8.2 For the platooning system in (8.1)–(8.2) with controller (8.43), adaptation law (8.44), and auxiliary system (8.45) under nonzero ITEs, the stability of vehicles can be guaranteed in the sense that the tracking errors $\bar{\varepsilon}_{ei}$ and $\bar{\varepsilon}_{\Pi i}$ are uniformly ultimately bounded and approach zero. Moreover, the string stability of the platoon can be guaranteed if $0 < q \leq 1$.

Proof: The proof is similar to Theorem 8.1, and along similar lines, we can also prove that the transformed errors $\bar{\varepsilon}_{\bar{e}i}, \bar{\varepsilon}_{\overline{\Pi}i}$ are uniformly ultimately bounded. Furthermore, the prescribed tracking performance and string stability can be guaranteed simultaneously.

Remark 8.5 Compared with the existing platoon control schemes [13–17], the prominent advantage is that the prescribed transient and steady-state performance can be achieved along with string stability for the vehicular platooning systems.

8.4 Simulation Studies

To evaluate the effectiveness of the proposed PPCC platoon control scheme, numerical simulations are performed for two scenarios with zero and nonzero ITEs. Comparisons are made with the one-layer PPC method in [2].

In the simulations, the control algorithms are applied to a platoon of five followers and one leader. The parameters of vehicles i ($i = 0, 1, \ldots, 5$) are: $l_i = 4$, $m_i = 1600$, $c_i = 0.33$, $\tau_i = 0.25$, and $h_i = 0.2$. The specific platoon velocity trajectory is given in Figure 8.3, and $D_i(t) = 0.1\sin(t)$. The desired distance between two consecutive vehicles is set as $d^* = 10$ m.

First, here the two PFs are considered the same with $\rho_i(t) = (2.5 - 0.06)\exp(-0.1t) + 0.06$ to simply verify the main results. The initial positions and velocities of each vehicle are given in Table 8.1. According to (8.3), here the ITEs are all zero. The controller parameters of the PPCC method and the one-layer PPC method in [2] are listed in Tables 8.2 and 8.3, respectively. The corresponding simulation results are given in Figures 8.4 and 8.5.

From Figures 8.4 and 8.5, it is obvious that both schemes can achieve string stability with prescribed performance guarantees, i.e. $|e_5| \le |e_4| \le \cdots \le |e_0|$ and $-\underline{\xi}_i\rho_i(t) < e_i(t) < \bar{\xi}_i\rho_i(t)$, $i = 0, 1, \ldots, 5$. From Figures 8.4d and 8.5b, it can be seen that when $t > 45$, the tracking errors with PPCC are much smaller. Furthermore, the computation time of the two methods is different (see Table 8.4), implying that the PPCC scheme is potentially more attractive for implementation. The nonzero ITEs are

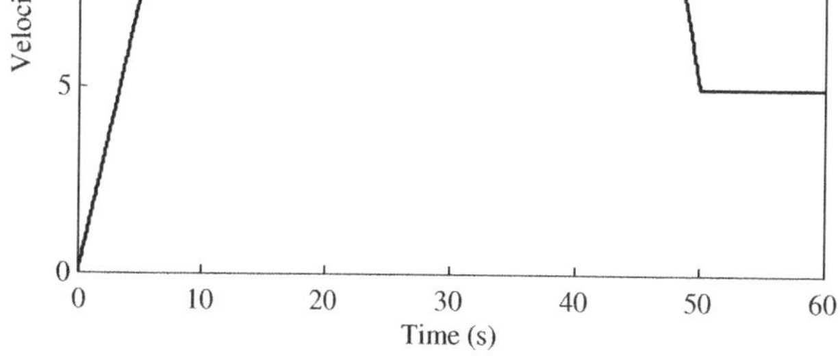

Figure 8.3 Profile of the platoon velocity.

Table 8.1 The parameters of $p_i(0)$ and $p_i(0)$ for vehicles ($i = 0, 1, \ldots, 5$).

i	5	4	3	2	1	0
$p_i(0)$(m)	0	14	28	42	56	70
$v_i(0)$(m/s)	0	0	0	0	0	0

Table 8.2 Controller parameters of vehicles ($i = 0, 1, \ldots, 5$).

q	λ	k_s	k_ϕ	β_i	σ_i	γ_i	$\underline{\xi}$	$\overline{\xi}_i$
0.9	35	2	200	10^{-5}	10^{-5}	0.01	0.9	0.9

Table 8.3 Controller parameters in [2].

q	λ	k_s	k_ϕ	β_i	σ_i	γ_i	$\underline{\xi}$	$\overline{\xi}_i$
0.95	208	165	260	10^{-8}	10^{-8}	0.01	0.9	0.9

Source: Adapted from Guo and Li [2].

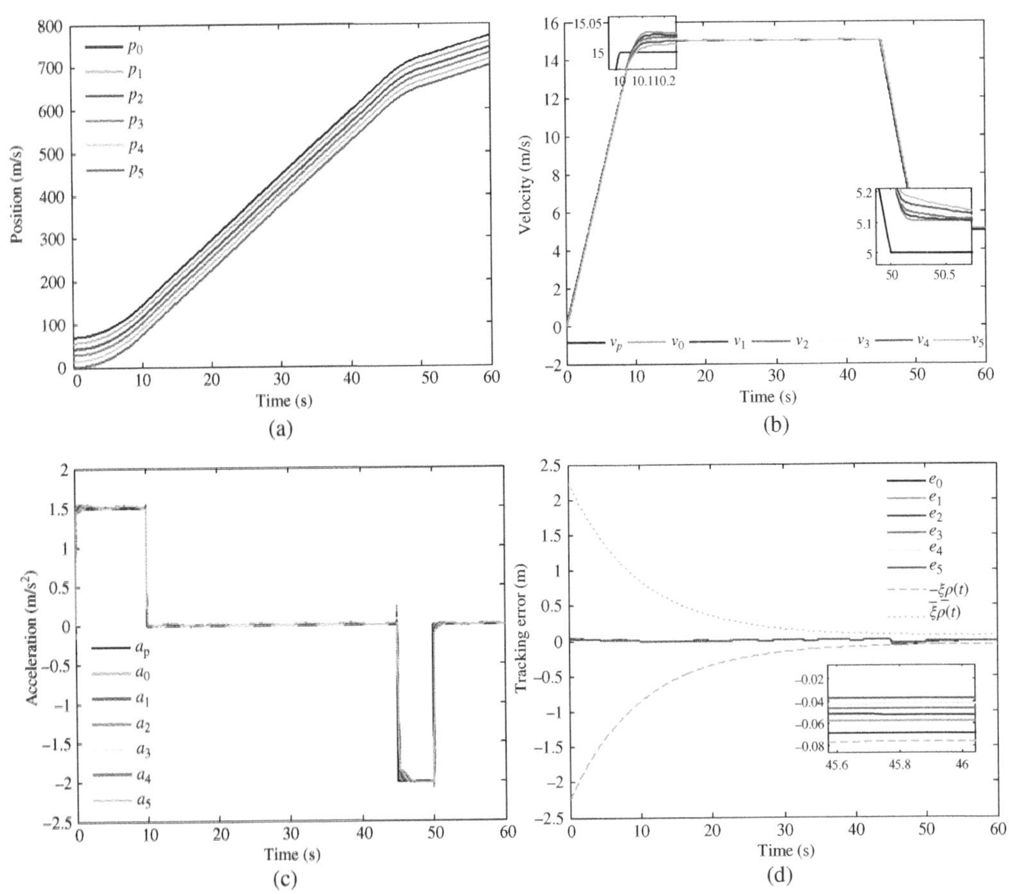

Figure 8.4 The results of PPCC with zero initial tracking error. (a) Position. (b) Velocity. (c) Acceleration. (d) Tracking error.

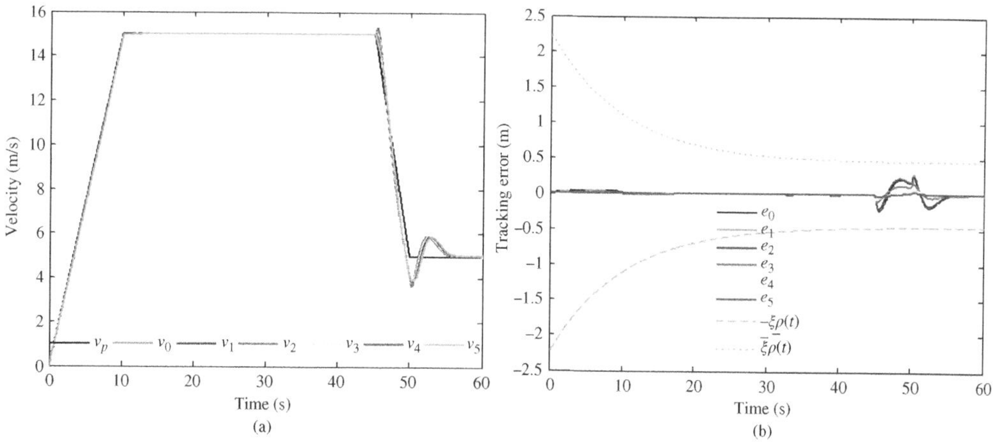

Figure 8.5 The results of PPC in [2] with zero ITEs. (a) Velocity. (b) Tracking error. *Source:* Adapted from Guo and Li [2].

Table 8.4 Comparison of computation time under zero ITEs.

	PPC in [2] (s)	The proposed PPCC (s)
First time	108	26
Second time	107	25
Third time	109	28
Average time	108	26

Table 8.5 The parameters of $p_i(0)$, $v_i(0)$, $a_i(0)$ for vehicles(i =0,1, ..., 5).

i	5	4	3	2	1	0
$p_i(0)$(m)	1	14	31	46	58	74
$v_i(0)$(m/s)	1.5	1.5	2	1	1.5	1
$a_i(0)$(m/s^2)	0	1.5	1	−2	1	0

given below in Table 8.5. The control parameters are listed in Table 8.2 except for $\lambda = 30$, $\alpha = 1$. The simulation results are illustrated in Figures 8.6 and 8.7.

The computation time of the two methods is given in Table 8.6. Note that the case of nonzero ITEs can be transformed into a zero ITEs case by introducing error transformation (8.39). Therefore, the simulation results of this scenario are similar to the previous subsection. That is to say, each vehicle in the platoon can track the platoon velocity successfully with prescribed performance guarantees whether the ITE is zero or not. Meanwhile, the string stability of the whole platoon can also be achieved in the presence of parameter uncertainties, disturbances, and actuator saturations.

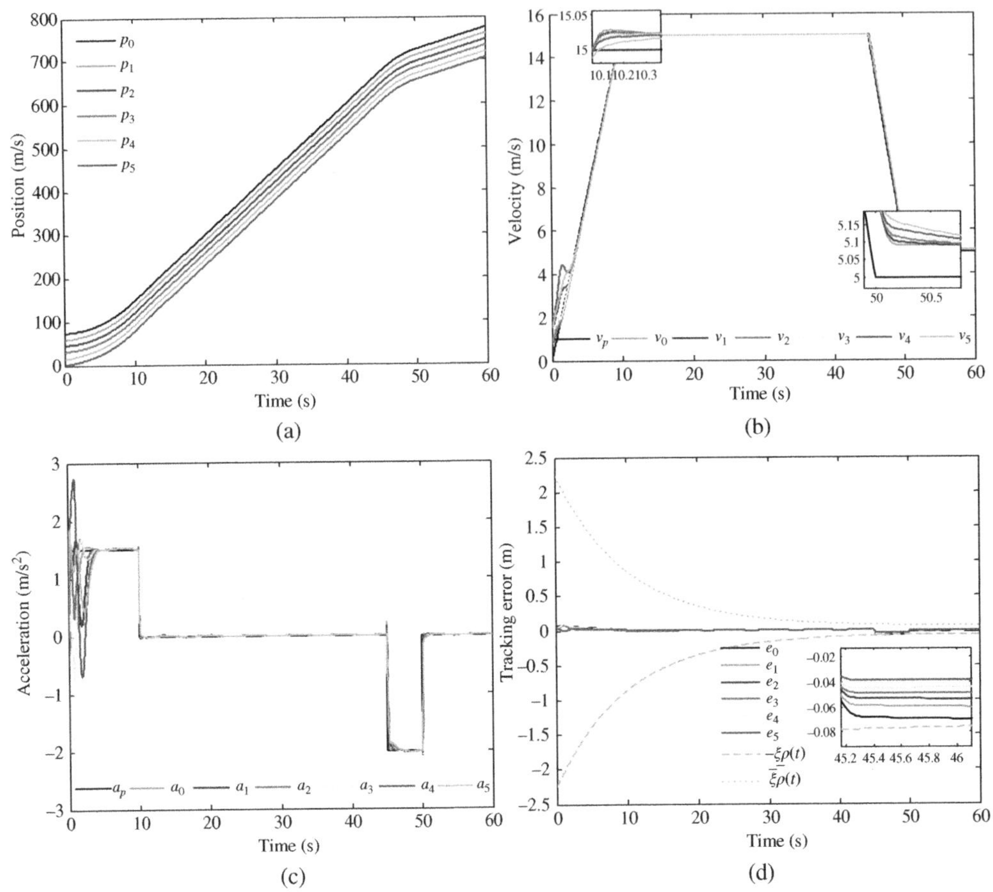

Figure 8.6 The results of PPCC with nonzero initial tracking error. (a) Position. (b) Velocity. (c) Acceleration. (d) Tracking error.

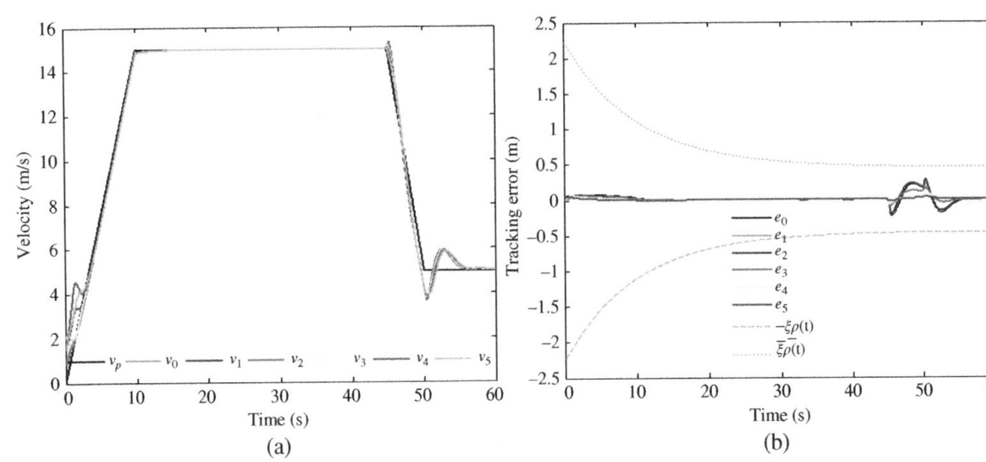

Figure 8.7 The results of PPC in [2] with nonzero ITEs. (a) Velocity. (b) Tracking error. *Source:* Adapted from Guo and Li [2].

Table 8.6 Comparison of computation time under nonzero ITEs.

	PPC in [2] (s)	The proposed PPCC (s)
First time	109	24
Second time	114	26
Third time	111	25
Average time of once	111	25

8.5 Conclusions and Future Work

This chapter proposes a novel distributed adaptive prescribed performance SMC scheme for nonlinear third-order vehicular platoons. Particularly, the presented concurrent PPC scheme can directly serve the platoon control goals. All the vehicles can track the desired intervehicle spacing and a common velocity with guaranteed string stability and prescribed performance even under nonzero ITEs. Simulation results show the satisfactory performance of the proposed method in both accuracy and rapidity.

References

1 Hu, Y., Chen, H., Gong, X. et al. (2019). Control-oriented modeling and robust nonlinear triple-step controller design for an air-feed system for polymer electrolyte membrane fuel cells. *Asian J. Control* 21 (4): 1811–1823.

2 Guo, G. and Li, D. (2019). Adaptive sliding mode control of vehicular platoons with prescribed tracking performance. *IEEE Trans. Veh. Technol.* 68 (8): 7511–7520.

3 Stanković, S.S., Stanojević, M.J., and Šiljak, D.D. (2000). Decentralized overlapping control of a platoon of vehicles. *IEEE Trans. Control Syst. Technol.* 8 (5): 816–832.

4 Ge, S.S. and Wang, C. (2004). Adaptive neural control of uncertain MIMO nonlinear systems. *IEEE Trans. Neural Netw.* 15 (3): 674–692.

5 Kwon, J.W. and Chwa, D. (2014). Adaptive bidirectional platoon control using a coupled sliding mode control method. *IEEE Trans. Intell. Transp. Syst.* 15 (5): 2040–2048.

6 Karayiannidis, Y., Papageorgiou, D., and Doulgeri, Z. (2016). A model-free controller for guaranteed prescribed performance tracking of both robot joint positions and velocities. *IEEE Robotics Autom. Lett.* 1 (1): 267–273.

7 Hu, X., Chen, H., Li, Z., and Wang, P. (2020). An energy-saving torque vectoring control strategy for electric vehicles considering handling stability under extreme conditions. *IEEE Trans. Veh. Technol.* 69: 1–10. https://doi.org/10.1109/TVT.2020.3011921.

8 Guo, G. and Li, D. (2018). PMP-based set-point optimization and sliding-mode control of vehicular platoons. *IEEE Trans. Comput. Social Syst.* 5 (2): 553–562.

9 Chen, M., Ge, S.S., and Choo, Y.S. (2009). Neural network tracking control of ocean surface vessels with input saturation. *Proc. IEEE Int. Conf. Autom. and Logistics*, Shenyang, China (5–7 August 2009), pp. 85–89.

10 Hu, Y., Gu, W., Zhang, H., and Chen, H. (2019). Adaptive robust triple-step control for compensating cogging torque and model uncertainty in a DC motor. *IEEE Trans. Syst. Man Cybern. Syst.* 49 (12): 2396–2405.

11 Polycarpou, M.M. and Ioannou, P.A. (1996). A robust adaptive nonlinear control design. *Automatica* 32 (3): 423–427.

12 Yan, M., Song, J., Yang, P., and Zuo, L. (2018). Neural adaptive sliding-mode control of a bidirectional vehicle platoon with velocity constraints and input saturation. *Complexity* 2018: 1–11.

13 Guo, X., Wang, J., Liao, F., and Teo, R.S.H. (2017). Distributed adaptive sliding mode control strategy for vehicle-following systems with nonlinear acceleration uncertainties. *IEEE Trans. Veh. Technol.* 66 (2): 981–991.

14 Yan, M., Song, J., Zuo, L., and Yang, P. (2017). Neural adaptive sliding-mode control of a vehicle platoon using output feedback. *Energies* 10 (11): 1906.

15 Guo, X., Wang, J., Liao, F., and Teo, R.S.H. (2018). CNN-based distributed adaptive control for vehicle-following platoon with input saturation. *IEEE Trans. Intell. Transp. Syst.* 19 (10): 3121–3132.

16 Zhu, Y. and Zhu, F. (2018). Distributed adaptive longitudinal control for uncertain third-order vehicle platoon in a networked environment. *IEEE Trans. Veh. Technol.* 67 (10): 9183–9197.

17 Rupp, A., Steinberger, M., and Horn, M. (2017). Sliding mode based platooning with non-zero initial spacing errors. *IEEE Control Syst. Lett.* 1 (2): 274–279.

9

Adaptive Sliding Mode Control with Prescribed Performance

9.1 Introduction

The prescribed performance control (PPC) method presented in Chapter 8 ignores the engine dynamics. It is more natural and practical to consider the engine dynamics, taking into account the saturation in the throttle and braking. On the other hand, the vehicle dynamics in Chapters 1–5 are based on feedback linearization of nonlinear vehicular dynamics that needs to be completely known. In a practical environment, however, the dynamics of a car is often nonlinear with uncertain parameters and external disturbances (due to wind gusts and road condition).

Motivated by these discussions, this chapter investigates the PPC with guaranteed string stability for uncertain third-order nonlinear vehicle platoons subject to actuator saturation. Two distributed adaptive sliding mode control schemes based on leader-predecessor (LP) and leader-bidirectional (LB) information topologies are proposed to ensure string stability and strong string stability, respectively, with the trajectory tracking errors of the vehicles retained within a prescribed region. The main contributions of this chapter in comparison with existing works are as follows:

1) Actuator saturation and exogenous disturbances in the third-order nonlinear vehicular platoons are addressed together with uncertain parameters. A smooth hyperbolic tangent function is used to approximate the actuator saturation nonlinearity. Adaptation laws are given to cope with the uncertain parameters and the disturbance.
2) Two distributed adaptive sliding mode control schemes based on LP and LB information topologies, respectively, are proposed, which not only achieve prescribed transient and steady-state tracking performance but also guarantee string stability of vehicle platoons.

The organization of this chapter is as follows. The problem description and system modeling are given in Section 9.2. Section 9.3 is the model transformation. Two adaptive sliding model control methods with string stability analysis are presented in Section 9.4. Section 9.5 is the simulation results, followed by the conclusions in Section 9.6.

9.2 Problem Formulation

Consider a vehicular platoon consisting of a lead vehicle (vehicle 0) and N following vehicles as shown in Figure 9.1, where p_i, v_i, and a_i are, respectively, the position, speed, and acceleration of vehicle i, $i \in \nu_N$, $\nu_N = \{1, 2, \ldots, N\}$. Assume that the vehicles are equipped with a road map, a global

Connected Vehicular Systems: Communication, Control, and Optimization, First Edition. Ge Guo and Shixi Wen.

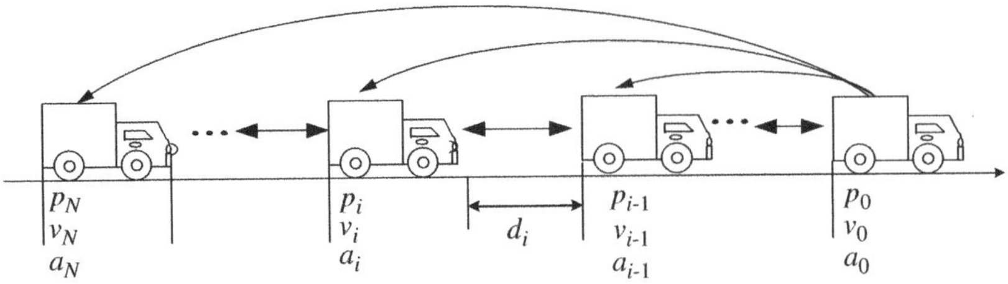

Figure 9.1 Platoon of heterogeneous vehicles.

positioning system (GPS), and onboard sensors. Each following vehicle periodically receives the state (position, speed, and acceleration) of the leader, and its nearest neighbors (immediate front or/and back vehicles). It is assumed that the communication channels are reliable.

The longitudinal dynamics of vehicle i is obtained by Newton's Second Law as:

$$\dot{p}_i(t) = v_i(t),\ i \in v_N \tag{9.1a}$$

$$\dot{v}_i(t) = a_i(t) = \frac{1}{m_i}\left[F_{ei}(t) - F_{ai}(t) - F_{gi}(t) - F_{ri}(t)\right] + y_i(t) \tag{9.1b}$$

where m_i denotes the i-th vehicle's mass, F_{ei} is the driving force generated by the engine, F_{ai} is the aerodynamic resistance and is given as $F_{ai} = c_i v_i^2$, $c_i = h_{ai} A_i C_{di}/2$ with h_{ai} being the air density, A_i denoting the vehicle cross-section area, and C_{di} being the air drag coefficient, $F_{ri} = r_i m_i g \cos\theta$ is the rolling resistance, with r_i being the rolling resistance coefficient, g being the gravity acceleration, and θ being the road slope. $F_{gi} = m_i g \sin\theta$ is the gravity force. y_i denotes the unknown external disturbances caused by wind gusts and rough road surfaces or others. Here, we assume the disturbance is bounded, i.e. $|y_i| \le \bar{y}_i$, at the same time, the time derivation of y_i is bounded.

The engine dynamics is modeled as a first-order saturated model given by

$$\dot{F}_{ei}(t) = -\frac{1}{\tau_i}F_{ei}(t) + \frac{1}{\tau_i}\mathrm{sat}(u_i(t)) \tag{9.1c}$$

where τ_i is the engine time constant, u_i represents the throttle/brake input, which is subject to saturation nonlinearity due to structural limitations as described by:

$$\mathrm{sat}(u_i(t)) = \begin{cases} u_{\max i}, & u_i(t) \ge u_{\max i} \\ u_i(t), & u_{\min i} < u_i(t) < u_{\max i} \\ u_{\min i}, & u_i(t) \le u_{\min i} \end{cases} \tag{9.1d}$$

with $u_{\max i}$ and $u_{\min i}$ being the upper and lower bounds, respectively.

Let $f_i = r_i m_i g \cos\theta + m_i g \sin\theta$, and $D_i = y_i/\tau_i + \dot{y}_i$. It is assumed that the disturbances are bounded as $|m_i \tau_i D_i| \le \varepsilon_{di}$ [1]. Then the longitudinal dynamics of the ith vehicle can be rewritten as:

$$\dot{a}_i(t) = \frac{1}{m_i\tau_i} sat(u_i(t)) - \frac{1}{\tau_i}a_i(t) - \frac{1}{m_i\tau_i}\left[c_i\left(v_i^2 + 2\tau_i v_i a_i\right) + f_i\right] + D_i(t) \tag{9.2}$$

It is worth noting that the vehicle model in (9.1a)–(9.1d) and (9.2) involves the engine dynamics and the saturation of the throttle/brake into the vehicle dynamics in a more natural and practical way than the existing models in [2, 3].

The intervehicle spacing is defined as:

$$d_i = p_{i-1} - p_i - l_i \tag{9.3}$$

where l_i is the length of the i-th vehicle. Let $\delta_i = d_i - d^*$ be the spacing error of the i-th following vehicle with respect to the desired safe spacing d^*. Here, we utilize the following lumped error as the tracking error:

$$e_i = k_1(d_i - d^*) + k_2(v_i - v_0) + k_3(p_i - p_i^*) \tag{9.4}$$

where p_i^* represents the desired trajectory of the i-th vehicle with $p_i^* = p_0 - id^* - (l_1 + l_2 + \cdots + l_i)$, p_0 and v_0 are position and speed of leader vehicle and with $\dot{p}_0(t) = v_0(t)$. k_1, k_2, and k_3 are the weight parameters.

This chapter aims at prescribed performance platoon control and the performance is that the tracking errors remain strictly within the following prescribed region [4]:

$$\begin{cases} -\xi_i \rho_i(t) < e_i(t) < \rho_i(t), e_i(0) \geq 0 \\ -\rho_i(t) < e_i(t) < \xi_i \rho_i(t), e_i(0) \leq 0 \end{cases} \tag{9.5}$$

where $0 < \xi_i < 1$ is a design parameter, $e_i(0)$ is the initial tracking error and satisfies $|e_i(0)| < \rho_i(0)$. The performance function $\rho_i(t)$ is assumed positive and strictly decreasing, $\lim_{t \to \infty} \rho_i(t) = \rho_{i\infty} > 0$, which is given by:

$$\rho_i(t) = (\rho_{0i} - \rho_{\infty i}) \exp(-\phi_i t) + \rho_{\infty i} \tag{9.6}$$

where ρ_{0i}, $\rho_{\infty i}$, and ϕ_i are strictly positive constants satisfying $\rho_{\infty i} < \rho_{0i}$ and $\rho_{0i} = \rho_{0i}(0)$. Note that ϕ_i is the descent speed of $\rho_i(t)$, which is related to the convergence rate of tracking error $e_i(t)$.

The objective of this chapter is to find an adaptive platooning control methodology to achieve the following goals, despite the presence of external disturbances and input saturation:

1) Each vehicle in the platoon can maintain the desired intervehicle distance (i.e. $d_i(t) \to d^*$) and track the leader's velocity (i.e. $v_i(t) \to v_0(t)$).
2) A prescribed performance of the tracking error is achieved by the vehicles in the platoon, i.e. the tracking error evolves strictly within a predefined region bounded by a decaying time function.
3) String stability of the platoon is guaranteed in either of the following senses.

Definition 9.1 (*String Stability*) [5] The origins of the tracking error systems with vehicle dynamics (9.2) are said to be string stable; if given any $\varepsilon > 0$, there exists a $\delta > 0$ such that $\|e_i(0)\|_\infty < \delta \Rightarrow \sup_i \|e_i(\cdot)\|\ \infty < \varepsilon$.

Definition 9.2 (*Strong String Stability*) [1] The origins of the tracking error systems with vehicle dynamics (9.2) are said to be strong string stable if $|e_N(t)| \leq |e_{N-1}| \leq \cdots \leq |e_1(t)|$.

Before proceeding further, we give the following lemma, which will be used later.

Lemma 9.1 [6] For a continuous function $V(t) \geq 0$, with $V(0)$ bounded. If $\dot{V}_i \leq -q_1 V(t) + q_2$, then $V(t)$ is bounded with $q_1 > 0$, q_2 being a constant.

9.3 Model Transformation

We use the following smooth hyperbolic tangent function to approximate the actuator saturation $\text{sat}(u_i(t))$ [2]:

$$g(u_i(t)) = \begin{cases} u_{\max i} & \tanh(u_i(t)/u_{\max i}), u_i(t) \geq 0 \\ u_{\min i} & \tanh(u_i(t)/u_{\min i}), u_i(t) < 0 \end{cases} \tag{9.7}$$

It is inevitable that there exists a difference $\Delta u_i(t)$ between $\text{sat}(u_i(t))$ and $g(u_i(t))$. Let $\text{sat}(u_i(t))$ in (9.2) be:

$$\text{sat}(u_i(t)) = g(u_i(t)) + \Delta u_i(t) \tag{9.8}$$

where the difference $\Delta u_i(t)$ is bounded by:

$$\begin{aligned} |\Delta u_i(t)| &= |\text{sat}(u_i(t)) - g(u_i(t))| \\ &\leq \max\{u_{\max i}(1 - \tanh(1)), u_{\min i}(\tanh(1) - 1)\} \end{aligned} \tag{9.9}$$

Note that, according to the mean-value theorem, we can express $g(u_i(t))$ in (9.7) as follows:

$$g(u_i(t)) = g(u^0) + \left.\frac{\partial g(u_i(t))}{\partial u_i(t)}\right|_{u_i(t) = u_{\mu i}} (u_i(t) - u^0) \tag{9.10}$$

where $u_{\mu i} = \mu u_i(t) + (1 - \mu)u^0 (0 < \mu < 1)$.

Considering $g(0) = 0$ for $u^0 = 0$, let $g_{ui}(t) = (\partial g(u_i(t)))/(\partial u_i(t))|_{u_i(t) = u_{\mu i}}$. Assume that there exists an unknown positive constant g_{mi} such that $0 < g_{mi} \leq g_{ui}(t) \leq 1$ [2]. Then we can have:

$$\text{sat}(u_i(t)) = g_{ui}(t)u_i + \Delta u_i(t) \tag{9.11}$$

Then the dynamics of the i-th vehicle can be rewritten as:

$$\begin{aligned} \dot{a}_i(t) = &-\frac{1}{\tau_i}a_i(t) - \frac{1}{m_i\tau_i}\left[c_i\left(v_i^2 + 2\tau_i v_i a_i\right) + f_i\right] + \frac{1}{m_i\tau_i}g_{ui}(t)u_i(t) \\ &+ \frac{1}{m_i\tau_i}\Delta u_i(t) + D_i(t) \end{aligned} \tag{9.12}$$

Here we use the following transformation to transform the original constrained tracking error into an equivalent unconstrained one, where the transformed error is:

$$s_i(t) = \psi_i^{-1}(e_i(t)/\rho_i(t)) \tag{9.13}$$

where $\psi_i^{-1}(\bullet)$ is the inverse function of $\psi_i(\bullet)$ which is a smooth, strictly increasing, and invertible function satisfying:

$$\begin{cases} \lim_{s_i \to -\infty} \psi_i(s_i) = -\xi_i \\ \lim_{s_i \to +\infty} \psi_i(s_i) = 1 \end{cases}, \quad \text{if } e_i(0) \geq 0 \tag{9.14a}$$

$$\begin{cases} \lim_{s_i \to -\infty} \psi_i(s_i) = -1 \\ \lim_{s_i \to +\infty} \psi_i(s_i) = \xi_i \end{cases}, \quad \text{if } e_i(0) \leq 0 \tag{9.14b}$$

Here, we choose $\psi_i(s_i)$ as [4]:

$$\begin{cases} \psi_i(s_i) = \dfrac{e^{s_i} - \xi_i e^{-s_i}}{e^{s_i} + e^{-s_i}}, & \text{if } e_i(0) \geq 0 \\ \psi_i(s_i) = \dfrac{\xi_i e^{s_i} - e^{-s_i}}{e^{s_i} + e^{-s_i}}, & \text{if } e_i(0) \leq 0 \end{cases} \tag{9.15}$$

Based on Equations (9.13) and (9.15), we can obtain:

$$\begin{cases} s_i(t) = \dfrac{1}{2}\ln\dfrac{e_i(t) + \xi_i\rho_i(t)}{\rho_i(t) - e_i(t)}, & \text{if } e_i(0) \geq 0 \\ s_i(t) = \dfrac{1}{2}\ln\dfrac{e_i(t) + \rho_i(t)}{\xi_i\rho_i(t) - e_i(t)}, & \text{if } e_i(0) \leq 0 \end{cases} \tag{9.16}$$

Thus, as long as the transformed error is bounded, the following facts hold true:

$$\begin{cases} -\xi_i < \psi_i(s_i) < 1, & \text{if } e_i(0) \geq 0 \\ -1 < \psi_i(s_i) < \xi_i, & \text{if } e_i(0) \leq 0 \end{cases} \tag{9.17}$$

Remark 9.1 For the convenience of analysis of PPC, here we give a general PPC scheme depending on the initial value of the tracking error. By the tracking error transformation, the tracking error will be forced into the allowable region as shown in Equation (9.5). There are other prescribed performances to make the control scheme more practical and effective. Details can be seen in [7–10].

9.4 Vehicles Tracking Controller Design

In this section, prescribed tracking performance platoon controllers are designed for the follower vehicles such that they all mutually synchronize and follow the leader vehicle, maintaining the desired intervehicle spacing. Furthermore, based on different definitions of string stability, we give two platoon control schemes to guarantee string stability of vehicle platoon including LP and LB information flow topologies.

In this subsection, a novel controller based on PPC is designed to guarantee the string stability of platoons defined in Definition 9.1. Unlike previous work, a new sliding surface with the transformed error is designed as:

$$S_i(t) = \dot{s}_i + \lambda s_i \tag{9.18}$$

where $\lambda > 0$ should satisfy the Hurwitz condition.

To design a control law to drive $S_i(t)$ converge to zero, consider the time derivative of $S_i(t)$:

$$\dot{S}_i(t) = \ddot{s}_i + \lambda\dot{s}_i \tag{9.19a}$$

Then, based on Equation (9.16), we can obtain $\dot{s}_i$ and $\ddot{s}_i$ as:

$$\dot{s}_i(t) = \frac{\partial\psi_i^{-1}}{\partial(e_i/\rho_i)}\frac{1}{\rho_i}\left(\dot{e}_i - \frac{e_i\dot{\rho}_i}{\rho_i}\right) = R_i\left(\dot{e}_i - \frac{e_i\dot{\rho}_i}{\rho_i}\right) \tag{9.19b}$$

$$\ddot{s}_i = \dot{R}_i\left(\dot{e}_i - \frac{e_i\dot{\rho}_i}{\rho_i}\right) + R_i\left(\ddot{e}_i - \frac{(\dot{e}_i\dot{\rho}_i + e_i\ddot{\rho}_i)\rho_i - e_i\dot{\rho}_i^{\,2}}{\rho_i^{\,2}}\right) \tag{9.19c}$$

where $R_i = \left(\partial\psi_i^{-1}/\partial\left(\dfrac{e_i}{\rho_i}\right)\right)\left(\dfrac{1}{\rho_i}\right)$, $R_i > 0$.

The controller to be designed is in the form as follows:

$$u_i = \frac{1}{R_ik_2g_{m\,i}}\left[-k_4S_i + R_ik_2c_i\left(v_i^2 + 2\tau_iv_ia_i\right) + R_ik_2f_i - R_ik_2\hat{\bar{\varepsilon}}_i\tanh\left(\frac{S_i}{\gamma_i}\right) - \tau_im_iT_i\right] \tag{9.20}$$

where k_4 and γ_i are positive constants, $\hat{\overline{\varepsilon}}_i$ is the estimation of $\overline{\varepsilon}_i$, $|\Delta u_i + \varepsilon_{di}| \leq \overline{\varepsilon}_i$, T_i is given by:

$$\begin{aligned} T_i(t) = R_i\left[k_1(a_{i-1} - a_i) + k_3(a_i - a_0) - \frac{k_2}{\tau_i}a_i\right] + \lambda\dot{s}_i \\ + \dot{R}_i\left(\dot{e}_i - \frac{e_i\dot{\rho}_i}{\rho_i}\right) - R_i\frac{(\dot{e}_i\dot{\rho}_i + e_i\ddot{\rho}_i)\rho_i - e_i\dot{\rho}_i^2}{\rho_i^2} \end{aligned} \tag{9.21}$$

Note that sometimes in practice, parameters c_i, m_i, and f_i in the vehicle model are unknown, we can use their estimated value, $\hat{c}_i$, $\hat{m}_i$ and $\hat{f}_i$ to design the controller as follows:

$$\begin{aligned} u_i = \frac{1}{R_i k_2 g_{m\,i}}\Big[- k_4 S_i + R_i k_2 \hat{c}_i\left(v_i^2 + 2\tau_i v_i a_i\right) \\ + R_i k_2 \hat{f}_i - R_i k_2 \hat{\overline{\varepsilon}}_i \tanh\left(\frac{S_i}{\gamma_i}\right) - \tau_i \hat{m}_i T_i\Big] \end{aligned} \tag{9.22}$$

with the following adaptation laws:

$$\dot{\hat{c}}_i(t) = -\eta_{ci}\left[\left(v_i^2 + 2\tau_i v_i a_i\right)R_i k_2 S_i + \sigma_{ci}\hat{c}_i\right] \tag{9.23a}$$

$$\dot{\hat{f}}_i(t) = -\eta f_i\left[R_i k_2 S_i + \sigma_{fi}\hat{f}_i\right] \tag{9.23b}$$

$$\dot{\hat{m}}_i(t) = \eta_{mi}[\tau_i T_i S_i - \sigma_{m\,i}\hat{m}_i] \tag{9.23c}$$

$$\dot{\hat{\overline{\varepsilon}}}_i(t) = \eta_{\varepsilon i}\left[R_i k_2 S_i \tanh\left(\frac{S_i}{\gamma_i}\right) - \sigma_{\overline{\varepsilon} i}\hat{\overline{\varepsilon}}_i\right] \tag{9.23d}$$

where η_{ci}, η_{fi}, η_{mi}, and η_{mi} are positive parameters, σ_{ci}, σ_{fi}, σ_{mi}, and $\sigma_{\overline{\varepsilon}i}$ are small constants.

Remark 9.2 In controller (9.22), a continuous hyperbolic tangent function is used for switching instead of a sign function or saturation function as usually done, which can mitigate the chattering phenomenon in sliding mode control. Moreover, a modification is introduced in adaptive laws (9.23) to improve the robustness, which prevents the estimated values of the adaptation parameters from drifting very large [11].

To sum up, our first result is given in the following theorem, which guarantees the bounded stability of individual vehicles and string stability of the whole platoon under the LP scheme.

Theorem 9.1 For a platoon of vehicles with dynamics (9.12), the control law in (9.22) and the updating laws in (9.23) can guarantee the stability of vehicles in the sense that the tracking error e_i and the transformed error s_i are uniformly ultimately bounded with the prescribed transient/steady state performance.

Proof: For stability analysis, choose the following Lyapunov function candidate:

$$V_i = \frac{m_i\tau_i}{2}S_i^2 + \frac{1}{2\eta_{ci}}\tilde{c}_i^2 + \frac{1}{2\eta_{fi}}\tilde{f}_i^2 + \frac{1}{2\eta_{mi}}\tilde{m}_i^2 + \frac{1}{2\eta_{\overline{\varepsilon}i}}\tilde{\overline{\varepsilon}}_i^2 \tag{9.24}$$

where $(\tilde{\cdot}) = (\hat{\cdot}) - (\cdot)$ is the estimation error. The time derivative of V_i can be derived as:

$$\dot{V}_i = m_i\tau_i S_i\dot{S}_i + \frac{1}{\eta_{ci}}\tilde{c}_i\dot{\hat{c}}_i + \frac{1}{\eta_{fi}}\tilde{f}_i\dot{\hat{f}}_i + \frac{1}{\eta_{mi}}\tilde{m}_i\dot{\hat{m}}_i + \frac{1}{\eta_{\overline{\varepsilon}i}}\tilde{\overline{\varepsilon}}_i\dot{\hat{\overline{\varepsilon}}}_i \tag{9.25}$$

Based on (9.19) and (9.22), it can be written as:

$$\dot{S}_i = \frac{R_i k_2 g_{ui}}{m_i \tau_i} \frac{1}{R_i k_2 g_{m\,i}} \left[-k_4 S_i + R_i k_2 \hat{c}_i \left(v_i^2 + 2\tau_i v_i a_i\right) + R_i k_2 \hat{f}_i - R_i k_2 \hat{\varepsilon}_i \tanh\left(\frac{S_i}{\gamma_i}\right) - \tau_i \hat{m}_i T_i \right]$$
$$- \frac{R_i k_2}{m_i \tau_i} \left[c_i \left(v_i^2 + 2\tau_i v_i a_i\right) + f_i \right] + \frac{R_i k_2}{m_i \tau_i} \left[\Delta u_i + m_i \tau_i D_i\right] + T_i \tag{9.26}$$

Then we can derive that:

$$m_i \tau_i S_i \dot{S}_i \le -k_4 S_i^2 + \tilde{c}_i \left(v_i^2 + 2\tau_i v_i a_i\right) R_i k_2 S_i + \tilde{f}_i R_i k_2 S_i - \tau_i T_i S_i \tilde{m}_i - R_i k_2 S_i \left(\hat{\bar{\varepsilon}}_i \tanh\left(\frac{S_i}{\gamma_i}\right) - \bar{\varepsilon}_i \right) \tag{9.27}$$

with the adaptive laws in (9.23), one can obtain that

$$\dot{V}_i \le -k_4 S_i^2 - R_i k_2 S_i \left(\hat{\bar{\varepsilon}}_i \tanh\left(\frac{S_i}{\gamma_i}\right) - \bar{\varepsilon}_i \right) - \sigma_{ci} \tilde{c}_i \hat{c}_i - \sigma_{fi} \tilde{f}_i \hat{f}_i - \sigma_{mi} \tilde{m}_i \hat{m}_i$$
$$- \sigma_{\bar{\varepsilon}i} \tilde{\bar{\varepsilon}}_i \hat{\bar{\varepsilon}}_i + R_i k_2 S_i \tilde{\bar{\varepsilon}}_i \tanh\left(\frac{S_i}{\gamma_i}\right) \tag{9.28}$$

Based on the property $0 \le |\kappa| - \kappa \tanh(\kappa/\gamma) \le k_\gamma \gamma$, $k_\gamma = 0.2785$ for $\forall k_\gamma > 0$ and $\kappa \in \mathfrak{R}$, $\mathfrak{R}$ denote real numbers [12], one can obtain the following inequality:

$$|S_i| - S_i \tanh\left(\frac{S_i}{\gamma_i}\right) \le k_\gamma \gamma_i, k_\gamma = 0.2785 \tag{9.29}$$

Further, we can obtain that

$$-\bar{\varepsilon}_i R_i k_2 S_i \tanh\left(\frac{S_i}{\gamma_i}\right) \le R_i k_2 \left(\bar{\varepsilon}_i k_\gamma \gamma_i - \bar{\varepsilon}_i |S_i|\right) \tag{9.30}$$

According to Young's inequality, we have:

$$-\sigma_{ci} \tilde{c}_i \hat{c}_i \le \frac{\sigma_{ci}}{2} c_i^2 - \frac{\sigma_{ci}}{2} \tilde{c}_i^2 \tag{9.31a}$$

$$-\sigma_{fi} \tilde{f}_i \hat{f}_i \le \frac{\sigma_{fi}}{2} f_i^2 - \frac{\sigma_{fi}}{2} \tilde{f}_i^2 \tag{9.31b}$$

$$-\sigma_{mi} \tilde{m}_i \hat{m}_i \le \frac{\sigma_{mi}}{2} m_i^2 - \frac{\sigma_{mi}}{2} \tilde{m}_i^2 \tag{9.31c}$$

$$-\sigma_{\bar{\varepsilon}i} \tilde{\bar{\varepsilon}}_i \hat{\bar{\varepsilon}}_i \le \frac{\sigma_{\bar{\varepsilon}i}}{2} \bar{\varepsilon}_i^2 - \frac{\sigma_{\bar{\varepsilon}u}}{2} \tilde{\bar{\varepsilon}}_i^2 \tag{9.31d}$$

Then we can have that:

$$\dot{V}_i \le -k_4 S_i^2 - \frac{\sigma_{ci}}{2} \tilde{c}_i^2 - \frac{\sigma_{fi}}{2} \tilde{f}_i^2 - \frac{\sigma_{mi}}{2} \tilde{m}_i^2 - \frac{\sigma_{\bar{\varepsilon}i}}{2} \tilde{\bar{\varepsilon}}_i^2 + \frac{\sigma_{ci}}{2} c_i^2 + \frac{\sigma_{fi}}{2} f_i^2$$
$$+ \frac{\sigma_{mi}}{2} m_i^2 + \frac{\sigma_{\bar{\varepsilon}i}}{2} \bar{\varepsilon}_i^2 + R_i k_2 \bar{\varepsilon}_i k_\gamma \gamma_i \le -\varsigma_i V_i + \chi_i \tag{9.32}$$

where

$$\varsigma_i = \frac{\min\left\{2k_4, \sigma_{ci}, \sigma_{fi}, \sigma_{mi}, \sigma_{\bar{\varepsilon}i}\right\}}{\max\left\{m_i\tau_i, 1/\eta_{ci}, 1/\eta_{fi}, 1/\eta_{mi}, 1/\eta_{\bar{\varepsilon}i}\right\}} \tag{9.33a}$$

$$\chi_i = \frac{\sigma_{ci}}{2}c_i^2 + \frac{\sigma_{fi}}{2}f_i^2 + \frac{\sigma_{mi}}{2}m_i^2 + \frac{\sigma_{\bar{\varepsilon}i}}{2}\bar{\varepsilon}_i^2 + R_i k_2 \bar{\varepsilon}_i k_\gamma \gamma_i \tag{9.33b}$$

Thus, according to Lemma 9.1, we know that V_i is ultimately bounded. Furthermore, the following inequality holds:

$$\begin{aligned} V_i(t) &\leq [V_i(0) - \chi_i/\varsigma_i]e^{-\varsigma_i t} + \chi_i/\varsigma_i \\ &\leq V_i(0) + \chi_i/\varsigma_i \end{aligned} \tag{9.34}$$

We now proceed to prove the string stability of the vehicular platoon. Consider the following global Lyapunov function:

$$V(t) = \sum_{i=1}^{N} V_i(t) \tag{9.35a}$$

whose time derivative can be obtained as

$$\dot{V} = -\varsigma \sum_{i=1}^{N} V_i(t) + \sum_{i=1}^{N} \chi_i = -\varsigma V + \chi \tag{9.35b}$$

where $\varsigma = \varsigma_i$, $\chi = \sum_{i=1}^{N} \chi_i$. In light of (9.32), it follows that

$$0 \leq V \leq [V(0) - \chi/\varsigma]e^{-\varsigma t} + \chi/\varsigma \tag{9.36}$$

This means that V is uniformly ultimately bounded, which along with the definitions, implies that $S_i(t)$ can converge to a small neighborhood of zero by choosing the appropriate parameters. Then, in view of (9.18), the transformed error $s_i(t)$ is uniformly ultimately bounded. Hence, based on (9.13) and (9.17), the tracking error is uniformly ultimately bounded, i.e. (9.5) is satisfied and the entire platoon system is string stable. The proof is completed.

Remark 9.3 The PPC scheme with LP information pattern is shown to be effective in coping with the input saturation, uncertain parameters, and external disturbances to achieve the prescribed tracking performance with guaranteed string stability in Definition 9.1.

In the forthcoming subsection, an LB control scheme is given to guarantee strong string stability as defined in Definition 9.2.

Since the sliding surface defined in (9.18) cannot guarantee strong string stability, to guarantee strong string stability, similar to [1], here we introduce a modified coupled sliding surface based on the transformed error for vehicle i:

$$\Pi_i(t) = qS_i(t) - S_{i+1} \tag{9.37}$$

where $i \in \nu_N$, $q > 0$ is a weighting factor. For the last vehicle, $S_{N+1} = 0$, and $s_{N+1} = \dot{s}_{N+1} = \ddot{s}_{N+1} = 0$. Then the relationship between $\Pi_i(t)$ and $S_i(t)$ is obtained as:

$$\boldsymbol{\Pi}(t) = \boldsymbol{QS}(t) \tag{9.38}$$

where

$$\boldsymbol{\Pi}(t) = [\Pi_1(t)\Pi_2(t)\cdots\Pi_N(t)]^T,$$

$$\mathbf{S}(t) = [S_1(t)S_2(t)\cdots S_N(t)]^T,$$

$$\mathbf{Q} = \begin{bmatrix} q & -1 & \cdots & 0 & 0 \\ 0 & q & -1 & \cdots & 0 \\ & & \vdots & & \\ 0 & 0 & \cdots & q & -1 \\ 0 & 0 & \cdots & 0 & q \end{bmatrix}$$

According to Lemma 9.1 of [1], the convergence of $\Pi_i(t)$ for all $i \in \nu_N$ can guarantee the same for $S_i(t)$. In what follows, we will give the main result guaranteeing not only individual vehicle stability but also strong string stability of the vehicular platoon. To this end, we introduce the following control law:

$$u_i = \frac{1}{qR_i k_2 g_{mi}}\Big[-k_4\Pi_i + qR_i k_2 \hat{c}_i(v_i^2 + 2\tau_i v_i a_i) + qR_i k_2 \hat{f}_i - qR_i k_2 \hat{\varepsilon}_i \tanh\left(\frac{\Pi_i}{\gamma_i}\right) - \tau_i \hat{m}_i P_i\Big] \tag{9.39}$$

where P_i is given by:

$$P_i = q\left[\lambda\dot{s}_i + \dot{R}_i\left(\dot{e}_i - \frac{e_i\dot{\rho}_i}{\rho_i}\right) - R_i\frac{(\dot{e}_i\dot{\rho}_i + e_i\ddot{\rho}_i)\rho_i - e_i\dot{\rho}_i^2}{\rho_i^2}\right] + qR_i\left[k_1(a_{i-1} - a_i) + k_3(a_i - a_0) - \frac{k_2}{\tau_i}a_i\right] - \ddot{s}_{i+1} - \lambda\dot{s}_{i+1}$$

and the adaptation laws of the parameters are:

$$\dot{\hat{c}}_i = -\eta_{ci}\left[(v_i^2 + 2\tau_i v_i a_i)qR_i k_2\Pi_i + \sigma_{ci}\hat{c}_i\right] \tag{9.40a}$$

$$\dot{\hat{f}}_i = -\eta_{fi}\left[qR_i k_2\Pi_i + \sigma_{fi}\hat{f}_i\right] \tag{9.40b}$$

$$\dot{\hat{m}}_i = \eta_{mi}[\tau_i P_i\Pi_i - \sigma_{mi}\hat{m}_i] \tag{9.40c}$$

$$\dot{\hat{\varepsilon}}_i = \eta_{\varepsilon i}\left[qR_i k_2\Pi_i \tanh\left(\frac{\Pi_i}{\gamma_i}\right) - \sigma_{\overline{\varepsilon} i}\hat{\overline{\varepsilon}}_i\right] \tag{9.40d}$$

The main result is given below.

Theorem 9.2 For a platoon of vehicles with dynamics (9.12), the controller u_i in (9.39) with parameters, the adaptation laws in (9.40) can guarantee the stability of vehicles in the sense that the tracking error e_i and the transformed error s_i are uniformly ultimately bounded with the prescribed transient/steady state performance. Meanwhile, the strong string stability of the vehicular platoon can be guaranteed if q satisfies $0 < q \leq 1$.

Proof: The proof of individual vehicle stability is similar to Section 9.3 and we can easily prove that the tracking error e_i and the transformed error s_i are uniformly ultimately bounded and $\Pi_i(t)$ (and $S_i(t)$) can converge to zero. What remains is to prove the strong string stability of the vehicular platoon, which is done in the Laplace domain as in [7]. Since $\Pi_i = qS_i - S_{i+1}$ can converge to zero in finite time, we can have:

$$q(\dot{s}_i + \lambda s_i) = \dot{s}_{i+1} + \lambda s_{i+1} \tag{9.41}$$

whose Laplace transform is obtained as follows:

$$q[(s+\lambda)E_i(s)] = (s+\lambda)E_{i+1}(s) \tag{9.42}$$

where $\overline{E}_i(s)$ is the Laplace transform of s_i. Therefore, the transfer function of the transformed error is:

$$G_i(s) = \frac{\overline{E}_{i+1}(s)}{\overline{E}_i(s)} = q \tag{9.43}$$

According to the definition in (9.13), the transformed error $s_i(t)$ has the same monotonicity with the tracking error $e_i(t)$. Thus, the string stability of the vehicular platoon is guaranteed when $0 < q \leq 1$. This completes the proof.

Remark 9.4 The difference between the two sliding mode control schemes is that vehicle i needs the information such as acceleration and velocity from the vehicle $i+1$ in order to guarantee strong string stability of vehicular platoons.

Remark 9.5 Note that string stability is of essential importance for platoon control. No theoretical analysis of string stability was given in the closely related works in [3, 13]. The SMC platooning method in [1, 2] can achieve strong string stability, but no prescribed performance is investigated. The method in this chapter involves a PPC scheme to guarantee strong string stability besides achieving both transient and steady-state performances of tracking errors. It is worth mentioning that here we use an indirect method to achieve prescribed performance and strong string stability. It is possible to address this issue using a direct method. For instance, one can introduce hierarchical performance functions with respect to tracking error e_i and use coupled sliding surface $\Pi_i = qS_i - S_{i+1}$ with $s_i = \dot{e}_i + \lambda e_i$. This can be investigated in the future.

9.5 Simulation Studies

In this section, the effectiveness of the proposed platoon control methods is evaluated via numerical simulations. We present two simulation studies: (i) Prescribed performance (PP) tracking control with the LP scheme; (ii) PP tracking control with the LB scheme.

In the simulations, a six-vehicle platoon with parameters assumed the same with [14]: $m_i = 1607$, $c_i = 0.414$, $f_i = 236.229$, $\tau_i = 0.25$, and $D_i(t) = 0.1\sin(t)$. The specific trajectories of the leader vehicle are given in Figure 9.2, and $p_0(0) = 60$. The desired distance between two consecutive vehicles is set as $d^* = 10$ m and the vehicle length is 2m. The initial positions and velocities of the vehicles in the platoon are chosen as follows: $p_i(0) = [0, 12, 24, 36, 48]$, and $v_i(0) = 0$, with zero initial tracking errors.

This subsection verifies the proposed LP controller in Theorem 9.1 and Definition 9.1. The controller parameters are listed in Table 9.1. The prescribed performance function used is $\rho(t) = (0.8 - 0.09)\exp(-0.5t) + 0.09$, with $\xi_i = 1$. Simulation results are given in Figures 9.3 and 9.4,

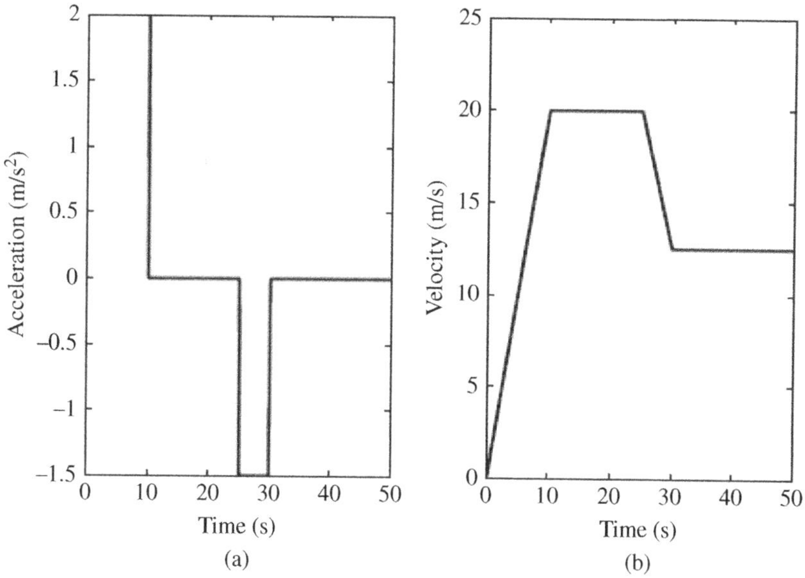

Figure 9.2 Profile of the leader vehicle: (a) Acceleration; (b) Velocity.

Table 9.1 Control parameters of vehicles (i = 1, 2, ..., 5).

k_1	k_2	k_3	k_4	λ	γ_i	η_i	σ_i	g_{mi}
0.1	0.1	1	65	20	0.01	0.001	0.001	0.5

where Figure 9.3 shows the platoon control results under the LP control scheme with prescribed performance, while Figure 9.4 shows those without a prescribed performance.

We also show that the method is effective for the case with nonzero initial tracking errors. The initial vehicle-spacing errors are set as $\delta_1 = \delta_3 = \delta_4 = 0.2$, $\delta_2 = 0.8$, and $\delta_5 = 0$. The initial velocities of the vehicles in the platoon are chosen as $v_i(0) = [1, 1.5, 2, 1, 2]$. The control parameters in this scheme are listed in Table 9.1 except that $k_2 = 0.5$, $k_4 = 75$, and $\lambda = 35$, and the prescribed performance function used in this case is $\rho(t) = (2 - 0.5)\exp(-0.5t) + 0.5$, with $\xi_i = 1$, which satisfies $|e_i(0)| < \rho(0)$. The simulation results are shown in Figure 9.5.

From Figures 9.3g and 9.5a, we can see that the tracking errors of all vehicles can achieve the prescribed region, guaranteeing string stability, whether the initial tracking errors are zero or not. Moreover, from Figures 9.4 and 9.5b, it can be obviously seen that the LP scheme without PPC cannot guarantee string stability.

We now proceed to demonstrate the effectiveness of the PP tracking control with the LB scheme. Simulation results are given in Figures 9.6–9.8 where Figure 9.6 shows the platoon control results under the LB control scheme with prescribed performance, Figure 9.7 shows those without a prescribed performance, and Figure 9.8 shows the tracking errors under the PP tracking control with

LB scheme and nonzero initial tracking errors. Note that in the simulations, the same prescribed performance function and control parameters are used as the previous subsection except that $k_4 = 155$, $\lambda = 75$, and $q = 0.9$ in Figures 9.6 and 9.7, and $k_2 = 0.5$, $k_4 = 145$, and $\lambda = 45$ in Figure 9.8.

It is seen from Figures 9.6–9.8 that the tracking errors of all vehicles can achieve strong string stability whether the initial tracking errors are zero or not, i.e. $|e_5| \leq |e_4| \leq \cdots \leq |e_1|$. At the same time, the LB scheme with PP has a better transient and steady-state tracking performance than the case without PP. Unlike the method in [3, 13], where string stability is not considered and hence their result cannot guarantee string stability, our method can achieve string stability of vehicular platoons, which is very important for platooning. In summary, the effectiveness of the proposed methods is well demonstrated, i.e. string stability of vehicular platoons can be achieved in the presence of input saturation, uncertain parameters, and disturbances with prescribed transient and steady-state tracking performance.

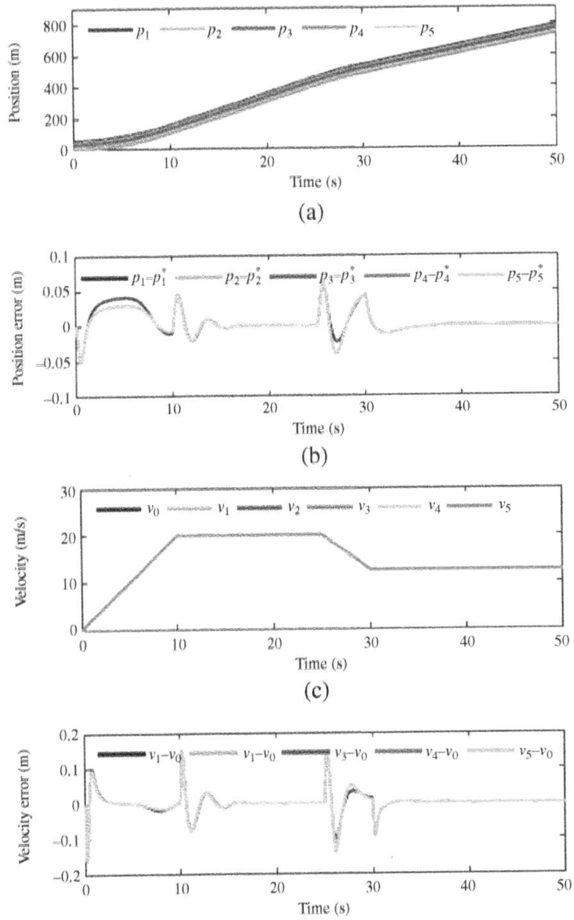

Figure 9.3 The results of PP tracking control with LP scheme: (a) Position; (b) position tracking error; (c) velocity; (d) velocity tracking error; (e) intervehicle distance between consecutive vehicles; (f) the distance error with the desired distance; (g) tracking error.

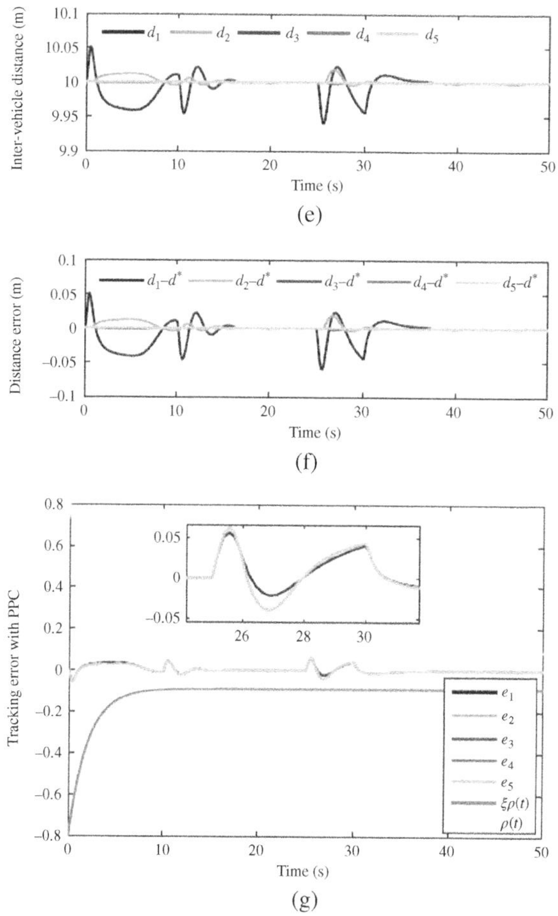

Figure 9.3 (Continued)

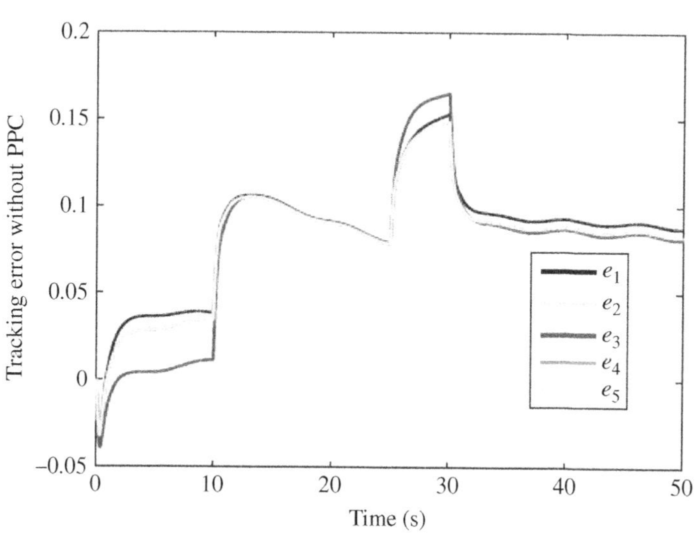

Figure 9.4 The errors of LP control without PP.

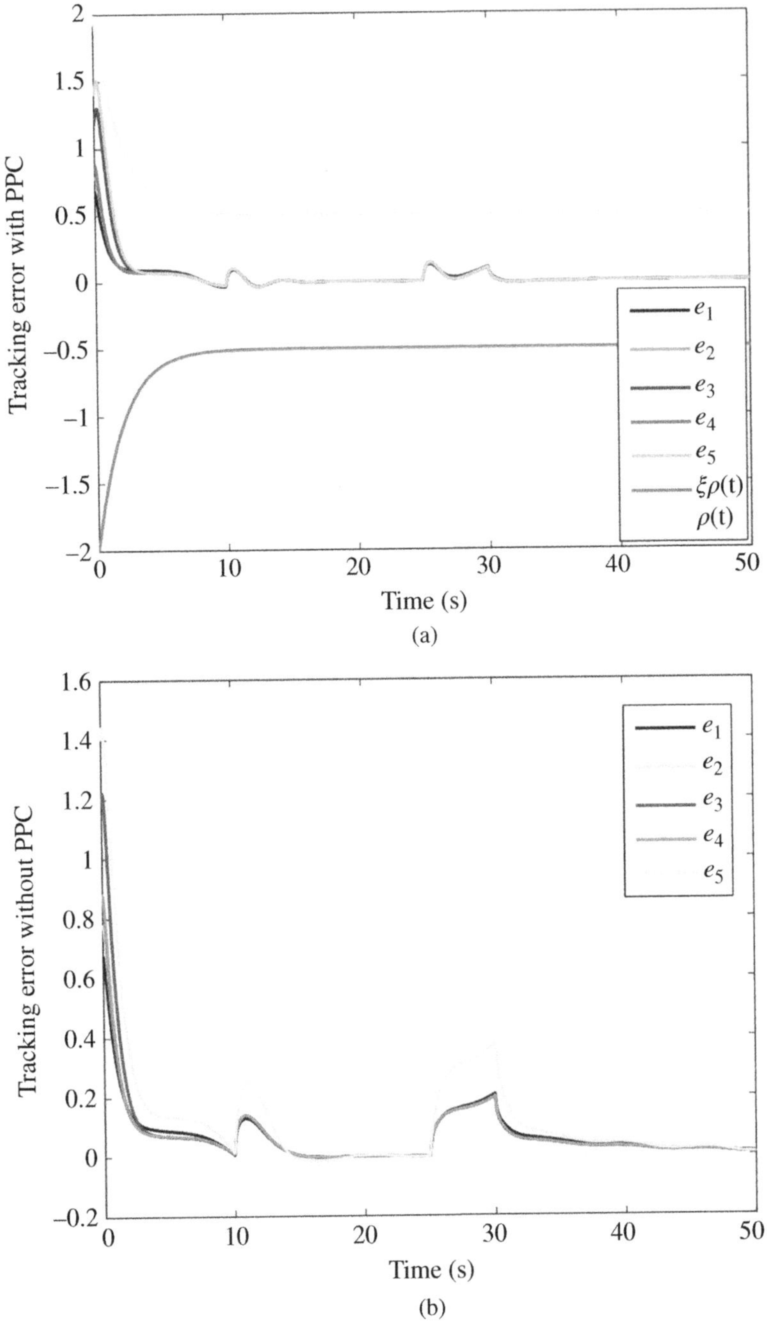

Figure 9.5 The tracking error results of error of the LP control method with nonzero initial tracking errors. (a) Tracking error with PPC. (b) Tracking error without PPC.

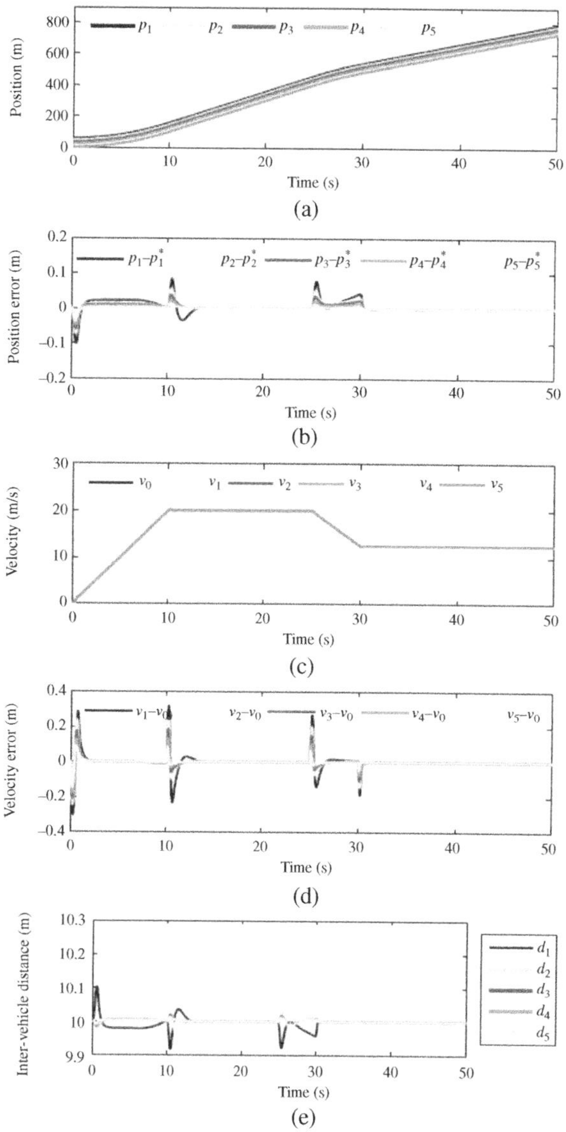

Figure 9.6 The results of PP tracking control with LB scheme: (a) Position, (b) position tracking error, (c) velocity, (d) velocity tracking error, (e) intervehicle distance between consecutive vehicles, (f) the distance error with the desired distance, (g) tracking error.

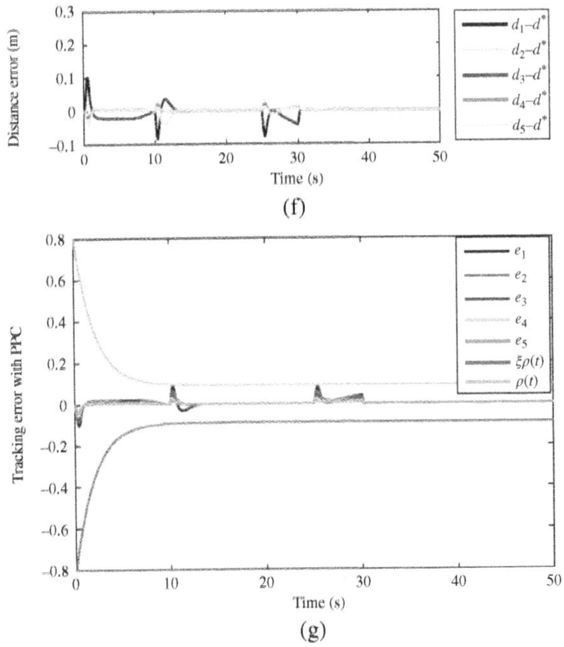

Figure 9.6 (Continued)

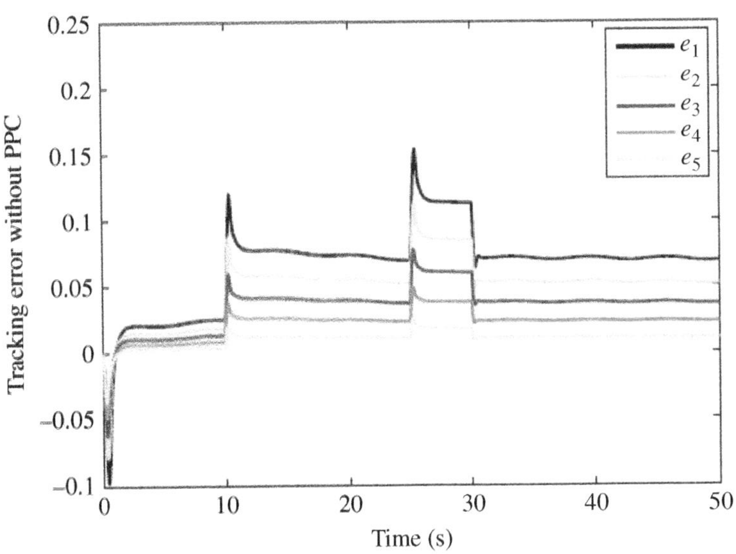

Figure 9.7 The tracking errors of the LB control method without PPC.

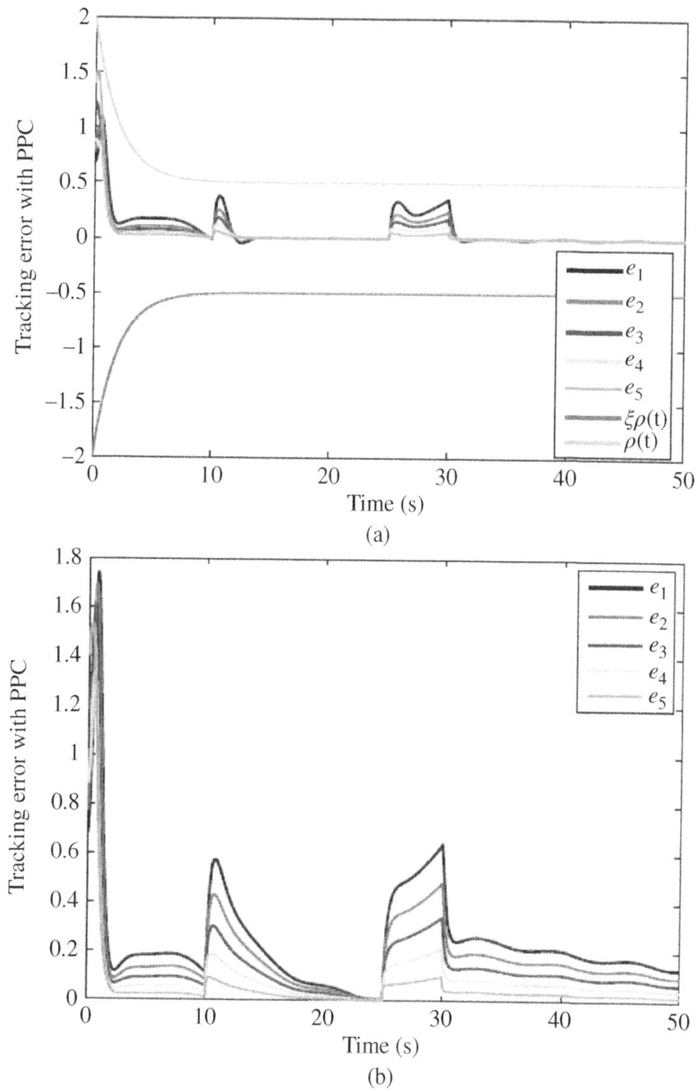

Figure 9.8 The tracking errors under the LB control method with nonzero initial tracking errors. (a) Tracking error with PPC. (b) Tracking error without PPC.

9.6 Conclusions and Future Work

In this chapter, the prescribed tracking performance control problem for vehicle platoons is studied in the presence of input saturation, uncertainty parameters, and external disturbances. Two adaptive sliding-mode platoon control schemes based on LP and LB information topologies are proposed. Under both schemes, the vehicles can follow each other and maintain the desired intervehicle spacing with guaranteed string stability and prescribed performance. Finally, simulation results show the effectiveness of the proposed platoon control schemes.

References

1 Kwon, J.W. and Chwa, D. (2014). Adaptive bidirectional platoon control using a coupled sliding mode control method. *IEEE Trans. Intell. Transp. Syst.* 15 (5): 2040–2048.

2 Guo, X., Wang, J., Liao, F., and Teo, R.S.H. (2018). CNN-based distributed adaptive control for vehicle-following platoon with input saturation. *IEEE Trans. Intell. Transp. Syst.* 19 (10): 3121–3132.

3 Bechlioulis, C.P., Dimarogonas, D.V., and Kyriakopoulos, K.J. (2014). Robust control of large vehicular platoons with prescribed transient and steady state performance. *Proc. IEEE Conf. Decision Control*, Los Angeles, CA (December 2014), pp. 3689–3694.

4 Bechlioulis, C.P. and Rovithakis, G.A. (2008). Robust adaptive control of feedback linearizable MIMO nonlinear systems with prescribed performance. *IEEE Trans. Autom. Control* 53 (9): 2090–2099.

5 Swaroop, D., Hedrick, J.K., and Choi, S.B. (2001). Direct adaptive longitudinal control of vehicle platoons. *IEEE Trans. Veh. Technol.* 50 (1): 150–161.

6 Ge, S.S. and Wang, C. (2004). Adaptive neural control of uncertain MIMO nonlinear systems. *IEEE Trans. Neural Netw.* 15 (3): 674–692.

7 Bechlioulis, C.P. and Rovithakis, G.A. (2014). A low-complexity global approximation-free control scheme with prescribed performance for unknown pure feedback systems. *Automatica* 50 (4): 1217–1226.

8 Liu, D. and Yang, G.H. (2017). Data-driven adaptive sliding mode control of nonlinear discrete-time systems with prescribed performance. *IEEE Trans. Syst. Man Cybern. Syst.* 1–7, https://doi.org/10.1109/TSMC.2017.2779564.

9 Liu, D. and Yang, G.H. (2018). Performance-based data-driven model-free adaptive sliding mode control for a class of discrete-time nonlinear processes. *J. Process Control* 68: 186–194.

10 Liu, D. and Yang, G.H. (2018). Prescribed performance model-free adaptive integral sliding mode control for discrete-time nonlinear systems. *IEEE Trans. Neural Netw. Learn. Syst.*, doi:https://doi.org/10.1109/TNNLS.2018.2881205.

11 Ioannou, P.A. and Kokotovic, P.V. (1984). Instability analysis and improvement of robustness of adaptive control. *Automatica* 20 (5): 583–594.

12 Polycarpou, M.M. and Ioannou, P.A. (1996). A robust adaptive nonlinear control design. *Automatica* 32 (3): 423–427.

13 Verginis, C.K., Bechlioulis, C.P., Dimarogonas, D.V., and Kyriakopoulos, K.J. (2018). Robust distributed control protocols for large vehicular platoons with prescribed transient and steady-state performance. *IEEE Trans. Control Syst. Technol.* 26 (1): 299–304.

14 Guo, G. and Li, D. (2018). PMP-based set-point optimization and sliding-mode control of vehicular platoons. *IEEE Trans. Comput. Soc. Syst.* 5 (2): 553–562.

Part III

Speed Trajectory Planning and Control

This part is concerned with speed trajectory planning and control for connected vehicles, with the aim to reduce fuel-time cost, improve road capacity, and guarantee road safety. There are five chapters in this part. **Chapter 10** studies the speed optimization and tracking control problem for heavy-duty truck platoons. The speed planning procedure is based on the fuel-time cost and derived for an average vehicle using receding dynamic programming. A back-stepping controller is given based on the nonlinear vehicle dynamics and the road slope, which can ensure string stability. **Chapter 11** gives a two-layered solution to the speed optimization and tracking control problem. The first layer gives the optimal team speed by averaging the fuel-time optimal speed of all vehicles obtained via Pontryagin's minimum principle. The second layer contains the distributed sliding-mode controllers to guarantee string stability and constant inter-vehicle spacing. **Chapters 12–14** investigate the same optimization and control problem all from the view of minimizing the inter-vehicle spacing errors but with different concerns. The solutions in these chapters provide distributed two-layered trajectory optimization and tracking control schemes. The differences between these solutions mainly lie in the controller in the lower layer, i.e. **Chapter 12** presents an adaptive sliding-mode controller that can guarantee string stability, **Chapter 13** gives a PID-type sliding-mode controller with a quadratic spacing policy that can guarantee string stability and traffic flow stability, and **Chapter 14** provides a fixed-time terminal sliding-mode controller that can robustly guarantee zero steady-state spacing errors and string stability.

Connected Vehicular Systems: Communication, Control, and Optimization, First Edition. Ge Guo and Shixi Wen.

10

Speed Planning and Tracking Control of Vehicles

10.1 Introduction

In Chapters 1–10, the attention is focused on platoon controller design to guarantee string stability and performance under a given reference speed. However, for a vehicular platoon, running under different reference speeds may yield different fuel consumptions even for a given road segment. Since vehicular platoons may operate with various lengths and the vehicles involved may have different sizes, weights, and dynamics, it is clearly unreasonable for all platoons to drive with the same reference speed under all road conditions and for whatever cars involved. Therefore, a properly designed speed set-point is of great importance to maximize the benefits of platooning, which should take into account the parameters and dynamics of the cars involved, the road slope, and the speed limits. Fortunately, it is now technologically feasible for the vehicles to have knowledge about the road profiles, other vehicles' states and traffic information via VANETs and smart sensing devices (such as GPS and cameras). Therefore, it is now feasible to plan the speed of vehicles en route at a tactical level so as to fully utilize the platooning benefits. For a given road segment, one can find an optimal speed profile by solving an optimization problem with dynamic programming (DP) [1, 2]. The optimization problem should be formulated carefully to avoid the high computational cost of the DP-based solution, and hence, to make it efficient enough for real-time implementation. Due to its toughness, fuel-efficient speed planning has been limited to single vehicles [3–6] yet. Little attention has been paid to platoon speed planning with regard to road slope and traffic conditions with the exception of the work in [7]. It still remains an open and challenging problem. The main contributions of this chapter are:

1) *Giving a hierarchical framework for truck convoy coordination.* The proposed framework is composed of two layers: a speed planning layer responsible for en route speed profile calculation in a fuel-optimal manner and a vehicle control layer for speed tracking with ensured string stability. Existing results usually just give a platoon control law, either ignoring speed planning or assuming a fixed speed profile.
2) *Proposing an en route speed planning algorithm.* The algorithm for speed planning is derived based on a combined fuel-time cost and receding DP, which makes the algorithm computationally less demanding than traditional DP-based algorithms and hence enables en route implementation for real-time vehicular platooning. An important factor that makes the method practical is the idea of planning the platoon speed with regard to an average vehicle instead of the platoon leader. As a consequence, the derived speed profile is far more fuel-efficient for heterogeneous vehicles with different weights and sizes.

Connected Vehicular Systems: Communication, Control, and Optimization, First Edition. Ge Guo and Shixi Wen.

3) *Providing a nonlinear vehicle control method.* The nonlinear vehicle controller is designed in the framework of discrete back-stepping control based on nonlinear vehicle dynamics instead of its linearized version in existing works. The nonlinear vehicle model factors in the road slope and heterogeneity of vehicles, making the new method more realistic than the linearization-based ones. Another significant contribution is that we give a criterion (including its proof) for string stability analysis in discrete time, which, to our knowledge, is the first time.

The rest of this chapter is organized as follows. Section 10.2 contains the system modeling and problem definition. In Section 10.3, the fuel-optimal speed planning algorithm is given. Section 10.4 contains the vehicle control design procedure, string stability analysis, and platoon control algorithm. Simulations and experimental studies are given in Section 10.5. Section 10.6 contains the conclusions.

10.2 Problem Formulations

Consider a platoon of N heterogeneous vehicles driving on a road as illustrated in Figure 10.1, where p_i, v_i, and a_i are, respectively, the position, speed, and acceleration of vehicle i ($i = 0, \ldots, N - 1$, with $i = 0$ being the leader). Assume that all the vehicles are equipped with a digital map (e.g. Google Map), GPS, wireless communication devices, and onboard sensors. The leader truck broadcasts its state (position, speed, and acceleration) via the communication network to all followers. The other trucks have knowledge about the road slope, its position, speed, and acceleration, the preceding vehicle's state, and the leader's state.

According to Newton's second law, the longitudinal dynamics of vehicle i can be represented as:

$$\dot{p}_i(t) = v_i(t) \tag{10.1}$$

$$m_i\dot{v}_i(t) = F_{e,i} - F_{g,i} - F_{r,i} - F_{air,i}(v_i(t), d_i(t)) \tag{10.2}$$

where m_i is its mass, $F_{e,i}$ is the engine force, $F_{g,i} = m_i g \sin(\theta(t))$ is the force due to gravity, with g being the gravity acceleration and θ the road slope, $F_{r,i} = c_r m_i g \cos(\theta(t))$ and $F_{air,i}(v_i, d_i) = (1/2)\rho c(d_i) A_i v_i^2$ are, respectively, the rolling resistance and aerodynamic resistance,

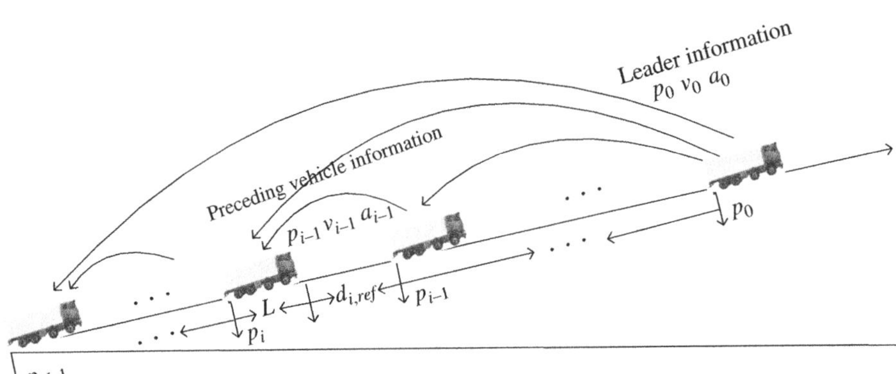

Figure 10.1 A vehicular platoon.

c_r is the rolling coefficient, ρ is the air density, A_i is the cross-section area of the vehicle, $c(d_i)$ is the air drag coefficient, which depends on the inter-vehicle distance d_i in the following way [7]:

$$c(d_i) = \begin{cases} c_d & i = 0 \\ \eta_i c_d & i \geq 1 \end{cases} \tag{10.3}$$

where c_d is the nominal drag coefficient and $0 < \eta_i < 1 - ((\alpha_i)/(\alpha_{i2} + d_i)) < 1$ presents the air drag reduction due to platooning with α_1 and α_2 being the parameters.

Note that, strictly speaking, the air drag coefficient is not a constant but varies as a function of airflow speed and direction, air density and viscosity, and vehicle size (i.e. cross-section area). While for cars at highway speed, the effect of the incoming flow is more or less the same. Therefore, the drag coefficient can often be treated as constant. When cars are running in close proximity to each other, the air drag may reduce considerably, which is also true for the leading car. Generally, for a platoon of vehicles, the drag reduction of the leading vehicle is not as obvious as that of the followers. Here we assume that the air drag of the leading car is equivalent to a solo-driving car, ignoring the air drag reduction of the leading vehicle as in [7, 8], namely $c(d_i) = c_d$.

Define the spacing error of the i th following vehicle as:

$$\Delta_i(t) = p_{i-1}(t) - p_i(t) - d_{i,r} \tag{10.4}$$

where $d_{i,r}$ is the desired inter-vehicle spacing. It is given as:

$$d_{i,r} = v_i\tau + \delta_d \tag{10.5}$$

where δ_d is the constant, and τ is the headway time. Then, the spacing error can be rewritten as:

$$\Delta_i(t) = p_{i-1}(t) - p_i(t) - v_i(t)\tau - \delta_d \tag{10.6}$$

Remark 10.1 The vehicle model in (10.1)–(10.2) is fully nonlinear and takes into account the variation in road slope and vehicle mass. However, in existing platoon control literature, the vehicle dynamics is usually simplified by feedback linearization [9–12], ignoring the effect of vehicle mass and road gradient on the system dynamics and, in turn, on the speed profile and fuel consumption.

Remark 10.2 The constant spacing policy is most commonly considered in the literature, which requires a constant distance between two successive vehicles [11, 12]. While, if the change in speed of the lead vehicle is the result of the disturbance, obviously, the effect of this disturbance is not suppressed throughout the string. The spacing policy we defined in (10.4) depending on the vehicle speed can inherently attenuate the disturbance effect on the platoon.

The fuel consumption of a vehicle depends on various factors, such as the engine speed, gear ratio, torque, temperature, and efficiency. For simplicity, here we adopt the following approximate fuel consumption model [13]:

$$f_{i,v}(t) = \begin{cases} b_0 + b_1 v_i(t) + b_2 v_i^2(t) + b_3 v_i^3(t) + \hat{a}_i(t)\left(q_0 + q_1 v_i(t) + q_2 v_i^2(t)\right) & a_i \geq 0 \\ 0 & a_i < 0 \end{cases} \tag{10.7}$$

where $f_{criusc} = b_0 + b_1 v_i(t) + b_2 v_i^2(t) + b_3 v_i^3(t)$ represents fuel consumed per second at a steady speed v_i, and $f_{accel} = \hat{a}_i(t)\left(q_0 + q_1 v_i(t) + q_2 v_i^2(t)\right)$ is the additional fuel consumption for acceleration $\hat{a}_i$ at speed v_i, the equivalent acceleration $\hat{a}_i$ of the vehicle i is defined as $\hat{a}_i = a_i + a_\theta$, with a_θ being the acceleration due to road slope, b_0, b_1, b_2, b_3 and q_0, q_1, q_2 are the fuel consumption parameters, which can be approximated using the data obtained from an engine torque-speed-efficiency map of a typical vehicle.

Note that diesel engines also consume fuel when idling [13], but the amount of fuel consumed is almost negligible. So, here we choose to ignore idle fuel consumption.

Remark 10.3 The fuel consumption of a real car is rather complicated. It is influenced by many factors, e.g. air resistance, weight, engine type/power, engine state of tune, tire size/thread/pressure, and road and weather conditions. Yes, a power train model like CMEM, FASTSim, and Autonomie might be better for fuel-efficiency control of an individual car since these microscopic real-time emission models are more sophisticated and subtler in calculating vehicle emissions and fuel consumption based on the driving cycle. However, this chapter is not concerned with the fuel efficiency of an individual car but the overall fuel economy of a fleet of cars in an intelligent transportation system. As the optimization problem involved in platoon speed planning is very sophisticated, we choose to use the relatively simpler regression-based fuel consumption model.

Remark 10.4 The fuel consumption model used here is an approximated function obtained by a curve-fitting process in [13]. Experiments have shown that it is very close to the practical fuel consumption. The curve-fitting process is based on the average mechanical power Pow_i of the vehicle, given as follows:

$$Pow_i = \left(\frac{1}{2}\rho c(d_i)A_i v_i^2 + m_i g\sin(\theta) + c_r m_i g\cos(\theta) + m_i a_i\right)v_i$$

The mechanical power corresponds to the fuel consumption rate $F_{i,rate}$ via the following equation:

$$F_{rate,i} = \frac{\varphi}{\kappa\psi}\left(LMD + \frac{Pow_i}{\eta_t\eta_{tf}}\right)$$

where φ is the fuel-to-air mass ratio, κ is the heating value of typical diesel fuel L, M, D, ψ, ψ, and η_{tf} are the engine friction factor, the engine speed, the engine displacement, a factor converting the fuel rate from grams per second to liters per second, the efficiency parameter for diesel engines, the drive train efficiency, respectively. Therefore, based on the engine characteristic map with regard to constant power and the engine efficiency curves with regard to torque and engine speed, one can obtain the fuel consumption rate for any vehicle state, which can lead to the model in (10.7).

Denote the platoon state as $x_p = [p_p, v_p]^T$, where p_p and v_p, respectively are the position and speed. For the platoon speed planning problem, we see all vehicles in the platoon as a whole and assume that $x_p = x_0 = x_1 = x_2 \cdots = x_{N-1}$. We then choose an average vehicle to plan the speed profile for the platoon. To this end, we define the platoon average engine force $F_{e,p}$ and mass m_p, respectively, as follows:

$$F_{e,p} = \sum_{i=0}^{N-1} F_{e,i}/N, \quad m_p = \sum_{i=0}^{N-1} m_i/N \tag{10.8}$$

with $F_{e,i}$ and m_i are the control input and mass of the ith vehicle, respectively. The dynamics of the platoon can be written as:

$$\dot{p}_p(t) = v_p(t) \tag{10.9}$$

$$m_p\dot{v}_p(t) = -c_r m_p g\cos(\theta(t)) - m_p g\sin(\theta(t)) - \frac{1}{2}\rho c_p A_p v_p^2(t) + F_{e,p}(t) \tag{10.10}$$

where $c_p = c_d$ for a solo-driving vehicle and the leading vehicle, otherwise, $c_p = c_d\sum_{i=1}^{N-1}\eta_i/N$ for the follower vehicles in the platoon.

Correspondingly, the platoon fuel consumption model can be written as:

$$f_{p,v}(t) = \begin{cases} b_0 + b_1 v_p(t) + b_2 v_p^2(t) + b_3 v_p^3(t) + \hat{a}_p(t)\left(q_0 + q_1 v_p(t) + q_2 v_p^2(t)\right) & a_p \geq 0 \\ 0, & a_p < 0 \end{cases} \tag{10.11}$$

Define the cost function as follows:

$$J = J_{p,f} + \beta J_{p,t} \tag{10.12}$$

which is the weighted sum of the fuel consumption $J_{p,f}$ and the travel time $J_{p,t}$ of the vehicular platoon, with $J_{p,f} = \int_{t_0}^{t_0+t} f_{p,v}(\lambda)d\lambda$ and $J_{p,t} = t$, β represents a trade-off weight.

Remark 10.5 Usually, the speed of the platoon is set as that of the leader [1], or planned with regard to the leader vehicle as in [7]. However, the leader-based speed planning is not fair and fuel-efficient for heterogeneous platoons with vehicles different in mass, size, and weight. So, we use an averaging method (introducing a vehicle with average mass, size, weight, and air drag) for platoon speed planning. The average platoon model is similar to the summation platoon model in [14], both are more fuel-efficient than the leader-based or any single vehicle-based platoon speed planning schemes.

This chapter aims to derive a speed planning mechanism and a cooperative control method for a group of heavy-duty trucks driving in a convoy. The objective of this study is to achieve the following goals:

1) To give a road-adaptive speed planning scheme that yields the reference speed profile v_p^* of the convoy of trucks in the sense that the overall fuel consumption is minimized. To be specific, v_p^* should be designed such that the cost function is minimized, namely min J.
2) To derive a cooperative control method for the group of trucks to track the given speed profile v_p^* with guaranteed string stability, i.e. avoid spacing error amplification along the string of vehicles due to the leader vehicle or other interference. Namely, the control law should guarantee $\lim_{t\to\infty}\left|v_i - v_p^*\right| = 0$ for any vehicle in the platoon and $|G_i(s)| = |G_i(j\omega)| \leq 1$ for any $\omega > 0$, where $|G_i(s)| = \Delta_i(s)/\Delta_{i-1}(s)$, $\Delta_i(s) = p_{i-1}(s) - p_i(s) - v_i(s)\tau - \delta_d/s$.

Note that the transfer function $|G_i(s)| = \Delta_i(s)/\Delta_{i-1}(s)$ in the frequency domain corresponds to an impulse response function $g_i(t)$ in the time domain, which is the inverse Laplace transform of $G_i(s)$. By definition of the impulse response function, we have that $\Delta_i(t) = \int_0^t g_i(t-r)\Delta_{i-1}(t)dr$, which means for unit impulse input $\Delta_{i-1}(t)$, one has $\Delta_i(t) = g_i(t)$. Therefore, the sting stability condition $|G_i(j\omega)| = |\Delta_i(j\omega)/\Delta_{i-1}(j\omega)| \leq 1$ for all $\omega > 0$ implies that $|g_i(t)| \leq 1$ or $|\Delta_i(t)| \leq |\Delta_{i-1}(t)|$, at any time instant t. Namely, the spacing error will not amplify backward along the string of vehicles.

10.3 Speed Planning

Fuel-efficient platoon control calls for en route fuel-optimal speed planning. Here, we give a speed planning methodology based on receding optimization. The optimization problem is formulated in a discrete space horizon on the basis of fuel consumption and travel time. Its solution is obtained using a fuel consumption model, the platoon model, and a receding DP (RDP) algorithm.

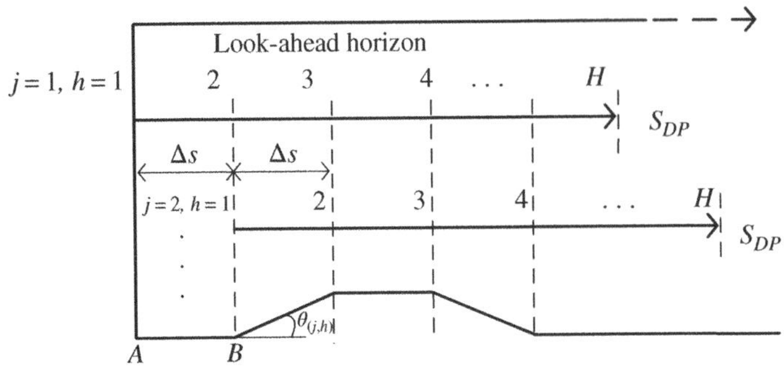

Figure 10.2 Illustration of the receding DP procedure.

Receding horizon optimization involves predicting the future behavior of the system with additional knowledge about disturbances such as road slopes. The optimization is accomplished through RDP in a procedure shown in Figure 10.2. Each look-ahead horizon S_{DP} (a segment of the total travel distance in a mission) is divided into H stages or sections with equal length Δs. The optimization is carried out within the look-ahead horizon backwardly from $h = H$ to $h = 1$, and the optimal solution for $h = 1$ is applied to the platoon control system for the present time cycle. Then the procedure is repeated at the beginning of the next cycle. Let j represent the jth cycle of the RDP optimization and (j,h) represents the hth stage of the jth cycle. For each stage, the slope of the road and the speed limits are assumed to be known.

Let $\Delta s = p_{(j,h+1)} - p_{(j,h)}$, where $p_{(j,h)}$ is the position of the hth space sample of the jth cycle, $v(p_{(j,h)})$ and $\theta(p_{(j,h)})$ are the speed and road slope, respectively, at the position $p_{(j,h)}$. Then, the platoon dynamics can be rewritten in discrete space form as:

$$\begin{aligned} v_p\left(p_{(j,h+1)}\right) = &\frac{\Delta s}{m_p v_p\left(p_{(j,h)}\right)}\left(-c_r m_p g\cos\left(\theta\left(p_{(j,h)}\right)\right) - m_p g\sin\left(\theta\left(p_{(j,h)}\right)\right)\right. \\ &\left.-\frac{1}{2}\rho A_p c_p v_p^2\left(p_{(j,h)}\right) + F_{e,p}\left(p_{(j,h)}\right)\right) + v_p\left(p_{(j,h)}\right) \end{aligned} \tag{10.13}$$

which will be referred to as $v_p\left(p_{(j,h+1)}\right) = f_p^s\left(v_p\left(p_{(j,h)}\right), F_{e,p}\left(p_{(j,h)}\right)\right)$ later in the RDP problem formulation. And the cost function at cycle j is written as:

$$J^{DP}(j) = J_{p,f}^{DP}(j) + \beta J_{p,t}^{DP}(j) \tag{10.14}$$

where $J_{p,f}^{DP}(j) = \sum_{h=1}^{H} f_{p,v}(h) \times \left(\Delta s / v_p\left(p_{(j,h)}\right)\right)$ and $J_{p,t}^{DP}(j) = \sum_{h=1}^{H} \Delta s / v_p\left(p_{(j,h)}\right)$.

Now, to address our objective *i*), we formulate the following optimization problem to be solved in the speed planning layer.

Problem 10.1

$$\min_{F_{e,p}(j)} J^{DP}(j), \quad j = 1, 2, \ldots \tag{10.15}$$

$$\text{subject to: } v_p\left(p_{(j,h+1)}\right) = f_p^s\left(v_p\left(p_{(j,h)}\right), F_{e,p}\left(p_{(j,h)}\right)\right) \tag{10.16}$$

$$F_{e,p\min} \le F_{e,p}(j,h) \le F_{e,p\max} \tag{10.17}$$

$$v_{\min}(j,h) \le v_p(j,h) \le v_{\max}(j,h) \tag{10.18}$$

$$v_{p,(j,1)} = v_p(t) \tag{10.19}$$

$$p_{p,(j,1)} = p_p(t) \tag{10.20}$$

for $h = 1, 2, \ldots, H$, where Equations (10.17) and (10.18) represent the input constraints and speed limits at road segment (j, h), respectively, (10.19) and (10.20) represent the initial conditions of the jth cycle. To solve Problem 10.1 with RDP, for $h = H - 1, H - 2, \ldots, 1$, the minimum cost is to be calculated as follows:

$$J_{(j,h)}^{DP}\left(v_{p,(j,h)}\right) = \min_{F_{e,p(j,h)} \in F_{e,ps}} \left\{\gamma_h\left(v_{p,(j,h)}, F_{e,p(j,h)}\right) + J_{(j,h+1)}^{DP}\left(v_{p,(j,h+1)}\right)\right\} \tag{10.21}$$

where $\gamma_h(v_{p,(j,h)}, F_{e,p(j,h)}) = f_{p,v}(h) \times (\Delta s / v_p(p_{(j,h)})) + \Delta s / v_p(p_{(j,h)})$ is the incremental cost at stage h and $J_{(j,h+1)}^{DP}\left(v_{p,(j,h+1)}\right)$ is the minimum cost from speed $v_{p,(j,h+1)}$ to an end speed $v_{p,(j,H)}$. At each stage h, the speed and driving force are quantized into N_{v_p} and $N_{F_{e,p}}$ levels with gaps δ_v and δ_F, respectively. Namely, $v_{p,(j,h)} = \left\{v_p^1, v_p^1 + \delta_v, \ldots, v_p^{N_{vp}}\right\}$ with $v_p^1 = v_{p,\min}$, $v_p^{N_{vp}} = v_{p,\max}$ and $F_{e,p(j,h)} = \left\{F_{e,p}^1, F_{e,p}^1 + \delta_F, \ldots, F_{e,p}^{N_{Fe,p}}\right\}$ with $F_{e,p}^1 = F_{e,p\min}$ and $F_{e,p}^{N_{Fe,p}} = F_{e,p\max}$. These quantization values will be used in (10.16) for recursive speed calculation for all stages h, namely $v_p\left(p_{(j,h+1)}\right) = f_p^S\left(v_p^i\left(p_{(j,h)}\right), F_{e,p}^{i,l}\left(p_{(j,h)}\right)\right)$, where $v_p^i\left(p_{(j,h)}\right) \in v_{p,(j,h)}$ and $F_{e,p}^{i,l}\left(p_{(j,h)}\right) \in F_{e,p}$ are the quantized speed of level i and driving force of level l at stage h, respectively.

We now give Algorithm 10.1 for platoon fuel-optimal speed planning.

Algorithm 10.1 RDP-based Speed Planning Algorithm

1) Initialize j: =1, δ_v and δ_F;
2) Measure states of the leader vehicle $v_0(j, 1)$ at the jth cycle, and let $v_p(j, 1) = v_0(j, 1)$;
3) Traversal the state search space $v_{p,(j,H)}$ to calculate $J_{(j,H)}$;
4) For $i = 1, 2, \ldots, N_{v_p}$, $l = 1, 2, \ldots, N_{F_{e,p}}$, solve Equation (10.20) from $h = H - 1$ to $h = 1$ to obtain the feasible stage decisions $F_{e,p}^{i,l}(j,h)$ for state $v_p^i(j,h)$;
5) Substitute the states $v_p(j, 1)$ into the results in step 4) to reverse deduce the ultimate optimal control sequence $F_{e,p}(j) = \left\{F_{e,p}^*(j,h)\right\}_{h=1}^{H}$;
6) Implement the first term of $F_{e,p}(j)$, i.e. $F_{e,p}(j,1) = F_{e,p}^*(j,1)$ for the system (10.15) and output the $v_p^*(j,2)$;
7) Set j: $= j + 1$ and go back to step 2).

10.4 Speed Tracking Controller Design

The vehicle trajectory tracking layer contains a local controller for each following vehicle. The aim of this section is to design the controller for each vehicle i in the platoon to track the calculated speed trajectory v_p^* with ensured string stability. Here, we first give a discrete back-stepping controller design procedure based on the nonlinear vehicle dynamic model. We then prove that the controller can guarantee string stability.

Rewriting Equation (10.2) as follows:

$$\dot{v}_i = \frac{1}{m_i}(F_{e,i} - F_{g,i} - F_{r,i} - F_{air,i}(v_i, d_i)) \tag{10.22}$$

where $F_{e,i}$ is the engine force of vehicle i, which satisfies:

$$\dot{F}_{e,i}(t) = -\frac{1}{\varsigma_i}F_{e,i}(t) + \frac{1}{\varsigma_i}u_i(t) \tag{10.23}$$

where u_i is the driving force (kg m/s^2) from the engine and ς_i denotes the time constant of the engine. From (10.2) and (10.23), we have:

$$F_{e,i}(t) = m_i\dot{v}_i + F_{a,i} + F_{g,i} + F_{r,i} \tag{10.24}$$

Differentiating both sides of (10.2) and using Equation (10.23) yields the vehicle dynamics as follows:

$$\begin{cases} \dot{p}_i = v_i \\ \dot{v}_i = a_i \\ \dot{a}_i = f_i(v_i, a_i) + g_i(v_i)u_i(t) \end{cases} \tag{10.25}$$

where $f_i(v_i, a_i)$ and $g_i(v_i)$ are given by:

$$f_i(v_i, a_i) = -\frac{1}{\varsigma_i}\left(\dot{v}_i(t) + \frac{\rho A_i c(d_i)}{2m_i}v_i^2(t) + c_r g\cos\theta(t) + g\sin\theta(t)\right)$$
$$+ c_r g\sin\theta(t) - g\cos\theta(t) - \frac{\rho A_i c(d_i)}{m_i}v_i(t)a_i(t)$$

$$g_i(v_i) = \frac{1}{\varsigma_i m_i}$$

It is worth noting that it is difficult to directly control the engine force based on (10.1)–(10.2) or (10.9)–(10.10). These models are useful just for speed planning but not for controller design. By introducing this model transformation to generate the model in (10.25), the controller design issue becomes more convenient. In this way, direct engine force control is avoided as the problem is now transformed into a problem of acceleration control, which is readily realizable via the throttle.

By discretizing the model in (10.25), we obtain the following nonlinear dynamic equations:

$$p_i(k+1) = p_i(k) + T_s v_i(k) \tag{10.26}$$
$$v_i(k+1) = v_i(k) + T_s a_i(k) \tag{10.27}$$
$$a_i(k+1) = a_i(k) + T_s[f_i(a_i(k), v_i(k)) + g_i u_i(k)] \tag{10.28}$$

where T_s is the sampling period. The forthcoming discussions are based on this discrete time vehicle model.

Before giving the backstepping controller design procedure, we define the following tracking errors:

$$z_{i1}(k) = 2p_i(k) - p_{id}(k) \tag{10.29}$$
$$z_{i2}(k) = 2v_i(k) - v_{id}(k) \tag{10.30}$$
$$z_{i3}(k) = 2a_i(k) - a_{id}(k) \tag{10.31}$$

where $p_{id}(k) = p_0(k) - \tau\sum_{j=1}^{j} v_{jj}(k) - i\delta_d + p_{i-1}(k) - v_i(k)\tau - \delta_d$ is the desired reference trajectory, $v_{id}(k)$ and $a_{id}(k)$ are the desired virtual control inputs to be designed.

Remark 10.6 Note that the error in (10.29) can be referred to as the sum of $p_i(k) + i\delta_d + \tau\sum_{j=1}^{i} v_{jj}(k) - p_0(k)$ and $p_i(k) + \delta_d + \tau v_i(k) - p_{i-1}(k)$, namely the position error of vehicle i and the leader and its preceding vehicle. This error makes sense because both errors play an important role in tracking performance and string stability [15].

In what follows, we give a three-stage back-stepping controller design procedure.

Step 1

From (10.26) and (10.29), we have:

$$z_{i1}(k+1) = 2(p_i(k) + T_s v_i(k)) - p_{id}(k+1) \tag{10.32}$$

where $p_{id}(k+1) = p_0(k+1) + p_{i-1}(k+1)$ is the reference trajectory.

Take $v_i(k)$ in (10.32) as the virtual control input and design its desired value as:

$$v_{id}(k) = \frac{1}{T_s}(-2p_i(k) + c_{i1}z_{i1}(k) + p_{id}(k+1)) \tag{10.33}$$

where $0 < c_{i1} < 1$.

By combining (10.30), (10.32), and (10.33), the following equation can be obtained:

$$z_{i1}(k+1) = c_{i1}z_{i1}(k) + T_s z_{i2}(k) \tag{10.34}$$

Step 2

From (10.27) and (10.30), we have:

$$z_{i2}(k+1) = 2(v_i(k) + T_s a_i(k)) - v_{id}(k+1) \tag{10.35}$$

Take $a_i(k)$ in (10.35) as the virtual control input and design its desired value as:

$$a_{id}(k) = \frac{1}{T_s}(-2v_i(k) + c_{i2}z_{i2}(k) + v_{id}(k+1)) \tag{10.36}$$

where $0 < c_{i2} < 1$. Then, combining (10.31), (10.35), and (10.36) yields the following equation:

$$z_{i2}(k+1) = c_{i2}z_{i2}(k) + T_s z_{i3}(k) \tag{10.37}$$

Step 3

From (10.28) and (10.31), we have:

$$z_{i3}(k+1) = 2(a_i(k) + T_s(f_i + g_i u_i(k))) - a_{id}(k+1) \tag{10.38}$$

Define the actual control input as:

$$u_i(k) = \frac{1}{T_s g_i}\left(-a_i(k) + c_{i3}z_{i3}(k) + \frac{1}{2}a_{id}(k+1) - T_s f_i + T_s z_{i3}(k)\right) \tag{10.39}$$

where $0 < c_{i3} < 1$. Then, combining (10.31), (10.38), and (10.39) yields the following equation:

$$z_{i3}(k+1) = 2c_{i3}z_{i3}(k) + 2T_s z_{i3}(k) \tag{10.40}$$

Next, we will show that the system in (10.29)–(10.31) under the back-stepping controller defined by (10.33), (10.36), and (10.39) is bounded, which is given by the following theorem and its proof.

Theorem 10.1 For system (10.29)–(10.31) with the controller in (10.33), (10.36), and (10.39), all errors $z_{i1}(k)$, $z_{i2}(k)$ and $z_{i3}(k)$ are semi-globally uniformly ultimately bounded.

Proof: Consider the following Lyapunov function:

$$L_i(k) = L_{i1}(k) + L_{i2}(k) + L_{i3}(k) \tag{10.41}$$

where

$$L_i(k) = L_{i1}(k) + L_{i2}(k) + L_{i3}(k) \tag{10.42}$$

$$L_{i1}(k) = z_{i1}^2(k) \tag{10.43}$$

$$L_{i2}(k) = z_{i2}^2(k) \tag{10.44}$$

$$L_{i3}(k) = z_{i3}^2(k) \tag{10.45}$$

Let the difference of the Lyapunov function be:

$$\Delta L_i(k) = L_i(k+1) - L_i(k) \tag{10.46}$$

Then, we have:

$$\begin{aligned}\Delta L_{i1}(k) &= z_{i1}(k+1)^2 - z_{i1}(k)^2 \\ &= (c_{i1}z_{i1}(k) + T_s z_{i2}(k))^2 - z_{i1}(k)^2 \\ &= (c_{i1}z_{i1}(k))^2 + (T_s z_{i2}(k))^2 + 2c_{i1}z_{i1}(k)\cdot T_s z_{i2}(k) - z_{i1}(k)^2\end{aligned} \tag{10.47}$$

Using the fact that $2ab \le \rho a^2 + (1/\rho)b^2$, where ρ is constant and positive, we have:

$$\begin{aligned}\Delta L_{i1}(k) &\le (c_{i1}z_{i1}(k))^2 + (T_s z_{i2}(k))^2 + \rho_{i1}(c_{i1}z_{i1}(k))^2 + \frac{1}{\rho_{i1}}(T_s z_{i2}(k))^2 - z_{i1}(k)^2 \\ &= -\left(1 - c_{i1}^2 - \rho_{i1}c_{i1}^2\right)z_{i1}(k)^2 + T_s^2\left(1 + \frac{1}{\rho_{i1}}\right)z_{i2}(k)^2 \\ &= -\tau_{11}L_{i1}(k) + \tau_{12}L_{i2}(k)\end{aligned} \tag{10.48}$$

where $\tau_{11} = 1 - c_{i1}^2 - \rho_{i1}c_{i1}^2$, $\tau_{12} = T_s^2(1 + ((1)/(\rho_{i1})))$.

Similarly, we can have:

$$\begin{aligned}\Delta L_{i2}(k) &= z_{i2}(k+1)^2 - z_{i2}(k)^2 = (c_{i2}z_{i2}(k) + T_s z_{i3}(k))^2 - z_{i2}(k)^2 \\ &\le -\tau_{21}L_{i2}(k) + \tau_{22}L_{i3}(k)\end{aligned} \tag{10.49}$$

$$\begin{aligned}\Delta L_{i3}(k) &= z_{i3}(k+1)^2 - z_{i3}(k)^2 = (2c_{i3}z_{i3}(k) + 2T_s z_{i3}(k))^2 - z_{i3}(k)^2 \\ &= -\tau_{31}L_{i3}(k)\end{aligned} \tag{10.50}$$

where $\tau_{21} = 1 - c_{i2}^2 - \rho_{i2}c_{i2}^2$, $\tau_{22} = T_s^2(1 + ((1)/(\rho_{i2})))$, and $\tau_{31} = 1 - (2c_{i3} + 2T_s)^2$. Therefore,

$$\Delta L_i(k) = \Delta L_{i1}(k) + \Delta L_{i2}(k) + \Delta L_{i3}(k) \le -\kappa_1 L_{i1}(k) - \kappa_2 L_{i2}(k) - \kappa_3 L_{i3}(k) \tag{10.51}$$

where $\kappa_1 = \tau_{11}$, $\kappa_2 = \tau_{21} - \tau_{12}$, $\kappa_3 = \tau_{31} - \tau_{22}$.

By properly selecting positive constants ρ_{ij} so that the following conditions hold:

$$\begin{cases}\kappa_1 = -\tau_{11} < 0 \\ \kappa_2 = -\tau_{21} + \tau_{12} < 0 \\ \kappa_3 = -\tau_{31} + \tau_{22} < 0\end{cases} \tag{10.52}$$

We can then have $\Delta L_i(k) \le -r_A L_i(k)$, where $r_A = \min\{\kappa_1, \kappa_2, \kappa_3\}$. From Ioannou and Sun [16], we know $L_i(k)$ is bounded. Therefore, all the errors $z_{i1}(k)$, $z_{i2}(k)$ and $z_{i3}(k)$ are semi-globally uniformly ultimately bounded. This completes the proof.

Remark 10.7 The proposed back-stepping controller takes nonlinear dynamics (such as road gradient) into full consideration, which is a great advantage over those linearization-based methods in [2–5]. In addition, [4] does not take into account the effects of the leader's information on the following vehicles in the controller design, which results in much longer time for platoon stabilization.

This section investigates string stability and controller design conditions to ensure the vehicles in the platoon travel with the desired spacing while avoiding spacing errors that enlarge along the platoon (called the slinky effect). Our aim is to find conditions achieving objective (ii), namely $|G_i(z)| = |G_i(e^{j\omega})| \le 1$ for any $\omega > 0$, where $G_i(z) = \Delta_i(z)/\Delta_{i-1}(z)$ is the Z-transform of $G_i(s)$.

The main result in this section is given by the following theorem.

Theorem 10.2 For the system in (10.26)–(10.28) under controller (10.33), (10.36), and (10.39), $|G_i(z)| = |G_i(e^{j\omega})| \le 1$ holds true for any $\omega > 0$.

Proof: For each following vehicle in the platoon under controller (10.39), from (10.28), we can have:

$$\begin{aligned} a_i(k+1) &= a_i(k) + T_s[f_i(a_i(k), v_i(k)) + g_i u_i(k)] \\ &= (T_s + c_{i3})z_3 + \frac{1}{2}a_d(k+1) \\ &= (T_s + c_{i3})\left[2a_i(k) + \frac{2v_i(k)}{T_s} - \frac{2c_{i2}}{T_s}v_i(k) - \frac{2c_{i2}}{T_s^2}p_i(k) + \frac{c_{i1}c_{i2}}{T_s^2}(2p_i(k) - p_{id}(k))\right. \\ &\quad \left. + \frac{c_{i2}}{T_s^2}p_{id}(k+1) - \frac{1}{T_s}v_{id}(k+1)\right] + \frac{1}{2}a_{id}(k+1) \end{aligned} \tag{10.53}$$

Using $p_{id}(k) = p_0(k) - \tau\sum_{jj=1}^{i} v_{jj}(k) - (i+1)\delta_d + p_{i-1}(k) - v_i(k)\tau$ and Equations (10.26)–(10.27), we can get:

$$\begin{aligned} &\Delta_i(k+3) - 2\Delta_i(k+2) + \Delta_i(k+1) \\ &\quad = (T_s + c_{i3})[2(\Delta_i(k+2) - 2\Delta_i(k+1) + \Delta_i(k)) \\ &\qquad + 2(1 - c_{i2})(\Delta_i(k+1) - \Delta_i(k)) - 2c_{i2}(1 - c_{i1})\Delta_i(k) \\ &\qquad + c_{i1}c_{i2}\Delta_{i-1}(k) - c_{i2}\Delta_{i-1}(k+1) - 2(c_{i1} - 2)\Delta_i(k+1) \\ &\qquad - c_{i1}\Delta_{i-1}(k+1) + \Delta_{i-1}(k+2)] + \frac{1}{2}T_s[(2c_{i1} - 2)\Delta_i(k+1) \\ &\qquad + c_{i1}\Delta_{i-1}(k+1) - \Delta_{i-1}(k+2)] \end{aligned} \tag{10.54}$$

By taking the Z transform for (10.54), we can obtain the transfer function (10.55), So, for any $\omega > 0$, $|G_i(z)| = 0.5 < 1$ holds, then this complements the proof.

$$\begin{aligned} G_i(z) &= \frac{\Delta_i(z)}{\Delta_{i-1}(z)} = \frac{(T_s + c_{i3})z^2 + [(T_s + c_{i3})(-c_{i2} - c_{i1}) + T_s c_{i1}/2]z + [(T_s + c_{i3})c_{i1}c_{i2} - T_s/2]}{z^3 - 2(1 + T_s + c_{i3})z^2 + [1 - (T_s + c_{i3})(-2c_{i2} - 2c_{i1} - T_s c_{i1} + T_s)]z - 2c_{i1}c_{i2}(T_s + c_{i3})} \\ &= \frac{1}{2} < 1 \end{aligned} \tag{10.55}$$

Remark 10.8 String stability ensures that the errors will not increase as they propagate along the vehicle stream. To our knowledge, in the existing results, string stability analysis is done in a continuous time domain [9–12]. This is the first time string stability conditions are given in a discrete time domain.

Algorithm 10.2 Platoon Control Algorithm

1) Given the desired platoon speed v_p^*, which is the output of Algorithm 10.1.
2) Initialize χ_0, ε_1, ε_2, $\mu_0 \leq \alpha\|F(\chi_0)\|$, $0 < \alpha < 1$, T_s, $k := 0$.
3) Compute $b_k = \Phi_{u_k}(\chi_k)$, if $\|b_k\| \leq \varepsilon_1$, then, output χ_k, go to step 6), else go to step d).
4) Compute $J_k = \Phi'_{u_k}$, where Φ'_{u_k} is the generalized inverse matrix of Φ_{μ_k}.
5) Solve $J_k\Delta\chi_k = -b_k$ to obtain $\Delta\chi_k$; let $\chi_{k+1} = \chi_k + \Delta\chi_k$, $\mu_{k+1} = 0.25\mu_k$; if $\|\Delta\chi_k\| \leq \varepsilon_2$, then output χ_{k+1}, go to step 6), otherwise, let $k := k+1$, $\chi_k := \chi_{k+1}$, $\mu_k := \mu_{k+1}$ and go to step 3).
6) Using parameters obtained in step 5) to calculate controller output via (10.38).

In determining the controller parameters, we need to solve the nonlinear inequalities in (10.52). Obviously, analytical solutions to inequality systems are very complicated. One can obtain numerical solutions using the mathematical method in [17] by transforming them into a set of nonlinear equations. To this end, combine the inequalities in (10.52) and conditions $-\rho_{i1} < 0$ and $-\rho_{i2} < 0$, denote them by

$$F(\chi) = [f^1(\chi), f^2(\chi), \ldots, f^5(\chi)]^T \leq 0,$$

where $\chi = [c_{i1}, c_{i2}, c_{i3}, \rho_{i1}, \rho_{i2}]^T$, $f^1(\chi) = -1 + c_{i1}^2 + \rho_{i1}c_{i1}^2$, $f^4(\chi) = -\rho_{i1}$, $f^5(\chi) = -\rho_{i2}$,

$$f^2(\chi) = -1 + c_{i2}^2 + \rho_{i2}c_{i2}^2 + T_s^2\left(1 + \frac{1}{\rho_{i1}}\right), f^3(\chi) = -1 + 4(c_{i3} + T_s)^2 + T_s^2\left(1 + \frac{1}{\rho_{i2}}\right).$$

Introducing the following so-called helper function $\phi(\mu, \chi)$ [18]:

$$\phi(\mu,\chi) = \begin{cases} \frac{1}{2}\chi^2 + \frac{1}{6}\mu^2, \chi \geq \mu \\ \frac{1}{12\mu}(\chi+\mu)^3, -\mu < \chi < \mu \\ 0, \ \chi \leq -\mu \end{cases} \tag{10.56}$$

where $\mu > 0$ and $\phi(\mu,\chi)$ is secondary continuous differentiable. Using the function in (10.56), the set of inequalities $F(\chi) \leq 0$ can be transformed into a group of equivalent nonlinear equations:

$$\phi(f^i(\chi), \mu) = 0, \quad (i = 1, 2, \ldots, 5) \tag{10.57}$$

Let $\Phi_\mu(\chi) = (\phi(f^1(\chi), \mu), \phi(f^2(\chi), \mu), \ldots, \phi(f^5(\chi), \mu))^T$. Then (10.57) can be written more concisely as $\Phi_\mu(\chi) = 0$. These sets of nonlinear equations can be solved using Newton's iteration method. Based on these discussions, we give Algorithm 10.2 for platoon control.

10.5 Simulation and Experiments

In this section, we will show how to apply the two-layer control structure to a platoon of vehicles, first in a virtual environment established using the System Build package in Matlab and then in a three-vehicle platoon of Arduino Cars in the laboratory.

(1) Speed planning

Consider a platoon of 10 heavy-duty vehicles (whose parameters are listed in Table 10.1) driving over the 3-km road as illustrated in Figure 10.3. The slope is composed of three parts, an uphill stretch of 300 m with a slope 3%, a flat stretch, and a downhill stretch of 300 m with a slope –3%. In the simulations, we use the following parameters in the fuel consumption model [15]:

$$b_0 = 0.1569, b_1 = 2.450 \times 10^{-2}, b_2 = -7.415 \times 10^{-4}, b_3 = 5.975 \times 10^{-5}, q_0 = 0.07224,$$
$$q_1 = 9.681 \times 10^{-2}, q_2 = 1.075 \times 10^{-3}.$$

The fuel-optimal speed profile of a platoon of vehicles obtained by Algorithm 10.1 is shown in Figure 10.4.

Table 10.1 Vehicle parameters.

Parameters	Value
Average mass (m_p)	$40t$
Rolling resistance coefficient (c_r)	3×10^{-3}
Vehicle cross-sectional area (A_p)	$10\ \text{m}^2$
Engine time constant (ς_i)	0.25 s
Air density (ρ)	$1.29\ \text{kg/m}^{-3}$
Air drag coefficient (c_p)	0.3
Speed limits	$v_{p,\max} = 25\ \text{m/s}$ $v_{p,\min} = 20\ \text{m/s}$

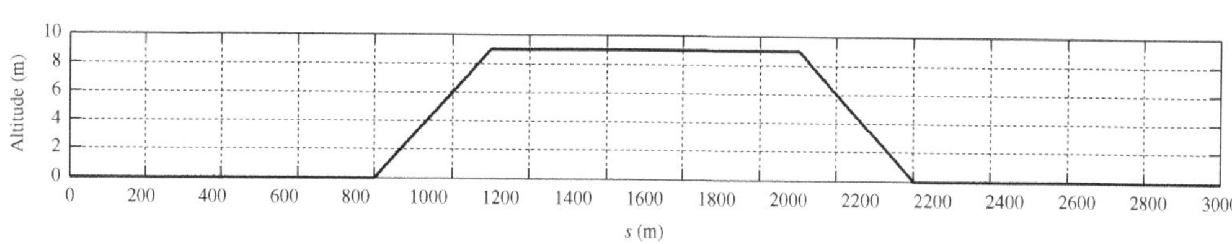

Figure 10.3 Scenario of road.

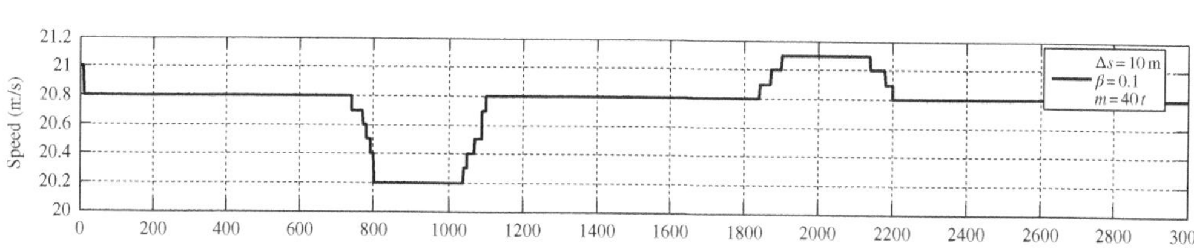

Figure 10.4 Fuel-efficient speed in the space domain.

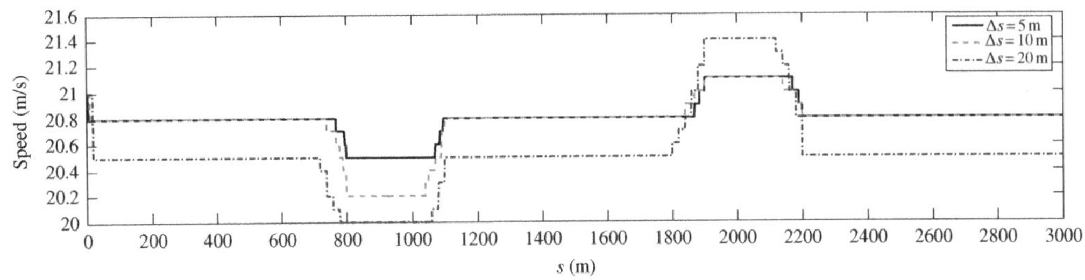

Figure 10.5 Fuel-efficient speed for different Δs.

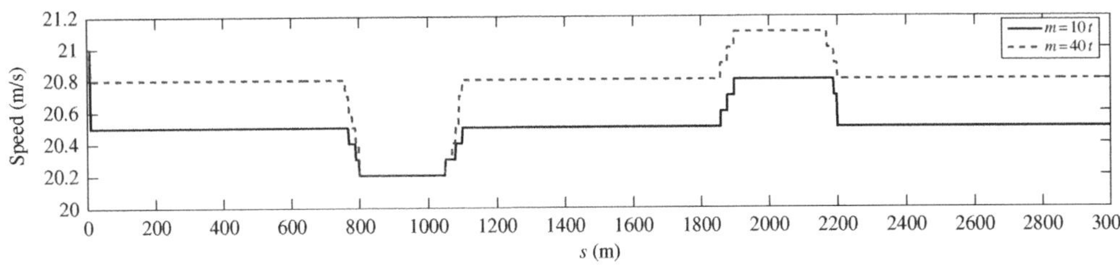

Figure 10.6 The profiles of the fuel-efficient speed for different masses.

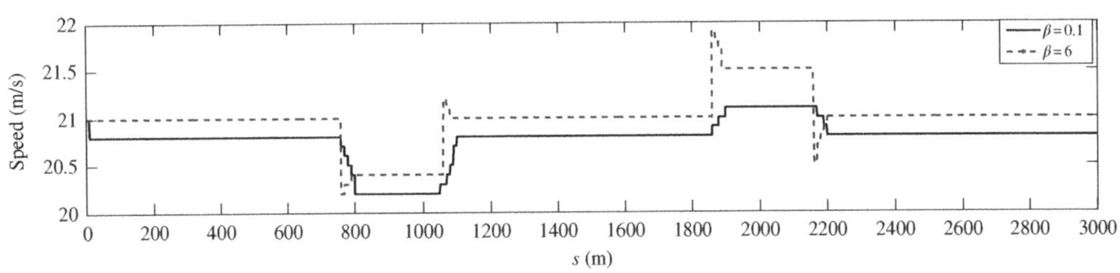

Figure 10.7 Fuel-efficient speed for different weight β.

The fuel-efficient speed trajectory is dependent on the step size in space. Figure 10.5 compares the trajectories with different spatial steps Δs ($\Delta s = 5$ m, $\Delta s = 20$ m, and $\Delta s = 10$ m). Although smaller step size may yield more delicate speed planning results, the simulation results show that when the space size decreases from 10 to 5 m, the planned speed trajectories are almost the same but the computation time increases by 40%. Therefore, as a compromise between computation time and fineness of the planned speed, we choose $\Delta s = 10$ m for all the following simulations.

The effect of the average mass ($m_p = 10t$ and $m_p = 40t$) and the trade-off weight ($\beta = 6$ and $\beta = 0.1$) on fuel-efficient speed planning is also studied. The speed profiles are shown in Figures 10.6 and 10.7, respectively. It is clearly seen in Figure 10.7 that when $\beta = 0.1$, the platoon of vehicles has minimum fuel consumption, while the optimal speed is much smaller than in the case with $\beta = 6$, where, in the latter case, the fuel consumption is much higher.

(2) Speed tracking

The presented control method is effective in speed tracking control of vehicular platoons in simulations (see Figure 10.8). The controller parameters are selected as $c_{i,1} = 0.7$, $c_{i,2} = 0.7$, $c_{i,3} = 0.03$, and $T_s = 0.02$ s according to Algorithm 10.2. Headway time τ and constant δ_d are set to be $\tau = 0.7$ s,

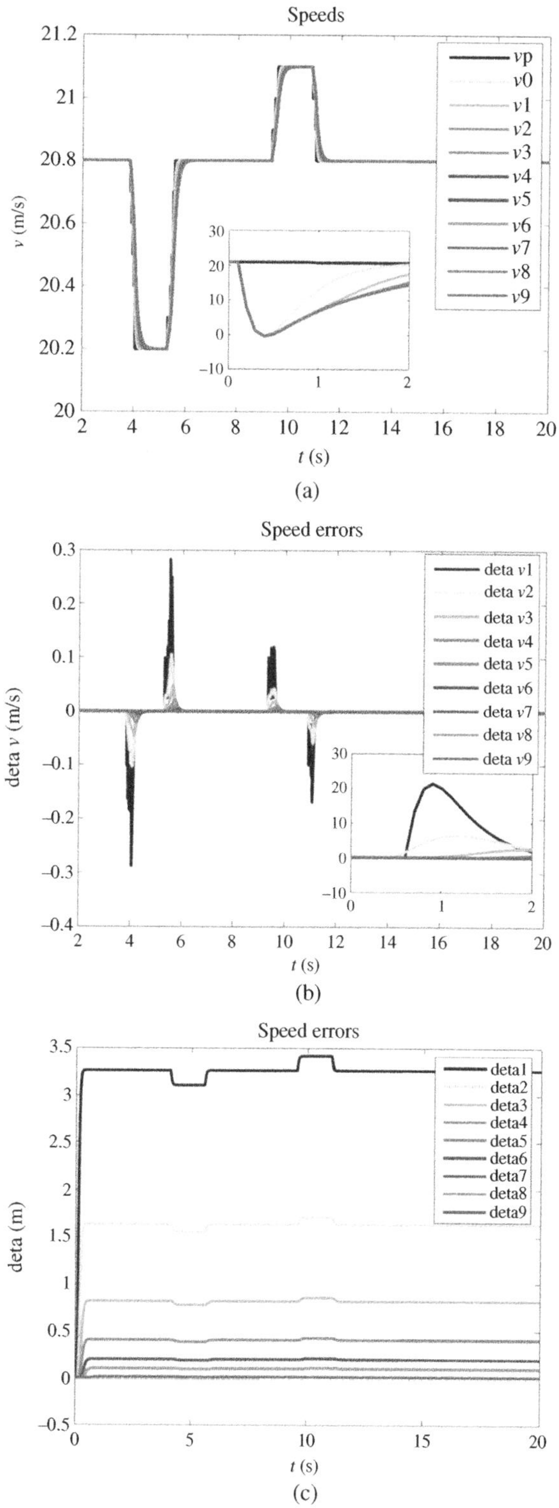

Figure 10.8 Ten-vehicle platoon track the reference speed planned in this chapter: Responses and spacing propagation characteristics: (a) speed tracking, (b) speed error, and (c) spacing error.

$\delta_d = 3$ m, respectively. From Figure 10.8a, b, we can see that all the vehicles can track the reference speed given by the speed planning layer. Figure 10.8c, d is the speed error and position error between neighboring vehicles.

Simulations have also shown that our method is superior to that of [9] where the reference speed is not planned but preset. Note that the superiority does not lie in the control performance (see Figure 10.8), especially if the speed reference in [9] is set exactly as the speed calculated by the planning algorithm in this chapter. Instead, the real difference shows up in the fuel consumption, namely the new method is more energy efficient. In the simulations of Figures 10.8 and 10.9, where

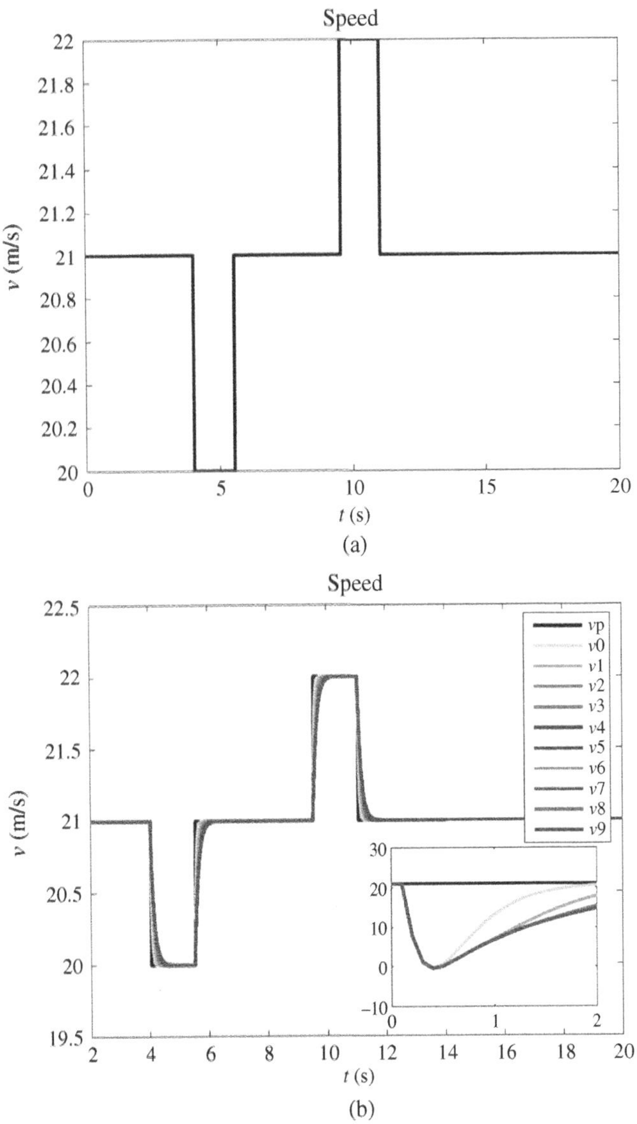

Figure 10.9 Ten-vehicle platoon track the reference speed set in advance: responses and spacing propagation characteristics: (a) reference speed, (b) speed tracking, (c) speed error, and (d) spacing error.

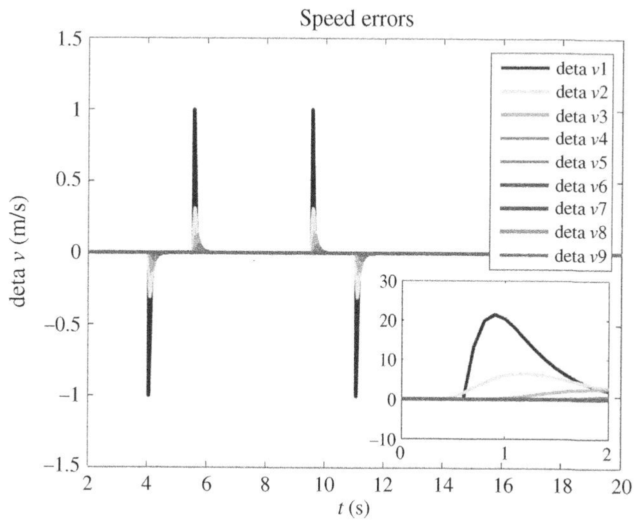

(c)

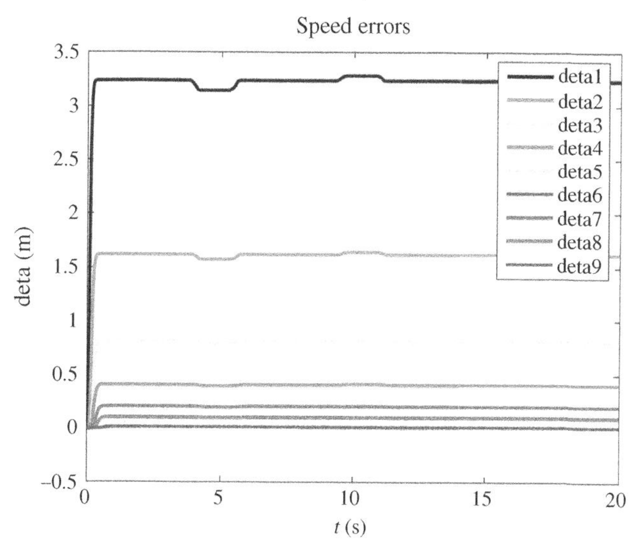

(d)

Figure 10.9 (Continued)

the same reference speed is used, the fuel consumption of the ten-car platoon in our method is 350.45 ml, 1.80% less than that is yield with the method 1 in [9], which is 356.86 ml. The advantage can be much greater if the reference speed profile in [9] is quite different from what is calculated by the planning algorithm here. For example, when a constant reference speed 23 m/s is used in [9], then the fuel consumption is higher than that of our method by 19.52%.

To verify the robustness of the designed nonlinear controller for heavy trucks (whose parameters are given in Table 10.1) to sudden external disturbances or noises, two cases are taken into consideration: a sudden extremum acceleration of $\pm 2\ \text{m/s}^2$ and a mild acceleration disturbance plus a wind gust disturbance

$$r(t) = 0.2a_p(t) + \varpi(t)$$

where $a_p(t)$ is the reference acceleration of the platoon and $\varpi(t)$ represents the wind gust, with

$$\varpi(t) = \begin{cases} 2 & 1\,\text{s} < t < 2\,\text{s} \\ 0 & \text{otherwise} \end{cases}$$

In the first case (see Figure 10.10a), although the maximum spacing error increased by 0.5 m, the entire platoon remains stable and the tracking performance is well as shown in Figure 10.10b. In the

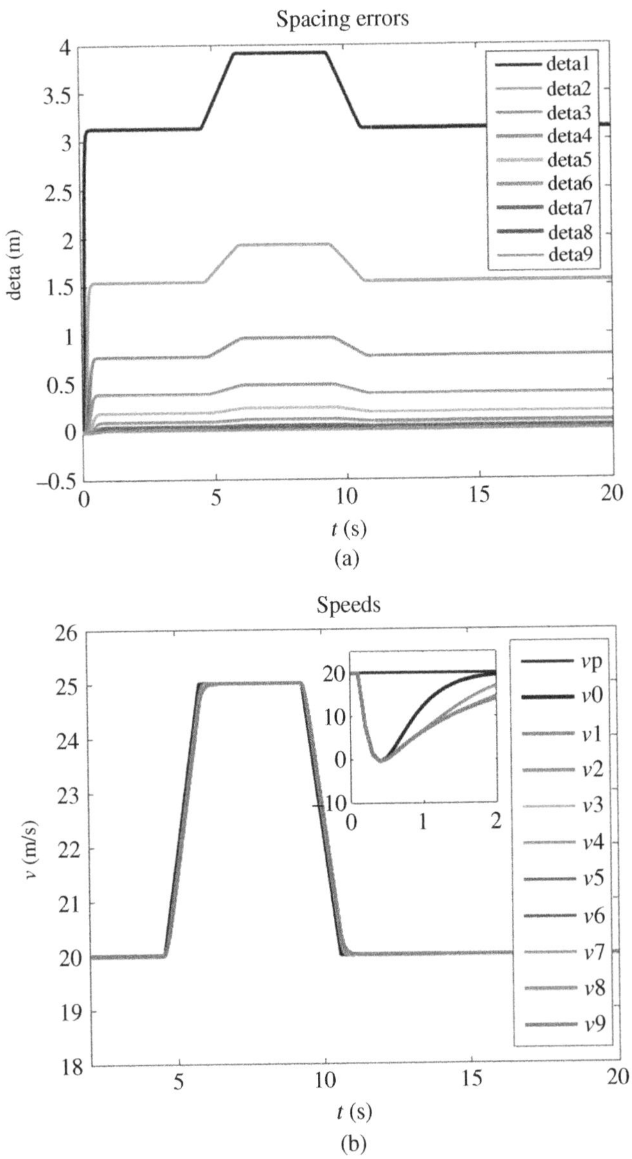

Figure 10.10 Maximum acceleration 2 m/s^2 for 10-vehicle platoon: responses and spacing propagation characteristics. (a) Spacing error, and (b) speed tracking.

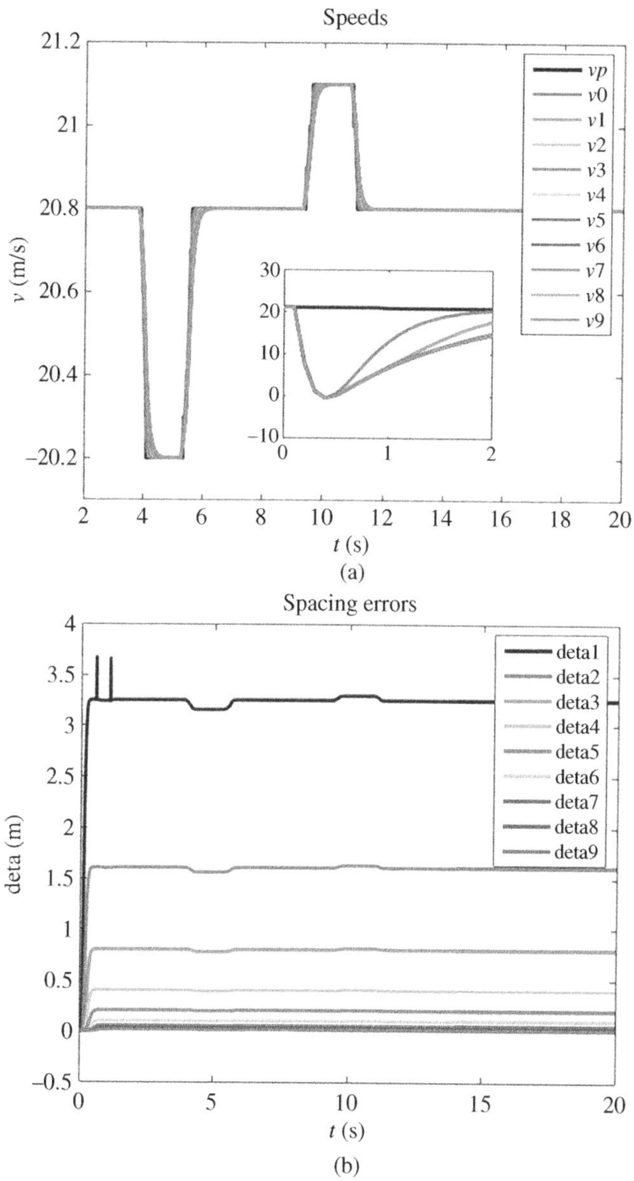

Figure 10.11 Lumped disturbance for 10-vehicle platoon: responses and spacing propagation characteristics. (a) Speed tracking, and (b) spacing error.

second case (see Figure 10.11), it can be seen that almost no effect of the disturbance on the tracking performance is seen (Figure 10.11a). Only the spacing error between vehicle 1 and vehicle 0 has slight fluctuation in the presence of gust disturbance, which then returns to the original value, and the spacing errors of other vehicles are basically unchanged (Figure 10.11b).

A similar simulation experiment is also conducted for a platoon of ten two-axis semitrucks for the two disturbance cases (Figure 10.12). The results are shown in Figure 10.13, from which we can see

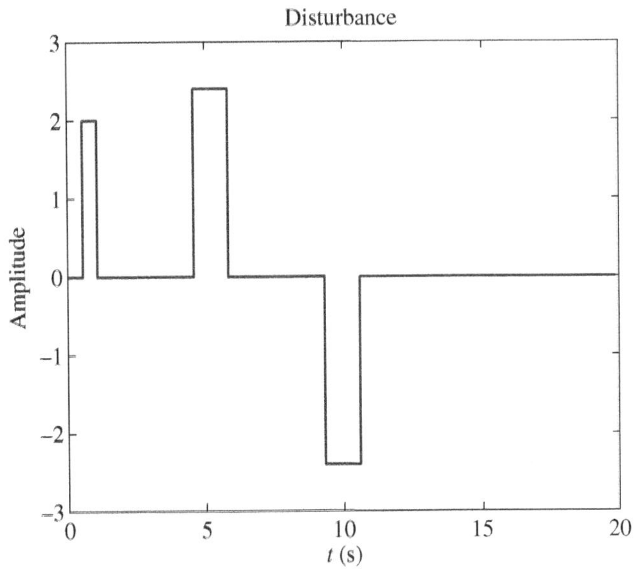

Figure 10.12 Lumped disturbance for two-axis semitrucks' platoon.

obviously that the nonlinear controller in our paper behaves well too. The semitruck parameters used in this simulation are: average mass $m_p = 16t$ rolling resistance coefficient $c_r = 0.015$, vehicle cross-sectional area $A_p = 5.4\ \text{m}^2$, engine time constant $\zeta_i = 0.25$ s, air density $\rho = 1.29\ \text{kg/m}^{-3}$, air drag coefficient $c_p = 0.3$, and speed limits $v_{p,\max} = 25$ m/s and $v_{p,\min} = 20\ \text{m/s}^2$.

We use three Arduino cars to verify the practicality and effectiveness of the designed nonlinear back-stepping controller (see Figure 10.14). The Arduino car parameters are given in Table 10.2. The desired spacing is set to be 15 cm. The longitudinal speed and spacing between a car and its preceding vehicle are detected by an increment encoder sensor installed on the shaft of the rear wheels and two infrared sensors, respectively. The sampling period is set to be $T_s = 0.1$ s. The controller parameters in the experiments are obtained according to Algorithm 10.2 as follows:

$$c_{i1} = 0.5, c_{i2} = 0.5, c_{i3} = 0.35$$

The corresponding closed-loop dynamics of vehicle i is

$$\begin{aligned} a_i(k+1) = {} & 0.15a_i(k) + 1.5v_i(k) + 22.5p_i(k) - 0.575(a_0(k) + a_{i-1}(k)) - 5.75(v_0(k) \\ & + v_{i-1}(k)) - 11.25(p_0(k) + p_{i-1}(k)) \end{aligned} \tag{10.58}$$

The experiment results are presented in Figure 10.16a, b in which the leader's speed profile is shown in Figure 10.15. Figure 10.16a shows that although speed fluctuation (not serious) appears at speed-switching instants, the speed can be stabilized within a short time and the platoon of Arduino cars achieves speed tracking control quite well. Figure 10.16b shows the spacing errors during speed tracking, where the spacing errors converge within an acceptable range, showing good

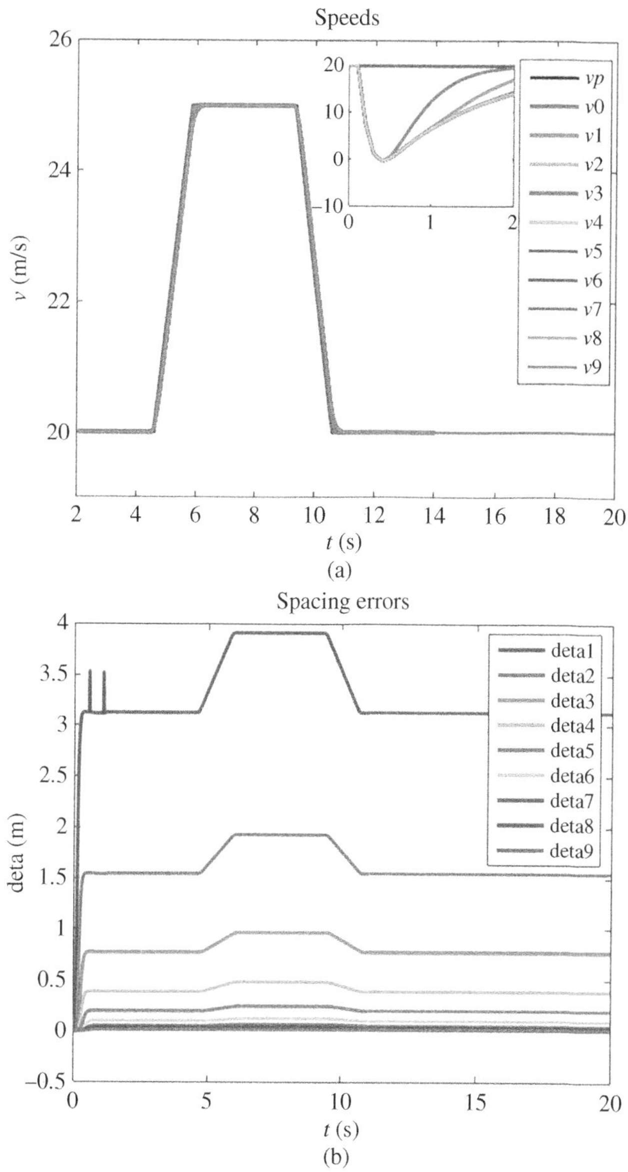

Figure 10.13 Lumped disturbance for two-axis semitrucks' platoon: responses and spacing propagation characteristics. (a) Speed tracking, and (b) spacing error.

robustness of the platoon. Note that, in part, the spacing errors are due to information delay and roughness of the runway.

10.6 Conclusions and Future Work

This chapter presents a hierarchical framework for fuel-efficient and safe platoon control. By exploiting the road slope, a planning algorithm for a fuel-optimal reference speed profile is obtained using the concept of average platoon modeling at the speed planning layer.

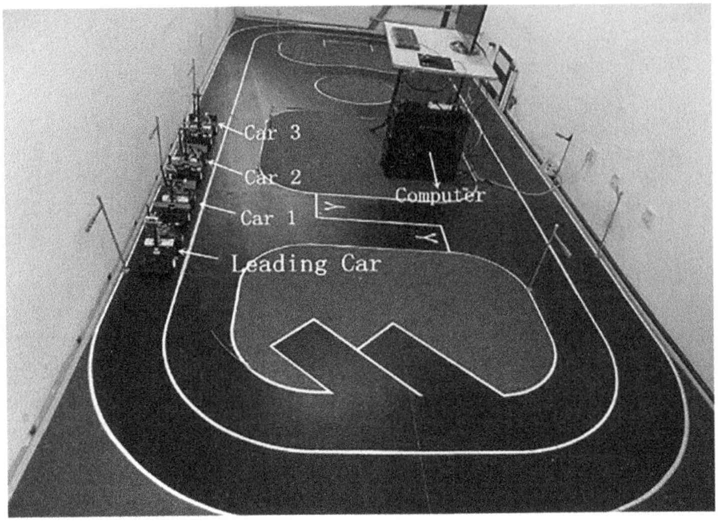

Figure 10.14 Arduino platoon in the experiment.

Table 10.2 Parameters of the Arduino cars.

The *i*th vehicle	m_i (kg)	A_i (cm^2)	$c(d_i)$
0	1.2	15	0.15
1	1.1	12	0.15
2	1.1	12	0.15

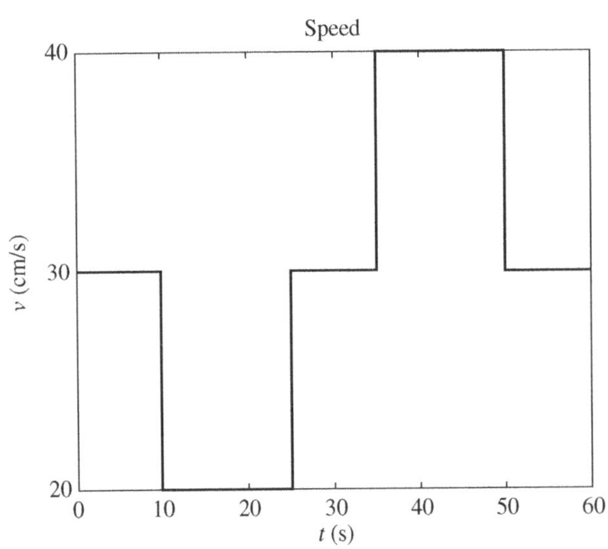

Figure 10.15 The leader's speed.

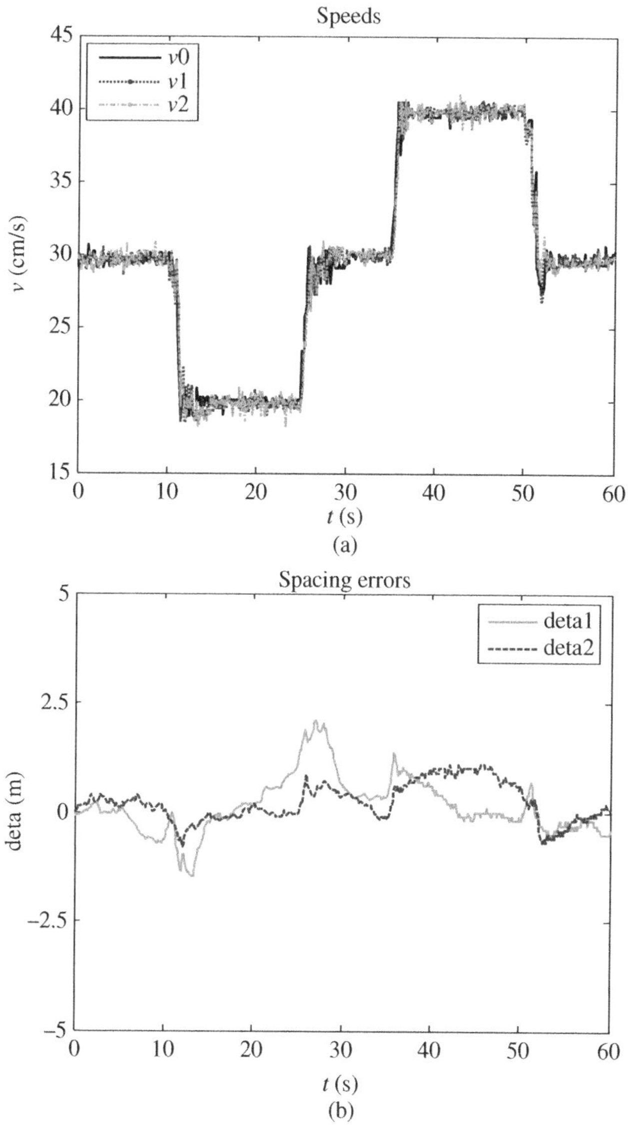

Figure 10.16 Arduino car experiments: (a) Speeds, and (b) spacing errors.

A nonlinear back-stepping control method is then derived for each vehicle at the speed tracking layer. A platoon control algorithm is given to guarantee all the vehicles to track the reference speed profile and achieve string stability. The two-layered planning and control method outperforms the existing single-layered platoon control methods with preset speed profiles thanks to the introduction of the en route speed planning algorithm. The advantages of the new method have been demonstrated in many simulations and laboratory experiment results conducted using laboratory-scale cars.

References

1 Deng, Q. and Ma, X. (2014). A fast algorithm for planning optimal platoon speeds on highway. *Proc. IFAC World Congr.*, Cape Town, South Africa, pp. 8073–8078.

2 Ma, X. (2013). Towards intelligent fleet management: local optimal speeds for fuel and emissions. *Proc. IEEE Int. Conf. Intell. Transp. Syst.*, Hague, The Netherlands (October 2013), pp. 2201–2206.

3 Gilbert, E.G. (1976). Vehicle cruise: improved fuel economy by periodic control. *Automatica* 12 (2): 159–166.

4 Schwarzkopf, A.B. and Leipnik, R.B. (1977). Control of highway vehicles for minimum fuel consumption over varying terrain. *Transp. Res.* 11 (4): 279–286.

5 Kamal, M.A.S., Mukai, M., Murata, J., and Kawabe, T. (2010). Ecological driver assistance system using model-based anticipation of vehicle-road traffic information. *IET Intell. Transp. Syst.* 4 (4): 244–251.

6 Larson, J., Liang, K.-Y., and Johansson, K.H. (2015). A distributed framework for coordinated heavy-duty vehicle platooning. *IEEE Trans. Intell. Transp. Syst.* 16 (1): 419–429.

7 Turri, V., Besselink, B., and Johansson, K.H. (2017). Cooperative look-ahead control for fuel-efficient and safe heavy-duty vehicle platooning. *IEEE Trans. Control Syst. Technol.* 25 (1): 12–28.

8 Alam, A. (2014). *Fuel-efficient heavy-duty vehicle platooning*. Ph.D. dissertation. School Elect. Eng. Autom. Control, KTH Roy. Inst. Technol., Stockholm, Sweden, pp. 1–18.

9 Xiao, L. and Gao, F. (2011). Practical string stability of platoon of adaptive cruise control vehicles. *IEEE Trans. Intell. Transp. Syst.* 12 (4): 1184–1194.

10 Naus, G.J.L., Vugts, R.P.A., Ploeg, J. et al. (2010). String-stable CACC design and experimental validation: a frequency-domain approach. *IEEE Trans. Veh. Technol.* 59 (9): 4268–4279.

11 Ghasemi, A., Kazemi, R., and Azadi, S. (2013). Stable decentralized control of a platoon of vehicles with heterogeneous information feedback. *IEEE Trans. Veh. Technol.* 62 (9): 4299–4308.

12 Guo, G. and Yue, W. (2014). Sampled-data cooperative adaptive cruise control of vehicles with sensor failures. *IEEE Trans. Intell. Transp. Syst.* 15 (6): 2404–2418.

13 Kamal, M.A.S., Mukai, M., Murata, J., and Kawabe, T. (2011). Ecological vehicle control on roads with up-down slopes. *IEEE Trans. Intell. Transp. Syst.* 12 (3): 783–794.

14 Koller, J.P.J., Colín, A.G.N., Besselink, B., and Johansson, K.H. (2015). Fuel-efficient control of merging maneuvers for heavy-duty vehicle platooning. *Proc. IEEE Int. Conf. Intell. Transp. Syst.* (September 2015), pp. 1702–1707.

15 Swaroop, D. and Hedrick, J.K. (1991). Constant spacing strategies for platooning in automated highway systems. *J. Dyn. Syst. Meas. Control.* 121 (3): 462–470.

16 Ioannou, P.A. and Sun, J. (1996). *Robust Adaptive Control*. Englewood Cliffs, NJ: Prentice-Hall.

17 Pshenichnyi, B.N. (1970). Newton's method for the solution of systems of equalities and inequalities. *Math. Notes Acad. Sci. USSR* 8 (5): 827–830.

18 Pshenichnyi, B.N. (1970). Newton's method for the solution of systems of equalities and inequalities. *Math Notes Acad. Sci. USSR* 8 (5): 827–830.

11

Analytical Solution for Speed Planning and Tracking Control

11.1 Introduction

The research topic presented in Chapter 10 is very important because online set-point optimization is an enabling tool that would upgrade vehicular platoon control to a new level. Note that the proposed speed-planning algorithm for connected vehicles is based on receding dynamic programming. However, the algorithm is not efficient enough for real-time implementation since it is a receding dynamic programming-based numerical solution and is computationally very expensive. This paper aims to give a new and computationally less demanding solution to the platoon speed planning and tracking control problem. We give an analytical method, using Pontryagin's minimum principle (PMP) [1] to solve the optimization problem involved, which yields an algorithm for platoon reference speed calculation with far less computations. We introduce a two-layered speed planning and platoon control architecture. In the upper layer, i.e. the speed-planning layer, the optimal speed trajectory of the platoon is calculated to minimize the fuel consumption and the travel time. The lower layer is responsible for speed tracking control, which achieves two objectives: (i) making each of the vehicles track the calculated optimal speed set point, and (ii) guaranteeing string stability of the entire platoon with a desired safe intervehicle spacing. To this end, a set of distributed sliding-mode controls are designed for the following vehicles, which are robust to external disturbance and parameter perturbation. The main contributions of this chapter in comparison with existing works are as follows.

1) *New vehicular platooning architecture is given*: It contains two layers, with a speed planner at the top layer and a speed tracking controller at the lower layer. The speed planner uses an averaged-speed scheme, which calculates the expected platoon speed profile in a fuel-time optimal manner. The lower layer controller makes each vehicle track the planned speed with guaranteed vehicle stability and platoon string stability. Unlike the existing results, the speed profile can vary with time and road conditions instead of being given and fixed regardless of the road conditions.
2) *Algorithms that are fuel-time-efficient for speed planning and control are derived*: The speed-planning algorithm is derived based on the combined fuel-time cost and PMP method. Unlike Chapter 10, which considers an averaged vehicle to plan the speed profile for the platoon, we choose to derive the optimal speed profile for each vehicle and then use the averaged optimal speed as the reference speed of the platoon. For large-size platoons with heterogeneous vehicles different in weight and size, the average speed profile is more flexible than the averaged vehicle scheme. In addition, the sliding mode controller given here is simpler and more practical for use

Connected Vehicular Systems: Communication, Control, and Optimization, First Edition. Ge Guo and Shixi Wen.

than that of Chapter 10, which contains three steps whose calculation involves virtual variables not easily tractable for computation.

The remainder of this chapter is arranged as follows. The problem description and system modeling are given in Section 11.2. Section 11.3 presents the platoon speed-planning problem and its solution based on PMP, yielding an algorithm for optimal speed trajectory generation. In Section 11.4, the design of the vehicle tracking controller and analysis of platoon string stability are given. Section 11.5 is the simulation results. Finally, the conclusions are given in Section 11.6.

11.2 System Modeling and Problem Formulation

Consider a vehicular platoon consisting of $n + 1$ vehicles driving on a freeway (see Figure 11.1), where s_i, v_i, and a_i are, respectively, the position, speed, and acceleration of vehicle i ($i = 0, 1, \ldots n$ with $i = 0$ being the leader). Assume that the vehicles are equipped with a road map, a global positioning system, wireless communication devices, and onboard sensors. Each of the following vehicles periodically receives the states (position, speed, and acceleration) of the leader and its immediate predecessor. The platoon of vehicles is allowed to stop by some of the service areas along the freeway during its mission trip. Therefore, depending on the service areas chosen as the docking sites, the route is divided into consecutive road segments as depicted in Figure 11.2.

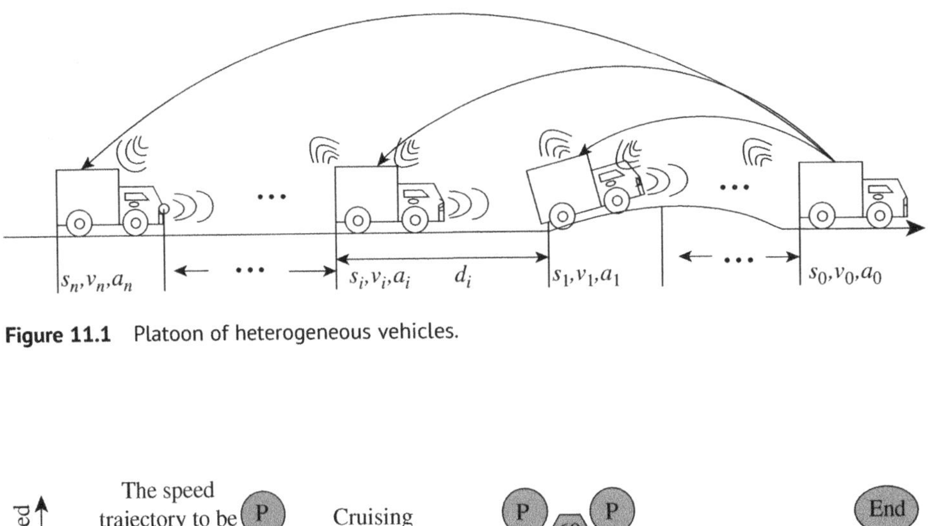

Figure 11.1 Platoon of heterogeneous vehicles.

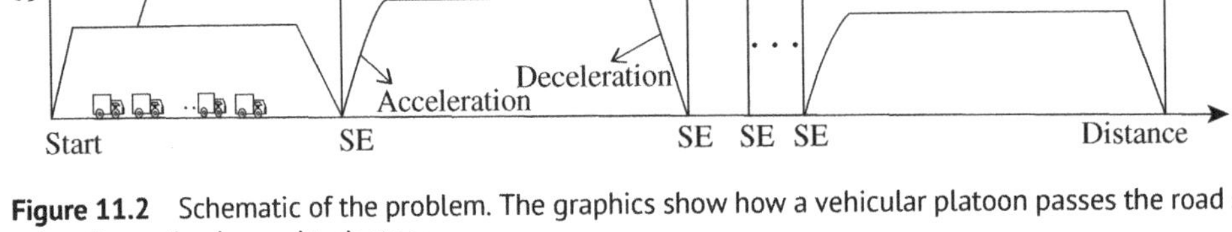

Figure 11.2 Schematic of the problem. The graphics show how a vehicular platoon passes the road sections using the optimal speed trajectory.

The longitudinal dynamics of the ith vehicle is given by the following equations:

$$\dot{s}_i(t) = v_i(t) \tag{11.1}$$

$$\begin{aligned}\dot{v}_i(t) &= \frac{1}{m_i}\left(F_{ei}(t) - F_{ai}(t) - F_{gi}(t) - F_{ri}(t)\right)\\ &= \frac{1}{m_i}\left(F_{e,i} - \frac{1}{2}\rho A_i C_{di} v_i^2 - m_i g \sin\theta - \mu_i m_i g \cos\theta\right)\end{aligned} \tag{11.2}$$

where m_i denotes the ith vehicle's mass, F_{ei} is the driving force generated by where m_i denotes the ith vehicle's mass, F_{ei} is the driving force generated by the engine, F_{ai} is the aerodynamic resistance and is given as $F_{ai} = (1/2)\rho A_i C_{di} v_i^2$ with ρ being the air density, A_i being the cross section area, C_{di} being the air drag coefficient, F_{gi} is the gravity force and is given by $F_{gi} = m_i g \sin\theta$, with g being the gravity acceleration, θ being the road slope, and F_{ri} represents the rolling resistance and is denoted by $F_{ri} = \mu_i m_i g \cos\theta$ with μ_i being the rolling resistance coefficient.

The spacing between two consecutive vehicles is defined as:

$$d_i = s_{i-1} - s_i - l_i \tag{11.3}$$

where l_i is the length of the ith vehicle. Let δ_i be the spacing error of the ith following vehicle with respect to the desired safe spacing d^*. We then have:

$$\delta_i = s_{i-1} - s_i - l_i - d^* \tag{11.4}$$

and

$$\dot{\delta}_i = v_{i-1} - v_i \tag{11.5}$$

which can be seen as the kinematic model of the platoon of vehicles.

We consider the following cost function:

$$J_i = J_{fi} + \beta J_{ti} \tag{11.6}$$

which is the weighted sum of the fuel consumption J_{fi} and the travel time J_{ti} of the vehicle, with $J_{ti} = \int_{t_{0,i}}^{t_{f,i}} \beta dt$, $\beta > 0$ is a trade-off weight. The total fuel consumption depends on the energy required to propel the vehicle and can be described as [2]

$$J_{fi} = K_{Ei}\int_{t_{0i}}^{t_{fi}} \phi F_{ei}(t) v_i(t) dt \tag{11.7}$$

where $t_{0,i}$ and $t_{f,i}$ denote, respectively, the initial and final time of the road section, K_{Ei} is an energy conversion constant depending on the energy density of oil and the engine combustion efficiency, and ϕ is an indicator function to prevent negative energy when the vehicle is braking

$$\phi = \leq \begin{cases} 1 & \text{if } F\,ei \geq 0 \\ 0 & \text{otherwise} \end{cases} \tag{11.8}$$

This chapter aims to derive a speed-planning mechanism and a cooperative control method for a platoon of vehicles. The objective of this chapter is to achieve the following goals:

1) To give an analytical speed-planning scheme that yields the reference speed profile $v_p^*(t)$ for a platoon in the sense that the overall cost of fuel consumption and travel time is minimized. Note that the optimization problem is solved by the PMP method and $v_p^*(t)$ is the average of the optimal velocity of each vehicle, which is different from Chapter 10.
2) To derive a cooperative control method for the vehicular platoon to track the given speed profile $v_p^*(t)$ with guaranteed string stability, i.e. the spacing errors do not amplify as it propagates

through the platoon due to the leader vehicle or other interference. Namely, the control law should guarantee $\lim_{t\to\infty}|v_i - v_p^*| = 0$ and $|G_i(j\omega)| \leq 1$ for any ω, where $G_i(s) = \delta_i(s)/\delta_{i-1}(s)$. Here, we utilize the idea of sliding mode control and propose a decentralized controller, which is simple and easy to implement.

We consider a two-layered planning and control architecture to achieve these objectives. For the speed-planning layer, the optimal speed profile is generated by utilizing the PMP. The generated fuel-efficient speed is taken as the reference speed for the trajectory tracking control layer, which is also responsible for guaranteeing the string stability with the desired intervehicle spacing.

11.3 Speed Optimization Based on PMP

Fuel-time-efficient platoon control calls for optimal speed planning or set-point optimization. Our idea is to obtain the optimal speed set-point for the platoon of vehicles under different road segments. The optimization is done at the departure service area of each road segment instead of doing it for the entire route at once. This is motivated by the fact that, depending on the service areas chosen as the docking sites, the route of the platoon can be divided into consecutive road segments. The speed set-point optimization problem can be formulated with regard to either the leader vehicle or an average-size vehicle in the platoon. We choose to derive the optimal speed set point for each vehicle i and then use the averaged optimal speed as the reference speed for the platoon. For heterogeneous vehicles with different weights and sizes, the average speed profile is far more reasonable than schemes based on the leader speed planning. The speed-planning methodology is based on the PMP approach. Here, we first give the optimization problem considering fuel consumption as well as travel time. Then, we discuss the solution procedure of the optimization problem and derive an algorithm for vehicles with simplified linear dynamics.

Due to complexity, the analytical solutions for nonlinear vehicle dynamics may not be feasible for implementation in real applications. However, it is possible to obtain an easily applicable algorithm for platoon speed planning if we use the linearized vehicle models in (11.1) and (11.2). Consider vehicle i in the platoon. The dynamics of vehicle i is expressed as

$$\begin{bmatrix} \dot{s}_i \\ \dot{v}_i \end{bmatrix} = \begin{bmatrix} v_i \\ C_1 F_{e,i} - \overline{C}_2 v_i - C_3 - C_4 + C_5 \end{bmatrix} \equiv f_i(s_i, v_i, F_{ei}) \tag{11.9}$$

where $C_1 = 1/m_i$, $\overline{C}_2 = 2C_2 v_{0,i}$, $C_2 = 0.5\rho A_i C_{di}/m_i$, $C_3 = g \sin\theta$, $C_4 = \mu_i g \cos\theta$ and $C_5 = C_2 {v_{0,i}}^2$, with $v_{0,i}$ being the initial speed of the ith vehicle. Namely, let the air drag term $C_2 v_i^2$ be approximated by a linear term $\overline{C}_2 v_i - C_5$.

To derive the optimal trajectory, the speed-planning problem can be formulated as the following optimization problem:

$$\min J_i = \int_{t_{0,i}}^{t_{f,i}} \phi K_E F_{ei} v_i + \beta dt \tag{11.10}$$

$$\text{s.t.}[\dot{s}_i \quad \dot{v}_i]^T = f_i(s_i, c_i, F_{ei}) \tag{11.11}$$

$$F_{ei} \in [F_{ei}, \text{ min}, F_{ei}, \text{ max}] \tag{11.12}$$

$$v_i \in [v_i, \text{ min}, v_i, \text{ max}] \tag{11.13}$$

$$s(t_{0,i}) = s_{0,i}, \quad v(t_{0,i}) = v_{0,i} \tag{11.14}$$

$$s(t_{f,i}) = s_{f,i}, \quad v(t_f) = v_{f,i} \tag{11.15}$$

where $[v_{i,\min}, v_{i,\max}]$ are the speed limits of road segments, s_0 and s_f are the given start and final positions, respectively, and v_0 and v_f are, respectively, the initial and final speeds.

We introduce the following Hamiltonian function (the subscript i is omitted for the sake of presentation simplicity):

$$H = \phi K_E F_e v + \beta + \lambda_1 v + \lambda_2 (C_1 F_e - \overline{C}_2 v - C_3 - C_4 + C_5) + \lambda_3 f_3[v(t)] \tag{11.16}$$

where λ_1, λ_2, and λ_3 are the three costates, and $f_3[v(t)]$ is defined as

$$f_3[v(t)] = g_1^2[v(t)]\,\mathrm{sgn}(-g_1) + g_2^2[v(t)]\,\mathrm{sgn}(-g_2) \tag{11.17}$$

which is a function of the speed constraints

$$g_1[v(t)] = v(t) - v_{\min} \geq 0 \tag{11.18}$$

$$g_2[v(t)] = v_{\max} - v(t) \geq 0 \tag{11.19}$$

and

$$\mathrm{sgn}(-g_j) = \begin{cases} 0, & g_j[v(t)] \geq 0 \\ 1, & g_i[v(t)] < 0 \; j = 1, 2 \end{cases} \tag{11.20}$$

The details of the aforementioned function can be seen in [3]. Let $\dot{x}_3 = f_3[v(t)]$. Then we have the following boundary condition:

$$x_3(t_0) = x_3(t_f) = 0 \tag{11.21}$$

Then, according to the definition of PMP, the necessary conditions of optimality are expressed as:

$$\frac{\partial H}{\partial F_e} = \phi K_E v + \lambda_2 C_1 = 0 \tag{11.22}$$

$$\dot{\lambda}_1 = -\frac{\partial H}{\partial s} = 0 \tag{11.23}$$

$$\begin{aligned} \dot{\lambda}_2 = -\frac{\partial H}{\partial v} &= -\phi K_E F_e - \lambda_1 + \lambda_2 \overline{C}_2 \\ &- 2\lambda_3[(v - v_{\min})\,\mathrm{sgn}(-g_1) - (v_{\max} - v)\,\mathrm{sgn}(-g_2)] \end{aligned} \tag{11.24}$$

$$\dot{\lambda}_3 = -\frac{\partial H}{\partial x_3} = 0 \tag{11.25}$$

As the vehicular platoon stops at both ends of such road segments, its speed profile in each road segment consists of three different periods: acceleration, cruising with constant speed, and deceleration before stopping at the destination docking site. Therefore, to yield the optimal speed and position trajectory using PMP, we consider the following bang-bang control law as in [9]:

$$F_e(t) = \begin{cases} F_{e,\max}, & \text{if } \sigma < 0 \\ F_{e,\sigma}, & \text{if } \sigma = 0 \\ F_{e,\min}, & \text{if } \sigma > 0 \end{cases} \tag{11.26}$$

where $\sigma = \phi K_E v + \lambda_2 C_1$ is known as the switching function. Here $\sigma < 0$, $\sigma > 0$ and $\sigma = 0$, respectively, correspond to the cases of acceleration, deceleration, and cruising with constant speed. For $\sigma < 0$, $\sigma > 0$, the control forces $F_e = F_{e,\max}$ and $F_e = F_{e,\min}$, respectively. In particular, for the case of $\sigma = 0$, i.e. the constant speed cruising case, the control force trajectory $F_{e,\sigma}$ is a singular arc. In this case, the time derivative of σ should satisfy that

$$\sigma = \phi K_E \dot{v} + \lambda_2 C_1 = 0 \tag{11.27}$$

This section focuses on solving the aforementioned problem for optimal speed trajectory. To simplify the problem, we do not consider the influence of other vehicles. Let $M = a, d$ and $M_\sigma = c$ be the driving modes where $M = a$ denotes the acceleration mode in which $F_e = F_{e,\max}$, $M = d$ denotes the deceleration mode in which $F_e = F_{e,\min}$, and $M_\sigma = c$ denotes the constant speed mode in which $F_e = F_{e,\sigma}$ and with the deceleration mode being the last phase. Let t_M^0, t_M^f, t_c^0, and t_c^f be the start time and end time of different driving modes, and s_M^0, s_M^f, v_M^0, and v_M^f be the corresponding initial position, speed and final position, and speed of these driving modes, respectively. Then, we can readily get the following results, which are very easy to implement in real applications.

For the start phase, it may be the acceleration mode or deceleration mode, namely $M = a, d$, the speed and position trajectories can be, respectively, obtained as

$$\begin{aligned} v^*(t) = &\left[-\frac{1}{\overline{C}_2}(C_1F_e - C_3 - C_4 + C_5) + v_M^0\right]e^{-\overline{C}_2(t - t_M^0)} \\ &+ \frac{1}{\overline{C}_2}(C_1F_e - C_3 - C_4 + C_5) \end{aligned} \tag{11.28}$$

$$\begin{aligned} s^*(t) = &-\frac{1}{\overline{C}_2}\left[v_M^0 - \frac{1}{\overline{C}_2}(C_1F_e - C_3 - C_4 + C_5)\right]e^{-\overline{C}_2(t - t_M^0)} + \frac{1}{\overline{C}_2}(C_1F_e - C_3 - C_4 + C_5)(t - t_M^0) \\ &+ \frac{1}{\overline{C}_2}\left[v_M^0 - \frac{1}{\overline{C}_2}(C_1F_e - C_3 - C_4 + C_5)\right] + s_M^0 \end{aligned} \tag{11.29}$$

And the costate trajectories can be written as

$$\lambda_1^*(t) = K_1 \tag{11.30}$$

$$\lambda_2^*(t) = K_2 e^{\overline{C}_2 t} + \frac{1}{\overline{C}_2}(\phi K_E F_e + K_1) \tag{11.31}$$

$$\lambda_3^*(t) = K_3 \tag{11.32}$$

For the constant speed mode, by taking (11.9) and (11.18) into (11.20) and using (11.16), we can easily obtain the optimal constant speed and the corresponding position as

$$\begin{aligned} v^*(t) = v_c = &\frac{\phi K_E(C_3 + C_4 - C_5) + K_1C_1}{-2\overline{C}_2\phi K_E - 2C_1K_3[\mathrm{sgn}(-g_1) + \mathrm{sgn}(-g_2)]} \\ &+ \frac{2C_1K_3[v_{\max}\mathrm{sgn}(-g_2)] + v_{\min} s\,\mathrm{sn}(-g_1)}{2\overline{C}_2\phi K_E + 2C_1K_3[\mathrm{sgn}(-g_1) + \mathrm{sgn}(-g_2)]} \end{aligned} \tag{11.33}$$

$$s^*(t) = v_c(t - t_c^0) + s(t_c^0) \tag{11.34}$$

Note that for (11.33), there are two cases about v_c because of the speed limits of road segments. In the case of without speed limits, the optimal constant speed is denoted by

$$v_c = -\frac{C_3 + C_4 - C_5}{2\overline{C}_2} - \frac{K_1C_1}{2\phi K_E C_2} \tag{11.35}$$

With speed limits, it becomes

$$v_c = \begin{cases} \dfrac{\phi K_E(C_3 + C_4 - C_5) + K_1C_1 - 2K_3v_{\min}}{-2\overline{C}_2\phi K_E - 2K_3}, & v_c < v_{\min} \\ \dfrac{\phi K_E(C_3 + C_4 - C_5) + K_1C_1 - 2K_3v_{\max}}{-2\overline{C}_2\phi K_E - 2K_3}, & v_c > v_{\max}. \end{cases} \tag{11.36}$$

Accordingly, the costate and the driving force can be calculated by

$$\lambda_2^* = -\frac{\phi K_E v_c}{C_1} \tag{11.37}$$

$$F_{e,\sigma} = \frac{1}{C_1}\left[\overline{C}_2 v_c + C_3 + C_4 - C_5\right] \tag{11.38}$$

For whatever acceleration or deceleration modes, we strive to derive a constant driving speed for the vehicle, namely we can have a constant speed $v_M^f = v_c$ at the mode end time t_M^f. In addition, at the transition instants between two phases, the corresponding speed must be equal, and the same is true for costate. Therefore, based on (11.21) and (11.25), we can obtain the mode end time t_M^f, with $t_M^f = t_c^0$, as

$$t_M^f = -\frac{1}{\overline{C}_2}\ln\left[\frac{\overline{C}_2 v_c - (C_1F_e - C_3 - C_4 + C_5)}{\overline{C}_2 v_M^0 - (C_1F_e - C_3 - C_4 + C_5)}\right] + t_M^0 \tag{11.39}$$

Similarly, for the last phase, the costate λ_2 at the time t_d^0 must be equal and with $t_d^0 = t_c^f$; then, based on (11.31) and (11.37), the parameter K_2 can be derived by different equations due to the different adjoining points. Furthermore, using the boundary conditions (11.17) and the Hamiltonian function at time $t = t_f$, we can have that

$$H(f_f) = 0 \tag{11.40}$$

by which it is possible to calculate the unknown parameters t_M^f, t_f, and K_1.

Finally, based on the aforementioned discussions, we can obtain the optimal speed trajectory for each vehicle $v_i^*(t)$. For the platoon of vehicles with different weights and sizes, we can adopt the following averaged optimal speed as the reference speed of the platoon:

$$v_p^*(t) = \frac{1}{n+1}\sum_{i=0}^{n} v_i^*(t) \tag{11.41}$$

Remark 11.1 The method is different from that of Chapter 10, although both papers consider speed planning and tracking control of heterogeneous vehicular platoons. Specifically speaking, Chapter 10 uses an averaged vehicle for speed planning and gives a numerical method, which needs more information on the vehicles and implies large computational and communicational costs for the upper layer controller. The speed-planning method in this chapter uses a speed-averaging method but not a vehicle-averaging method. Also, it is an analytical scheme that leads to an algorithm more flexible and efficient because the optimal speed profile of each vehicle can be calculated independently in a distributed manner. Moreover, the presented scheme can be easily extended to the cases of platoon merging and splitting.

Based on the earlier given results, we now give the speed-planning algorithm for vehicular platoons. See Algorithm 11.1.

Algorithm 11.1 PMP-Based Optimal Speed-Planning Algorithm

Input: Parameters of the current road segment and each of the vehicles including speed limits $[v_{\min}, v_{\max}]$, K_E, C_1, C_2, C_3, C_4, C_5, $F_{e,\max}$, $F_{e,\min}$ and t_0, s_0, s_f, v_0, v_f.
Output: The averaged optimal speed trajectory $v_p^*(t)$.
Step 0: Set $\text{sgn}(-g_j) = 0, j = 1, 2$ and $F_e = F_{e,\max}$.
Step 1: Calculate K_1, according to the conditions at $t = t_f$ using Equations (11.28), (11.31), and (11.40).
Step 2: Calculate the constant speed v_c according to Equation (11.35).
Step 3: Determine the driving mode for the first phase according to the initial speed v_0 and v_c. If $v_0 < v_c$, use acceleration mode, i.e. $M = a$, $v_0 = v_a^0$; if $v_0 > v_c$, choose the deceleration mode, $M = d$, $v_0 = v_d^0$;
if $v_0 = v_c$, choose the constant speed mode.
Step 4: Determine whether to trigger speed constraints.
If $v_c \in [v_{\min}, v_{\max}]$, go to **Step 6**
Else if $v_c \notin [v_{\min}, v_{\max}]$, go to **Step 5**
Step 5: Calculate the constant speed v_c according to Equation (11.36).
Step 6: Calculate t_M^f according to Equation (11.39).
Step 7: Calculate $s\left(t_M^f\right)$ according to Equation (11.29).
Step 8: Calculate t_c^f according to s_f.
Step 9: Calculate $s\left(t_c^f\right)$ according to Equation (11.34).
Step 10: Calculate the final time t_f.
Step 11: Calculate the optimal trajectory $v^*(t)$ and $s^*(t)$ of each vehicle.
Step 12: Calculate the optimal platoon speed trajectory $v_p^*(t)$ by averaging using Equation (11.41).

11.4 Speed Tracking Control and String Stability

This section aims to design decentralized control laws for the follower vehicles such that they all mutually synchronize and follow the given platoon speed trajectory (represented by the leader), maintaining a desired and constant intervehicle spacing and guaranteeing platoon string stability, i.e. achieving the control objective 2).

Using $\dot{v}_i = a_i$ and based on (11.2), we can get the following equation:

$$F_{ei} = m_i a_i + \frac{\rho_i A_i C_{di}}{2} v_i^2 + m_i g \sin\theta + \mu_i m_i g \cos\theta \tag{11.42}$$

which, by taking the time derivative at both sides, can yield the following equation:

$$\dot{a}_i = \frac{1}{m_i}\left(\dot{F}_{ei}(t) - \rho_i A_i C_{di} v_i(t) a_i(t) - m_i g \cos\theta + \mu_i m_i g \sin\theta\right) \tag{11.43}$$

In (11.43), F_{ei} is the driving force generated by the engine, whose dynamics can be given by the following equation [4]:

$$\dot{F}_{ei}(t) = -F_{ei}(t)/\tau_i + c_i(t)/\tau_i \tag{11.44}$$

where c_i is the engine input and $\tau_i > 0$ is the time constant of the engine.

Based on (11.42)–(11.44), the vehicle dynamics can be written as

$$\begin{cases} \dot{s}_i = v_i \\ \dot{v}_i = a_i \\ \dot{a}_i = f_i(v_i, a_i) + g_i(v_i)c_i \end{cases} \tag{11.45}$$

where $f_i(v_i, a_i)$ and $g_i(v_i)$ are given by

$$f_i(v_i, a_i) = -\frac{1}{\tau_i}\left(a_i + \frac{\rho_i A_i C_{di}}{2m_i} v_i^2 + g\sin\theta + \mu_i g\cos\theta\right) - \frac{\rho_i A_i C_{di}}{m_i} v_i a_i - g\cos\theta + \mu_i g\sin\theta$$

$$g_i(v_i) = \frac{1}{\tau_i m_i}.$$

$$v^*(t) = \left[-\frac{1}{\overline{C}_2}(C_1F_e - C_3 - C_4 + C_5) + v_M^0\right]e^{-\overline{C}_2\left(t - t_M^0\right)} + \frac{1}{\overline{C}_2}(C_1F_e - C_3 - C_4 + C_5)$$

$$s^*(t) = -\frac{1}{\overline{C}_2}\left[v_M^0 - \frac{1}{\overline{C}_2}(C_1F_e - C_3 - C_4 + C_5)\right]e^{-\overline{C}_2\left(t - t_M^0\right)} + \frac{1}{\overline{C}_2}(C_1F_e - C_3 - C_4 + C_5)\left(t - t_M^0\right)$$

$$+ \frac{1}{\overline{C}_2}\left[v_M^0 - \frac{1}{\overline{C}_2}(C_1F_e - C_3 - C_4 + C_5)\right] + s_M^0$$

$$\lambda_1^*(t) = K_1$$

$$\lambda_2^*(t) = K_2 e^{\overline{C}_2 t} + \frac{1}{\overline{C}_2}(\phi K_E F_e + K_1)$$

$$\lambda_3^*(t) = K_3$$

$$v^*(t) = v_c = \frac{\phi K_E(C_3 + C_4 - C_5) + K_1C_1}{-2\overline{C}_2\phi K_E - 2C_1K_3[\mathrm{sgn}(-g_1) + \mathrm{sgn}(-g_2)]}$$

$$+ \frac{2C_1K_3[v_{\max}\mathrm{sgn}(-g_2)] + v_{\min} s\,\mathrm{sn}(-g_1)}{2\overline{C}_2\phi K_E + 2C_1K_3[\mathrm{sgn}(-g_1) + \mathrm{sgn}(-g_2)]}$$

$$s^*(t) = v_c\left(t - t_c^0\right) + s\left(t_c^0\right)$$

$$v_c = -\frac{C_3 + C_4 - C_5}{2\overline{C}_2} - \frac{K_1C_1}{2\phi K_E C_2}$$

$$v_c = \begin{cases} \dfrac{\phi K_E(C_3 + C_4 - C_5) + K_1C_1 - 2K_3v_{\min}}{-2\overline{C}_2\phi K_E - 2K_3}, & v_c < v_{\min} \\ \dfrac{\phi K_E(C_3 + C_4 - C_5) + K_1C_1 - 2K_3v_{\max}}{-2\overline{C}_2\phi K_E - 2K_3}, & v_c > v_{\max}. \end{cases}$$

$$\lambda_2^* = -\frac{\phi K_E v_c}{C_1}$$

$$F_{e,\sigma} = \frac{1}{C_1}\left[\overline{C}_2 v_c + C_3 + C_4 - C_5\right]$$

$$t_M^f = -\frac{1}{\overline{C}_2}\ln\left|\frac{\overline{C}_2 v_c - (C_1F_e - C_3 - C_4 + C_5)}{\overline{C}_2 v_M^0 - (C_1F_e - C_3 - C_4 + C_5)}\right| + t_M^0$$

$$H\left(t_f\right) = 0$$

$$v_p^*(t) = \frac{1}{n+1}\sum_{i=0}^{n} v_i^*(t)$$

$$F_{ei} = m_i a_i + \frac{\rho_i A_i C_{di}}{2} v_i^2 + m_i g \sin\theta + \mu_i m_i g \cos\theta$$

$$\dot{a}_i = \frac{1}{m_i}\left(\dot{F}_{ei}(t) - \rho_i A_i C_{di} v_i(t) a_i(t) - m_i g \cos\theta + \mu_i m_i g \sin\theta\right)$$

$$\dot{F}_{ei}(t) = -F_{ei}(t)/\tau_i + c_i(t)/\tau_i \begin{cases} \dot{s}_i = v_i \\ \dot{v}_i = a_i \\ \dot{a}_i = f_i(v_i, a_i) + g_i(v_i) c_i \end{cases}$$

$$f_i(v_i, a_i) = -\frac{1}{\tau_i}\left(a_i + \frac{\rho_i A_i C_{di}}{2m_i} v_i^2 + g \sin\theta + \mu_i g \cos\theta\right) - \frac{\rho_i A_i C_{di}}{m_i} v_i a_i - g \cos\theta + \mu_i g \sin\theta$$

$$g_i(v_i) = \frac{1}{\tau_i m_i}$$

The aforementioned nonlinear model can be further transformed into a linear one by introducing the following feedback linearization control law:

$$\begin{aligned} c_i = {} & m_i u_i + \tau_i \rho_i A_i C_{di} v_i \dot{v}_i + \tau_i m_i g \cos\theta - \tau_i \mu_i m_i g \sin\theta \\ & + \frac{1}{2}\rho_i A_i C_{di} v_i^2 + m_i g \sin\theta + \mu_i m_i g \cos\theta \end{aligned} \tag{11.46}$$

where u_i is the additional control input to be designed. After introducing (11.46), the third equation in (11.45) becomes

$$\dot{a}_i = -a_i/\tau_i + u_i + \tau_i \tag{11.47}$$

Based on the intervehicle spacing and the platoon dynamics in (11.3)–(11.5), the platoon model can be described by the following differential equations:

$$\begin{bmatrix} \dot{\delta}_i \\ \dot{v}_i \\ \dot{a}_i \end{bmatrix} = \begin{bmatrix} v_{i-1} - v_i \\ a_i \\ -a_i/\tau_i + u_i/\tau_i \end{bmatrix} \tag{11.48}$$

which serves as the basis for speed-tracking control design and string stability analysis.

Here, we assume the leader's trajectory is the generated optimal trajectory, i.e. $v_0 = v_p^*$. Define the tracking error e_i of the ith following vehicle as:

$$\dot{e}_i = k_1(d_i - d^*) + k_2(v_i - v_0) + k_3(s_i - s_i^*) \tag{11.49}$$

where s_i^* represents the desired trajectory of ith vehicle with $s_i^* = s_0 - id^* - (l_1 + l_2 + \cdots + l_i)$, and k_1, k_2, and k_3 denote the error weight parameters.

To steer this error to zero by using the sliding mode control approach, we will choose the sliding mode surface, such that:

$$S_i = \dot{e}_i + ce_i \tag{11.50}$$

where $c > 0$ which is satisfied with the condition of Hurwitz. Taking the time derivative of S_i, then

$$\dot{S}_i = k_1(\dot{v}_{i-1} - \dot{v}_i) + k_2\left(-\frac{a_i}{\tau_i} + \frac{u_i}{\tau_i}\right) + k_3(\dot{v}_i - \dot{v}_0) + c\dot{e}_i \tag{11.51}$$

In order to ensure a finite-time convergence of the sliding surface to zero, the following dynamic will be imposed on the sliding surface S_i:

$$\dot{S}_i = -\eta S_i \tag{11.52}$$

where $\eta > 0$. According to Equations (11.49) and (11.50), the control law is calculated by:

$$\begin{aligned} u_i &= \frac{\tau_i}{k_2}\left[-\eta S_i - k_1(\dot{v}_{i-1} - \dot{v}_i) + \frac{k_2}{\tau_2}a_i - k_3(\dot{v}_i - \dot{v}_0) - c\dot{e}_i\right] \\ &= \frac{\tau_i}{k_2}\Big\{-(\eta + c)k_1(v_{i-1} - v_i) - k_1(\dot{v}_{i-1} - \dot{v}_i) - k_2(\eta + c)(\dot{v}_i - \dot{v}_0) \\ &\quad - [k_3(\eta + c) + \eta c k_2](v_i - v_0) - \eta c k_3(s_i - s_i^*) - \eta c k_1(d_i - d^*) + \frac{k_2}{\tau_i}a_i\Big\} \end{aligned} \tag{11.53}$$

Theorem 11.1 When the control law in (11.53) is employed in the vehicles described by Equation (11.48), the stability of the vehicles can be guaranteed in the sense that the sliding surface and the tracking error converge to zero asymptotically.

Proof: Firstly, choose the following Lyapunov function candidate:

$$V_i = \frac{1}{2}S_i^2 \tag{11.54}$$

Using (11.49), the time derivative of the Lyapunov function for each vehicle can be obtained as:

$$\dot{V}_i = S_i\dot{S}_i = S_i\left[k_1(\dot{v}_{i-1} - \dot{v}_i) - \frac{k_2}{\tau_i}a_i + \frac{k_2}{\tau_i}\frac{\tau_i}{k_2}u_i + k_3(\dot{v}_i - \dot{v}_0) + c\dot{e}_i\right] \tag{11.55}$$

Substituting the control law (11.53) into (11.55) yields

$$\begin{aligned} \dot{V}_i = S_i\dot{S}_i = S_i\Big\{&k_1(\dot{v}_{i-1} - \dot{v}_i) - \frac{k_2}{\tau_i}a_i + \frac{k_2}{\tau_i}\frac{\tau_i}{k_2}\left[-\eta S_i - k_1(\dot{v}_{i-1} - \dot{v}_i) + \frac{k_2}{\tau_i}a_i - k_3(\dot{v}_i - \dot{v}_0) - c\dot{e}_i\right] \\ &+ k_3(\dot{v}_i - \dot{v}_0) + c\dot{e}_i\Big\} = -\eta S_i^2 \end{aligned} \tag{11.56}$$

Since $\eta > 0$, $\dot{V}_i$ is negative definite and the condition for Lyapunov stability is satisfied. The tracking error can reach the sliding mode $S_i = 0$ within a finite time. The proof of the theorem is completed.

In this subsection, the string stability of the vehicular platoon is conducted and, based on the string stability conditions obtained, a platoon control algorithm is given.

Combining the longitudinal vehicle dynamic model in (11.48) and the proposed control law (11.53), and using some basic algebraic calculations, we can write the spacing error dynamics in the time domain as follows:

$$\begin{aligned} &k_2\dddot{\delta}_i + [k_3 + (c + \eta)k_2 - k_1]\ddot{\delta}_i + [(c + \eta)(k_3 - k_1) + c\eta k_2]\dot{\delta}_i + c\eta(k_3 - k_1)\delta_i \\ &= -k_1\ddot{\delta}_{i-1} - k_1(c + \eta)\dot{\delta}_{i-1} - c\eta k_1\delta_{i-1} \end{aligned} \tag{11.57}$$

Taking the Laplace transformation, the system dynamics in (11.57) can be written in the frequency domain as follows:

$$G_i(s) = \frac{\delta_i(s)}{\delta_{i-1}(s)} = \frac{\text{num}}{\text{den}} \tag{11.58}$$

where

$$\begin{aligned} \text{num} &= -k_1 s^2 - k_1(c+\eta)s - c\eta k_1 \\ \text{den} &= k_2 s^3 + [k_3 + (c+\eta)k_2 - k_1]s^2 + [(c+\eta)(k_3-k_1) + c\eta k_2]s + c\eta(k_3 - k_1) \end{aligned}$$

Note that the denominator of $G_i(s)$ is Hurwitz if and only if

$$\begin{cases} k_2 > 0 \\ \eta > 0, c > 0 \\ k_3 - k_1 > 0 \end{cases} \tag{11.59}$$

To ensure that the spacing error does not increase as it propagates through the platoon, it suffices to guarantee that

$$|G_i(j\omega)| \leq 1 \forall \omega \tag{11.60}$$

with

$$|G_i(j\omega)| = \sqrt{\frac{A^2 + B^2}{C^2 + D^2}} \tag{11.61}$$

where

$$\begin{aligned} A &= k_1\omega^2 - c\eta k_1, B = -k_1(c+\eta)\omega \\ C &= -(k_3 + ck_2 + \eta k_2 - k_1)\omega^2 + c\eta(k_3 - k_1) \\ D &= -k_2\omega^2 + [(c+\eta)(k_3 - k_1) + c\eta k_2]\omega. \end{aligned}$$

Algorithm 11.2 Platoon Control Algorithm

Step 1: Given the optimal platoon speed trajectory v_p^* using Algorithm 11.1.
Step 2: Design the feedback-linearization controller in (11.46), which is a routine procedure.
Step 3: Determine parameters c and η according to Theorem 11.1. Using these parameters, Equation (11.59) and inequalities (11.62) in Theorem 11.2, find conditions for the controller gains k_1, k_2, and k_3.
Step 4: Calculate the control input $u_i(t)$ in (11.53) using parameters obtained in Step 3.
Step 5: Apply the tracking controller $u_i(t)$ to the platoon system.

Theorem 11.2 The transfer function of the spacing errors satisfies $|\delta_i(j\omega)/\delta_{i-1}(j\omega)| \leq 1$ for any $\omega > 0$ if the following conditions are satisfied:

$$\begin{cases} (a).k_3^2 - 2k_1k_3 \geq 0 \\ (b).c > 0, \quad \eta > 0 \end{cases} \tag{11.62}$$

Proof: Based on the condition in (11.60), we can have

$$a_1\omega^6 + a_2\omega^4 + a_3\omega^2 + a_4 \geq 0 \tag{11.63}$$

with

$$\begin{aligned}
a_1 &= k_2^2 \\
a_2 &= [k_3 + (c+\eta)k_2 - k_1]^2 - 2k_2[(c+\eta)(k_3-k_1) + c\eta k_2] - k_1^2 \\
a_3 &= [(c+\eta)(k_3-k_1) + c\eta k_2] - 2c\eta(k_3-k_1)[k_3 + (c+\eta)k_2 - k_1] + 2k_1^2 c\eta - k_1^2(c+\eta)^2 \\
a_4 &= [c\eta(k_3-k_1)]^2 - (c\eta k_1)^2
\end{aligned} \tag{11.64}$$

Since $a_1 > 0$ and $a_4 > 0$, we can see that $|\delta_i(j\omega)/\delta_{i-1}(j\omega)| \leq 1$ for any $\omega > 0$ if $a_2 > 0$ and $a_3 > 0$, which are equivalent to the conditions in (11.62). The proof is completed.

Finally, we are in a position to give the platoon control algorithm. See Algorithm 11.2 here.

11.5 Simulation Studies

In this section, the proposed speed planning and platoon control method is tested via simulations done with the MATLAB toolbox. Consider a platoon of four identical vehicles with one leader running on a straight lane, where the parameters of the vehicles are the same with [5], as listed in Table 11.1. The vehicles are carefully spaced in the simulation such that there are no initial spacing errors. We first evaluate the optimal reference speed trajectories for different roads. Second, vehicle tracking performance is studied.

In this section, Algorithm 11.1 is tested by considering different cases. First, according to Algorithm 11.1, the driving mode should be determined. For this, the simulation is performed on a 10-km-flat highway segment under different initial velocities. The simulation results are shown in Figure 11.3. Second, simulations are performed on a 20-km road divided by two service

Table 11.1 Parameters of the vehicle model.

Parameter	Description	Value
l	Length of the vehicle	2 m
m	Mass of the vehicle	1607 kg
ρ	The air density	1.19854 kg/m^3
A	The cross-section area	2.250842 m^2
C_d	The air drag coefficient	0.3
μ	The rolling resistance coefficient	0.015
g	The gravity acceleration	9.81 m/s^2

areas into three road segments with different slopes and speed limits. Using the presented method, the optimal speed trajectories are calculated, as shown in Figures 11.4 and 11.5.

In this section, we analyze the performance of the vehicle tracking for one road segment. The optimal reference trajectory is calculated by Algorithm 11.1 as shown in Figure 11.3, where the leader vehicle starts with $v_0 = 5$ m/s and decelerates to $v_f = 5$ m/s. The initial positions and initial

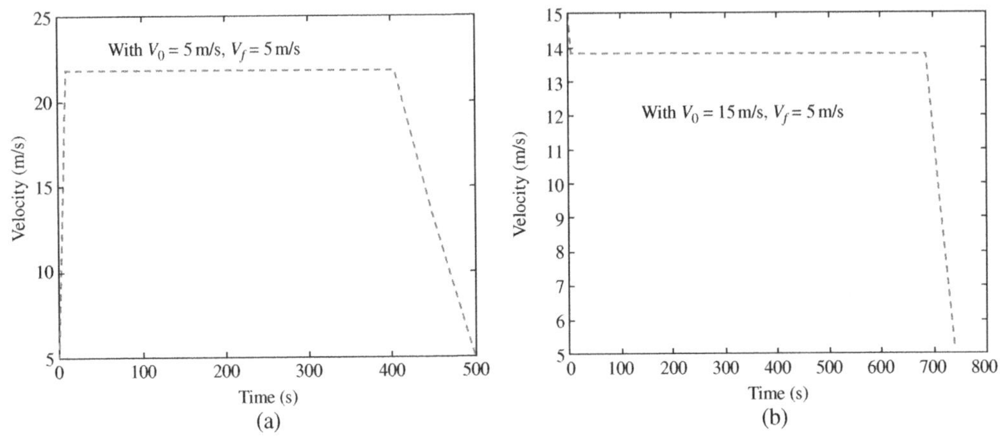

Figure 11.3 Optimal speed trajectory. (a) Initial velocity 5 m/s. (b) Initial velocity 15 m/s.

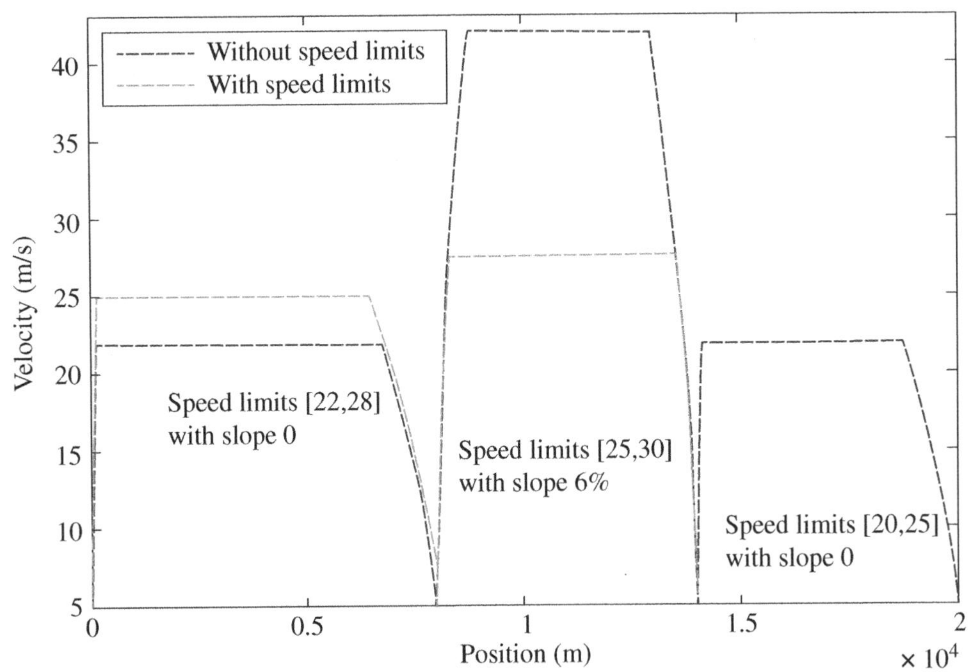

Figure 11.4 Optimal speed trajectory including an uphill.

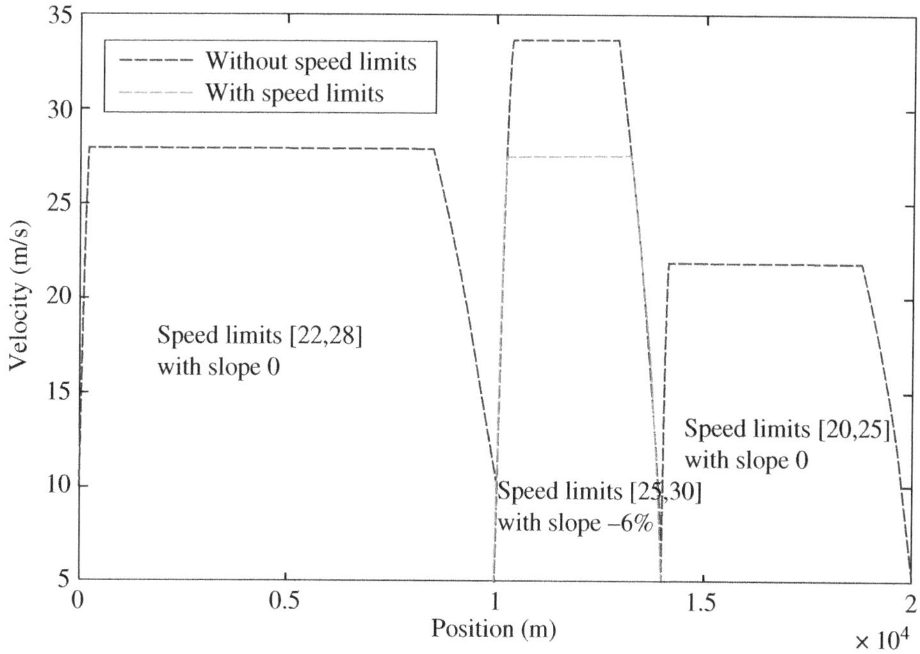

Figure 11.5 Optimal speed trajectory including a downhill.

velocities of four vehicles are chosen as follows: $s(t_0) = [0\ 12\ 24\ 36]$ and $v(t_0) = [5\ 5\ 5\ 5]$. The desired intervehicle spacing is chosen as $d^* = 10\ m$. Based on the conditions (11.59) and (11.62), we can set $c = 0.5$, $\eta = 0.5$, $k_1 = -1$, $k_2 = 1.5$, and $k_3 = 3$. The simulation results are shown in Figures 11.6–11.8.

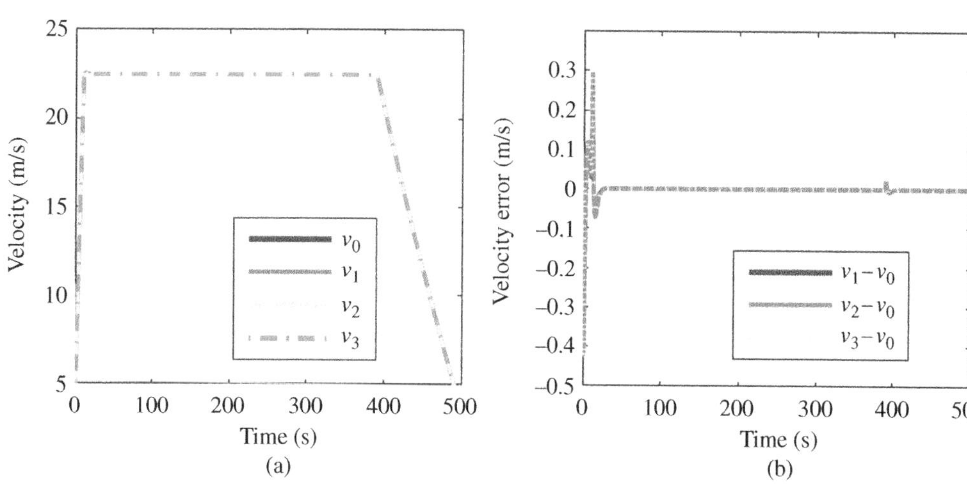

Figure 11.6 Speed trajectory tracking. (a) Speed trajectory of all vehicles. (b) Speed errors of followers.

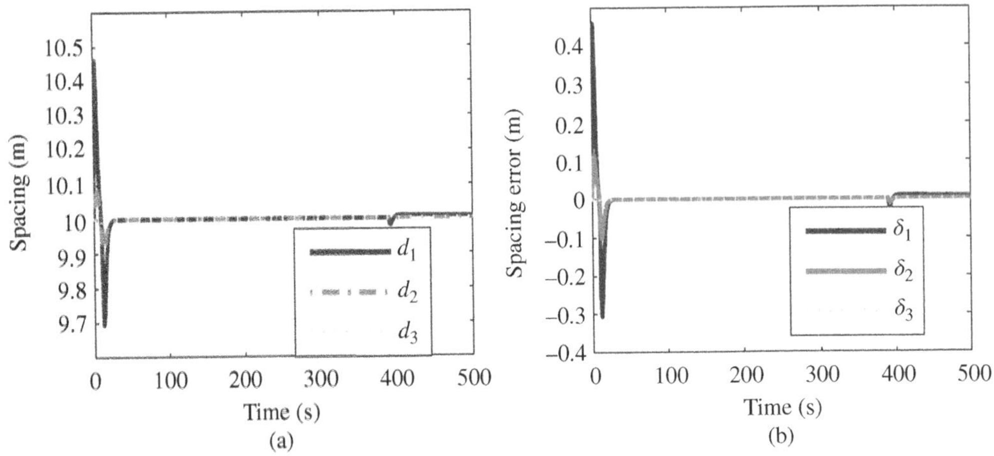

Figure 11.7 Vehicle spacing keeping. (a) Intervehicle spacing. (b) Spacing errors of followers.

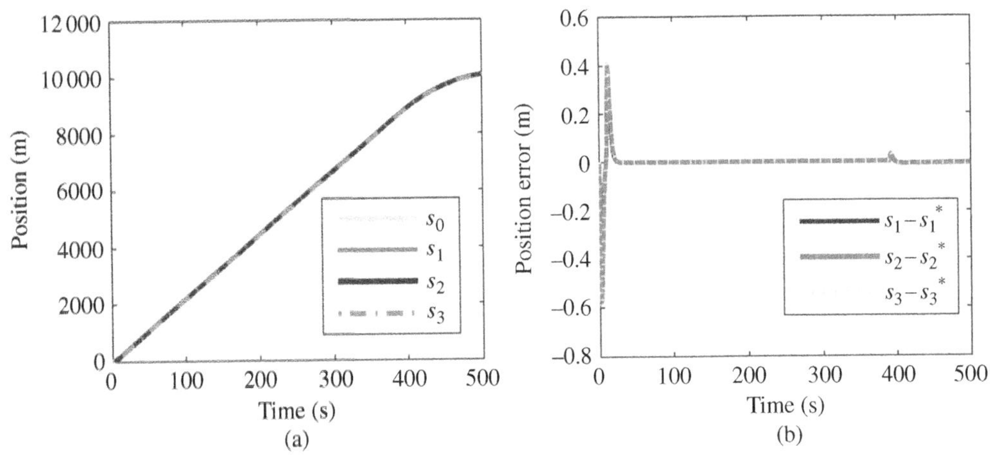

Figure 11.8 Position tracking. (a) Position of vehicles. (b) Position errors of followers.

11.6 Conclusions and Future Work

This chapter studies the platooning problem of road freight transport trucks for fuel efficiency. Introducing a two-layer control structure, we first derive the optimal speed trajectory for the vehicular platoon by using the PMP approach, considering a joint cost of fuel consumption and travel time. Second, we give a sliding mode bang-bang control algorithm for speed tracking, which runs in the tracking layer in a distributed manner and guarantees the string stability of the platoon. Simulations are given to verify the effectiveness of the method.

References

1 Kirk, D.E. (2012). *Optimal Control Theory: An Introduction (Dover Books on Electrical Engineering)*. Mineola, NY: Dover.

2 Liang, K.-Y., Mårtensson, J., and Johansson, K.H. (2016). Heavy-duty vehicle platoon formation for fuel efficiency. *IEEE Trans. Intell. Transp. Syst.* 17 (4): 1051–1061.

3 Li, C.J. and Ma, G.F. (2011). *Optimal Control*. Beijing: Science Press.

4 Ali, A., Garcia, G., and Martinet, P. (2015). The flatbed platoon towing model for safe and dense platooning on highways. *IEEE Intell. Transp. Syst. Mag.* 7 (1): 58–68.

5 Lim, H., Su, W., and Mi, C.C. (2017). Distance-based ecological driving scheme using a two-stage hierarchy for long-term optimization and short-term adaptation. *IEEE Trans. Veh. Technol.* 66 (3): 1940–1949.

12

Speed Planning and Sliding-Mode Control to Reduce Intervehicle Spacing

12.1 Introduction

In this chapter, we are interested in a joint problem of vehicle speed planning and optimal tracking control. This important problem is studied in Chapter 10, which presents a framework for receding horizon speed planning and back-stepping platoon control. In the meanwhile, Chapter 11 gives an analytical solution to the same problem based on Pontryagin's minimum principle and sliding mode control. Being optimal in the sense of overall cost, however, these two results are often accompanied by turbulence of spacing errors in the transient state. As is well known, intervehicle spacing errors are crucial for keeping the platoons compact [1, 2].

To cooperatively minimize the intervehicle spacing errors via distributed control is very challenging. The main difficulties lie in that: (i) Each follower can communicate only with its neighboring vehicles. In an unreliable vehicular ad-hoc network (VANET) environment, it is highly possible that the state of the leader and the neighbors could not be successfully received for the vehicle to update its control input. As a result, control methods assuming specific reliable information flows (i.e. leader predecessor following or LPF, predecessor following or PF, and bidirectional following or PSF) may not be able to guarantee string stability. (ii) A heterogeneous vehicular platoon comprises vehicles with different sizes, parameters, and uncertain dynamics. Moreover, in the cascaded platoon control system, the intervehicle spacing errors of one follower could affect the others [3].

This chapter investigates the problem of minimizing intervehicle spacing errors based on distributed cooperative control. To deal with random information flow in VANETs and uncertain vehicle dynamics, a two-layered hierarchical control framework is proposed for each follower. Specifically, the upper layer aims to provide a desired trajectory for each follower, which is implemented by the distributed continuous-time convex optimization technique. We prove that the desired trajectory is unique for each follower. Moreover, consensus with the leader's running trajectory is achieved when intervehicle spacing errors are minimized. In the second layer, an adaptive sliding mode (ASM) controller is designed to track the desired trajectory obtained in the first layer. By analyzing the dynamics of the tracking errors on the sliding surface, a useful condition is derived to determine the parameters of the ASM controller so that both internal and string stability are satisfied. The contributions of this study are summarized as follows:

1) Since the optimal trajectory provided by the first layer is in consensus with the leaders when the intervehicle spacing errors are minimized, the string stability of a platoon can be guaranteed without directly using the information from the leader or the nearest neighbors. Zheng et al.

Connected Vehicular Systems: Communication, Control, and Optimization, First Edition. Ge Guo and Shixi Wen.

[4] and Seiler et al. [5] find that the more followers have the information from the leader, the better controlled the performance will be.

2) By using only information exchange with neighboring vehicles, at any time, the cooperative minimization on intervehicle spacing is implemented, in which each follower is allowed uncertain dynamics with different sizes.
3) By establishing a two-layered hierarchical control framework, the two aspects of communication and control are decoupled in the control synthesis of the platoon. In this way, it is convenient to introduce an advanced control technique (i.e. ASM controller) in the second layer to improve the control performance of the platoon.

The remainder of the chapter is organized as follows. Section 12.2 describes the vehicle longitudinal dynamics, the objectives of platoon optimization control, and the introduction of the two-layered hierarchical control framework. Section 12.3 outlines the tracking optimization procedure in the first layer. Section 12.4 provides the design of the ASM tracking controller in the second layer. Section 12.5 illustrates the effectiveness of the proposed control method using numerical simulations, while Section 12.6 presents concluding remarks.

Notation $\mathbb{R}^n$ denotes the n-dimensional Euclidean space and diag{·} is a block-diagonal matrix. The matrix inequality $A > (\geq B)$ means that matrix A−B is positive (semi-) definite. In addition, $\|x\|_p$ denotes the p-norm of the vector x, A^T is the transpose of a matrix A, $\otimes$ is the matrix Kronecker product, and $\lambda_{\min}(A)$ is the minimal eigenvalue of a matrix A. Let 1_N and 0_N denote the column vectors of N ones and zeros, respectively, where I_N denotes the $N \times N$ identity matrix. sgn(·) is the signum function defined componentwise and |·| is the absolute value.

12.2 Problem Statement

Consider a heterogeneous platoon of $N + 1$ vehicles driving on a horizontal road (see Figure 12.1) in a VANET environment. Denote z_i, v_i, and a_i as the position, velocity, and acceleration of the $i-th$ ($i = 0, 1, 2..., N$) vehicle, respectively. In the platoon, the leader is indexed by 0 and the followers are indexed from 1 to N in order. Each follower can communicate with the neighbors to exchange state information (i.e. z_i, v_i, and a_i). It is assumed that intervehicle communication is fixed and perfect.

In this study, the intervehicle communication connections among the followers are described by a directed graph $\mathcal{G}_N = (\mathcal{V}_N, \mathcal{E}_N)$, where $\mathcal{V}_N = \{1, 2, ..., N\}$ is the vertices set and $\mathcal{G}_N = (\mathcal{V}_N, \mathcal{E}_N)$ is the edges set. An edge defined as $\varepsilon_{ij} \triangleq (i, j) \in \mathcal{E}_N$ implies that a directed communication link exists from vehicle j to vehicle i, for instance, $(i, j) \in \mathcal{E}_N$. A path is a sequence of connected edges in a graph. $\mathcal{A} = [a_{ij}]_{N\times N}$ represents an adjacency matrix of graph $\mathcal{G}_N$, where $a_{ii} = 0$ for all i, and $a_{ij} = 1$ if $\varepsilon_{ij} \in \mathcal{E}_N$, otherwise, $a_{ij} = 0$. Graph $\mathcal{G}_N$ is connected if there is a path between every pair of nodes. Let

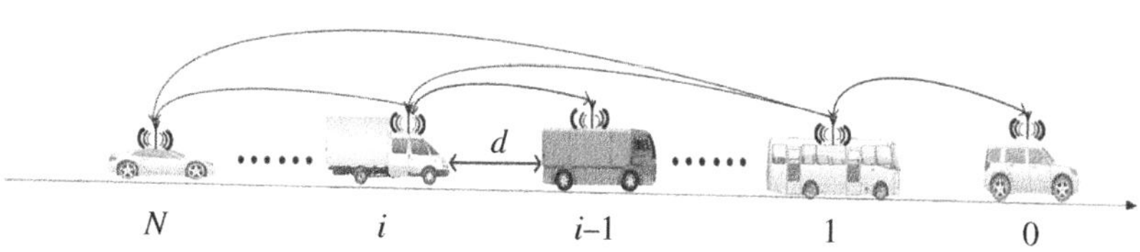

Figure 12.1 Platoon control.

$D = [d_{ij}]_{N \times |\varepsilon_N|}$ be the incidence matrix associated with $\mathcal{G}_N$, where $d_{ij} = -1$ if the edge ε_j leaves follower i, $d_{ij} = 1$ if it enters the follower i, and $d_{ij} = 0$ otherwise. The Laplacian matrix L describes the directional connection among followers, defined as $L = [l_{ij}]_{N \times N}$, $l_{ii} = -\sum_{j=1, j \neq i}^{N} l_{ij}$. If follower i can receive the information of i, $l_{ij} = -1$; otherwise, $l_{ij} = 0$. Note that $L \triangleq DD^T$. We define a neighbor set of nodes i in the followers as $\mathbf{N}_i = \{j \in \mathcal{V}_N, (i,j) \in \mathcal{E}_N\}$. We define a neighbor set of nodes i in the followers as

$$\mathbf{N}_i = \{j \in \mathcal{V}_N, (i,j) \in \mathcal{E}_N\} \tag{12.1}$$

An augmented directed graph $\mathcal{G}_{N+1} = (\mathcal{V}_{N+1}, \mathcal{E}_{N+1})$ is defined to describe the communication between the leader and the followers with a set of $N+1$ vertices $\mathcal{V}_{N+1} = \{0, 1, 2, \ldots, N\}$. The pinning matrix $\mathcal{P} = \text{diag}\{p_1, p_2, \ldots, p_N\}$ denotes the connection between the leader and the follower, where $p_i = 1$ if follower i is connected with the leader; otherwise, $p_i = 0$. The leader-accessible set of nodes i is defined as

$$\mathbf{P}_i \begin{cases} \phi & \text{if} \quad p_i = 0 \\ \{0\} & \text{if} \quad p_i = 1 \end{cases} \tag{12.2}$$

where ϕ denotes the empty set and set $\{0\}$ only has the index number 0. Depending on the connection status p_i, set $\mathbf{P}_i$ is either an empty set or the set including the index number of the leader only.

The incidence matrix associated with $\mathcal{G}_{N+1}$ is $\overline{D} = [d_{ij}]_{N \times |\varepsilon_{N+1}|}$. Then, we have $L + \mathcal{P} \triangleq \overline{DD}^T$. These notations can also be adapted for time-varying graphs, where $\mathcal{G}_N(t)/\mathcal{G}_{N+1}(t)$, $L(t)$, $D(t)/\overline{D}(t)$, and $\mathcal{P}(t)$ are the undirected graph, the Laplacian matrix, the incidence matrix, and the pinning matrix at time t, respectively.

This study focuses on undirected communication topology. The communication topology is considered undirected if the information flow among followers is undirected, which means the edges (i, j) and (j, i) are equivalent.

Throughout this study, we make the following assumption [6]:

Assumption 12.1 The augmented graph $\mathcal{G}_{N+1}$ contains at least one spanning tree rooted at the leader.

This assumption implies that the leader is globally reachable in $\mathcal{G}_{N+1}$. Namely each follower can obtain the leader's information directly or indirectly, which is a prerequisite to guarantee the internal stability of a platoon. Generally, the internal stability indicates the asymptotic stability of the platoon with respect to the eigenvalues of the closed-loop system, considering the intervehicle communication topology, which is defined as:

Definition 12.1 A platoon with linear time-invariant dynamics is internal stability if and only if its closed-loop control system is asymptotically stable [7].

Lemma 12.1 For any undirected connected topology, $\lambda_{\min}(L) = 0$ with $\mathbf{1}_N$ as the corresponding eigenvector. Moreover, if Assumption 12.1 holds, $\lambda_{\min}(L) = 0$ is a simple eigenvalue, and all the eigenvalues of $L + \mathcal{P}$ are greater than zero, that is, $\lambda_{\min}(L + \mathcal{P}) > 0$ [8].

According to the feedback linearization technique, the following linearized third-order model is used to describe the longitudinal behavior of each vehicle in a platoon [9, 10]

$$\begin{cases} \dot{z}_i(t) = v_i(t) \\ \dot{v}_i(t) = a_i(t) \\ \dot{a}_i(t) = -\varsigma_i a_i(t) + \varsigma_i u_{i2}(t) \end{cases} \tag{12.3}$$

where $\varsigma_i = 1/\theta_i$ with θ_i denoting the uncertain time constant of drivetrain dynamics (TCDD), and with $u_{i2}(t)$ as the control input for the follower i.

Define the following spacing error for each follower

$$\delta_i(t) = z_i(t) - z_{i-1}(t) - d_{i,i-1} \tag{12.4}$$

where constant $d_{i,\,i-1} > 0$ denotes the desired distance between vehicle i and vehicle $i-1$. Here we adopt a constant spacing policy for the platoon, that is, $d_{i,\,i-1} = d$. To determine the desired inter-vehicle distance d, the length of the individual vehicles should be considered.

To minimize intervehicle space errors in a cascaded platoon control system, we define the following optimization and control problem:

$$\min_{z_i(t)} J(t) = \sum_{i=1}^{N} \delta_i^2(t) \tag{12.5a}$$

$$\text{s.t.} \begin{cases} \lim_{t\to\infty} |v_i(t) - v_0(t)| = 0 \\ \lim_{t\to\infty} |\delta_i(t)| = 0 \end{cases} \tag{12.5b}$$

Here we assume that the leader runs a constant speed trajectory, that is, $a_0 = 0$, $z_0 = v_0 t$. It should be noted that the cost function for the platoon defined in (12.5a) is convex.

A two-layered hierarchical control framework is proposed to solve the distributed optimization control problem defined in (12.5a) and (12.5b). As shown in Figure 12.2, trajectory optimization and tracking control for the platoon are implemented in different layers. Specifically, the first layer is for tracking optimization. Using its own and the neighbor's measured position and velocity, a group of time-varying reference positions $z_i^*(t)$ and velocity $v_i^*(t)$ is determined to satisfy the control objectives defined in (12.5a) and (12.5b). In the second layer, an adaptive tracking controller is designed to track $z_i^*(t)$ and $v_i^*(t)$ so that internal and string stability are satisfied under the effect of uncertain parameter ς_i in vehicle dynamics.

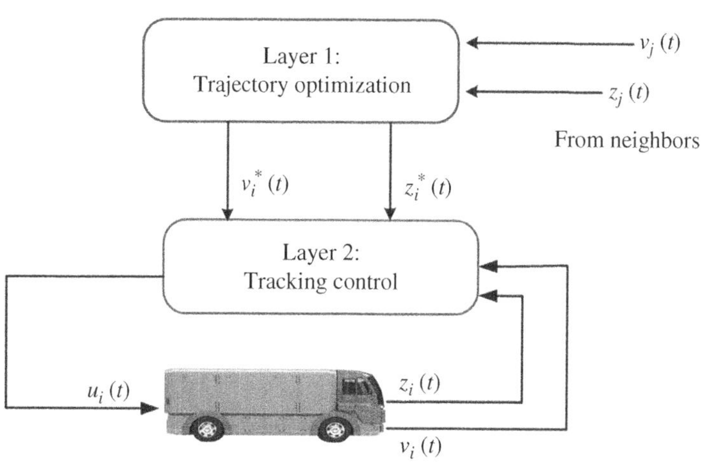

Figure 12.2 Two-layered hierarchical control framework for platoon.

12.3 Intervehicle Spacing Optimization

In this section, we consider the problem of trajectory optimization for the vehicular platoon in the upper layer. For this purpose, consider the following kinematic equation for the reference position and velocity.

$$\begin{cases} \dot{z}_i^*(t) = v_i^*(t) \\ \dot{v}_i^*(t) = u_{i1}(t) \end{cases} \tag{12.6}$$

Here we aim to design control input $u_{i1}(t)$ by using the information of nodes in the set $\mathbf{I}_i = \mathbf{N}_i \cup \mathbf{P}_i$ so that trajectory $z_i^*(t)$ and $v_i^*(t)$ determined by (12.6) are the feasible solutions for the optimization problem defined in (12.5a) and (12.5b). According to the set $\mathbf{I}_i$, we know whether or not the leader is a neighbor of the ith follower.

Then, we obtain the following results for trajectory optimization in the first layer.

Theorem 12.1 Consider a heterogeneous platoon with uncertain vehicle dynamics. It is assumed that Assumption 12.1 holds. For positive coefficients μ, α, γ, and ξ, if $(\gamma/(\alpha\xi)) < \lambda_{\min}(L + \mathcal{P})$ holds, the controller (12.7a) designed for the first-layer guarantee that the reference position $z_i^*(t)$ and velocity $v_i^*(t)$ are the feasible trajectories for the optimization problem defined in (12.5a) and (12.5b). The gain $\beta_{ij}(t)$ and internal signal $\phi_i(t)$ in controller (12.7a) are given as

$$\begin{aligned} u_{i1}(t) = &-\sum_{j\in\mathbf{I}_i}\mu\left(z_i(t) - z_j(t) - d_{i,j}\right) + \alpha\left(v_i(t) - v_j(t)\right) \\ &-\sum_{j\in\mathbf{I}_i}\beta_{ij}(t)\,\mathrm{sgn}\left(\gamma\left(z_i(t) - z_j(t) - d_{i,j}\right) + \xi\left(v_i(t) - v_j(t)\right)\right) + \phi_i(t) \end{aligned} \tag{12.7a}$$

$$\dot{\beta}_{ij}(t) = \left|\gamma\left(z_i(t) - z_j(t) - d_{i,j}\right) + \xi\left(v_i(t) - v_j(t)\right)\right| \tag{12.7b}$$

$$\phi_i(t) = -2\left(z_i(t) - z_j(t) - d_{i,j}\right) - \left(v_i(t) - v_j(t)\right) \tag{12.7c}$$

where $\beta_{ij}(0) = \beta_{ji}(0) \geq 0$.

Proof: We define the following tracking errors in the first layer

$$\begin{cases} \hat{z}_i^*(t) = z_i^*(t) - z_0(t) - d_{i,0} \\ \hat{v}_i^*(t) = v_i^*(t) - v_0(t) \end{cases} \tag{12.8}$$

where $\hat{z}_i^*(t)$ and $\hat{v}_i^*(t)$ denote the spacing error and velocity error between the ith follower and the leader, respectively. By introducing tracking errors $\hat{z}_i^*(t)$ and $\hat{v}_i^*(t)$, it is clear that

$$\begin{cases} \hat{z}_i^*(t) - \hat{z}_j^*(t) = z_i^*(t) - z_j^*(t) - d_{i,j} \\ \hat{v}_i^*(t) - \hat{v}_j^*(t) = v_i^*(t) - v_j^*(t) \end{cases}$$

According to (12.8), the dynamics of the tracking errors can be derived as

$$\begin{cases} \dot{\hat{z}}_i^*(t) = \hat{v}_i^*(t) \\ \dot{\hat{v}}_i^*(t) = u_{i1}(t) \end{cases} \tag{12.9}$$

Here we use the fact that the acceleration of the leader is $a_0(t) = 0$.

Using the tracking controller designed in the second layer, the position and velocity of each follower are satisfied so that $z_i(t) \to z_i^*(t)$ and $v_i(t) \to v_i^*(t)$ as time $t \to \infty$. For this reason, we replace position $z_i(t)$ and velocity $v_i(t)$ in controller $u_{i1}(t)$ with $z_i^*(t)$ and $v_i^*(t)$, respectively. Then, controller $u_{i1}(t)$ in (12.7a) can be rewritten as in (12.10a) with varying gain $\beta_{ij}(t)$ and internal signal $\phi_i(t)$ given as

$$
\begin{aligned}
u_{i1}(t) = &-\sum_{j\in\mathbf{I}_i}\mu\left(\hat{z}_i^*(t)-\hat{z}_j^*(t)\right)+\alpha\left(\hat{v}_i^*(t)-\hat{v}_j^*(t)\right)\\
&-\sum_{i\in\mathbf{I}_i}\beta_{ij}(t)\operatorname{sgn}\left(\gamma\left(\hat{z}_i^*(t)-\hat{z}_j^*(t)\right)+\xi\left(\hat{v}_i^*(t)-\hat{v}_j^*(t)\right)\right)+\phi_i(t)
\end{aligned}
\tag{12.10a}
$$

$$
\dot{\beta}_{ij}(t)=\left|\gamma\left(\hat{z}_i^*(t)-\hat{z}_j^*(t)\right)+\xi\left(\hat{v}_i^*(t)-\hat{v}_j^*(t)\right)\right| \tag{12.10b}
$$

$$
\phi_i(t)=-2\left(\hat{z}_i^*(t)-\hat{z}_j^*(t)\right)-\left(\hat{v}_i^*(t)-\hat{v}_j^*(t)\right) \tag{12.10c}
$$

According to $\hat{z}_i^*(t)$ and $\hat{v}_i^*(t)$ defined in (12.8), the optimization problem described in (12.5a) and (12.5b) can be transformed into the following optimization consensus problem

$$
\min_{\hat{z}_i^*(t)} J=\sum_{i=1}^{N}\left[\hat{z}_i^*(t)-\hat{z}_{i-1}^*(t)\right]^2 \tag{12.11a}
$$

$$
\text{s.t.}\begin{cases}\hat{z}_i^*(t)=\hat{z}_j^*(t)\\ \hat{v}_i^*(t)=\hat{v}_j^*(t)\end{cases} \tag{12.11b}
$$

To prove that $\hat{z}_i^*(t)$ and $\hat{v}_i^*(t)$ are the feasible solutions for the optimization problem in (12.11a) and (12.11b), we required the following definition and assumption.

Definition 12.2 Defining $a'_{ij}=a_{ij}\beta_{ij}(t)$ and $p'_i=p_i\beta_{i0}(t)$. A new Laplacian matrix $L'=[l'_{ij}]_{N\times N}$ and pinning matrix $\mathcal{P}'=\operatorname{diag}\{p'_1,p'_2,\ldots,p'_n\}$ are introduced, where $l'_{ii}=-\sum_{j=1,j\neq i}^{N}l'_{ij}$ and $l'_{ij}=-a'^2_{ij}$. Since $l'_{ij}=l'_{ji}$, Laplacian matrix L' is symmetric. Similar to the definition of D and $\overline{D}$, $D'=\left[d'_{ij}\right]_{N\times|\varepsilon_N|}$ and $\overline{D}'=\left[d'_{ij}\right]_{N\times|\varepsilon_{N+1}|}$ are the two new incidence matrices associated with the Laplacian matrix L', where $d'_{ij}=-a'_{ij}$ if the edge ε_{ij} leaves node i, $d'_{ij}=a'_{ij}$ if it enters the nodes i, and $d_{ij}=0$ otherwise. Note that $L'\triangleq D'D'^T$ and $L'+\mathcal{P}'\triangleq\overline{D}'\overline{D}'^T$.

Assumption 12.2 For internal signal $\phi_i(t)$ defined in (12.7c), a positive constant $\overline{\phi}$ exists so that $|\phi_i(t)-\phi_j(t)|\le\overline{\phi}$, $\forall i,\ j\in\{1,2,\ldots,N\}$. It is clear that the boundness of $|z_i(t)-z_j(t)-d_{i,j}|$ and $|v_i(t)-v_j(t)|$ can guarantee the boundness of $\overline{\phi}$.

Define $\overline{Z}(t)=\left[\hat{z}_1^*(t)\ \hat{z}_i^*(t)\ \cdots\ \hat{z}_N^*(t)\right]^T$ and $\overline{V}(t)=\left[\hat{v}_1^*(t)\ \ \hat{v}_2^*(t)\ \ \cdots\ \ \hat{v}_N^*(t)\right]^T$ and $\Phi(t)=\left[\phi_1(t)\ \ \phi_2(t)\ \ \cdots\ \ \phi_N(t)\right]^T$. First, we prove that spacing error $\hat{z}_i^*(t)$ and velocity error $\hat{v}_i^*(t)$ can reach a consensus as $t\to\infty$.

Taking controller (12.10a) into tracking error dynamic (12.9), a compact form of a closed-loop platoon control system can be obtained as

$$
\begin{cases}\dot{\overline{Z}}(t)=\overline{V}(t)\\ \dot{\overline{V}}(t)=-\mathcal{G}\left(\mu\overline{Z}(t)+\alpha\overline{V}(t)\right)-\Omega(t)+\Phi(t)\end{cases} \tag{12.12}
$$

where $\mathcal{G}=L+\mathcal{P}$ and $\Omega(t)=\overline{\mathcal{D}}'\operatorname{sgn}\left(\overline{\mathcal{D}}'^T\left(\gamma\overline{Z}(t)+\xi\overline{V}(t)\right)\right)$. To simplify the analysis for the platoon control system (12.12), we define the following error vector:

$$
\begin{cases}\hat{Z}(t)=\Pi\overline{Z}(t)\\ \hat{V}(t)=\Pi\overline{V}(t)\end{cases}
$$

where matrix $\Pi \triangleq I_N - (1/N)\mathbf{1}_N\mathbf{1}_N^T$ due to the fact that matrix Π has one simple zero eigenvalue with $\mathbf{1}_N$ as its right eigenvector and has 1 as its other eigenvalue with multiplicity $N-1$. Then, $\hat{Z}(t)=0$ and $\hat{V}(t)=0$ if and only if $\hat{z}_i^*(t)=\hat{z}_j^*(t)$ and $\hat{v}_i^*(t)=\hat{v}_j^*(t)$, respectively. As a result, we can rewrite the platoon control system (12.12) as

$$\begin{cases} \dot{\hat{Z}}(t) = \hat{V}(t) \\ \dot{\hat{V}}(t) = -\mathcal{G}(\mu\hat{Z}(t) + \alpha\hat{V}(t)) - \Omega'(t) + \Pi\phi(t) \end{cases} \tag{12.13}$$

where $\Omega'(t) = \overline{\mathcal{D}}'\,\mathrm{sgn}\left(\overline{\mathcal{D}}^T\left(\gamma\hat{Z}(t) + \xi\hat{V}(t)\right)\right)$

Then, we define the following Lyapunov function for the derived closed-loop platoon control system (12.13):

$$W_1(t) = \frac{1}{2}\begin{bmatrix}\hat{Z}(t)\\ \hat{V}(t)\end{bmatrix}^T H \begin{bmatrix}\hat{Z}(t)\\ \hat{V}(t)\end{bmatrix} + \frac{1}{4}\sum_{i=1}^{N}\sum_{j\in\mathbf{I}_i}\left(\beta_{ij}(t) - \overline{\beta}\right)^2 \tag{12.14}$$

where $H = \begin{bmatrix}(\alpha\gamma+\mu\xi)\mathcal{G} & \gamma I_N\\ \gamma I_N & \xi I_N\end{bmatrix}$ and $\overline{\beta} > 0$ to be selected. To prove the positive definiteness of matrix H, we define matrix $\hat{H}$ as

$$\hat{H} = \begin{bmatrix}(\alpha\gamma+\mu\xi)\lambda_{\min}(\mathcal{G})I_N & \gamma I_N\\ \gamma I_N & \xi I_N\end{bmatrix}.$$

It is clear that $\hat{H} \le H$. Since it is true that

$$(\alpha\gamma+\mu\xi)\lambda_{\min}(\mathcal{G}) - \frac{\gamma^2}{\xi} > 0$$

Based on Schur's complement, we know that $\hat{H} > 0$. Namely, $0 \le \hat{H} \le H$.

Taking the time derivative of $W_1(t)$, we obtain

$$\begin{aligned}
\dot{W}_1(t) &= \begin{bmatrix}\hat{Z}(t)\\ \hat{V}(t)\end{bmatrix}^T H \begin{bmatrix}\hat{V}(t)\\ \dot{\hat{V}}(t)\end{bmatrix} + \frac{1}{2}\sum_{i=1}^{N}\sum_{j\in\mathbf{I}_i}\left(\beta_{ij}(t) - \overline{\beta}\right)\dot{\beta}_{ij}(t)\\
&= -\gamma\mu\hat{Z}^T(t)\mathcal{G}\hat{Z}(t) + \hat{V}^T(t)(\gamma I_N - \alpha\xi\mathcal{G})\hat{V}(t)\\
&= -\gamma\mu\hat{Z}^T(t)\mathcal{G}\hat{Z}(t) + \hat{V}^T(t)(\gamma I_N - \alpha\xi\mathcal{G})\hat{V}(t)\\
&\quad - \frac{1}{2}\sum_{i=1}^{N}\sum_{j\in\mathbf{I}_i}\beta_{ij}(t)\left|\gamma\left(\hat{z}_i^*(t) - \hat{z}_j^*(t)\right) + \xi\left(\hat{v}_i^*(t) - \hat{v}_j^*(t)\right)\right|\\
&\quad + \frac{1}{N}\sum_{i=1}^{N}\sum_{j=1}^{N}\left[\gamma\left(\hat{z}_i^*(t) - \hat{z}_j^*(t)\right) + \xi\left(\hat{v}_i^*(t) - \hat{v}_j^*(t)\right)\right]\phi_i(t)\\
&\quad + \frac{1}{2}\sum_{i=1}^{N}\sum_{j\in\mathbf{I}_i}\left(\beta_{ij}(t) - \overline{\beta}\right)\dot{\beta}_{ij}(t)
\end{aligned} \tag{12.15}$$

Substituting $\dot{\beta}_{ij}(t)$ (12.10b) into (12.15) yields

$$\begin{aligned}
\dot{W}_1(t) \leq & -\gamma\mu\hat{Z}^T(t)\mathcal{G}\hat{Z}(t) + \hat{V}^T(t)(\gamma I_N - \alpha\xi\mathcal{G})\hat{V}(t) \\
& - \frac{1}{2}\sum_{i=1}^{N}\sum_{j\in\mathbf{I}_i}\beta_{ij}(t)\left|\gamma\left(\hat{z}_i^*(t) - \hat{z}_j^*(t)\right) + \xi\left(\hat{v}_i^*(t) - \hat{v}_j^*(t)\right)\right| \\
& + \frac{1}{2N}\sum_{i=1}^{N}\sum_{j=1}^{N}\left|\gamma\left(\hat{z}_i^*(t) - \hat{z}_j^*(t)\right) + \xi\left(\hat{v}_i^*(t) - \hat{v}_j^*(t)\right)\right|\cdot\left(\phi_i(t) - \phi_j(t)\right) \\
& \frac{1}{2}\sum_{i=1}^{N}\sum_{j\in\mathbf{I}_i}\left(\beta_{ij}(t) - \overline{\beta}\right)\left|\gamma\left(\hat{z}_i^*(t) - \hat{z}_j^*(t)\right) + \xi\left(\hat{v}_i^*(t) - \hat{v}_j^*(t)\right)\right| \\
\leq & -\gamma\mu\hat{Z}^T(t)\mathcal{G}\hat{Z}(t) + \hat{V}^T(t)(\gamma I_N - \alpha\xi\mathcal{G})\hat{V}(t) \\
& + \left(\frac{(N-1)\overline{\phi}}{4} - \frac{\overline{\beta}}{2}\right)\sum_{i=1}^{N}\sum_{j\in\mathbf{I}_i}\left(\beta_{ij}(t) - \overline{\beta}\right)\left|\gamma\left(\hat{z}_i^*(t) - \hat{z}_j^*(t)\right)\right| \\
& + \left(\frac{(N-1)\overline{\phi}}{4} - \frac{\overline{\beta}}{2}\right)\sum_{i=1}^{N}\sum_{j\in\mathbf{I}_i}\left(\beta_{ij}(t) - \overline{\beta}\right)\left|\xi\left(\hat{v}_i^*(t) - \hat{v}_j^*(t)\right)\right|
\end{aligned} \tag{12.16}$$

where the last inequality is derived based on Assumptions 12.1 and 12.2. Note that $\gamma - \alpha\xi\lambda_{\min}(\mathcal{G}) < 0$, then $\gamma I_N - \alpha\xi\mathcal{G} < 0$. As a result, the term $\hat{V}^T(t)(\gamma I_N - \alpha\xi\mathcal{G})\hat{V}(t) \leq 0$.

Selecting $\overline{\beta}$ so that $\overline{\beta} > \left(\left((N-1)\overline{\phi}\right)/2\right)$, we obtain

$$\begin{aligned}
\dot{W}_1(t) & \leq \left(\frac{(N-1)\overline{\phi}}{4} - \frac{\overline{\beta}}{2}\right)\sum_{i=1}^{N}\sum_{j\in\mathbf{I}_i}\left|\gamma\left(\hat{z}_i^*(t) - \hat{z}_j^*(t)\right)\right| + \xi\left(\hat{v}_i^*(t) - \hat{v}_j^*(t)\right) \\
& = \left(\frac{(N-1)\overline{\phi}}{2} - \overline{\beta}\right)\left(\gamma\hat{Z}(t) + \xi\hat{V}(t)\right)^T\Omega'(t) \\
& \leq \left(\frac{(N-1)\overline{\phi}}{2} - \overline{\beta}\right)\left\|\overline{D}^T\left(\gamma\hat{Z}(t) + \xi\hat{V}(t)\right)\right\|_1 \\
& \leq \left(\frac{(N-1)\overline{\phi}}{2} - \overline{\beta}\right)\sqrt{\lambda_{\min}(\mathcal{G})}\left\|\gamma\hat{Z}(t) + \xi\hat{V}(t)\right\|_2 < 0
\end{aligned} \tag{12.17}$$

where $\mathcal{G} = \overline{D}\overline{D}^T$ is used in the last inequality. Since $W_1(t) \geq 0$ and $\dot{W}_1(t) \leq 0$, we have $\hat{Z}(t)$, $\hat{V}(t) \in \mathcal{L}_\infty$. By integrating both sides of (12.17), one can obtain $\hat{Z}(t)$, $\hat{V}(t) \in \mathcal{L}_2$. According to Barbalat's Lemma [11], it is concluded that $\hat{Z}(t)$ and $\hat{V}(t)$ will converge asymptotically to zero. Consequently, $\hat{z}_i^*(t) = \hat{z}_j^*(t)$ and $\hat{v}_i^*(t) = \hat{v}_j^*(t)$ as $t \to \infty$.

We now illustrate that the optimization goal (12.11a) can be achieved. For this purpose, we define the following Lyapunov function

$$W_2(t) = \frac{1}{2}\sum_{i=1}^{N}\sum_{j\in\mathbf{I}_i}\left(\hat{z}_i^*(t) - \hat{z}_j^*(t)\right)^2 + \frac{1}{2}\sum_{i=1}^{N}\sum_{j\in\mathbf{I}_i}\left(\hat{z}_i^*(t) - \hat{z}_j^*(t) - \dot{v}_i^*(t)\right)^2 \tag{12.18}$$

The time derivative of $W_2(t)$ along with the system defined by (12.9) with controller (12.10a) is obtained as

$$\dot{W}_2(t) = \sum_{i=1}^{N}\sum_{j\in\mathbf{I}_i}\left(\hat{z}_i^*(t) - \hat{z}_j^*(t)\right)\left(\hat{v}_i^*(t) - \hat{v}_j^*(t)\right) + \sum^{N}\sum\left(\hat{z}_i^*(t) - \hat{z}_j^*(t) - \dot{v}_i^*(t)\right)\left(\hat{v}_i^*(t) - \hat{v}_j^*(t)\right) \tag{12.19}$$

By summing both sides of the closed-loop tracking system (12.9) with controller (12.10a), we have $\sum_{i=1}^{N}\dot{\hat{v}}_i^*(t) = \sum_{i=1}^{N}\phi_i(t)$. Substituting the value $\phi_i(t)$ (12.10c) into (12.19), one can derive

$$\dot{W}_2(t) = \sum_{i=1}^{N}\sum_{j\in \mathbf{I}_i} -\left(\hat{v}_i^*(t) - \hat{v}_j^*(t)\right)^2 \tag{12.20}$$

Therefore, $\dot{W}_2(t) < 0$ for $\sum_{i=1}^{N}\sum_{j\in \mathbf{I}_i}\left(\hat{v}_i^*(t) - \hat{v}_j^*(t)\right) \neq 0$. As $W_2(t) > 0$ and $\dot{W}_2(t) < 0$, we conclude that $\sum_{i=1}^{N}\sum_{j\in \mathbf{I}_i}\left(\hat{z}_i^*(t) - \hat{z}_j^*(t)\right) \in \mathcal{L}_\infty$, which means $\sum_{i=1}^{N}\sum_{j\in \mathbf{I}_i}\left(\hat{v}_i^*(t) - \hat{v}_j^*(t)\right) \in \mathcal{L}_\infty$. By integrating both sides of (12.20), one can derive $\sum_{i=1}^{N}\sum_{j\in \mathbf{I}_i}\left(\hat{v}_i^*(t) - \hat{v}_j^*(t)\right) \in \mathcal{L}_2$. Based on Barbalat's Lemma [11], it is concluded that $\sum_{i=1}^{N}\sum_{j\in \mathbf{I}_i}\left(\hat{v}_i^*(t) - \hat{v}_j^*(t)\right)$ will converge to zero asymptomatically. This means that the time derivative of the platoon cost function $J(t)$ will converge to zero asymptomatically. In addition, the platoon cost function $J(t)$ is a continuously differentiable convex function. Hence, it is minimized at $\hat{z}_i^*(t)$ as time $t \to \infty$ [12]. Moreover, the optimal solution $\hat{z}_i^*(t)$ is unique according to [13]. According to (12.8), it is clear that $z_i^*(t)$ and $v_i^*(t)$ are the feasible trajectory for the optimization problem defined in (12.5a) and (12.5b).

Remark 12.1 In the proposed controller $u_{i1}(t)$, the summation term $\sum_{j\in \mathbf{I}_i}\mu\left(z_i(t) - z_j(t) - d_{i,j}\right) + \alpha\left(v_i(t) - v_j(t)\right)$ and $\sum_{j\in \mathbf{I}_i}\beta_{ij}(t)\,\mathrm{sgn}\left(\gamma\left(z_i(t) - z_j(t) - d_{i,j}\right) + \xi\left(v_i(t) - v_j(t)\right)\right)$ contribute to the consensus problem (12.5b) via local-information feedback. As the signum function switches instantaneously at zero, it can compensate for the effect of the inconsistent time-varying optimization signals among the vehicles such that (12.5b) holds. Since the cost function $J(t)$ is time-varying, varying gain β_{ij} has the ability to adjust its value according to spacing error $z_i - z_j - d_{i,j}$ and velocity error $v_i - v_j$. Additionally, the internal signal $\phi_i(t)$ places a bound on the cost function $J(t)$ and its changing rate so that the optimization problem defined in (12.5a) and (12.5b) can be solved.

12.4 Sliding-Mode Controller Design

In this section, an ASM controller $u_{i2}(t)$ is derived in the second layer to track the reference trajectory $z_i^*(t)$ and $v_i^*(t)$, which can reject the effect of uncertain TCDDs. By analyzing the dynamics of the tracking errors on the sliding surface, we provide the condition to design the parameter of an ASM controller so that both internal and string stability are satisfied.

In the second layer, position tracking error and velocity tracking error are defined as

$$\begin{cases} \tilde{z}_i(t) = z_i(t) - z_i^*(t) \\ \tilde{v}_i(t) = v_i(t) - v_i^*(t) \end{cases} \tag{12.21}$$

Then, we use the following sliding surface to track the optimal position $z_i^*(t)$ and velocity $v_i^*(t)$

$$s_i(t) = \tilde{a}_i(t) + K\tilde{e}_i(t) \tag{12.22}$$

where tracking error vector $\tilde{e}_i(t) = \left[\tilde{z}_i(t) \quad \tilde{v}_i(t)\right]^T$, $\tilde{a}_i(t) = a_i(t) - u_{i1}(t)$ and $K = \left[k_1 \quad k_2\right]$ is the design parameter of the sliding motion.

To guarantee the speed and smoothness of convergence for the designed ASM controller, the following exponential reaching law is selected [14].

$$\dot{s}_i(t) = -\kappa s_i(t) \tag{12.23}$$

where constant $\kappa > 0$ determines the exponential convergence speed of $s_i(t)$.

We are now able to present the results for the tracking control.

Theorem 12.2 Consider the tracking control in the second layer. Based on the ASM control technique [14, 15], we derive the following controller:

$$u_{i2}(t) = a_i(t) - \frac{K\dot{\tilde{e}}_i}{\hat{\varsigma}_i} - \frac{\kappa s_i}{\hat{\varsigma}_i} \tag{12.24}$$

where the parameter adaptive law is given by

$$\dot{\hat{\varsigma}}_i = -\frac{K\dot{\tilde{e}}_i(t)s_i(t)}{\hat{\varsigma}_i q} \tag{12.25}$$

then the sliding surface $s_i(t)$ of each follower will exponentially converge to zero. Here the constant $q > 0$ is designed in accordance with the requirement of adaptive speed.

Proof: To prove the asymptotical convergence of $s_i(t)$, we define the following Lyapunov function:

$$\mathcal{W}_i(t) = \frac{1}{2}s_i^2(t) + \frac{1}{2}q(\hat{\varsigma}_i - \varsigma_i)^2 \tag{12.26}$$

Taking the time derivative of $\mathcal{W}_i(t)$ yields

$$\dot{\mathcal{W}}_i(t) = s_i(t)\dot{s}_i(t) + q(\hat{\varsigma}_i - \varsigma_i)\dot{\hat{\varsigma}}_i \tag{12.27}$$

According to (12.22), it is clear that

$$\dot{s}_i(t) = \dot{\tilde{a}}_i(t) + K\dot{\tilde{e}}_i(t) \tag{12.28}$$

Substituting the ASM controller given in (12.24) and (12.25) into (12.28) using dynamics (12.3) and $u_{i1}(t) = 0$, one can derive that

$$\dot{s}_i(t) = \left(1 - \frac{\varsigma_i}{\hat{\varsigma}_i}\right)K\dot{\tilde{e}}_i(t) - \frac{\varsigma_i \kappa s_i(t)}{\hat{\varsigma}_i} \tag{12.29}$$

Taking (12.25) and (12.29) into (12.27), $\dot{\mathcal{W}}_i(t)$ can be derived a

$$\dot{\mathcal{W}}_i(t) = -\frac{\varsigma_i}{\hat{\varsigma}_i}\kappa s_i^2(t) \tag{12.30}$$

Since uncertain parameter ς_i is positive, $(\varsigma_i/\hat{\varsigma}_i) > 0$ leading to $\dot{\mathcal{W}}_i(t) < 0$, if $s_i(t) \neq 0$. It is clear that $s_i(t)$ will asymptotically converge to zero according to Lyapunov stability theory.

To analyze the tracking error dynamics on the sliding surface, we consider the following tracking error system:

$$\dot{\tilde{e}}_i(t) - A\tilde{e}_i(t) + B\tilde{a}_i(t) \tag{12.31}$$

where $A = \begin{bmatrix} 0 & 1 \\ 0 & 0 \end{bmatrix}$, and $B = \begin{bmatrix} 0 \\ 1 \end{bmatrix}$

After the platoon dynamics reach the sliding surface, that is $s_i(t) = 0$, according to (12.22), we obtain

$$\tilde{a}_i(t) = -K\tilde{e}_i(t) \tag{12.32}$$

Taking (12.32) into (12.31), we can derive the following tracking error dynamics in the second layer.

$$\dot{\tilde{e}}_i(t) = (A - BK)\tilde{e}_i(t) \tag{12.33}$$

Theorem 12.3 Considering the tracking error dynamics (12.33) on the sliding surface, the tracking errors $\tilde{e}_i(t)$ asymptotically converge to zero under the ASM controller in (12.24) and (12.25) with parameter K determined by

$$K = B^T P^{-1} \tag{12.34}$$

where positive matrix $P > 0$ satisfies the following linear matrix inequality (LMI)

$$AP + PA^T - 2BB^T < 0 \tag{12.35}$$

Proof: We chose the following Lyapunov function for tracking error system (12.3)

$$\mathfrak{W}_i(t) = \tilde{e}_i^T(t) P \tilde{e}_i(t) \tag{12.36}$$

The time derivative of $\mathfrak{W}_i(t)$ along the trajectory of system (12.33) yields

$$\dot{\mathfrak{W}}_i(t) = 2\tilde{e}_i^T(t) P \dot{\tilde{e}}_i(t) = 2\tilde{e}_i^T(t)(A - BK)^T P \tilde{e}_i(t) \tag{12.37}$$

Taking controller gain (12.34) into (12.37), according to LMI (12.35), one can derive

$$\dot{\mathfrak{W}}_i(t) < 0 \tag{12.38}$$

which means tracking errors $\tilde{e}_i(t)$ asymptotically converge to zero.

According to Theorem 12.3, we know that tracking errors are stable on the sliding surface. Then, we have $\dot{u}_{i1}(t) = 0$ for a platoon with a constant speed trajectory. In the string stability analysis, we assume that $\tilde{z}_0(t) = 0$, $\tilde{a}_0(t) = 0$ and $\tilde{e}_0(t) = 0$.

According to the spacing error $\delta_i(t)$ defined in (12.4), we obtain

$$\ddot{\delta}_i(t) = \dot{a}_i(t) - \dot{a}_{i-1}(t) = \dot{\tilde{a}}_i(t) - \dot{\tilde{a}}_{i-1}(t) \tag{12.39}$$

Since $\dot{\tilde{a}}_i(t) = \dddot{\tilde{z}}_i(t)$, (12.39) can be represented by

$$\dddot{\delta}_i(t) = \ddot{\tilde{a}}_i(t) - \ddot{\tilde{a}}_{i-1}(t) = \dddot{\tilde{z}}_i(t) - \dddot{\tilde{z}}_{i-1}(t) \tag{12.40}$$

When platoon dynamics reach the sliding surface, according to (12.32), it is clear that

$$\dot{\tilde{a}}_i(t) - \dot{\tilde{a}}_{i-1}(t) = -K\dot{\tilde{e}}_i(t) + K\dot{\tilde{e}}_{i-1}(t) \tag{12.41}$$

Combined with (12.40) and (12.41), according to basic algebraic calculations, we obtain

$$\dddot{\tilde{z}}_i(t) - \dddot{\tilde{z}}_{i-1}(t) = -k_1\dot{\tilde{z}}_i(t) - k_2\ddot{\tilde{z}}_i(t) + k_1\dot{\tilde{z}}_{i-1}(t) + k_2\ddot{\tilde{z}}_{i-1}(t) \tag{12.42}$$

Applying the Laplace transformation for (12.42) under zero initial condition, one can derive that

$$\frac{\tilde{z}_i(s)}{\tilde{z}_{i-1}(s)} = 1 \tag{12.43}$$

According to (12.40), we give the following relationship between spacing errors $\delta_i(t)$ and $\delta_{i+1}(t)$.

$$\dddot{\delta}_{i+1}(t) - \dddot{\delta}_i(t) = \dddot{\tilde{z}}_{i+1}(t) - \dddot{\tilde{z}}_i(t) - \left(\dddot{\tilde{z}}_i(t) - \dddot{\tilde{z}}_{i-1}(t)\right) \tag{12.44}$$

Under zero initial condition, taking the Laplace transformation for both sides of (12.44) with (12.43), it is clear that

$$\delta_{i+1}(s) - \delta_i(s) = 0 \tag{12.45}$$

which means that the transformation function of the spacing error for the platoon satisfies

$$G_i(s) = \frac{\delta_{i+1}(s)}{\delta_i(s)} = 1, \quad i = 1, 2, \ldots, N.$$

According to [3], it is clear that the string stability is satisfied for the platoon with the proposed ASM controller.

12.5 Simulation Studies

In this section, numerical simulations are presented to verify the effectiveness of the proposed two-layered hierarchical platoon control method. According to [15], a realistic and simplified model of nonlinear vehicle dynamics is used, which is represented by

$$a_i(t) = [F_{d,i}(t) - F_{r,i}(t)]/m_i \tag{12.46}$$

where m_i is the mass of vehicle. $F_{r,\,i}(t)$ and $F_{d,\,i}(t)$ are actual resistance force and driving/braking acting on vehicles, respectively, given by [15]

$$F_{r,i}(t) = \phi_i'[v_i + v_w]^2 + m_i g[f_i \cos(\rho) + \sin(\rho)] \tag{12.47a}$$

$$\dot{F}_{d,i}(t) = [u_i(t) - F_{d,i}(t)]/\theta_i \tag{12.47b}$$

Here ϕ_i' is the aerodynamic drag coefficient, g is the acceleration due to gravity, f_i is the coefficient of rolling resistance, v_w is the wind speed, ρ is the road slope, and $u_i(t)$ is the engine input for vehicle i.

Taking the time derivative of both sides of (12.46) along with (12.47a) and (12.47b), one can derive that

$$\dot{a}_i(t) = -\varsigma_i a_i(t) + \frac{\varsigma_i}{m_i}\left[u_i(t) - F_{r,i}(t) - \frac{1}{\varsigma_i}\dot{F}_{r,i}(r)\right] \tag{12.48}$$

where $\dot{F}_{r,i}(t) = 2\phi_i'[v_i + v_w]a_i$.

By using the feedback linearization control technique, the controller $u_i(t)$ in (12.48) is derived as

$$u_i(t) = m_i u_{i2}(t) + F_{r,i}(t) + \frac{1}{\varsigma_i}\dot{F}_{r,i}(t) \tag{12.49}$$

Taking controller (12.49) into vehicle dynamics (12.48) yields the linearized dynamics $\dot{a}_i(t) - -\varsigma_i a_i(t) + \varsigma_i u_{i2}(t)$ as shown in the third equation of (12.3).

In simulations, the desired intervehicle spacing is set to $d_{i,\,i-1} = 5m$. An uncertain parameter ς_i is considered $\varsigma_i = \bar{\varsigma}_i + \Delta\varsigma_i$, where $\Delta\varsigma_i$ is the white noise with mean zero and variance 0.01. Vehicle mass m_i, parameter $\bar{\varsigma}_i$, and aerodynamic drag coefficient ϕ_i' are listed in Table 12.1. The coefficient of rolling resistance is set to $f_i = 0.02$. The wind speed and road slope are represented by the following sine function [15], respectively:

Table 12.1 Vehicle parameters used in the simulation.

Index	1	2	3	4
$m_i(\times 10^3$ kg)	2.81	2.90	2.27	2.09
$\bar{\varsigma}_i$(s)	0.50	0.25	0.60	0.35
ϕ_i'	0.29	0.30	0.25	0.35

$$\begin{aligned} v_m(t) &= 0.25\sin(\pi t/5) \\ \rho(z) &= 0.01\sin(\pi z/150 + \pi) \end{aligned} \tag{12.50}$$

where π is the circumference ratio and z denotes the position. Here, it is assumed that all followers are aware of the road slope. To measure vehicle platoons' control performance, we define the following cost function:

$$\begin{cases} J_1(t) = \lim\limits_{T\to\infty} \int_0^T \sum\limits_{i=1}^{4} \left(z_i(t) - z_i^*(t)\right)^2 dt \\ J_2(t) = \lim\limits_{T\to\infty} \int_0^T \sum\limits_{i=1}^{4} (z_{i-1}(t) - z_i(t) - d_{i,i-1})^2 dt \end{cases} \tag{12.51}$$

In simulations, the profile of acceleration and velocity for the leader is given in Figure 12.3, which is usually considered a disturbance in a platoon [6]. According to parameters listed in Table 12.2, we can derive the trajectory optimization controller $u_{i1}(t)$ and tracking controller $u_{i2}(t)$ according to Theorems 12.1 and 12.2, respectively. Using the Matlab LMI toolbox, the parameter K in the ASM controller is obtained as $K = [-0.46 \quad -1.38]$. To verify the effectiveness of the proposed algorithm, we consider the case for a platoon with BPF, leader-BPF (LBPF), and two-BPF

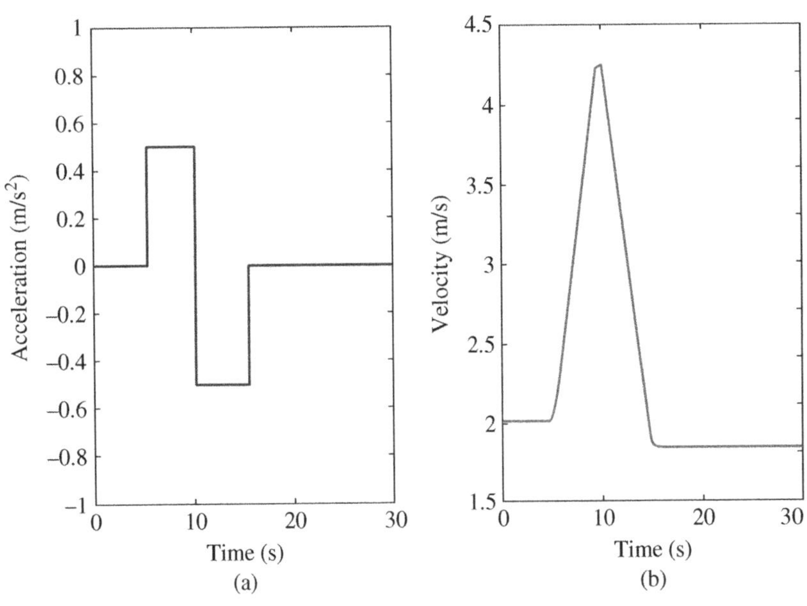

Figure 12.3 Acceleration (a) and velocity (b) of the leader.

Table 12.2 The value of parameters used in the proposed control algorithm.

μ	α	γ	ξ	κ	q
0.5	0.5	0.5	0.5	5	2

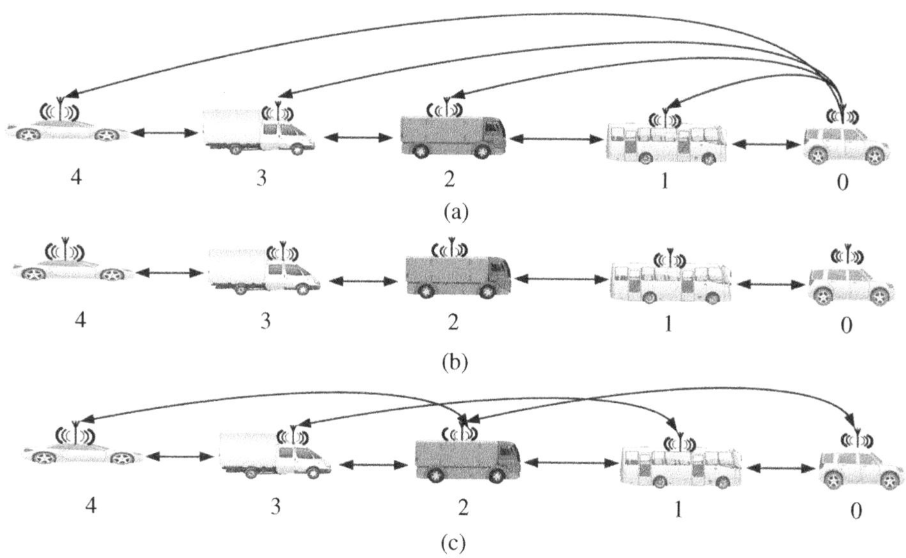

Figure 12.4 (a) LBPF, (b) BPF, and (c) TBPF communication topologies for vehicle platoons.

(TBPF) communication topologies that are shown in Figure 12.4. From Figure 12.4, it is observed that the leader is globally reachable in the union graph of each communication topology.

With the proposed two-layered control framework, Figures 12.5–12.7 present simulation results for a five-vehicles platoon under LBPF, BPF and TBPF communication topologies, respectively. The control performance of the first layer can be found in Figures #(a) and #(b) with symbol # ∈ {12.5, 12.6, 12.7}. As $\hat{z}_i^*$ and $\hat{v}_i^*$ defined in (12.8) asymptotically converge to zero, desired trajectory z_i^* and v_i^* can reach a consensus with the leaders. Note that not all the followers are directly communicating with the leader under BPF and TBPF communication topologies. As a result, the first layer has the potential to minimize the spacing error and strengthen string stability. The tracking control performance of the proposed ASM controller is shown in Figure #(c)–(d). It is clear that the ASM controller can track the desired trajectory z_i^* and v_i^* rapidly and accurately without the effect of uncertain TCDDs. As shown in Figures #(e) and #(f), by using the two-layered hierarchical control framework, each follower can maintain the desired intervehicle distance and guarantee the consistent velocity with the leaders. Since spacing error δ_i and velocity error $v_i - v_{i-1}$ are not amplified with the vehicle index, string stability is satisfied for the platoon subject to disturbances such as uncertain TCDDs and acceleration/deceleration operation of the leader. For heterogeneous vehicular platoons considered here, as shown in Figure 12.8, the resistance force $F_{r,i}(t)$ of each follower is

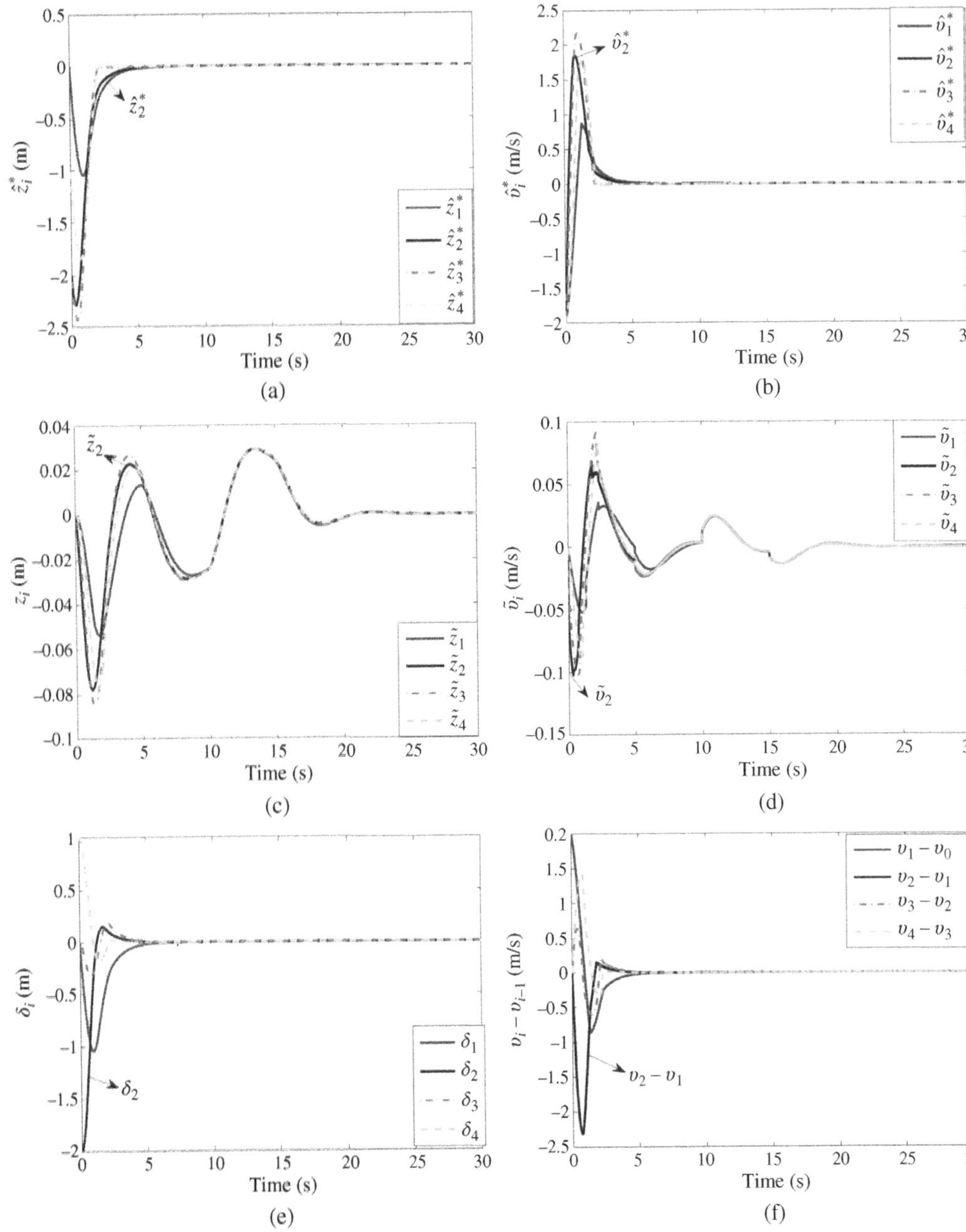

Figure 12.5 Control performance of the proposed hierarchical control method for the five-vehicle platoon under LBPF communication topology: (a) $\hat{z}_i^*$, (b) $\hat{v}_i^*$, (c) $\tilde{z}_i$, (d) $\tilde{v}_i$, (e) δ_i, and (f) $v_i - v_{i-1}$.

different. Those results validate the effectiveness of the proposed hierarchical platoon control method.

Contrarily, in the presence of a random jerk level, we investigate the behavior of the platoon under the proposed control method, for example,

$$\overline{a}_0 = a_0 \pm \vartheta \tag{12.52}$$

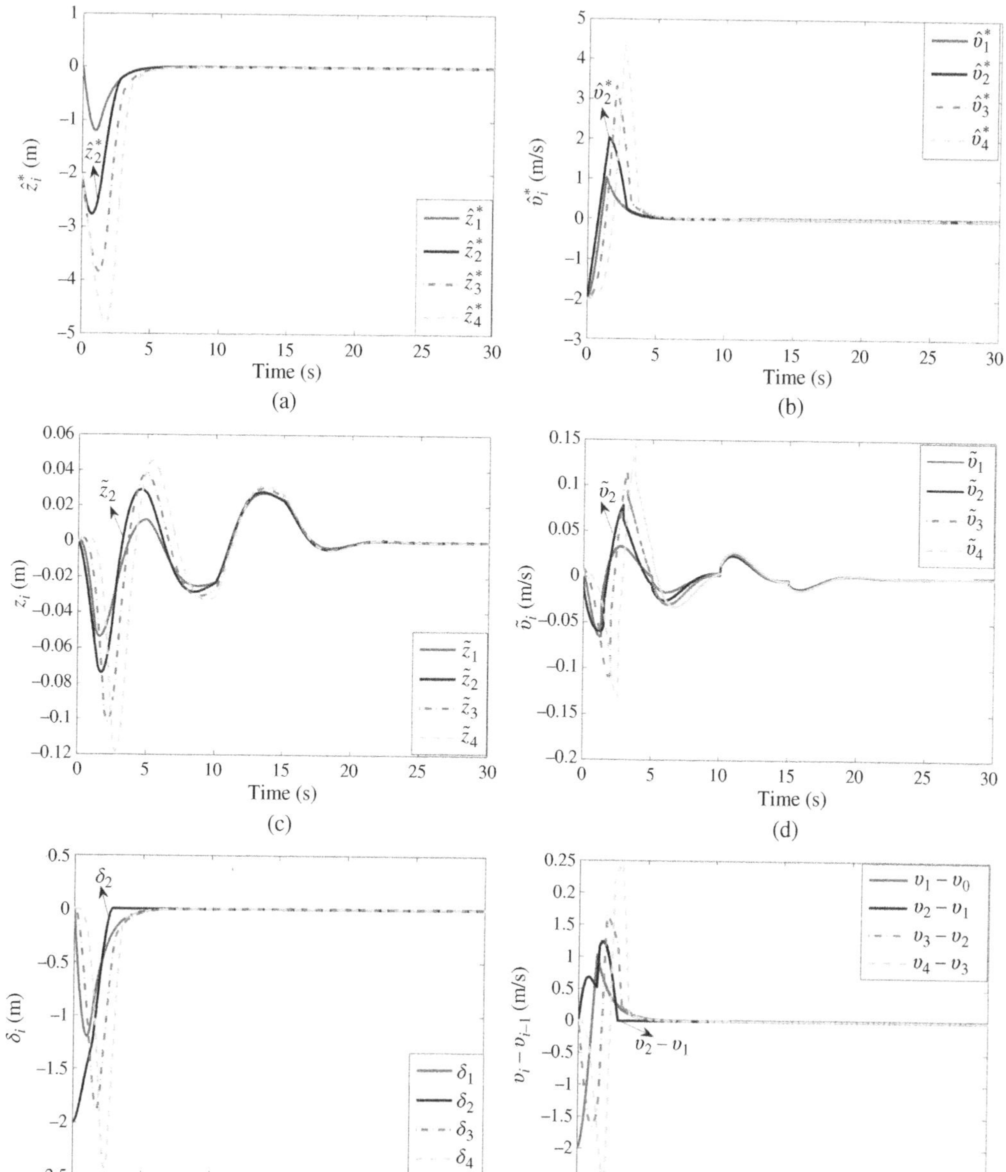

Figure 12.6 Control performance of the proposed hierarchical control method for the five-vehicle platoon under BPF communication topology: (a) $\hat{z}_i^*$, (b) $\hat{v}_i^*$, (c) $\tilde{z}_i$, (d) $\tilde{v}_i$, (e) δ_i, and (f) $v_i - v_{i-1}$.

where ϑ denotes the jerk level which randomly takes value from $[-0.25, 0.25]$. Under such circumstances, the profile of acceleration and velocity for the leader is shown in Figure 12.9. As per Figure 12.10, the control performance of the platoon is clearly degraded. Compared to Figure # (e)–(f), # $\in$ {12.5, 12.6, 12.7}, tracking errors in Figure 12.10 have oscillation with slower

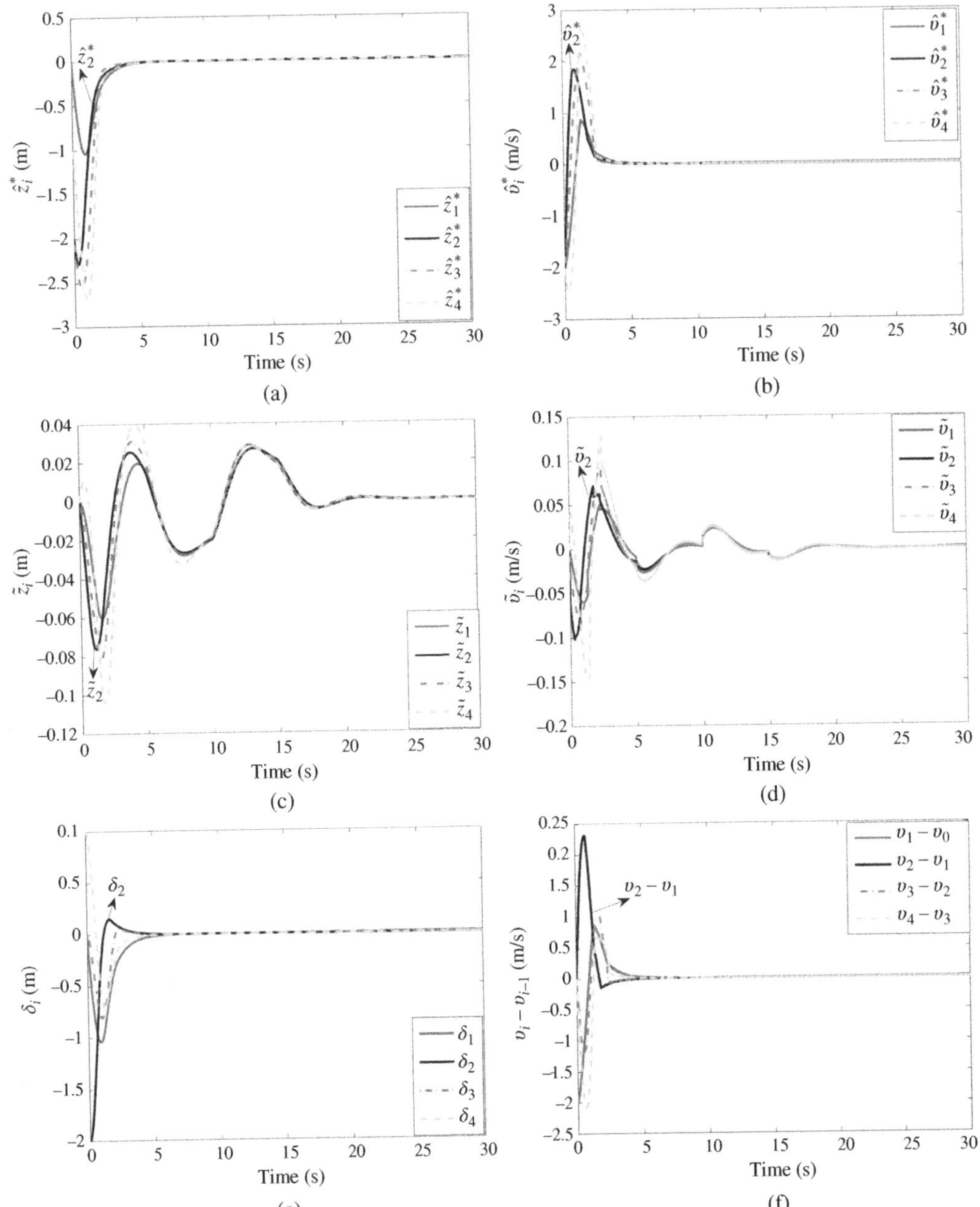

Figure 12.7 Control performance of the proposed hierarchical control method for the five-vehicle platoon under TBPF communication topology: (a) $\hat{z}_i^*$, (b) $\hat{v}_i^*$, (c) $\tilde{z}_i$, (d) $\tilde{v}_i$, (e) δ_i, and (f) $v_i - v_{i-1}$.

convergence of tracking performance, which affect the riding comfort. By comparing the cost function listed in Tables 12.3 and 12.4, it is clear that the random jerk level leads to a greater cost function $J_1(t)$ and $J_2(t)$ as defined in (12.51), since tracking errors are affected by the behavior of the leader. Specifically, the frequent acceleration/deceleration of the leader arises from the random jerk

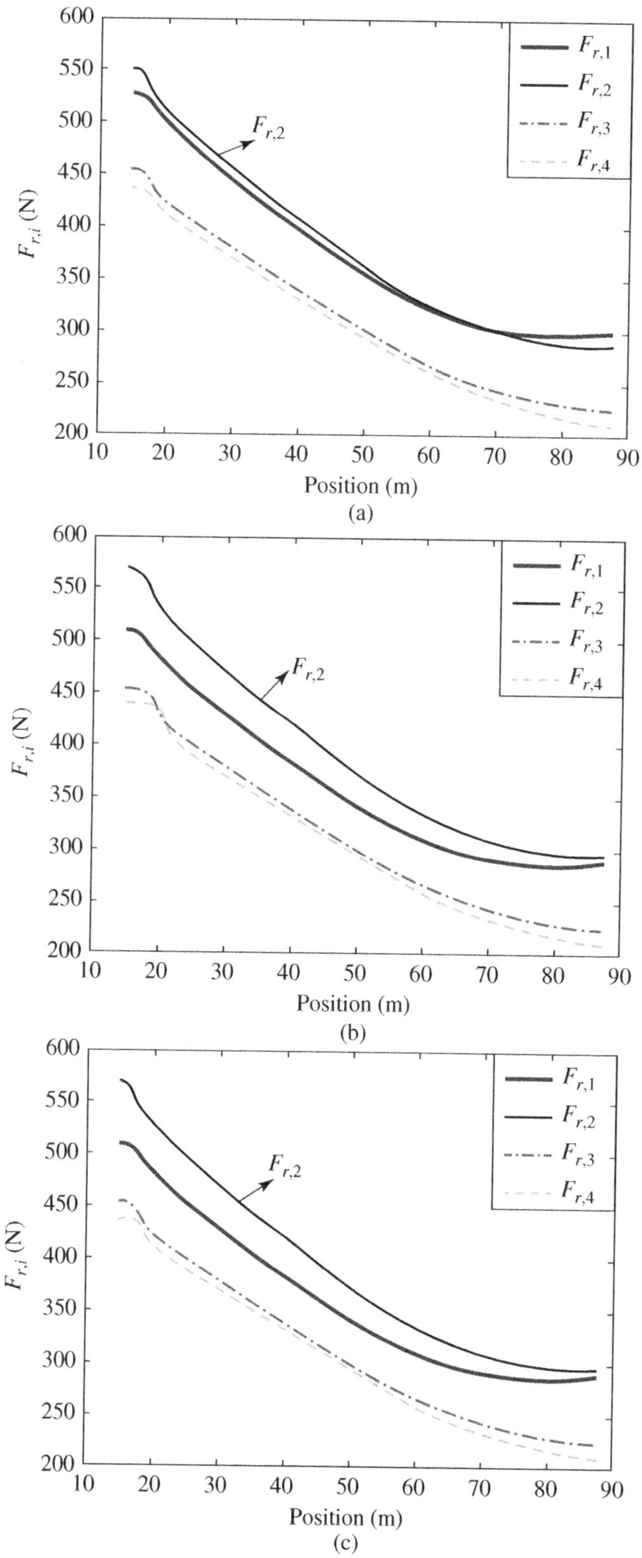

Figure 12.8 The resistance force $F_{r,i}(t)$ of each follower: (a) platoon with LBPF communication topology, (b) platoon with BPF communication topology, (c) platoon with TBPF communication topology.

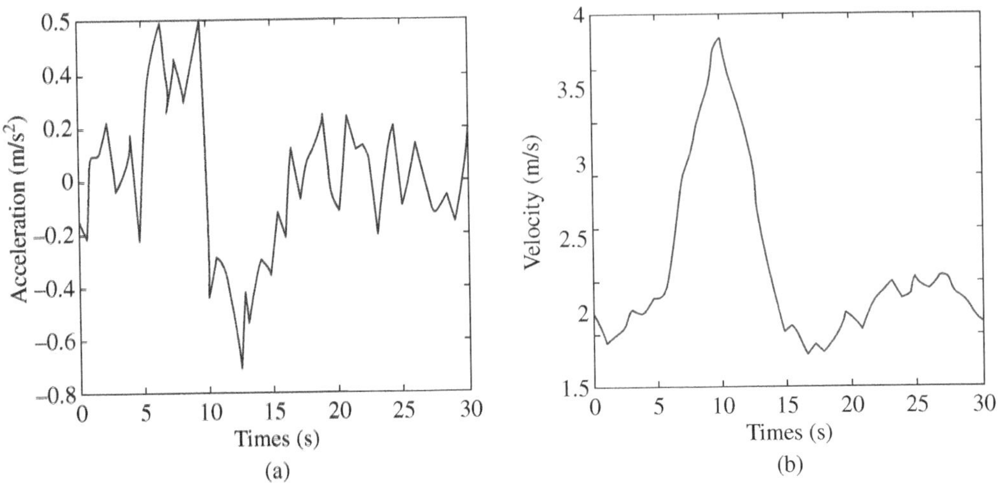

Figure 12.9 (a) Acceleration and (b) velocity of the leader with a random jerk.

level. However, the proposed hierarchical control method ignores the effect of the random jerk level.

In this section, the results of the proposed hierarchical platoon control method have been compared with the H_∞ feedback controller designed in [6]. In Ref. [6] and in this study, both vehicular platoon control models are similar since the same third-order vehicle dynamics (12.3) and undirected topology are involved. Here, the BPF communication topology is used in comparative simulation studies. According to [6], the H_∞ feedback controller is designed by

$$u_i(t) = -ck^T \sum_{j \in \mathbf{I}_i} (\hat{x}_i - \hat{x}_j) \tag{12.53}$$

where $\hat{x}_i = [z_i - z_0 - d_{i,0} \quad v_i - v_0 \quad a_i - a_0]^T$ with the coupling strength $c = 25$, controller gain $k = [2.122 \quad 3.425 \quad 2.501]$ and $d_{i,0} = i^*5$ m. It is important to note that the obtained H_∞ controller in (12.53) depends on equal and certain TCDDs, that is, $\theta_i = 0.25$.

First, we present comparative results to illustrate the advantages of trajectory optimization operated in the first layer. Certain TCDDs are involved in this simulation with the same scenario as in Figures 12.5–12.7, for example, $\theta_i = 0.25$. In comparative simulations, each follower has a different control framework. Without trajectory optimization, Figure 12.11c, d gives the spacing and velocity error of the platoon using the H_∞ controller (12.53) only. In contrast, Figure 12.11a, b gives the control performance of the platoon under a hierarchical control framework. Note that the H_∞ controller of [6] is used in place of the ASM controller in the second layer to track the desired trajectory $z_i^*(t)$ and $v_i^*(t)$, which is described by

$$u_{i2}(t) = -ck^T \hat{x}_i' \tag{12.54}$$

where the tracking error vector is $\hat{x}_i' = [z_i - z_i^* \quad v_i - v_i^* \quad a_i - u_{i1}]^T$. As listed in Table 12.5, the cost function $J_2(t)$ in Figure 12.11a is smaller than in Figure 12.11c since each follower is tracking the optimization trajectory provided by the first layer. This is consistent with the convergence speed of the tracking errors. Evidently, the convergence speed of the tracking errors in Figure 12.11a, b is faster than in Figure 12.11c, d. Consequently, the proposed trajectory optimization technique in the first layer is useful to reduce spacing error in the platoon.

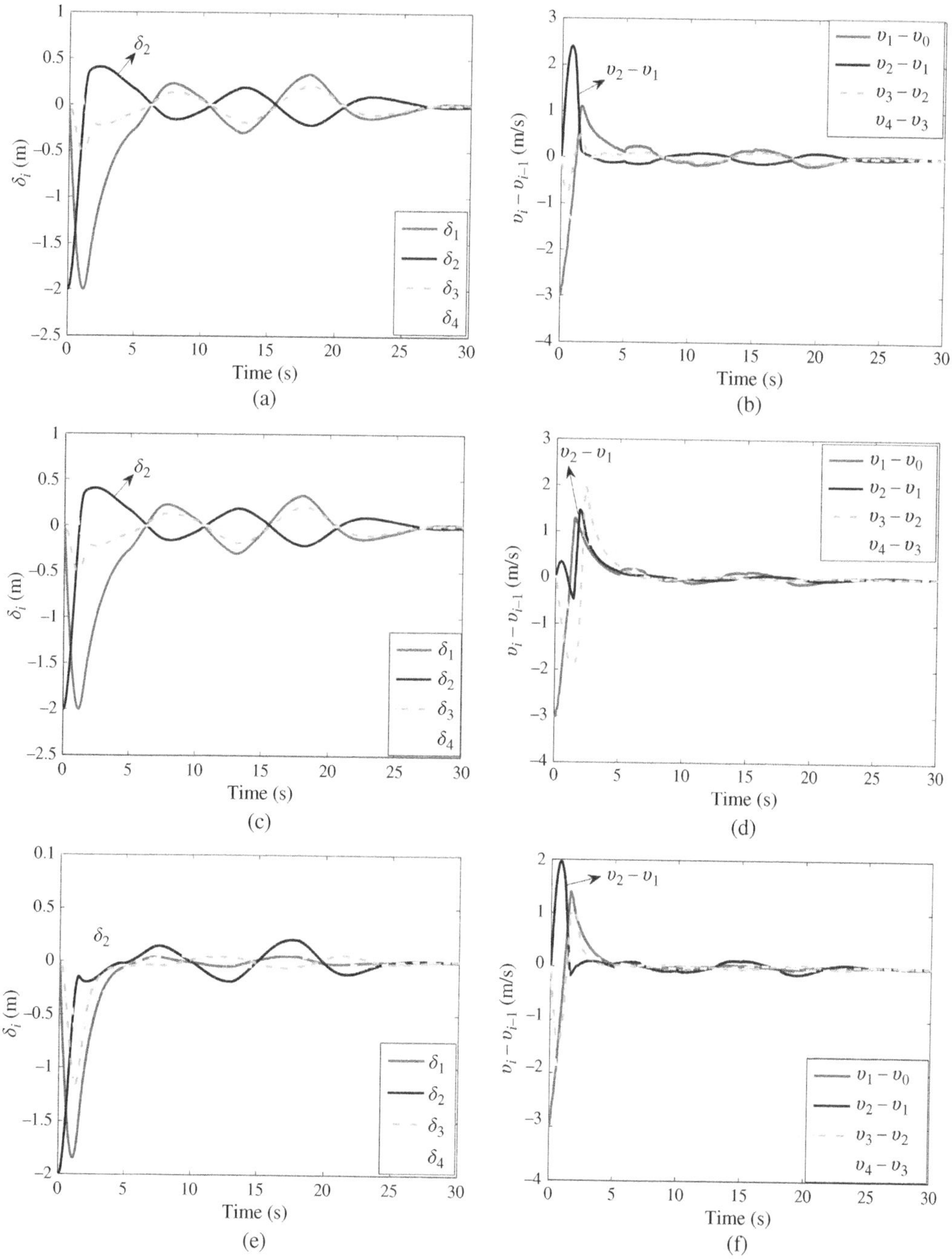

Figure 12.10 Spacing error δ_i and velocity error $v_i - v_{i-1}$ of the proposed hierarchical control method for the platoon with jerk under LBPF, BPF, and TBPF communication topologies, respectively. (a) Spacing error, (b) velocity error, (c), spacing error, (d) velocity error, (e), spacing error, (f) velocity error.

Table 12.3 Cost function for the platoon under LBPF, BPF, and TBPF communication topology.

Cost function \ $\mathcal{G}$	LBPF	BPF	TBPF
$J_1(t)$	0.05	0.06	0.053
$J_2(t)$	3.81	11.41	4.5

Table 12.4 Cost function for a platoon with random jerk level.

$\mathcal{G}$ / Cost function	LBPF	BPF	TBPF
$J_1(t)$	0.21	0.39	0.26
$J_2(t)$	13.61	20.20	17.72

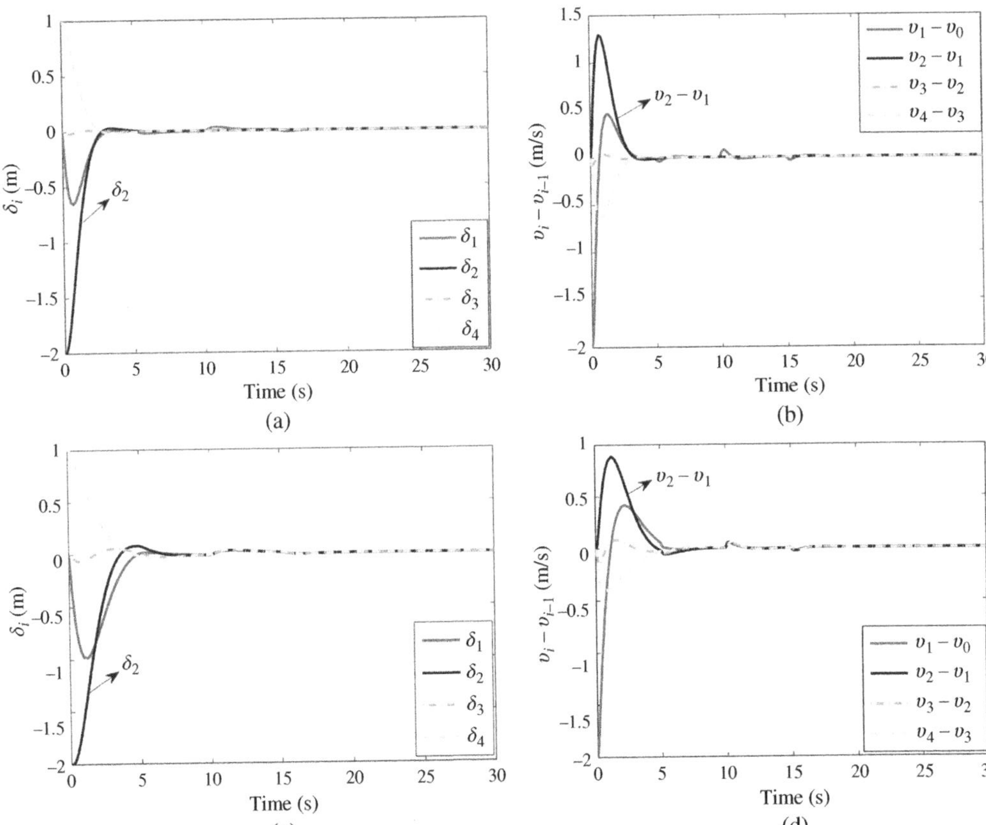

Figure 12.11 Comparative simulation results of a platoon with certain TCDDs θ_i: (a) δ_i and (b) $v_i - v_{i-1}$ of platoon using the proposed hierarchical control framework; (c) δ_i and (d) $v_i - v_{i-1}$ of platoon using the H_∞ controller in [6] without tracking optimization.

Table 12.5 Cost function $J_2(t)$ of comparative simulation studies.

	Figure 12.11a	Figure 12.11c	Figure 12.12a	Figure 12.12c
$J_2(t)$	10.23	18.76	11.53	23.57

Subsequently, comparative simulation results are presented to illustrate the effectiveness of the proposed ASM controller. Here uncertain TCDDs are considered, which are represented by $\theta_i = 0.25 + \Delta\theta_i$, where $\Delta\theta_i$ is the white noise with mean zero and variance 0.01. In simulations, each follower has a hierarchical control framework which is composed of trajectory optimization and tracking control in two layers. However, the tracking controller is different in the second layer, where the proposed ASM controller and H_∞ controller (12.54) are involved, respectively. Corresponding simulation results for each follower are demonstrated in Figure 12.12a–d, respectively. From Figure 12.12c, d, it is clear that the control performance of the H_∞ controller is degraded. Compared to the results in Figure 12.11a, b, the platoon has a bigger spacing error and slower convergence of the tracking performance. In contrast, the control performance of the platoon with the proposed ASM controller is hardly affected by uncertain TCDDs θ_i. As shown in Figure 12.12a, b, tracking errors of each follower can rapidly be stabilized to zero. This means that the proposed ASM controller can effectively estimate and compensate for disturbances caused by uncertain TCDDs. As listed in Table 12.5, the cost function $J_2(t)$ in Figure 12.12a is much smaller than in Figure 12.12c. It is clear that the control performance of the proposed method outperforms the H_∞ method in [6], especially for vehicle platoons subject to uncertain TCDDs.

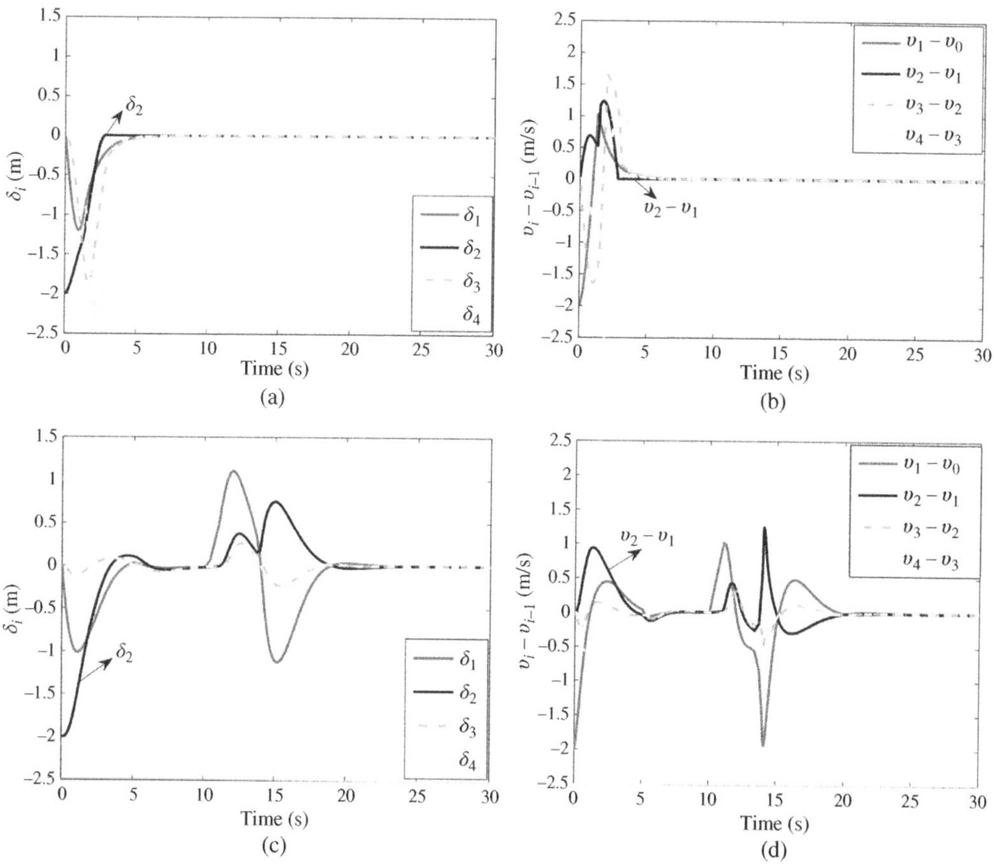

Figure 12.12 Comparative simulation results of the platoon subject to uncertain TCDDs θ_i using different tracking controllers in the second layer: (a) δ_i and (b) $v_i - v_{i-1}$ of the platoon using ASM controller; (c) δ_i and (d) $v_i - v_{i-1}$ of each follower using the H_∞ controller. Source: Adapted from Zheng et al. [6].

By comparing control performance of the platoon under different communication topologies, we found that the more followers obtain information from the leader, the better the control performance. Under the LBPF communication topology, as shown in Figure 12.5a, b, the trajectory of each follower has the fastest speed to reach consensus with the leaders because each follower can directly communicate with the leader. As a result, this platoon has the smallest spacing error, which is observed in simulation results shown in Figure #(e)–(f) $\# \in \{12.5, 12.6, 12.7\}$ with the cost function $J_2(t)$ listed in Table 12.3. Due to the fact that the eigenvalues of topology Laplacian matrix can influence the control performance of vehicle platoons. In contrast, a platoon with BPF topology has larger tracking errors since the smaller eigenvalues of the topology Laplacian matrix will reduce the effect of feedback control. Since the minimum eigenvalue of LBPF topology Laplacian matrix is always "1" (see Figure 12.13a), the control performance of the platoon affected by LBPF topology is better compared to the TBPF topology. For LBPF, BPF, and TBPF communication topologies, as shown in Figure 12.13b, a larger size of platoon leads to larger eigenvalues of the topology Laplacian matrix.

Due to the high mobility of vehicles, the condition of the wireless channel varies dynamically in VANETs [16]. To verify the robustness of the proposed control method, we consider the case of a platoon with switching communication topologies [17]. As shown in Figure 12.14, S_1 and S_2 are two different switching sequences, where values "0", "1", and "2" of the y-axis denote LBPF, BPF, and TBPF communication topologies, respectively. During the time interval [0, 30 s], the number of switching times in S_1 and S_2 are 3 and 16, respectively. However, as listed in Table 12.6, the total activation time of each communication topology in S_1 and S_2 is equal during this time interval. With switching communication topologies, spacing error $\delta_i(t)$ and velocity error $v_i(t) - v_{i-1}(t)$ are shown in Figure 12.15. Compared to Figure 12.15a–d, it is observed that the control performance of the platoon under the switching sequence S_1 is better than the switching sequence S_2. By computation, the average dwell time (ADT) of S_1 (ADT = 7.5 s) is more extensive than S_2 (ADT = 1.8 s). Here the ADT is equal to the total time divided by the switching times between different communication

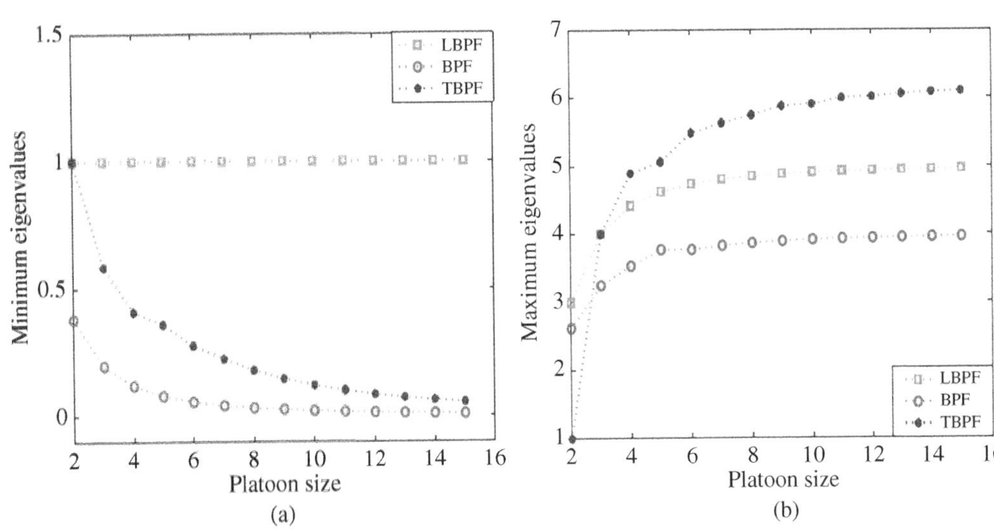

Figure 12.13 Minimum eigenvalues (a) and maximum eigenvalues (b) of the Laplacian matrix of communication topologies LBPF, BPF, and TBPF, respectively.

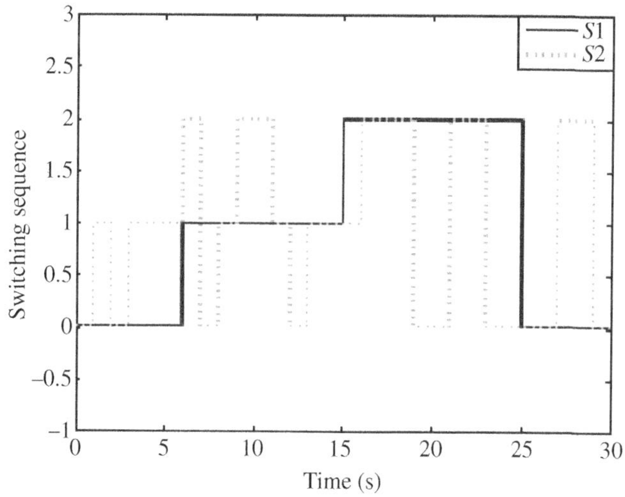

Figure 12.14 Vehicle platoons with switching communication topologies.

Table 12.6 Total activation time of LBPF, BPF, and TBPF communication topology in switching sequence S_1 and S_2.

$\mathcal{G}$ / Time (s)	LBPF	BPF	TBPF
S_1	12	9	10
S_2	12	9	10

topologies. According to switched system control theory [18], the control performance of the switched platoon control system has this relation with the ADT.

12.6 Conclusions and Future Work

We proposed a distributed cooperative optimization and control method for a platoon, which aims to cooperatively bring intervehicle spacing between any two successive vehicles as close as possible to the desired value. Based on the techniques of distributed continuous-time convex optimization and ASM control, a two-layered hierarchical control framework is established for tracking trajectory optimization and control. It is confirmed that the established hierarchical control framework does not only stabilize tracking errors but can also guarantee string stability for heterogeneous platoons with uncertain vehicle dynamics.

In the simulations, it is observed that the control performance of the proposed hierarchical control framework may be degraded under frequently switching communication topologies. Furthermore, existing communication constraints such as communication delay, packet dropouts, and

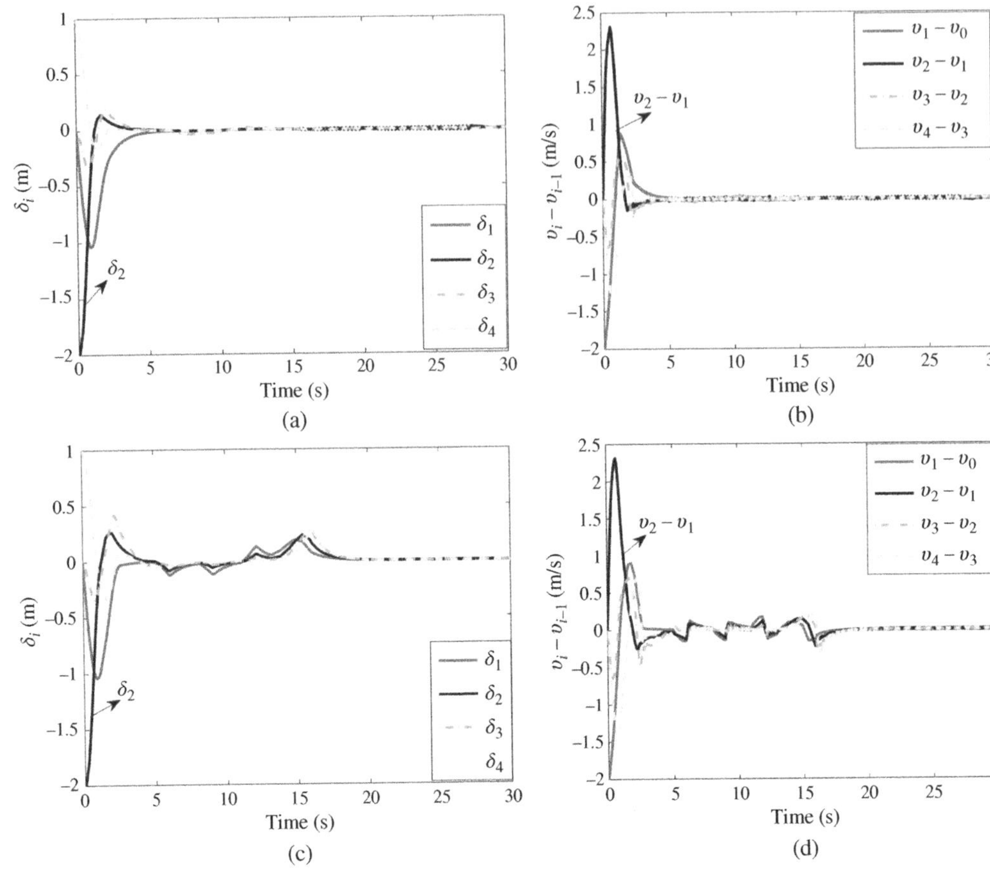

Figure 12.15 Control performance of the proposed two-layered hierarchical controller under switching communication topologies: (a) $\delta_i(t)$ and (b) $v_i(t) - v_{i-1}(t)$ of the platoon under switching sequence S_1; (c) $\delta_i(t)$ and (d) $v_i(t) - v_{i-1}(t)$ of the platoon under switching sequence S_2.

medium access constraints are inevitable in VANETs. In future research, it will be interesting to improve the tracking trajectory optimization approach to relieve the effect of communication constraints. Another interesting topic to consider is the effect of actuator delay (e.g. the fueling and braking delay) and random jerk level in the proposed ASM controller.

References

1 Naus, G.J.L., Vugts, R.P.A., Ploeg, J. et al. (2010). String-stable CACC design and experimental validation: a frequency-domain approach. *IEEE Trans. Veh. Technol.* 59 (9): 4268–4279.

2 Chen, Q., Zhou, Y., Ahn, S. et al. (2022). Robustly string stable longitudinal control for vehicle platoons under communication failures: a generalized extended state observer-based control approach. *IEEE Trans. Intell. Veh.* 8: 1–1.

3 Swaroop, D. and Hedrick, J.K. (1996). String stability of interconnected systems. *IEEE Trans. Autom. Control* 41 (3): 349–357.

4 Zheng, Y., Li, S.E., Li, K., and Wang, L.-Y. (2016). Stability margin improvement of vehicular platoon considering undirected topology and asymmetric control. *IEEE Trans. Control Syst. Technol.* 24 (4): 1253–1265.

5 Seiler, P., Pant, A., and Hedrick, K. (2004). Disturbance propagation in vehicle strings. *IEEE Trans. Autom. Control* 49 (10): 1835–1842.

6 Zheng, Y., Li, S.E., Li, K., and Ren, W. (2018). Platooning of connected vehicles with undirected topologies: robustness analysis and distributed H-infinity controller synthesis. *IEEE Trans. Intell. Transp. Syst.* 19 (5): 1353–1364.

7 Zheng, Y., Li, S.E., Wang, J. et al. (2016). Stability and scalability of homogeneous vehicular platoon: Study on the influence of information flow topologies. *IEEE Trans. Intell. Transp. Syst.* 17 (1): 14–26.

8 Olfati-Saber, R. and Murray, R.M. (2004). Consensus problems in networks of agents with switching topology and time-delays. *IEEE Trans. Autom. Control* 49 (9): 1520–1533.

9 Guo, G. and Yue, W. (2014). Sampled-data cooperative adaptive cruise control of vehicles with sensor failures. *IEEE Trans. Intell. Transp. Syst.* 15 (6): 2404–2418.

10 Guo, G. and Wen, S. (2016). Communication scheduling and control of a platoon of vehicles in VANETs. *IEEE Trans. Intell. Transp. Syst.* 17 (6): 1551–1563.

11 Slotine, J. and Li, W. (1991). *Applied Nonlinear Control.* Upper Saddle River, NJ: Prentice-Hall.

12 Bazaraa, M.S., Sherali, H.D., and Shetty, C.M. (2005). *Nonlinear Programming: Theory and Algorithms.* Hoboken, NJ: Wiley.

13 Boyd, S. and Vandenbergher, L. (2004). *Convex Optimization.* Cambridge: Cambridge Univ. Press.

14 Baek, J., Jin, M., and Han, S. (2016). A new adaptive sliding-mode control scheme for application to robot manipulators. *IEEE Trans. Ind. Electron.* 63 (6): 3628–3637.

15 Gao, F., Hu, X., Li, S.E. et al. (2018). Distributed adaptive sliding mode control of vehicular platoon with uncertain interaction topology. *IEEE Trans. Ind. Electron.* 65 (8): 6352–6361.

16 Fu, F. and van der Schaar, M. (2012). Structure-aware stochastic control for transmission scheduling. *IEEE Trans. Veh. Technol.* 61 (9): 3931–3945.

17 Wen, S. and Guo, G. (2020). Sampled-data control for connected vehicles with Markovian switching topologies and communication delay. *IEEE Trans. Intell. Transp. Syst.* 21 (7): 2930–2942.

18 Liberzon, D. (2013). *Switching in Systems and Control.* Boston, MA: Birkhäuser.

13

Trajectory Planning and PID-Type Sliding-Mode Control to Reduce Intervehicle Spacing

13.1 Introduction

To address the joint problem of vehicle speed planning and optimal tracking control, Chapter 12 proposes a distributed optimal platoon control method that minimizes a cost defined by intervehicle spacing errors. The result is built on a strict constant spacing (CS) policy, which leaves little room for speed regulation to deal with the dynamic coupling among vehicles, resulting in degraded speed tracking and high communication cost [1, 2]. Besides, the simple sliding mode surface and reaching law used in Chapter 12 are subject to poor convergence and anti-chattering performance that is not good for string instability [3–5].

This chapter revisits the problem in Chapter 12, aiming at improvements in at least two aspects. First, we introduce a quadratic spacing policy based on the leader speed (i.e. the expected platoon speed) to leave room for speed tracking, so as to ensure string stability and traffic flow stability. Second, a new PID-type sliding-mode controller with a double high-power reaching law is presented, resulting in significantly improved platoon control performance and anti-chattering capacity. The main contributions are as follows:

1) By introducing a realistic team speed-dependent quadratic spacing policy, a distributed speed optimization method is derived for vehicle speed planning based on the minimization of intervehicle spacing errors.
2) A PID-type sliding-mode controller with a double high-power reaching law is presented, which is less sensitive to the chattering of speed and spacing with a higher rate of convergence. In addition to performance and string stability, this method can also guarantee traffic flow stability, which is not addressed in Chapter 12.

The remainder of the paper is organized as follows. Section 13.2 describes the vehicle longitudinal dynamics and the intervehicle spacing policy, the objectives of platoon optimization control. Section 13.3 describes the distributed trajectory optimization for a platoon. Section 13.4 provides the design of the tracking controller. Section 13.5 illustrates the effectiveness of the proposed control method using numerical simulations, while Section 13.6 presents concluding remarks.

Notations $\mathbb{R}^n$ be the n-dimensional Euclidean space. Denote $1_N = [1, 1, ..., 1]^T \in \mathbb{R}$ and $0_N = [0, 0, ..., 0]^T \in \mathbb{R}^N$ $I_N \in \mathbb{R}^{N\times N}$ is the identity matrix of size $N \times N$ p-norm of vector x is denoted by $\|x\|_p$. For a matrix A, A^T is the transpose matrix and $\lambda_{\min}(A)$ is the minimal eigenvalue of A. diag$\{\cdot\}$ is defined as the diagonal matrix.

Connected Vehicular Systems: Communication, Control, and Optimization, First Edition. Ge Guo and Shixi Wen.

The matrix inequality $A > B(A \geq B)$ means that the matrix $A - B$ is positive-define (semidefinite). The Kronecker product is denoted by $\otimes$ and $\text{sgn}(\cdot)$ is the signum function.

13.2 Problem Description

Consider a collection of $N+1$ vehicles driving as a platoon on the road (see Figure 13.1), with the leader indexed by 0. Each follower can exchange information (e.g. position, velocity, and acceleration) with its neighbors via the communication topology described here.

The communication among followers constitutes a directed graph, $\mathcal{G}_N = (\mathcal{V}_N, \mathcal{E}_N)$, where the vertices set $\mathcal{V}_N = \{1, 2, ..., N\}$ represents vehicles and the edges set $\mathcal{E}_N \subseteq \mathcal{V}_N \times \mathcal{V}_N$ denotes communication link. Define the edge $\varepsilon_{ij} \triangleq (i, j) \in \mathcal{E}_N$ which means that there exists an information flow from vehicle j to vehicle i. The adjacency matrix of $\mathcal{G}_N$ graph is denoted by $A = [a_{ij}]_{N \times N}$, where $a_{ii} = 0$, $a_{ij} = 1$ if $\varepsilon_{ij} \in \mathcal{E}_N$, otherwise, $a_{ij} = 0$ The Laplacian matrix is defined as $\mathcal{L} = [l_{ij}]_{N \times N}$, where $l_{ij} = -a_{ij}$ for $i \neq j$ and $l_{ii} = \sum_{j=1}^{N} a_{ij}$. The incidence matrix is denoted by $D = \left[d_{ij}\right]_{N \times |\mathcal{E}_N|}$, where $d_{ij} = -1$ if the edge ε_j leaves node i, $d_{ij} = 1$ if it enters the node i, and $d_{ij} = 0$ otherwise. Note that $L \triangleq DD^T A$ path is a sequence of connected edges and $\mathcal{G}_N$ is connected if there is a path between any two distinct nodes. We assume an undirected communication topology among the followers and the neighbor set of node i is

$$\boldsymbol{N}_i = \{j \in \mathcal{V}_N, (i,j) \in \mathcal{E}_N\}$$

The leader and all followers together form an augmented directed graph $\mathcal{G}_{N+1} = (\mathcal{V}_{N+1}, \mathcal{E}_{N+1})$ with vertices $\mathcal{V}_{N+1} = \{0, 1, ..., N\}$. Let $P = \text{diag}\,\{p_1 p_2, ..., p_N\}$ be the pinning matrix, with $p_i = 1$ when vehicle i communicates with the leader, otherwise, $p_i = 0$. Note that, for $\mathcal{G}_{N+1}$, $\mathcal{L} + \mathcal{P} \triangleq \overline{D}\overline{D}^T$ where $\overline{D} = \left[d_{ij}\right]_{N \times |\mathcal{E}_{N+1}|}$ is the incident matrix. The set indicating if the leader is reachable for vehicle i is denoted by

$$\boldsymbol{P}_i = \begin{cases} \emptyset & \text{if} \quad p_i = 0 \\ \{0\} & \text{if} \quad p_i = 1 \end{cases} \tag{13.1}$$

where $\emptyset$ is the empty set. Then, the set of neighbors of vehicle i is $\boldsymbol{I}_i = \boldsymbol{N}_i \cup \boldsymbol{P}_i$.

Assumption 13.1 The augmented graph $\mathcal{G}_{N+1} = (\mathcal{V}_{N+1}, \mathcal{E}_{N+1})$ has at least one spanning tree rooted in the leader.

Lemma 13.1 For an undirected connected graph, $\lambda_{\min}(\mathcal{L}) = 0$ with $\mathbf{1}_N$ being the associated eigenvector. For graphs satisfying Assumption 13.1, $\lambda_{\min}(\mathcal{L}) = 0$ is a simple eigenvalue, and $\lambda_{\min}(\mathcal{L}) = 0$ [6].

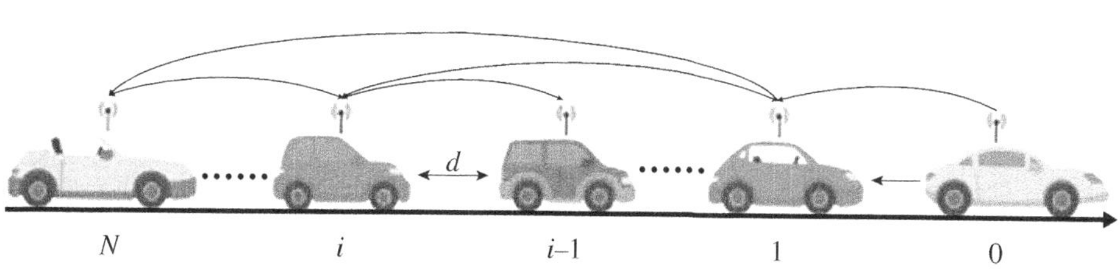

Figure 13.1 Platoon communications and control system.

Let $x_i(t)$, $v_i(t)$ and $a_i(t)$ be, respectively, position, velocity, and acceleration of vehicle $i = 0, 1, 2, \ldots, N$. Assume that the leader executes a constant team speed trajectory, namely, $a_0 = 0$, $x_0 = v_0 t$. For the follower vehicles, we consider the following longitudinal model

$$\begin{cases} \dot{x}_i(t) = v_i(t) \\ \dot{v}_i(t) = a_i(t) \\ \dot{a}_i(t) = -\xi_i a_i(t) + \xi_i u_{i2}(t) \end{cases} \tag{13.2}$$

where $\xi_i = 1/\theta_i$, with θ_i being the time constant in the drivetrain dynamics of vehicle i, and u_{i2} is the control input.

Let $d_{i,\,i-1}(t) > 0$ be the desired distance between i and $i-1$. We introduce the following quadratic spacing strategy, which is based on the leader speed (i.e. the expected platoon speed):

$$d_{i,i-1}(t) = hv_0^2(t) + cv_0(t) + l_i \tag{13.3}$$

where $h > 0$, c are coefficients, $l_i > 0$ denotes the expected intervehicle spacing when stationary, wherein the vehicle length L_i is considered. These parameters can be designed to ensure traffic flow stability, as will be given later.

The objective of this chapter is to develop a distributed speed planning and platoon control framework that solves the following optimization problem:

$$\min_{x_i(t)} J(t) = \sum_{i=1}^{N} \delta_i^2(t) \tag{13.4a}$$

$$\text{s.t.} \begin{cases} \lim\limits_{t \to \infty} |v_i(t) - v_0(t)| = 0 \\ \lim\limits_{t \to \infty} |\delta_i(t)| = 0 \end{cases} \tag{13.4b}$$

where δ_i is the spacing error for follower vehicle i, which is defined by

$$\delta_i(t) = x_i(t) - x_{i-1}(t) + d_{i,i-1}(t) + L_{i-1} \tag{13.5}$$

And the controller of all the following vehicles should meet the following criteria:

1) *Internal stability [7]*: The speed and acceleration of all followers can track the leader with convergent spacing errors.
2) *String stability [7]*: $|\delta_N(s)| \leq |\delta_{N-1}(s)| \leq \cdots \leq |\delta_1(s)|$, or, equivalently, the error transfer function $G_i(s) = \delta_{i+1}(s)/\delta_i(s)$ satisfies $|G_i(s)| \leq 1$, where $\delta_i(s)$ is the Laplace transform of $\delta_i(t)$.
3) *Traffic flow stability [5, 8]*: The gradient inequality $\partial Q/\partial P > 0$ should be guaranteed, where Q and P are the flow rate and density of road traffic, respectively.

In order to achieve these objectives, a two-layered framework is proposed (see Figure 13.2), with trajectory optimization and tracking control in different layers.

Remark 13.1 The optimal control method in Chapters 10 and 11 is based on CS policy and is aimed at saving fuel and time, but pays little attention to the smoothness and stability of traffic flow. Unlike their centralized methods, we give a fully distributed framework based on a quadratic spacing policy to strengthen traffic flow stability with decreased cost of calculation and communication.

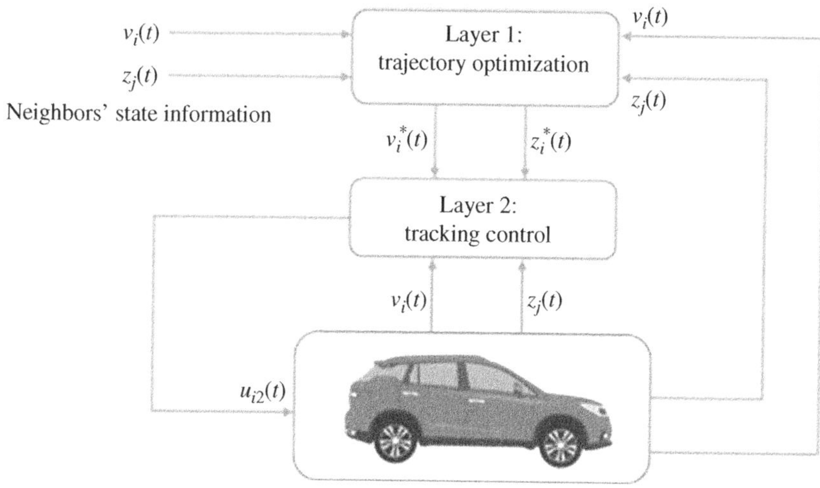

Figure 13.2 The proposed two-layered control structure for the platoon.

13.3 Distributed Trajectory Optimization

Let the reference trajectories of the position and speed for the following vehicle i be given by

$$\begin{cases} \dot{x}_i^*(t) = v_i^*(t) \\ \dot{v}_i^*(t) = u_{i1}(t) \end{cases} \tag{13.6}$$

We aim to find a controller $u_{i1}(t)$ to generate the reference trajectories $x_i^*(t)$ and $v_i^*(t)$ that solve the distributed optimization problem in (13.4a) and (13.4b). To this end, introduce the following reference position errors and reference velocity errors:

$$\begin{cases} \hat{x}_i^*(t) = x_i^*(t) - x_0(t) + d_{i,0}(t) \\ \hat{v}_i^*(t) = v_i^*(t) - v_0(t) \end{cases} \tag{13.7}$$

by which we can have that

$$\begin{cases} \hat{x}_i^*(t) - \hat{x}_j^*(t) = x_i^*(t) - x_j^*(t) + d_{i,j}(t) \\ \hat{v}_i^*(t) - \hat{v}_j^*(t) = v_i^*(t) - v_j^*(t) \end{cases} \tag{13.8}$$

Taking the time derivative of (13.7), we can obtain that

$$\begin{cases} \dot{\hat{x}}_i^*(t) = \hat{v}_i^*(t) + d\left(\sum_{i=1}^{N} \left(hv_0^2 + cv_0 + l_i\right)\right)/dt \\ \hat{v}_i^*(t) = v_i^*(t) - v_0(t) \end{cases} \tag{13.9}$$

by which and (13.6), we can have

$$\begin{cases} \dot{\hat{x}}_i^*(t) = \hat{v}_i^*(t) \\ \dot{\hat{v}}_i^*(t) = u_{i1}(t) \end{cases} \tag{13.10}$$

Now, according to (13.7) and (13.8), the optimization problem in (13.4a) and (13.4b) can be equivalently written as follows:

$$\min_{\hat{x}_i^*(t)} J = \sum_{i=1}^{N} \left[\hat{x}_i^*(t) - \hat{x}_{i-1}^*(t)\right]^2 \tag{13.11a}$$

$$\text{s.t.}\begin{cases}\lim\limits_{t\to\infty}\hat{x}_i^*(t)=\hat{x}_j^*(t)\\ \lim\limits_{t\to\infty}\hat{v}_i^*(t)=\hat{v}_j^*(t)\end{cases} \tag{13.11b}$$

Based on these discussions, the problem now becomes as designing $u_{i1}(t)$ so that the optimization problem in (13.11) is solved with (13.10). To this end, we have the following result.

Theorem 13.1 For a vehicular platoon with topologies satisfying Assumption 13.1, if $((\rho)/(\tau\varsigma)) < \lambda_{\min}(L+P)$ holds for positive parameters η, τ, ρ, and ς, then the controller that solves the optimization problem in (13.4a) and (13.4b) is given here, which generates the reference trajectory $x_i^*(t)$ and $v_i^*(t)$ for the group of vehicles:

$$\begin{aligned} du_{i1}(t) = &-\sum_{j\in I_i}\left(\eta\left(x_i(t)-x_j(t)+d_{i,j}(t)\right)+\tau\left(v_i(t)-v_j(t)\right)\right)\\ &-\sum_{j\in I_i}\left\{\chi_{ij}(t)\,sgn\left(\rho\left(x_i(t)-x_j(t)+d_{i,j}(t)\right)+\varsigma\left(v_i(t)-v_j(t)\right)\right)\right\}+\psi_i(t)\end{aligned} \tag{13.12}$$

where the controller gain $\chi_{ij}(t)-\chi_{ji}(t)\geq 0$ and internal signal $\psi_i(t)$ are given by

$$\dot{\chi}_{ij}(t) = \left|\rho\left(x_i(t)-x_j(t)+d_{i,j}(t)\right)+\varsigma\left(v_i(t)-v_j(t)\right)\right| \tag{13.13}$$

$$\psi_i(t) = -2\left(x_i(t)-x_j(t)+d_{i,j}(t)\right)-\left(v_i(t)-v_j(t)\right) \tag{13.14}$$

Proof: The proof is completed by successively using the distributed optimization algorithm with a time-varying cost function [9].

To prove that the reference signals $\hat{x}_i^*$ and $\hat{v}_i^*$ generated by the system (13.6) with $u_{i1}(t)$ are the solution to the optimization problem (13.4a), it is only necessary to prove that the system (13.10) applying $u_{i1}(t)$ solves problem (13.11).

The lower layer of the tracking controller makes that $x_i(t)\to x_i^*(t)$ and $v_i(t)\to v_i^*(t)$ as time $t\to\infty$. So, the position $x_i(t)$ and velocity $v_i(t)$ in controller (13.12)–(13.14) can be replaced by $x_i^*(t)$ and $v_i^*(t)$ which yields:

$$\begin{aligned} u_{i1}(t) = &-\sum_{j\in I_i}\left(\eta\left(\hat{x}_i^*(t)-\hat{x}_j^*(t)\right)+\tau\left(\hat{v}_i^*(t)-\hat{v}_j^*(t)\right)\right)\\ &-\sum_{j\in I_i}\chi_{ij}(t)\left(\rho\left(\hat{x}_i^*(t)-\hat{x}_j^*(t)\right)+\varsigma\left(\hat{v}_i^*(t)-\hat{v}_j^*(t)\right)\right)+\psi_i(t)\end{aligned} \tag{13.15}$$

$$\dot{\chi}_{ij}(t) = \left|\rho\left(\hat{x}_i^*(t)-\hat{x}_j^*(t)\right)+\varsigma\left(\hat{v}_i^*(t)-\hat{v}_j^*(t)\right)\right| \tag{13.16}$$

$$\psi_i(t) = -2\left(\hat{x}_i^*(t)-\hat{x}_j^*(t)\right)-\left(\hat{v}_i^*(t)-\hat{v}_j^*(t)\right) \tag{13.17}$$

First, we prove that $x_i^*(t)$ and $v_i^*(t)$ can reach a consensus as $t\to\infty$.

Define $a'_{ij}=a_{ij}\chi_{ij}(t)$ and $p'_i=p_i\chi_{i0}(t)$. A new Laplacian matrix is denoted by $\mathcal{L}'=\left[l'_{ij}\right]_{N\times N}$, where for $i\neq j$ and $l'_{ii}=\sum_{j=1}^{N}a'_{ij}$ The pinning matrix is $\mathcal{P}=\mathrm{diag}\{p'_1,p'_2,\ldots,p'_N\}$. Define the incidence matrices $D'=\left[d'_{ij}\right]_{N\times|\mathcal{E}_N|}$ and $\overline{D'}=\left[d'_{ij}\right]_{N\times|\mathcal{E}_{N+1}|}$, where $d'_{ij}=-a_{ij}$ when the edge ε_{ij} leaves node i, $d'_{ij}=a_{ij}$ when ε_{ij} enters the nodes i and $d'_{ij}=0$ otherwise. Notice that $\mathcal{L}'=D'D'^T$ and $\mathcal{L}+\mathcal{P}\triangleq\overline{D}'\overline{D}'^T$.

Let the column vectors $\overline{X}(t)=\left[\hat{x}_1^*(t),\hat{x}_2^*(t),\ldots,\hat{x}_N^*(t)\right]^T$, $\overline{V}(t)=\left[\hat{v}_1^*(t),\hat{v}_2^*(t),\ldots,\hat{v}_N^*(t)\right]^T$ and define vector $\Psi(t)=[\psi_1(t),\psi_2(t),\ldots,\psi_N(t)]^T$

The closed-loop system (13.10) and (13.15) can be written in a compact form

$$\begin{cases} \dot{\overline{X}}(t) = \overline{V}(t) \\ \dot{\overline{V}}(t) = -G\left(\eta\overline{X}(t) + \tau\overline{V}(t)\right) - \Theta(t) + \Psi(t) \end{cases} \tag{13.18}$$

where $\begin{cases} \dot{\overline{X}}(t) = \overline{V}(t) \\ \dot{\overline{V}}(t) = -G\left(\eta\overline{X}(t) + \tau\overline{V}(t)\right) - \Theta(t) + \Psi(t) \end{cases}$, $\Theta(t) = \overline{D}'\left(\overline{D}^T\left(\rho\overline{X}(t) + \varsigma\overline{V}(t)\right)\right)$.

Utilizing the following linear transformation

$$\begin{cases} \hat{X}(t) = \Gamma\overline{X}(t) \\ \hat{V}(t) = \Gamma\overline{V}(t) \end{cases} \tag{13.19}$$

where matrix $\Gamma \triangleq I_N - (1/N)\boldsymbol{1}_N\boldsymbol{1}_N^T$, the system (13.18) can be rewritten as

$$\begin{cases} \dot{\hat{X}}(t) = \hat{V}(t) \\ \dot{\hat{V}}(t) = -G\left(\eta\hat{X}(t) + \tau\hat{V}(t)\right) - \Theta'(t) + \Gamma\Psi(t) \end{cases} \tag{13.20}$$

where $\Theta'(t) = \overline{D}'\,\mathrm{sgn}\left(\overline{D}^{\mathrm{T}}\left(\rho\hat{X}(t) + \varsigma\hat{V}(t)\right)\right)$.

Consider the following Lyapunov function:

$$W_1(t) = \frac{1}{2}\begin{bmatrix} \hat{X}(t) \\ \hat{V}(t) \end{bmatrix}^T M \begin{bmatrix} \hat{X}(t) \\ \hat{V}(t) \end{bmatrix} + \frac{1}{4}\sum_{i=1}^{N}\sum_{j\in I_i}\left(\chi_{ij}(t) - \overline{\chi}\right)^2 \tag{13.21}$$

where $\overline{\chi} > 0$ to be chosen and

$$M = \begin{bmatrix} (\tau\rho + \eta\varsigma)G & \rho I_N \\ \rho I_N & \varsigma I_n \end{bmatrix}$$

Define matrix $\hat{M}$ as

$$\hat{M} = \begin{bmatrix} (\tau\rho + \eta\varsigma)\mathcal{G} & \rho I_N \\ \rho I_N & \varsigma I_n \end{bmatrix}$$

It is obvious that $\hat{M} \le M$. Due to $\lambda_{\min}(G) > (\rho/\tau\varsigma)$, one has

$$(\tau\rho + \eta\varsigma)\lambda_{\min}(G) - \frac{\rho^2}{\varsigma} > 0$$

By Schur's complement lemma, it can be concluded that $\hat{M} \ge 0$, then we have $0 \le \hat{M} \le M$. Therefore, M is positive definite and $W_1(t) \ge 0$. The time derivation of $W_1(1)$ along (13.40) is

$$\begin{aligned} \dot{W}_1(t) = {} & \begin{bmatrix} \hat{X}(t) \\ \hat{V}(t) \end{bmatrix}^T M \begin{bmatrix} \hat{X}(t) \\ \hat{V}(t) \end{bmatrix} + \frac{1}{2}\sum_{i=1}^{N}\sum_{j\in I_i}\left(\chi_{ij}(t) - \overline{\chi}\right)\dot{\chi}_{ij}(t) = -\rho\eta\hat{X}^T(t)G\hat{X}(t) + \hat{V}^T(t)(\rho I_N - \tau\varsigma G)\hat{V}(t) \\ & -\frac{1}{2}\sum_{i=1}^{N}\sum_{j\in I_i}\chi_{ij}(t)\left|\rho\left(\hat{x}_i^*(t) - \hat{x}_j^*(t)\right) + \varsigma\left(\hat{v}_i^*(t) - \hat{v}_j^*(t)\right)\right| \\ & + \frac{1}{N}\sum_{i=1}^{N}\sum_{j=1}^{N}\left[\rho\left(\hat{x}_i^*(t) - \hat{x}_j^*(t)\right) + \varsigma\left(\hat{v}_i^*(t) - \hat{v}_j^*(t)\right)\right]\psi_i(t) + \frac{1}{2}\sum_{i=1}^{N}\sum_{j\in I_i}\left(\chi_{ij}(t) - \overline{\chi}\right)\dot{\chi}_{ij}(t) \end{aligned} \tag{13.22}$$

Then, suppose that $|\psi_i(t) - \psi_j(t)|$ is bounded, namely, there exists a constant $\overline{\psi} > 0$ so that $|\psi_i(t) - \psi_j(t)| \leq \overline{\psi}$. It is obvious that this assumption is trivial and the boundness of $|x_i(t) - x_j(t) + d_{i,j}(t)|$ and $|v_i(t) - v_j(t)|$ can guarantee the boundness of $\overline{\psi}$.

Since $\rho - \tau\varsigma\lambda_{\min}(G) < 0$, then $\rho I_N - \tau\varsigma G < 0$ and $\hat{V}^T(t)(\rho I_N - \tau\varsigma G)\hat{V}(t) \leq 0$. Substituting (13.16) into (13.22), it follows from Assumption 13.1 and the boundness hypothesis here that

$$\begin{aligned}\dot{W}_1(t) \leq & -\rho\eta\hat{X}^T(t)G\hat{X}(t) + \hat{V}^T(t)(\rho I_N - \tau\varsigma G)\hat{V}(t) + \left(\frac{(N-1)\overline{\psi}}{4} - \frac{\overline{\chi}}{2}\right) \\ & \times \sum_{i=1}^{N}\sum_{j\in I_i}(\chi_{ij}(t) - \overline{\chi})\left|\rho(\hat{x}_i^*(t) - \hat{x}_j^*(t))\right| \times \sum_{i=1}^{N}\sum_{j\in I_i}(\chi_{ij}(t) - \overline{\chi})\left|\varsigma(\hat{v}_i^*(t) - \hat{v}_j^*(t))\right| \\ \leq & \left(\frac{(N-1)\overline{\psi}}{4} - \frac{\overline{\chi}}{2}\right)(\rho\hat{X}(t) + \varsigma\hat{V}(t))^T\Theta'(t) \leq \left(\frac{(N-1)\overline{\psi}}{4} - \frac{\overline{\chi}}{2}\right)\left\|\overline{D}^T(\rho\hat{X}(t) + \varsigma\hat{V}(t))\right\|_1 \\ \leq & \left(\frac{(N-1)\overline{\psi}}{4} - \frac{\overline{\chi}}{2}\right)\sqrt{\lambda_{\min}(G)}\left\|\rho\hat{X}(t) + \varsigma\hat{V}(t)\right\|_2 < 0 \end{aligned} \tag{13.23}$$

The second inequality is derived by selecting suitable $\overline{\chi}$ which satisfy $\overline{\chi} > ((N-1)\overline{\psi})/2$. The penultimate inequality is derived by $\mathcal{G} \triangleq \overline{DD}^T$. Based on Barbalat's Lemma [10], one has that $\hat{X}(t)$ and $\hat{V}(t)$ will converge to zero asymptotically, namely, $\hat{x}_i^*(t) = \hat{x}_j^*(t)$ and $\hat{v}_i^*(t) = \hat{v}_j^*(t)$.

The following shows that the optimization problem (13.11) is solved. Consider the following Lyapunov function:

$$W_2(t) = \frac{1}{2}\sum_{i=1}^{N}\sum_{j\in I_i}(\hat{x}_i^*(t) - \hat{x}_j^*(t))^2 + \frac{1}{2}\sum_{i=1}^{N}\sum_{j\in I_i}(\hat{x}_i^*(t) - \hat{x}_j^*(t) - \dot{v}_i^*(t))^2 \tag{13.24}$$

Taking the time derivative of (13.24), we have

$$\begin{aligned}\dot{W}_2(t) = & \sum_{i=1}^{N}\sum_{j\in I_i}(\hat{x}_i^*(t) - \hat{x}_j^*(t))(\hat{v}_i^*(t) - \hat{v}_j^*(t)) \\ & + \sum_{i=1}^{N}\sum_{j\in I_i}(\hat{x}_i^*(t) - \hat{x}_j^*(t) - \dot{v}_i^*(t))(\hat{v}_i^*(t) - \hat{v}_j^*(t))\end{aligned} \tag{13.25}$$

Invoking $\sum_{i=1}^{N}\dot{\hat{v}}_i^*(t) = \sum_{i=1}^{N}\psi_i(t)$ and (13.17), we can obtain

$$\dot{W}_2(t) = \sum_{i=1}^{N}\sum_{j\in I_i}(\hat{v}_i^*(t) - \hat{v}_j^*(t))^2 \tag{13.26}$$

Thus, $\dot{W}_2(t) < 0$, which yields $\sum_{i=1}^{N}\sum_{j\in I_i}(\hat{v}_i^*(t) - \hat{v}_j^*(t))$ will converge to zero asymptotically. It follows that the derivative of the cost function $J(t)$ in the optimization problem (13.12) will converge asymptotically to zero. Furthermore, since $J(t)$ is convex, it is minimized to $\hat{x}_i^*(t)$ as time goes to infinity [11]. By [12], we can conclude that $\hat{x}_i^*(t)$ are the unique solutions of the optimization problem (13.11). This completes the proof.

Remark 13.2 Similar to Chapter 12, the proposed quadratic spacing policy (13.3) can transform the reference platoon dynamics into a standard double integrator by introducing reference position and velocity errors. Using this spacing policy, the resulting method can ensure traffic flow stability within a critical traffic density (as shown later).

13.4 PID-Type Sliding-Mode Controller Design

In this section, we design the sliding-mode controller by introducing an improved PID-type sliding-mode surface, which is valued by its exceptional properties in tracking and eliminating steady-state spacing errors. The controller $u_{i2}(t)$ is designed to track the reference position $x_i^*(t)$ and speed $v_i^*(t)$, which guarantees individual stability, string stability, and traffic flow stability.

The position-tracking errors and velocity-tracking errors are defined as

$$\begin{cases} \tilde{x}_i(t) = x_i(t) - x_i^*(t) \\ \tilde{v}_i(t) = v_i(t) - v_i^*(t) \end{cases} \tag{13.27}$$

Define the total tracking errors

$$e_i(t) = \tilde{x}_i(t) + \tilde{v}_i(t) \tag{13.28}$$

It follows that

$$\begin{cases} \dot{e}_i(t) = \dot{\tilde{x}}_i(t) + \dot{\tilde{v}}_i(t) = \tilde{v}_i(t) + \tilde{a}_i(t) \\ \ddot{e}_i(t) = \dot{\tilde{v}}_i(t) + \dot{\tilde{a}}_i(t) = a_i(t) - u_{i1}(t) + \dot{a}_i(t) - \dot{u}_{i1}(t) \end{cases} \tag{13.29}$$

where $\tilde{a}_i(t) = a_i(t) - u_{i1}(t)$.

Introduce the following PID-type sliding mode surface:

$$s_i(t) = k_p e_i(t) + k_i \int_0^t e_i^{q/p}(\tau) d\tau + k_d \dot{e}_i(t) \tag{13.30}$$

where k_p, k_i, and k_d are positive real numbers, p and q denote the odd positive integer values with $q < p$. We now give the following result on the tracking controller design.

Theorem 13.2 If the controller $u_{i2}(t)$ of each follower is designed as follows, then the sliding mode surface $s_i(t)$ can converge to zero complying with reaching law (13.33):

$$u_{i2}(t) = -\frac{1}{\xi_i} a_i(t) + a_i(t) - \frac{k_p}{\xi_i k_d} \dot{e}_i(t) - \frac{k_i}{\xi_i k_d} e_i^{q/p}(t) - \frac{K_1}{\xi_i k_d} s_i^{\alpha}(t) - \frac{K_2}{\xi_i k_d} s_i^{\beta}(t) \tag{13.31}$$

where ξ_i is determined by the engine time constant, K_1 and K_2 are both positive real numbers, and $\alpha, \beta > 0$ are odd numbers.

Proof: The time derivative of the sliding mode surface $s_i(t)$ is given as follows:

$$\dot{s}_i(t) = k_p \dot{e}_i(t) + k_i \dot{e}_i^{q/p}(t) + k_d(a_i(t) - \xi_i a_i(t) + \xi_i u_{i2}(t)) \tag{13.32}$$

Substituting (13.31) into (13.32), we obtain

$$\dot{s}_i(t) = -K_1 s_i^{\alpha}(t) - K_2 s_i^{\beta}(t) \tag{13.33}$$

Consider the following Lyapunov function

$$V(t) = \frac{1}{2} s_i^2(t) \tag{13.34}$$

whose time derivative is

$$V'(t) = s_i(t)\dot{s}_i(t) - K_1 s_i^{\alpha+1}(t) - K_2 s_i^{\beta+1}(t) \le 0 \tag{13.35}$$

which implies that the sliding-mode surface $s_i(t)$ converges to zero. Here, (13.32) is the chosen reaching law to ensure the convergence speed and smoothness of the designed controller.

According to (13.29), when the sliding mode surface is reached, we have the following tracking error dynamnics:

$$\dot{e}_i(t) = -\frac{k_p}{k_d}e_i(t) - \frac{k_i}{k_d}\int_0^t e_i^{q/p}(v)dv \tag{13.36}$$

which is asymptotically convergent as shown by the following result.

Theorem 13.3 For mutually odd parameters p and q, if $0 < q < p$ then position tracking errors $\tilde{x}_i(t)$ and velocity tracking errors $\tilde{v}_i(t)$ can converge to zero asymptotically on the sliding mode surface with the sliding-mode controller in (13.30).

Proof: For the tracking error system (13.35), consider the following Lyapunov function

$$V_0(t) = \frac{1}{2}e_i^2(t) \tag{13.37}$$

whose time derivative of $V_0(t)$ along system (13.36) is

$$\dot{V}_0(t) = e_i(t)\left(-\frac{k_p}{k_d}e_i(t) - \frac{k_i}{k_d}\int_0^t e_i^{q/p}(v)dv\right) \tag{13.38}$$

Since p and q are mutually odd with $0 < q < p$, we have that

$$\dot{V}_0(t) = -\frac{k_p}{k_d}e_i^2(t) - \frac{k_i}{k_d}\int_0^t \left(e_i^2(v)\right)^{((q/p\,+\,1)/2)}dv \leq 0 \tag{13.39}$$

which means that the tracking errors $e_i(t)$ will converge asymptotically to zero, i.e.

$\tilde{x}_i(t) + \tilde{v}_i(t) \to 0$. Thus $\dot{\tilde{x}}_i(t) = \tilde{v}_i(t) = -\tilde{x}_i(t)$, which is asymptotically stable. So, the position tracking errors $\tilde{x}_i(t)$ and the velocity tracking errors $\tilde{v}_i(t)$ can be stabilized asymptotically.

Remark 13.3 In comparison with the methods in Chapter 12, by introducing the term with fractional-order q/p to reduce integral saturation, the PID-type sliding-mode controller proposed in this chapter has better performance in convergence speed and control accuracy, as well as better robustness and anti-disturbance ability.

For vehicle platoon with a constant pacesetter's velocity, $u_{i1}(t) = 0$ should be known as well as $\tilde{x}_0(t) = 0$, $\tilde{v}_0(t) = 0$ and $e_0 = 0$ hold.

By definition, the intervehicle spacing errors $\delta_i(t)$ (13.5) satisfy that

$$\dddot{\delta}_i(t) = \dddot{x}_i(t) - \dddot{x}_{i-1}(t) = \dot{a}_i(t) - \dot{a}_{i-1}(t) \tag{13.40}$$

According to (13.2) and the designed controller $u_{i2}(t)$, we have that

$$\dot{a}_i(t) = -a_i(t) - \frac{k_p}{k_d}\dot{e}_i(t) - \frac{k_i}{k_d}e_i^{q/p}(t) \tag{13.41}$$

which, together with (13.40), leads to the following spacing error dynamics

$$\begin{aligned}\dddot{\delta}_i(t) &= \dot{a}_i(t) - \dot{a}_{i-1}(t)\\ &= \left(-a_i(t) - \frac{k_p}{k_d}\dot{e}_i(t) - \frac{k_i}{k_d}e_i^{q/p}(t)\right) - \left(-a_{i-1}(t) - \frac{k_p}{k_d}\dot{e}_{i-1}(t) - \frac{k_i}{k_d}e_{i-1}^{q/p}(t)\right)\end{aligned} \tag{13.42}$$

On the sliding mode surface, we have $\dot{s}_i(t) = 0$. So, $k_p \dot{e}_i(t) + k_i e_i^{q/p}(t) + k_d \ddot{e}_i(t)$ namely

$$e_i^{q/p}(t) = -\frac{k_p}{k_i}\dot{e}_i(t) - \frac{k_d}{k_i}\ddot{e}_i(t) \tag{13.43}$$

Thus, (13.42) can be rewritten as

$$\dddot{\delta}_i(t) = -(a_i(t) - a_{i-1}(t)) - \frac{k_p}{k_d}(\dot{e}_i(t) - \dot{e}_{i-1}(t)) + \frac{k_p}{k_d}(e_i(t) - e_{i-1}(t)) + \ddot{e}_i(t) - \ddot{e}_{i-1}(t) \tag{13.44}$$

By definitions given earlier, we have that

$$\ddot{\delta}_i(t) - \delta_i(t) = \ddot{d}(t) - d(t) - \frac{k_d}{k_p}\ddot{d}(t) - \frac{k_d}{k_p}\dddot{d}(t)$$

which, by Laplace transform, is equivalent to

$$\delta_i(s) = \frac{-(k_p/k_d)\cdot s^3 d_{i,i-1}(s) + (1 - (k_p/k_d))\cdot s^2 d_{i,i-1}(s) - d_{i,i-1}(s)}{s^2 - 1}$$

where s is a complex variable. Therefore, we can have the following transfer function for $i = 1, 2, \ldots, N$

$$G_i(s) = \frac{\delta_{i+1}(s)}{\delta_i(s)} = \frac{-(k_p/k_d)\cdot s^3 d_{i+1,i}(s) + (1 - (k_p/k_d))\cdot s^2 d_{i+1,i}(s) - d_{i+1,i}(s)}{-(k_p/k_d)\cdot s^3 d_{i,i-1}(s) + (1 - (k_p/k_p))\cdot s^2 d_{i,i-1}(s) - d_{i,i-1}(s)}$$

According to the spacing policy, the intervehicle spacing in the platoon is equal in the s domain, i.e. $d_{i,\,i-1}(s) = d_{i+1,\,i}(s)$. So, we have that $G_i(s) = 1$ for all vehicles, namely, string stability can be guaranteed.

The condition and critical traffic density for traffic flow stability are given below.

Theorem 13.4 For the vehicular platoon system in (13.2), if $h \geq c^2/4l$ and $-2\sqrt{h(l+L)} \leq c \leq 0$, then the quadratic spacing policy (13.3) and the control method can guarantee traffic flow stability. Furthermore, the critical traffic density is given by

$$P_{\lim} = \frac{h}{c\sqrt{h(l+L)} + 2h(l+L)} - \frac{h}{c\sqrt{h(l+L)} - 2h(l+L)}$$

Proof: If $h \geq c^2/4l$ holds, it follows that $d_{i,\,i-1}(t) > 0$ for $\forall v_0(t) > 0$. In the steady state of road traffic, it is natural that the speed and separation of each vehicle are identical, i.e. $d_{i,\,i-1}(t) = d_{steady}(t)$ and $v_i(t) = v(t)$, for all i. Hence, according to [8], one can have the traffic density in steady-state distance

$$d_{steady}(t) = hv^2(t) + cv(t) + l + L \tag{13.45}$$

the steady-state traffic density

$$P = \frac{1}{d_{steady}} = \frac{1}{hv^2(t) + cv(t) + l + L} \tag{13.46}$$

According to the flow and density formula $Q = Pv$ and Equation (13.46), be obtained the following traffic flow rate can be obtained:

$$Q(P) = P\left(-\frac{c}{2h} + \frac{\sqrt{c^2 - 4h(l + L - (1/P))}}{2h}\right)$$

Taking partial derivatives can lead to

$$\frac{\partial Q}{\partial P} = -\frac{c}{2h} + \frac{\sqrt{c^2 - 4h(l + L - (1/P))}}{2h} - \frac{1}{P\sqrt{c^2 - 4h(l + L - (1/P))}}$$

From $\frac{\partial Q}{\partial P} = 0$, we can get the upper limit of traffic flow density $P_{\lim}$ as

$$P_{\lim} = \frac{h}{c\sqrt{h(l + L)} + 2h(l + L)} - \frac{h}{c\sqrt{h(l + L)} - 2h(l + L)} \tag{13.47}$$

Note that $P_{\lim} \geq 0$ for $-2\sqrt{h(l + L)} \leq c \leq 0$.

Remark 13.4 Unlike the result in Chapter 12, the presented method can guarantee traffic flow stability given that the traffic density is below a critical value, besides individual vehicle stability and string stability, by introducing a quadratic spacing policy.

13.5 Simulation Results

This section shows the implementation and performance of the presented platoon control method in numerical simulations. Consider a platoon of five vehicles, wherein the leader is denoted as 0 and the followers are 1–4 whose parameters are shown in Table 13.1.

In the simulations, the initial positions of the five vehicles are $x(0) = [0, -5, -12, -17, -12]$ m. The velocity and acceleration of the leader are given in Figure 13.3, and the followers all have zero initial velocities with nonzero initial spacing errors. Communication topologies considered in the simulations include bidirectional-predecessor following (BPF), two-bidirectional-predecessor following (LBPF), and two bidirectional-predecessor following (TBPF) (see Figure 13.4), all of which satisfy Assumption 13.1. According to Theorems 13.1, 13.2, and 13.3, the parameters of controller $u_{i1}(t)$ and tracking controller $u_{i2}(t)$ are listed in Table 13.2.

The trajectory optimization and tracking control results of these vehicles under topologies BPF, LBPF, and TBPF are given in Figures 13.5–13.7, respectively. In these figures, (a) and (b) show that the reference trajectory (position and velocity) of all vehicles can achieve consensus with that of the leader; (c) and (d) mean that with the method in this chapter, all vehicles can track their reference trajectory; (e) and (f) show that the spacing errors and velocity errors between neighboring vehicles

Table 13.1 Physical parameters used in the experiments.

Vehicle index	1	2	3	4
$\xi_i(s^{-1})$	0.5	0.25	0.60	0.35
L_i(m)	5.0	4.6	4.0	4.7

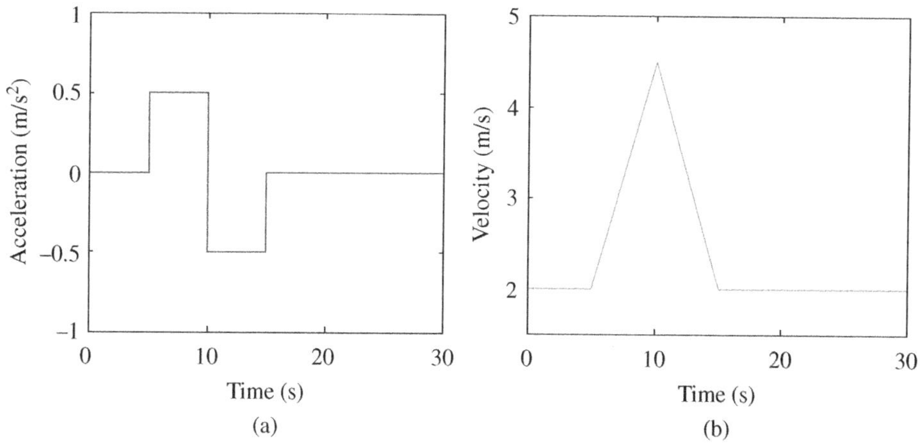

Figure 13.3 Acceleration (a) and velocity (b) of the leader.

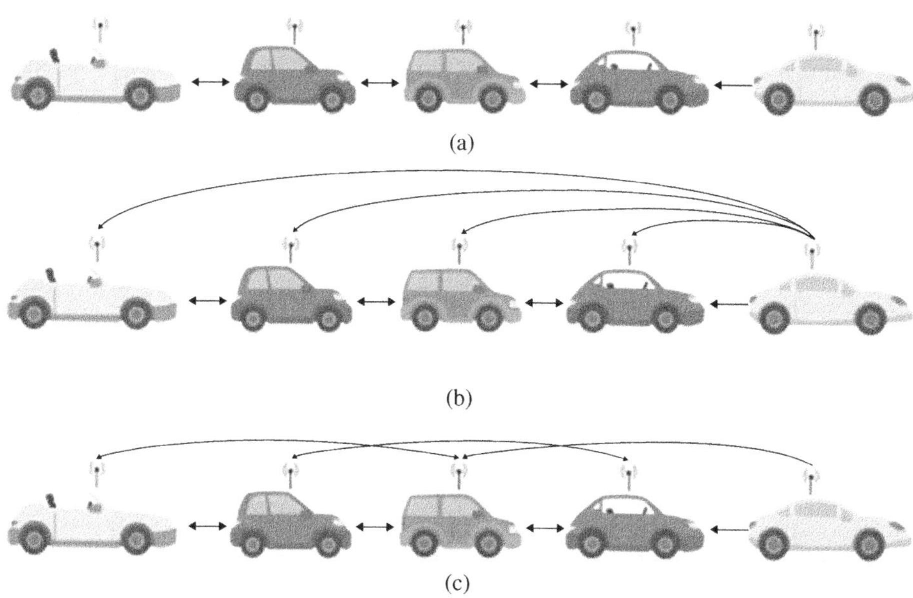

Figure 13.4 Three communication topologies for platoons. (a) BPF, (b) LBPF, (c) TBPF.

Table 13.2 Parameters used in the devised control structure.

Parameter name	Simulation values	Parameter name	Simulation values
η	0.5	q	3
τ	0.5	a	5
ρ	0.5	β	7
ς	0.5	K_1	5
p	5	K_2	40

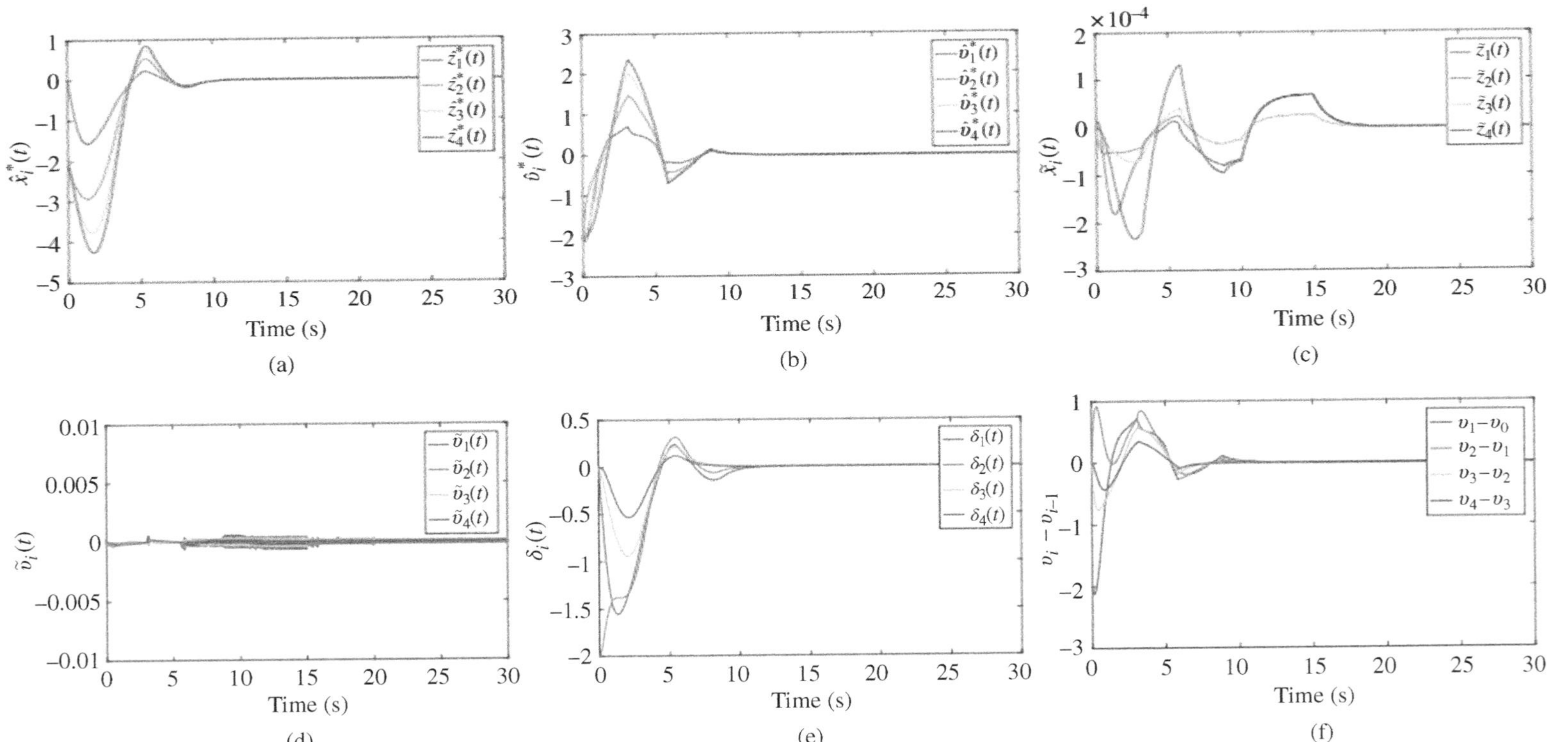

Figure 13.5 Simulation results of the platoon control system with the devised control framework under BPF. (a) Reference position errors $\hat{x}_i^*(t)$, (b) reference velocity errors $\hat{v}_i^*(t)$, (c) position tracking errors $\tilde{x}_i(t)$, (d) velocity tracking errors $\tilde{v}_i(t)$, (e) spacing errors $\delta_i(t)$, (f) velocity errors $v_i(t) - v_{i-1}(t)$.

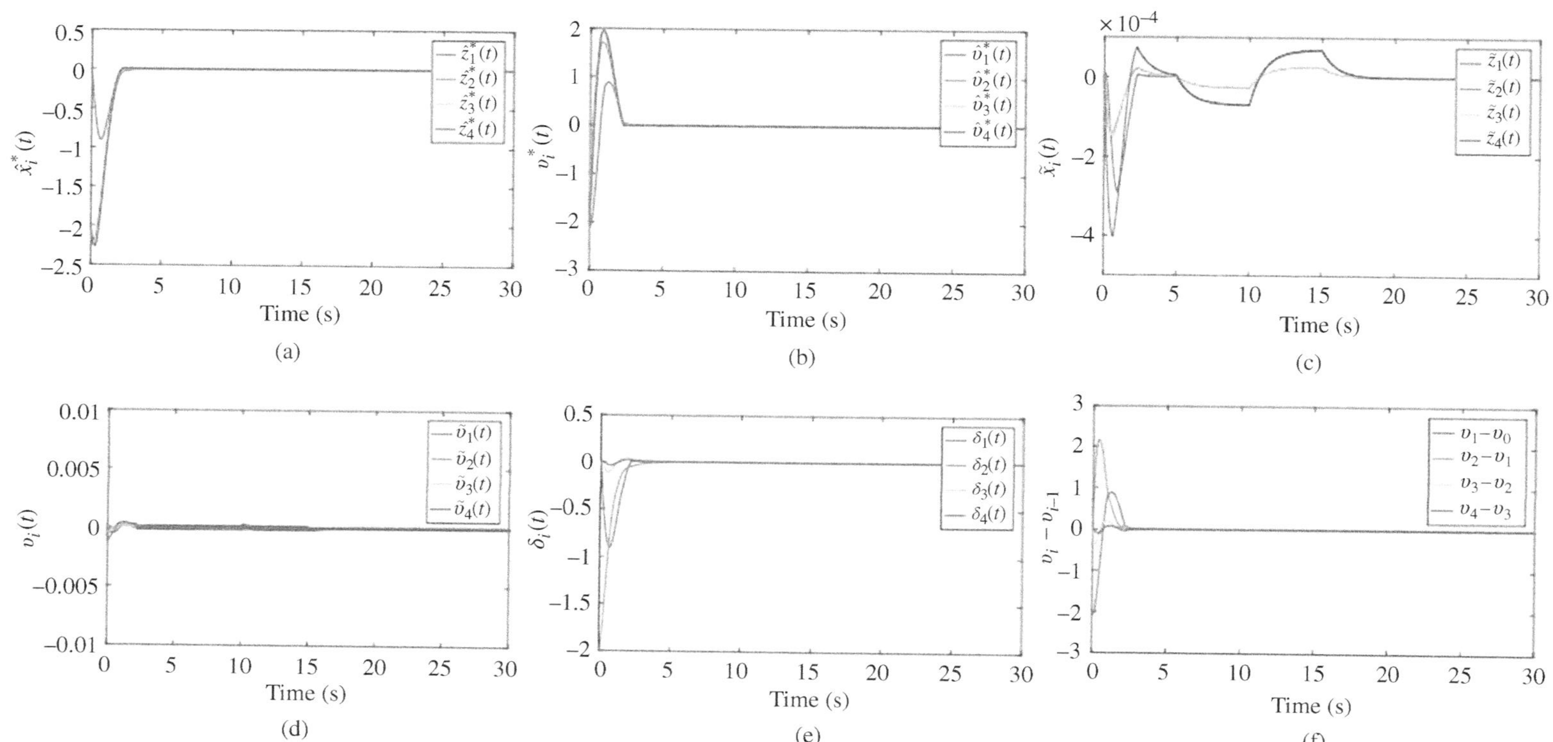

Figure 13.6 Simulation results of the platoon under LBPF with the devised control framework. (a) Reference position errors $\hat{x}_i^*(t)$, (b) reference velocity errors $\hat{v}_i^*(t)$, (c) position tracking errors $\tilde{x}_i(t)$, (d) velocity tracking errors $\tilde{v}_i(t)$, (e) spacing errors $\delta_i(t)$, (f) velocity errors $v_i(t) - v_{i-1}(t)$

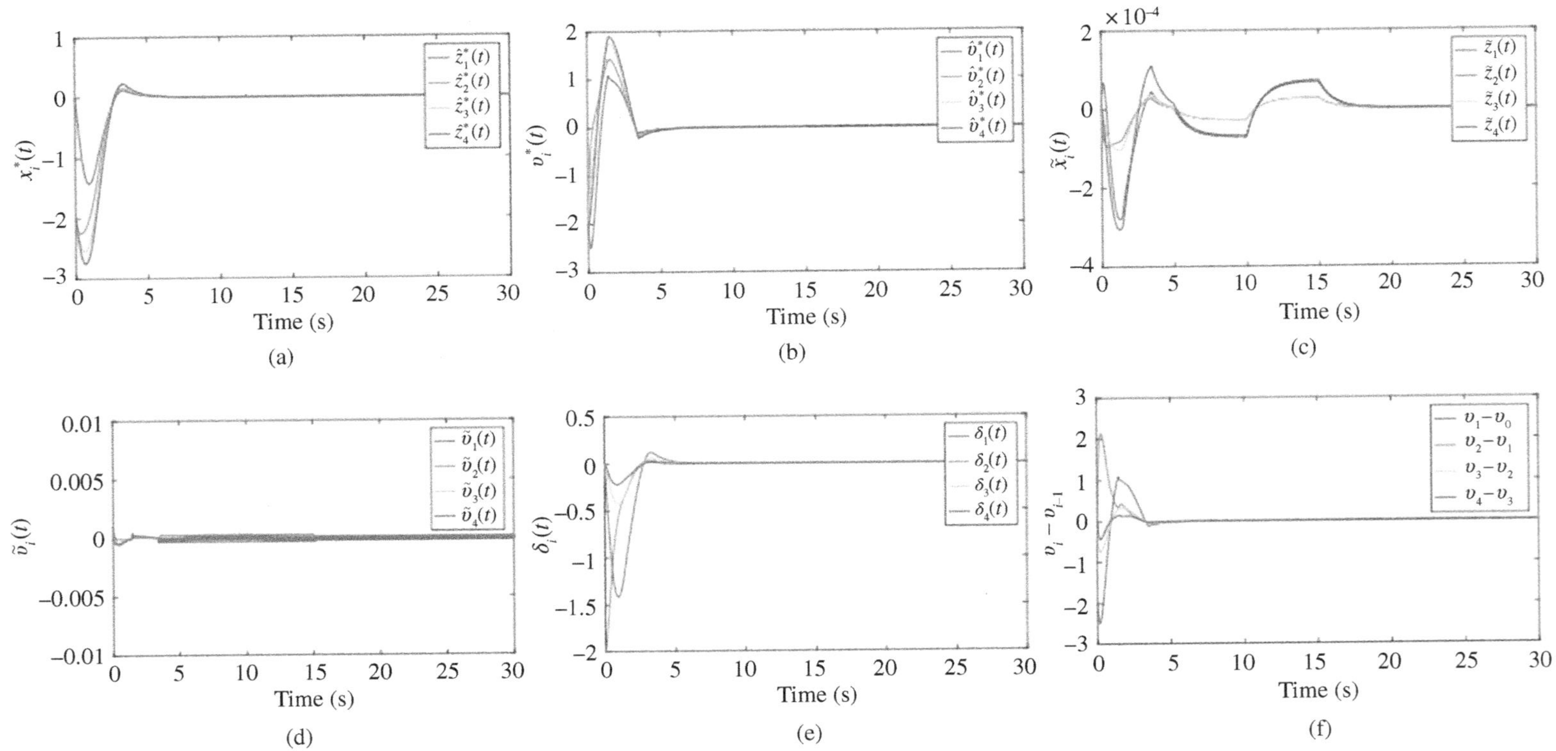

Figure 13.7 Simulation results of the platoon control system with the devised control framework under TBPF. (a) Reference position errors $\hat{x}_i^*(t)$, (b) reference velocity errors $\hat{v}_i^*(t)$, (c) position tracking errors $\tilde{x}_i(t)$, (d) velocity tracking errors $\tilde{v}_i(t)$, (e) spacing errors $\delta_i(t)$, (f) velocity errors $v_i(t) - v_{i-1}(t)$

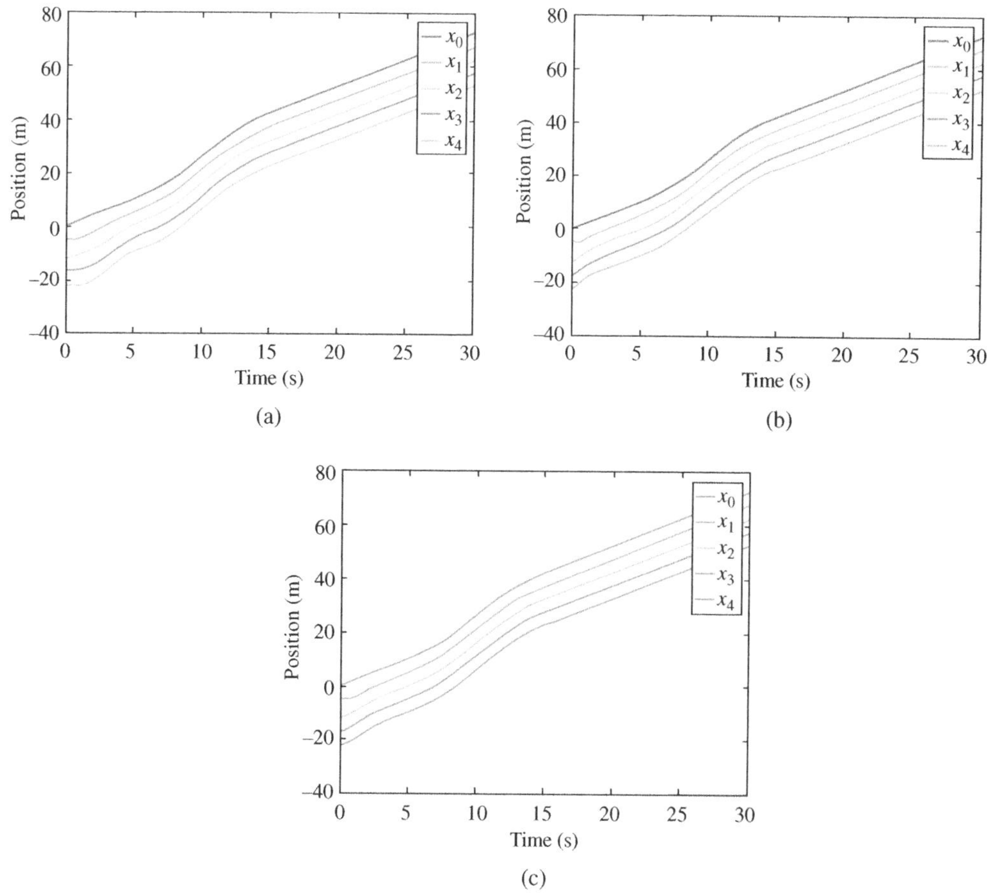

Figure 13.8 Time-position diagram of the five-vehicle platoon under BPF (a), LBPF (b), and TBPF (c) communication topologies.

can both converge to zero. The real-time positions of these vehicles under three topologies are shown in Figure 13.8. From these results, it is clearly seen that string stability and smooth traffic flow are well-guaranteed.

The proposed method is compared with two benchmark methods in [13] and Chapter 12, based on the following performance measures:

$$\begin{cases} J_1(t) = \lim\limits_{T\to\infty} \int_0^T \sum\limits_{i=1}^{4} \delta_i^2(t)dt \\ J_2(t) = \lim\limits_{T\to\infty} \int_0^T \sum\limits_{i=1}^{4} \tilde{x}_i^2(t)dt \end{cases} \tag{13.48}$$

where $J_1(t)$ is the overall spacing errors, and $J_2(t)$ represents the tracking performance.

Figure 13.9 shows the actual velocity, the spacing errors, and velocity errors of the platoon under the LBPF communication topology. It can be seen that the intervehicle spacing errors and speed-tracking errors of the proposed method are smaller and converge faster without obvious chattering. The corresponding performance measures are shown in Table 13.3.

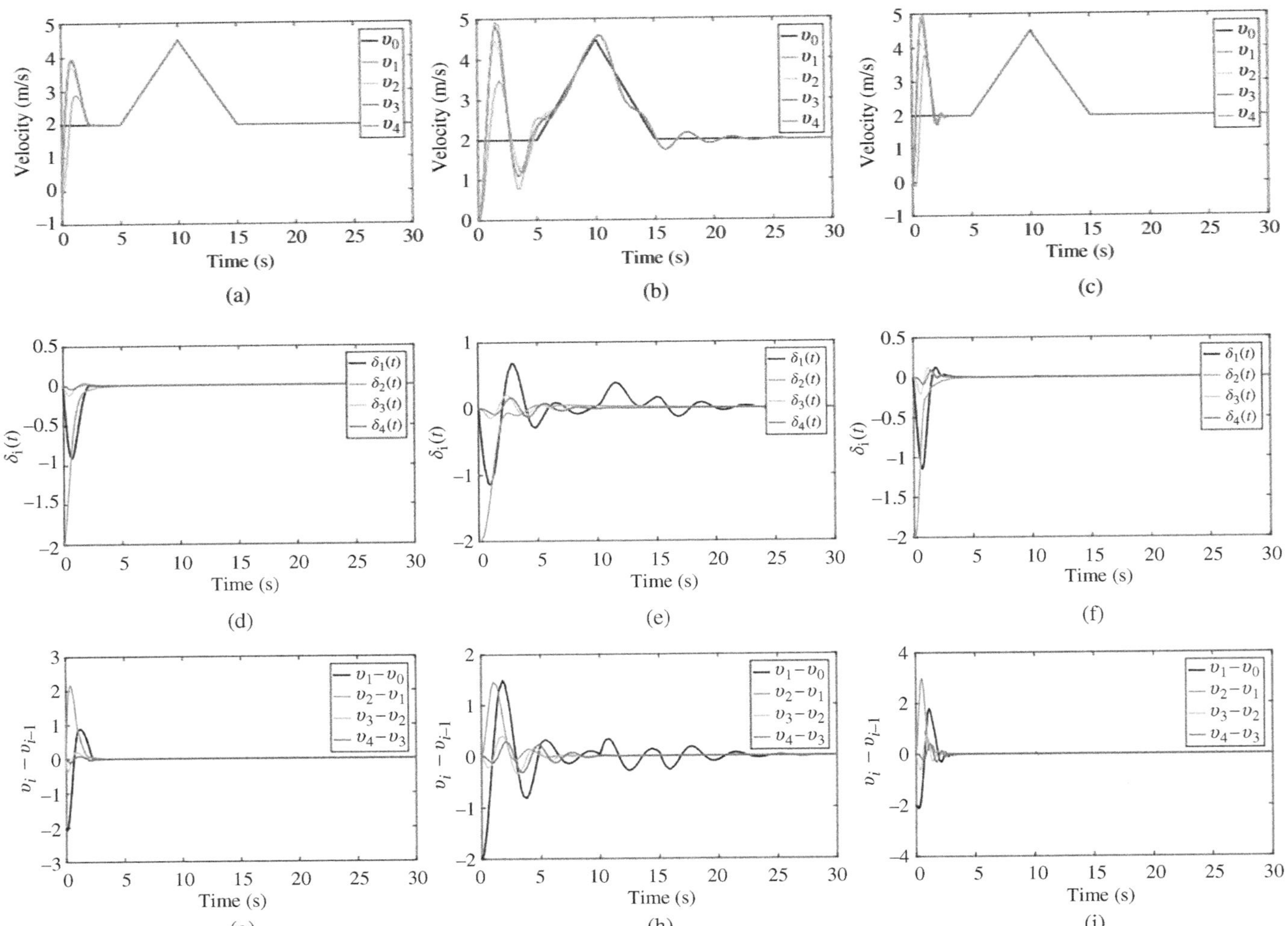

Figure 13.9 Comparisons of platoon control performance under LBPF among the proposed method (left), (middle), and Chapter 12 (right). (a) Vehicles' velocityv_i, (b) vehicles' velocityv_i, (c) vehicles' velocityv_i, (d) spacing errors $\delta_i(t)$, (e) spacing errors $\delta_i(t)$, (f) spacing errors $\delta_i(t)$, (g) velocity errors $v_i(t) - v_{i-1}(t)$, (h) velocity errors $v_i(t) - v_{i-1}(t)$, (i) velocity errors $v_i(t) - v_{i-1}(t)$. Source: Adapted from Hossain et al. [14].

Table 13.3 Performance measures for the platoon with different control under LBPF.

Controller	Proposed	Fiengo et al.	Chapter 12
$J_1(t)$	2.422	5.914	2.702
$J_2(t)$	3.095×10^{-7}	—	0.762
Setting time (s)	3.783	18.810	4.978

Then, we compare the performance of our method with that of Chapter 12 and in the existence of jerks in the leader acceleration

$$a_{jerk}(t) = a_0(t) + \sigma(t) \tag{13.49}$$

where $a_0(t)$ denotes the original leader acceleration and $\sigma(t)$ indicates the jerk, which takes arbitrary values in $[-0.15, 0.15]$. The acceleration and velocity of the leader in this case are shown in Figure 13.10.

The performance comparisons are shown in Figure 13.11, from which it can be seen that the method in Chapter 12 cannot deal with the acceleration jerk. The performance of our method is clearly much better, with faster convergent tracking errors and no oscillations. The performance measures under LBPF communication topology are given in Table 13.4, which show the clear advantage of our method. Figure 13.11 depicts the interspacing errors and velocity errors between neighbors.

Finally, a comparison with the centralized control scheme [14] is provided to demonstrate the advantages of the distributed framework in larger-size platoons. Figure 13.12 depicts the position of a 20-vehicle platoon and the interspacing and velocity errors between neighbors. It is obviously seen that our method performs better and faster.

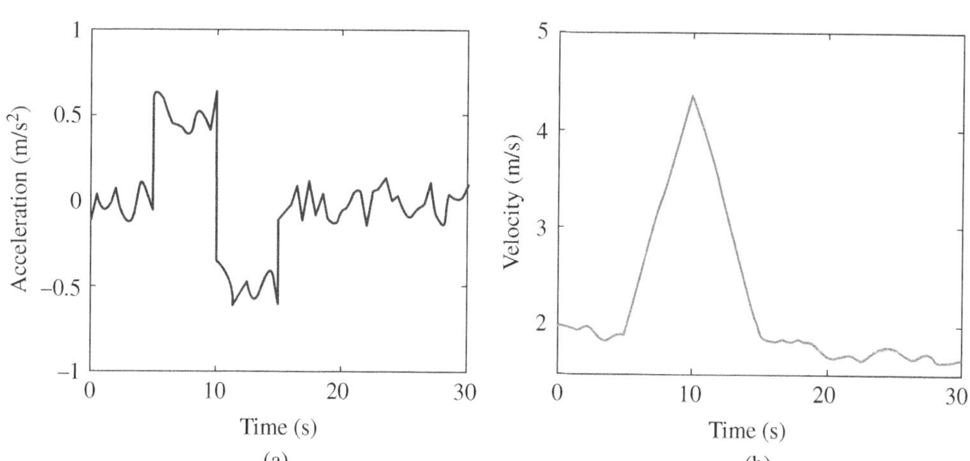

Figure 13.10 Acceleration (a) and velocity (b) of the leader with a random jerk.

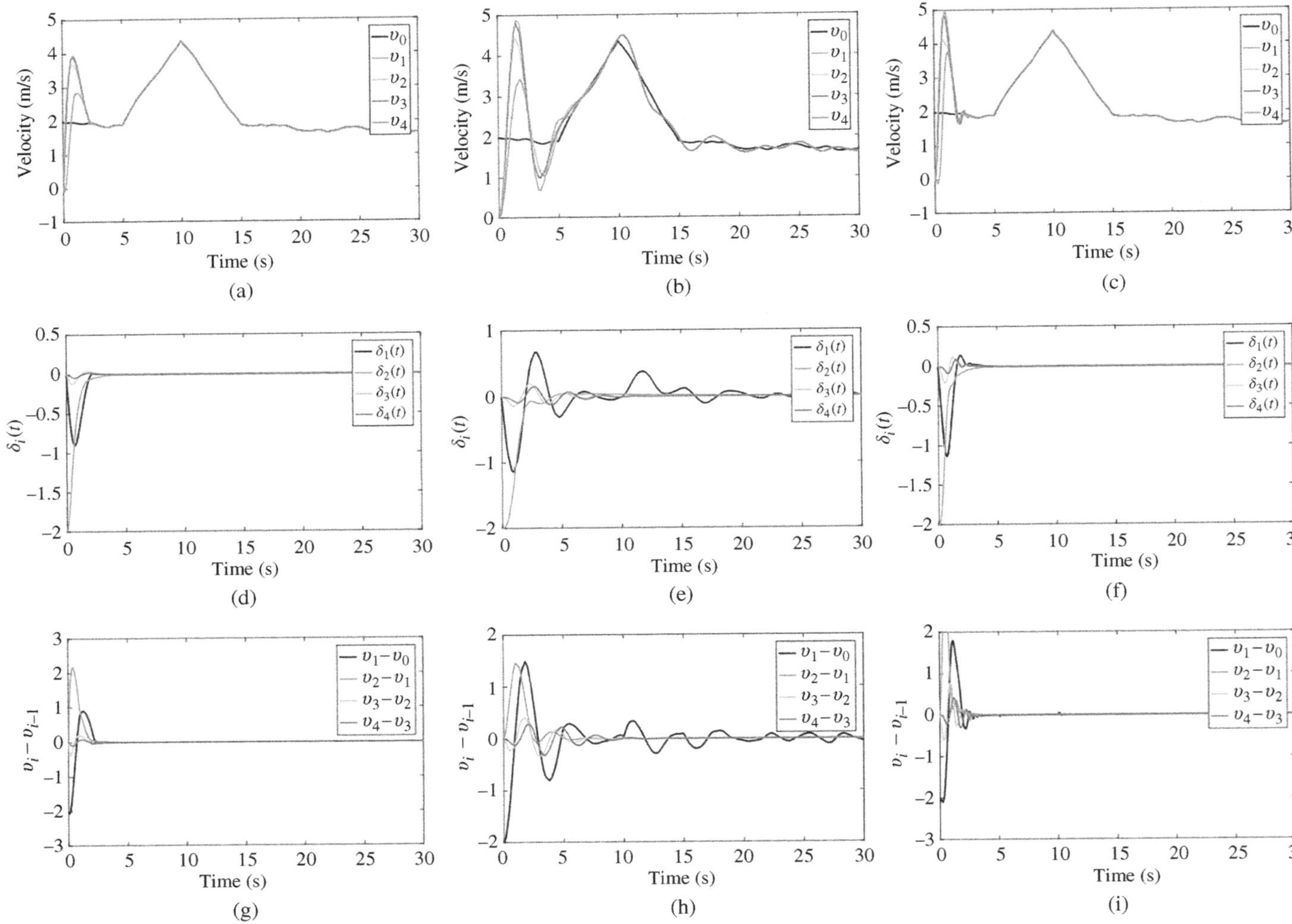

Figure 13.11 Comparisons of platoon control performance with random jerk under LBPF among the proposed method (left), (middle), and Chapter 12 (right). (a) Vehicles' velocityv_i, (b) vehicles' velocityv_i, (c) vehicles' velocityv_i, (d) spacing errors δ_i, (e) spacing errors δ_i, (f) spacing errors δ_i, (g) velocity errors $v_i - v_{i-1}$, (h) velocity errors $v_i - v_{i-1}$, (i) velocity errors $v_i - v_{i-1}$. *Source:* Adapted from Hossein et al. [14].

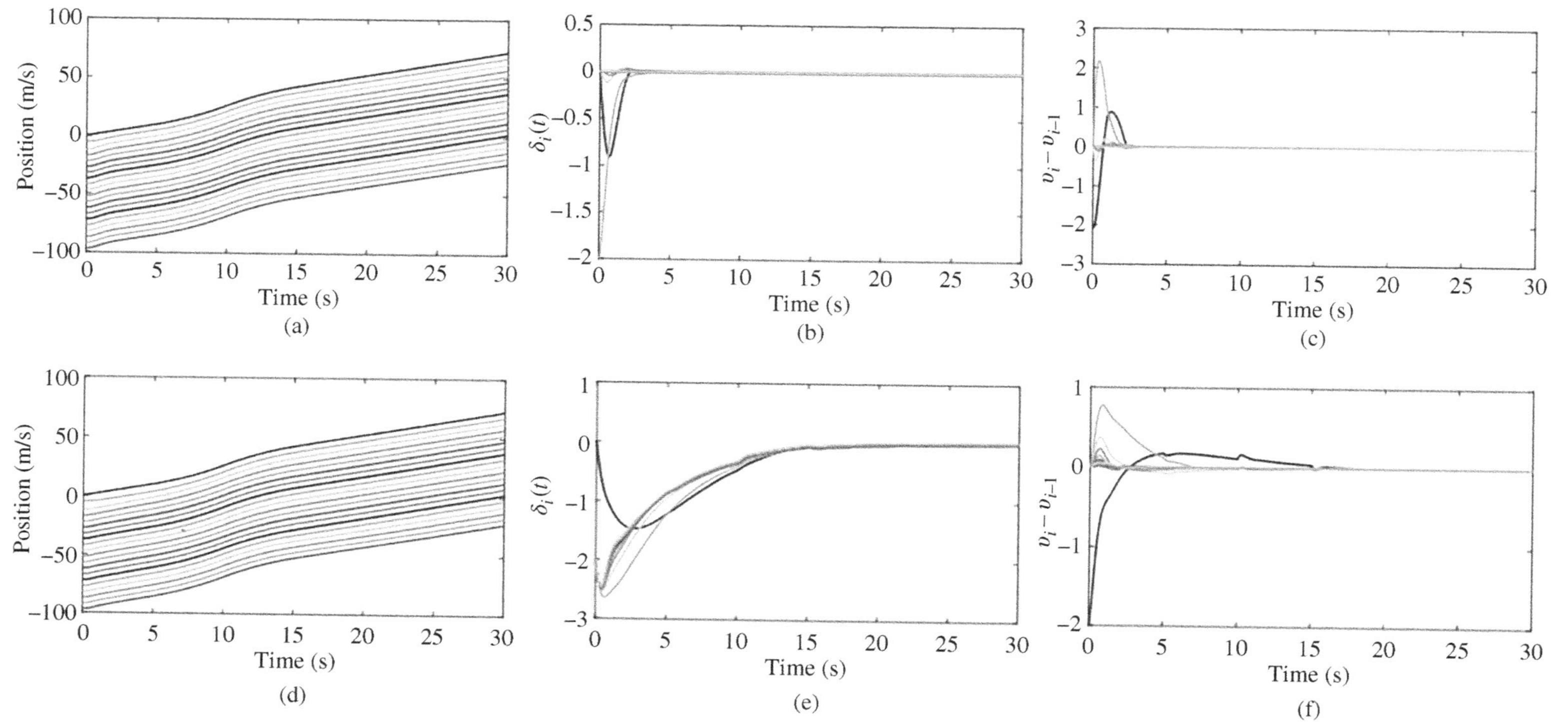

Figure 13.12 Simulation results of the 20-vehicle platoon under LBPF with the proposed method (top) and the centralized control method in [8] (bottom). (a) Position $x_i(t)$, (b) spacing errors $\delta_i(t)$, (c) velocity errors $v_i - v_{i-1}$, (d) position $x_i(t)$, (e) spacing errors $\delta_i(t)$, (f) velocity errors $v_i - v_{i-1}$. *Source:* Adapted from Fiengo et al. [13].

Table 13.4 Performance measures the platoon with random jerk using different controllers under LBPF.

Controller	Proposed	Fiengo et al.	Wen et al. [5]
$J_1(t)$	2.410	5.911	2.709
$J_2(t)$	3.079×10^{-7}	—	0.762
Settling time (s)	4.647	27.82	5.071

13.6 Conclusions and Future Work

In this chapter, a distributed optimal platoon control framework is developed. The optimal trajectory for each following vehicle is derived in the context of distributed convex optimization based on a spacing error cost function. Then, a novel PID-type sliding-mode controller with a double high-power reaching law is designed for the vehicles to track the optimal trajectory. The presented methodology can achieve smooth and stable platoon control, allowing for flexible speed regulation to guarantee steady global traffic flow.

References

1 Zheng, Y., Li, S.E., Li, K., and Wang, L.-Y. (2016). Stability margin im provement of vehicular platoon considering undirected topology and asymmetric control. *IEEE Trans. on Control Systems Tech.* 24 (4): 1253–1265.

2 Zheng, Y., Eben Li, S., Wang, J. et al. (2016). Stability and scalability of homogeneous vehicular platoon: study on the influence of information flow topologies. *IEEE Trans. Intel. Transp. Syst.* 17 (1): 14–26.

3 Ali, A., Garcia, G., and Martinet, P. (2015). The flatbed platoon towing model for safe and dense platooning on highways. *IEEE Intel. Transp. Syst. Magazine* 7 (1): 58–68.

4 Guo, G., Li, P., and Hao, L.-Y. (2022). A new quadratic spacing policy and adaptive fault-tolerant platooning with actuator saturation. *IEEE Trans. Intel. Transp. Syst.* 23 (2): 1200–1212.

5 Darbha, S. and Rajagopal, K. (1999). Intelligent cruise control systems and traffic flow stability. *Transp. Res. Part C: Emerging Tech.* 7 (6): 329–352.

6 Olfati-Saber, R. and Murray, R. (2004). Consensus problems in networks of agents with switching topology and time-delays. *IEEE Trans. Autom. Control* 49 (9): 1520–1533.

7 Swaroop, D. and Hedrick, J. (1996). String stability of interconnected systems. *IEEE Trans. Autom. Control* 41 (3): 349–357.

8 Sungu, H.E., Inoue, M., and Imura, J. (2015). Nonlinear spacing policy based vehicle platoon control for local string stability and global traffic flow stability. *2015 European Control Conference (ECC)*, Linz, Austria (July 2015), pp. 3396–3401.

9 Rahili, S. and Ren, W. (2017). Distributed continuous-time convex optimization with time-varying cost functions. *IEEE Trans. Autom. Control* 62 (4): 1590–1605.

10 Slotine, J. and Li, W. (1991). *Applied Nonlinear Control.* Upper Saddle River, NJ: Prentice-Hall.

11 Bazaraa, M.S., Sherali, H.D., and Shetty, C.M. (2005). *Nonlinear Programming: Theory and Algorithms.* Hoboken, NJ: Wiley.

12 Boyd, S. and Vandenberghe, L. (2004). *Convex Optimization*. Cambridge, UK: Cambridge University Press.

13 Fiengo, G., Lui, D.G., Petrillo, A. et al. (2019). Distributed robust PID control for leader tracking in uncertain connected ground vehicles with V2V communication delay. *IEEE/ASME Trans. Mechatr.* 24 (3): 1153–1165.

14 Hossein, C., Ali, G., and Ali, N. (2019). Centralized and decentralized distributed control of longitudinal vehicular platoons with nonuniform communication topology. *Asian Journal of Control* 21 (6): 2691–2699.

14

Trajectory Planning and Fixed-Time Terminal Sliding-Mode Control

14.1 Introduction

For the joint problem of vehicle speed planning and optimal tracking control considered in Chapters 12 and 13, according to the proposed control algorithm, vehicles can asymptotically track the leader speed when the time approaches infinity, which is not realistic in practice. This chapter revisits the problem with the aim to advance in at least two aspects. First, based on a cost defined by the accumulated intervehicle spacing errors, we introduce a novel distributed trajectory optimization method [1, 2]. Second, we propose a new terminal sliding mode control (TSMC) method for fixed-time platoon tracking control [3, 4], which is advantageous with alleviated chattering, zero steady-state spacing errors and string stability.

The contributions in comparison with the state-of-the-art results are given below:

1) A novel distributed trajectory optimization algorithm is proposed to minimize the intervehicle spacing errors, which is derived by means of consensus seeking of all vehicles with the platoon leader.
2) A fixed-time terminal sliding mode controller is given based on the bi-limit homogeneous property [3], which can make all the vehicles in the platoon track the optimal trajectory quickly and robustly with guaranteed string stability.

The remainder of this chapter is organized below. Section 14.2 gives problem description. The distributed vehicle trajectory optimization is in Section 14.3. Section 14.4 is the trajectory tracking control design procedure. Section 14.5 is numerical simulation. Finally, the conclusions and future work are addressed in Section 14.6.

Notations Define $\mathbb{R}^n$ as the n-dimensional Euclidean space. Denote $1^N = [1, 1, \ldots, 1]^T \in \mathbb{R}^N$, $0_N = [0, 0, \ldots, 0]^T \in \mathbb{R}^N$, $I_N \in \mathbb{R}^{N\times N}$ is the identity matrix of size $N \times N$. Let $\|p\|_q$ be q-norm of vector p. For a matrix B, B^T is its transpose matrix, $\lambda_{\min}(B)$ and $\lambda_{\max}(B)$ are defined as the minimal and maximum eigenvalues, diag{·} is defined as the diagonal matrix, and $\sigma_{\max}(B) = \sqrt{\lambda_{\max}(B^T B)}$. The set V_N contains {1, 2, ..., N}. $\lceil p \rfloor^\eta = |p|^\eta \operatorname{sgn}(p)$. The Kronecker product is denoted by $\otimes$ and $\operatorname{sgn}(\cdot)$ is the signum function, and the nonlinear function $h(\cdot)$ is designed as $h(a) = \begin{cases} \dfrac{a}{\|a\|} & \|a\| \neq 0 \\ 0 & \|a\| = 0 \end{cases}$.

Connected Vehicular Systems: Communication, Control, and Optimization, First Edition. Ge Guo and Shixi Wen.

14.2 Problem Formulation

The communication among the followers constitutes an undirected graph $\mathcal{G}_N = (\mathcal{V}_N, \mathcal{E}_N)$, where $\mathcal{V}_N = \{1, 2, ..., N\}$ is the vertices set and $\mathcal{E}_N \subseteq \mathcal{V}_N \times \mathcal{V}_N$ is the set of edges. An edge is defined as $\mathcal{E}_{ij} \triangleq (i, j) \in \mathcal{E}_N$, $i, j \in \mathcal{V}_N$, which illustrates there is an information flow from vehicle j to vehicle i. The adjacency matrix of graph $\mathcal{G}_N$ is expressed by $\mathcal{A} = [a_{ij}]_{N \times N}$, where $a_{ii} = 0$ for all i, and $a_{ij} = 1$ if $\mathcal{E}_{ij} \in \mathcal{E}_N$, otherwise, $a_{ij} = 0$. The Laplacian matrix is defined as $L = [l_{ij}]_{N \times N}$, where $l_{ij} = -a_{ij}$ for $i \neq j$ and $l_{ii} = \sum_{j=1}^{N} a_{ij}$. $D = [d_{ij}]_{N \times |\varepsilon_N|}$ is the incidence matrix, where $d_{ij} = -1$ if the edge E_j leaves node i, $d_{ij} = 1$ if it enters i, and $d_{ij} = 0$ otherwise. Note that $L \triangleq DD^T$. The neighbour set of node i is denoted by $N_i = \{j \in \mathcal{V}_N, (i, j) \in \mathcal{E}_N\}$.

Lemma 14.1 [5] Suppose an undirected graph $\mathcal{G}$ is connected, and its Laplacian matrix L is irreducible and the following conditions hold

1) $L1_N = 0$;
2) $\xi^T L = 0$ where $\xi = (\xi_1, \xi_2, ...\xi_N)^T$;
3) there is a positive-definite diagonal matrix $\Xi = \text{diag}(\xi_1, \xi_2, ...\xi_n) > 0$, and $\hat{L}$ is a symmetric matrix with $\sum_{j=1}^{N} \hat{l}_{ij} = \sum_{j=1}^{N} \hat{l}_{ji} = 0$, $\hat{L} = (1/2)(\Xi L + L^T \Xi)$.

Definition 14.1 [6] Define $a(L) = \min_{x^T\xi = 0, x \neq 0} \frac{x^T \hat{L} x}{x^T \Xi x}$, $b(L) = \max_{x^T\xi = 0, x \neq 0} \frac{x^T \hat{L} x}{x^T \Xi x}$. If $\Xi = \eta I_N$, then $a(L) = \lambda_2(L)$, i.e. the second smallest eigenvalue of L.

Lemma 14.2 [6] If a matrix $P \in R^{n \times n}$ is positive-definite and $Q \in R^{n \times n}$ is symmetric, then for the vector $x \in R^n$, we can draw the conclusion:

$$\lambda_{\min}(P^{-1}Q) x^T P x \leq x^T Q x \leq \lambda_{\max}(P^{-1}Q) x^T P x.$$

Consider a nonlinear dynamic system:

$$\dot{x} = f(x, t), f(0) = 0, x_0 \triangleq x(0) \tag{14.1}$$

where $x(t) \in R^n$ is the state vector and $f(x)$: $R^n \to R^n$ is the vector field function.

Definition 14.2 [7] For a scalar function $V(x)$: $R^n \to R^n$ or vector field $f(x)$: $R^n \to R^n$, it is said to be homogeneous in the bilimit if it is homogeneous in the 0-limit and the ∞-limit.

Lemma 14.3 [8] (Lyapunov stability theorem at fixed time) For system (14.1), if there exists a Lyapunov function $V(x)$: $U \to R^n$

$$\dot{V}(x) \leq -\varpi V^s(x) - \theta V^r(x) \tag{14.2}$$

where the coefficients $\varpi > 0$, $\theta > 0$, $0 < s < 1$, and $r > 1$, then the equilibrium of system (14.1) is fixed-time stable. The convergence time has a definite upper bound satisfying:

$$T \leq T_{\max} = \frac{1}{\varpi(1-s)} + \frac{1}{\theta(r-1)} \tag{14.3}$$

Consider a collection of vehicles with a leader and N follower vehicles, as seen in Figure 14.1, where the dynamics of vehicle is described by the following model

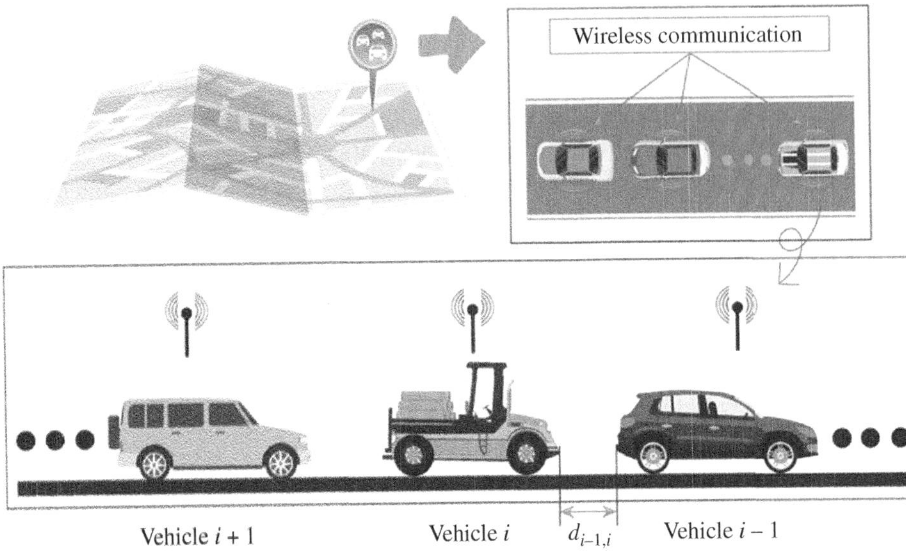

Figure 14.1 A heterogeneous vehicle platoon.

$$\begin{aligned}\dot{x}_i(t) &= v_i(t)\\ \dot{v}_i(t) &= a_i(t)\\ \dot{a}_i(t) &= \frac{u_i(t)}{\tau_i} + \mu_i\end{aligned} \tag{14.4}$$

where $x_i(t)$, $v_i(t)$, and $a_i(t)$ are, respectively, the position, velocity, and acceleration of vehicle i, $u_i(t)$ represents throttle/brake input, μ_i denotes the unknown uncertainty such as rough pavement and parameter uncertainties or others, and τ_i is the engine time constant. Suppose that the leading vehicle executes a constant speed trajectory, that is, $a_0 = 0$, $x_0 = v_0 t$.

Define spacing error for each follower as follows

$$\delta_i(t) = x_{i-1}(t) - x_i(t) - d_{i-1,i} \tag{14.5}$$

where $d_{i-1,i} > 0$ denotes the expected intervehicle spacing when stationary. Here, we introduce a constant spacing policy (CSP) strategy [9], i.e. $d_{i-1,i} = d$, which is a constant and contains the vehicle length and desired intervehicle distance.

Next, let us make the following assumptions.

Assumption 14.1 In a platoon system, the leading vehicle only communicates and transmits information with the following vehicles, and the information topology formed by all the following vehicles is an undirected connected graph.

Assumption 14.2 The disturbance of the platoon system and its first derivative are bounded:

$$|\mu_i| \le l_d, |\dot{\mu}_i| \le k_d$$

where l_d, $k_d > 0$ are known normal constants.

The paper aims at designing a distributed speed planning and fixed-time platoon tracking control framework that solves the following optimization problem:

$$\min_{x_i(t)} J(t) = \frac{1}{2}\sum_{i=1}^{N} \delta_i^2(t) \tag{14.6}$$

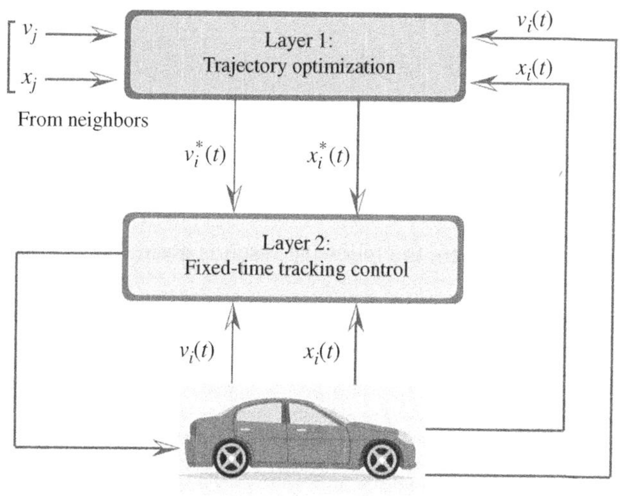

Figure 14.2 The proposed two-layer control structure for a platoon.

$$\text{s.t.}\begin{cases} \lim_{t\to\infty} |v_i(t) - v_0(t)| = 0 \\ \lim_{t\to\infty} |\delta_i(t)| = 0 \end{cases} \tag{14.7}$$

As mentioned earlier, string stability is critical to vehicle platooning, which is defined here.

Definition 14.3 [10] (String stability) The origins of the vehicular tracking error systems are defined to be string stable, if given any $\Delta > 0$. There is $\partial > 0$ such that $\|e_i(0)\| < \partial \Rightarrow$ $\sup_i \|e_i(\cdot)\|_\infty < \Delta$.

To sum up, this chapter mainly focuses on a layered control framework as shown in Figure 14.2 which meets the following requirements:

1) The first layer proposes a set of reference position $x_i^*(t)$ and velocity $v_i^*(t)$ to satisfy these control objectives proposed in (14.6) and (14.7) by using the measured position and velocity.
2) The lower layer designs a fixed-time terminal sliding mode tracking controller to track $x_i^*(t)$ and $v_i^*(t)$, which not only achieve stability of each vehicle but also achieve string stability of the whole platoon.

14.3 Vehicles Trajectory Optimization

First, we construct the following reference trajectories of the position and speed for the vehicle follower i.

$$\begin{aligned} \dot{x}_i^*(t) &= v_i^*(t) \\ \dot{v}_i^*(t) &= u_{i1}(t) \end{aligned} \tag{14.8}$$

In what follows, we design control laws $u_{i1}(t)$, which can make $x_i^*(t)$ and $v_i^*(t)$ determined by (14.8), the effective trajectory for the optimization problem proposed in (14.6) and (14.7).

To this end, the control law of the first layer is designed as

$$\begin{aligned} u_{i1}(t) = & -\sum_{j=1}^{N} a_{ij}\left(\mu\left(x_i(t)-x_j(t)-d_{i,j}\right)+\partial\left(v_i(t)-v_j(t)\right)\right) \\ & -\beta h\left(\sum_{j=0}^{N} a_{ij}\left(\lambda\left(x_i(t)-x_j(t)-d_{i,j}\right)-\zeta\left(v_i(t)-v_j(t)\right)\right)\right)\phi_i(t) \end{aligned} \quad (14.9)$$

$$\phi_i(t) = -\left(x_i(t)-x_{i-1}(t)-d_{i,j}\right)-\left(v_i(t)-v_{i-1}(t)\right) \quad (14.10)$$

where μ, ∂, β, λ, and ζ are controller parameters. Then, the following result is given.

Theorem 14.1 For heterogeneous platoon system with external disturbance, if the controller parameters μ, ∂, β, λ, and ζ are all positive, and $a(L) > \max\left\{\frac{\lambda}{\partial\zeta}, \frac{\sigma_{\max}(L)\overline{\xi}}{\beta\underline{\xi}}\|\Phi\|\right\}$ holds with $\underline{\xi} = \min\{\xi_i\}$ and $\overline{\xi} = \max\{\xi_i\}$, then the system in (14.8) with controller (14.9) and (14.10) can solve the optimization problem in (14.6) and (14.7) by generating the reference trajectory $x_i^*(t)$ and $v_i^*(t)$ for the group of vehicles.

Proof: Define the following reference position spacing errors and reference velocity spacing errors as:

$$\begin{aligned} \hat{x}_i^*(t) &= x_i^*(t)-x_0(t)-d_{i,0} \\ \hat{v}_i^*(t) &= v_i^*(t)-v_0(t) \end{aligned} \quad (14.11)$$

by which we can have that

$$\begin{aligned} \hat{x}_i^*(t)-\hat{x}_j^*(t) &= x_i^*(t)-x_j^*(t)-d_{i,j} \\ \hat{v}_i^*(t)-\hat{v}_j^*(t) &= v_i^*(t)-v_j^*(t) \end{aligned} \quad (14.12)$$

Based on (14.11) and (14.12), Equation (14.11) can be equivalently written as:

$$\begin{aligned} \dot{\hat{x}}_i^*(t) &= \hat{v}_i^*(t) \\ \dot{\hat{v}}_i^*(t) &= u_{i1}(t) \end{aligned} \quad (14.13)$$

Then, Equations (14.9) and (14.10) can be given by:

$$\begin{aligned} u_{i1} = & \ \phi_i(t)-\sum_{j=0}^{N} a_{ij}\left(\mu\left(\hat{x}_i^*(t)-\hat{x}_j^*(t)\right)+\partial\left(\hat{v}_i^*(t)-\hat{v}_j^*(t)\right)\right) \\ & -\beta h\left(\sum_{j=0}^{N} a_{ij}\left(\lambda\left(\hat{x}_i^*(t)-\hat{x}_j^*(t)\right)-\zeta\left(\hat{v}_i^*(t)-\hat{v}_j^*(t)\right)\right)\right) \end{aligned} \quad (14.14)$$

$$\phi_i(t) = -\left(\hat{x}_i^*(t)-\hat{x}_{i-1}^*(t)\right)-\left(\hat{v}_i^*(t)-\hat{v}_{i-1}^*(t)\right) \quad (14.15)$$

Now, according to this analysis, the optimization problem in this part can be equivalently written as follows:

$$\min_{\hat{x}_i^*(t)} J = \frac{1}{2}\sum_{i=1}^{N}\left[\hat{x}_{i-1}^*(t)-\hat{x}_i^*(t)\right]^2 \quad (14.16)$$

$$\text{s.t.}\begin{cases} \hat{x}_i^*(t) = \hat{x}_j^*(t) \\ \hat{v}_i^*(t) = \hat{v}_j^*(t) \end{cases} \quad (14.17)$$

Next, we will prove that the $\hat{x}_i^*(t)$ and $\hat{v}_i^*(t)$ can achieve consistency as $t \to \infty$.

Let $\Phi(t) = (I_N - 1_N\xi^{\mathrm{T}})\phi(t)$, $\phi(t) = \left(\phi_1^{\mathrm{T}}(t), \phi_2^{\mathrm{T}}(t), \cdots, \phi_N^{\mathrm{T}}(t)\right)^{\mathrm{T}}$, $\overline{X}(t) = \left[\hat{x}_1^*(t), \hat{x}_2^*(t)\cdots, \hat{x}_N^*(t)\right]^T$, $\overline{V}(t) = \left[\hat{v}_1^*(t), \hat{v}_2^*(t)\cdots, \hat{v}_N^*(t)\right]^T$.

Substituting (14.14) into (14.13), the closed-loop platoon system can be written in the following compact form:

$$\begin{aligned}\dot{\overline{X}}(t) &= \overline{V}(t)\\ \dot{\overline{V}}(t) &= -L\left(\mu\overline{X}(t) + \partial\overline{V}(t)\right) + \Phi(t) + \beta h\left[-L\left(\lambda\overline{X}(t) + \zeta\overline{V}(t)\right)\right]\end{aligned} \tag{14.18}$$

In order to simplify system (14.18), let us define:

$$\begin{cases}\hat{X}(t) = \Pi\overline{X}(t)\\ \hat{V}(t) = \Pi\overline{V}(t)\end{cases} \tag{14.19}$$

where $\Pi \triangleq I_N - (1/N)1_N 1_N^T$. Note that $\hat{x}_i^*(t) = \hat{x}_j^*(t)$, $\hat{v}_i^*(t) = \hat{v}_j^*(t)$ if and only if $\hat{X}(t) = 0$, $\hat{V}(t) = 0$. Utilizing the above coordinate transformation, the system (14.18) is converted to:

$$\begin{aligned}\dot{\hat{X}}(t) &= \hat{V}(t)\\ \dot{\hat{V}}(t) &= -L\left(\mu\hat{X}(t) + \partial\hat{V}(t)\right) + \Pi\Phi(t) + \beta h\left[-L\left(\lambda\hat{X}(t) + \zeta\hat{V}(t)\right)\right]\end{aligned} \tag{14.20}$$

Let $y = \left(\hat{X}^{\mathrm{T}}(t), \hat{V}^{\mathrm{T}}(t)\right)^{\mathrm{T}}$, one has $\dot{y}(t) = \Psi + Ay(t)$, where

$\Psi = \begin{pmatrix} 0_N \\ \Pi\Phi(t) + \beta h\left[-L\left(\gamma\hat{X}(t) + \zeta\hat{V}(t)\right)\right]\end{pmatrix}$ and $A = \begin{pmatrix} 0_N & I_N \\ -\mu L & -\partial L\end{pmatrix}$.

Consider the following candidate Lyapunov function

$$W(t) = \frac{1}{2}y^T(t)Py(t) \tag{14.21}$$

where $P = \begin{pmatrix} (\partial\lambda + \mu\zeta)\hat{L} & \lambda\Xi \\ \lambda\Xi & \zeta\Xi\end{pmatrix}$, $\hat{L} = \frac{1}{2}\left(\Xi L + L^{\mathrm{T}}\Xi\right)$, $\Xi = \mathrm{diag}(\xi_1, \xi_2, \ldots, \xi_N)$, ξ is the positive left eigenvector of the matrix related with zero eigenvalues, and meet $\xi^{\mathrm{T}}1_N = 1$. According to Definition 14.1, we can obtain that:

$$\begin{aligned}W(t) &= \frac{\partial\lambda + \mu\zeta}{2}\hat{X}^{\mathrm{T}}(t)\hat{L}\hat{X}(t) + \lambda\hat{X}^{\mathrm{T}}(t)\Xi\hat{V}(t) + \frac{\zeta}{2}\hat{V}^{\mathrm{T}}(t)\Xi\hat{V}(t)\\ &\geq \frac{1}{2}y^{\mathrm{T}}(t)\left(\hat{P}\otimes\Xi\right)y(t)\end{aligned} \tag{14.22}$$

where $\hat{P} = \begin{pmatrix} ((\partial\lambda + \mu\zeta)a(L)) & \lambda \\ \lambda & \zeta\end{pmatrix}$, by Schur Complement [11], invoking $P \geq \hat{P}\otimes\Xi$, we can find $W(t) \geq 0$, since $\hat{X}^{\mathrm{T}}(t)\Xi L\hat{X}(t) = \hat{X}^{\mathrm{T}}(t)L^{\mathrm{T}}\Xi\hat{X}(t) = \hat{X}^{\mathrm{T}}(t)\hat{L}\hat{X}(t)$, one has $(\partial\lambda + \mu\zeta)\hat{X}^{\mathrm{T}}(t)\hat{L}\hat{X}(t) = \partial\lambda\hat{X}^{\mathrm{T}}(t)\Xi L\hat{X}(t) + \mu\zeta\hat{X}^{\mathrm{T}}(t)L^{\mathrm{T}}\Xi\hat{X}(t)$, the Lyapunov function (14.21) can be transformed as:

$$W(t) = \frac{1}{2}y^T(t)\tilde{P}y(t) \tag{14.23}$$

where $\tilde{P} = \begin{pmatrix} \partial\lambda\Xi L + \mu\zeta L^T\Xi & \lambda\Xi \\ \lambda\Xi & \zeta\Xi\end{pmatrix}$.

By applying $\hat{L} = (1/2)\left(\Xi L + L^T\Xi\right)$, we can have:

$$\frac{1}{2}\left(\tilde{P}A + A^T\tilde{P}^T\right) = \begin{pmatrix} -\lambda\mu\hat{L} & O_N \\ O_N & \lambda\Xi - \partial\zeta\hat{L}\end{pmatrix} \tag{14.24}$$

where

$$\begin{aligned}\tilde{P}A &= \begin{pmatrix} \partial\lambda\Xi L + \mu\zeta L^T\Xi & \lambda\Xi \\ \lambda\Xi & \zeta\Xi \end{pmatrix}\begin{pmatrix} O_N & I_N \\ -\mu L & -\partial L \end{pmatrix} \\ &= \begin{pmatrix} -\lambda\mu\Xi L & \mu\zeta L^T\Xi \\ -\mu\zeta\Xi L & \lambda\Xi - \partial\zeta\Xi L \end{pmatrix}\end{aligned} \tag{14.25}$$

The time derivative of $W(t)$ should satisfy that

$$\begin{aligned}\dot{W}(t) &= y^T(t)\tilde{P}[Ay(t) + \Psi] \\ &= \frac{1}{2}y^T(t)\left(\tilde{P}A + A^T\tilde{P}^T\right)y(t) + y^T(t)\tilde{P}\Psi \\ &\leq -\lambda\mu a(L)\hat{X}^{\mathrm{T}}(t)\Xi\hat{X}(t) - (\partial\zeta a(L) - \lambda)\Xi\hat{V}^{\mathrm{T}}(t) \\ &\quad + \left(\lambda\hat{X}(t) + \zeta\hat{V}(t)\right)^T\Xi \times \left[\Phi + \beta h(-L(\lambda\hat{X}(t) + \zeta\hat{V}(t))\right] \\ &= -V^{\mathrm{T}}(t)(P_1\otimes\Xi)\hat{V}(t) + G\end{aligned} \tag{14.26}$$

where $P_1 = \begin{pmatrix} \lambda\mu a(L) & 0 \\ 0 & \partial\zeta a(L) - \lambda \end{pmatrix} > 0,$

$$G = \beta\left(\lambda\hat{X}^{\mathrm{T}}(t) + \zeta\hat{V}^{\mathrm{T}}(t)\right)^T\Xi \times h(-L)\left(\lambda\hat{X}^{\mathrm{T}}(t) + \zeta\hat{V}^{\mathrm{T}}(t)\right) + \left(\lambda\hat{X}^{\mathrm{T}}(t) + \zeta\hat{V}^{\mathrm{T}}(t)\right)^T\Xi\Phi.$$

Let $z = \left(z_1^{\mathrm{T}}, z_2^{\mathrm{T}}, \cdots z_N^{\mathrm{T}}\right) = \lambda\hat{X}^{\mathrm{T}}(t) + \zeta\hat{V}^{\mathrm{T}}(t)$, so

$$z^{\mathrm{T}}\xi = \lambda\hat{X}^{\mathrm{T}}(t)\xi + \zeta\hat{V}^{\mathrm{T}}(t)\xi = 0, z^{\mathrm{T}}\hat{L}z \geq a(L)z^{\mathrm{T}}\Xi z.$$

We can obtain that:

$$\|Lz\| \leq \sqrt{z^T(L^TL)z} \leq \sigma_{\max}(L)\|z\| \tag{14.27}$$

Based on these discussions, we have:

$$\begin{aligned}G &= -\beta z^T\Xi hLz + z^T\Xi\Phi \\ &= -\beta\frac{z^T\Xi Lz}{\|Lz\|} + z^T\Xi\Phi \leq -\beta a(L)\frac{z^T\Xi z}{\|Lz\|} + \overline{\xi}\|z\|\|\Phi\| \\ &= -\left(\frac{\beta a(L)\xi}{\sigma_{\max}(L)} - \overline{\xi}\|\Phi\|\right)\|z\|\end{aligned} \tag{14.28}$$

It is implied that if $a(L) > \left((\sigma_{\max}(L)\overline{\xi})/(\beta\underline{\xi})\right)\|\Phi\|$, one has $G \leq 0$, by which we can have that:

$$\dot{W}(t) \leq -y^T(t)(P_1\otimes\Xi)y(t) \tag{14.29}$$

Based on $\hat{X}(t)\hat{L}\hat{X}(t) \leq b(L)\hat{X}^{\mathrm{T}}(t)\Xi\hat{X}(t)$, we can obtain:

$$W(t) \leq \frac{1}{2}y^T(t)(P_2\otimes\Xi)y(t) \tag{14.30}$$

where $P_2 = \begin{pmatrix} (\partial\lambda + \mu\zeta)b(L) & \lambda \\ \lambda & \zeta \end{pmatrix} > 0$, and by using Lemma 14.2, we can obtain:

$$\dot{W}(t) \leq -\lambda_{\min}(P_1)\underline{\xi}y^T(t)y(t) = -\delta_1 y^T(t)y(t) \tag{14.31}$$

$$W(t) \leq \lambda_{\max}(P_2)\bar{\xi}y^T(t)y(t) = \delta_2 y^T(t)y(t) \tag{14.32}$$

From (14.31) and (14.32), it follows that

$$W(t) \leq -\Theta_1 y^T(t)y(t) \leq -\frac{\Theta_1}{\Theta_2}W(t) \tag{14.33}$$

where $\Theta_1 = \min\{\lambda\mu a(L), (\partial\zeta a(L) - \lambda)\}$ and $\Theta_2 = \left(\sigma + \zeta + \sqrt{(\sigma-\zeta)^2 + 4\lambda^2}\right)\bar{\xi}/4, \sigma = (\partial\lambda + \mu\zeta)b(L)$. Furthermore, the following inequality holds:

$$W(t) \leq W(0)\exp(-\Theta_1 t/\Theta_2) \rightarrow 0, t \rightarrow \infty \tag{14.34}$$

which implies that $\hat{x}_i^*(t) = \hat{x}_j^*(t)$, $\hat{v}_i^*(t) = \hat{v}_j^*(t)$. We next prove that the goal (14.6) can be achieved. Choose the following Lyapunov function:

$$\tilde{W}(t) = \frac{1}{2}\left[\sum_{i=1}^{N}\xi_i 2\left(\hat{x}_i^* - \hat{x}_{i-1}^*\right)\right]^2 + \frac{1}{2}\left[\sum_{i=1}^{N}\xi_i\left(\hat{v}_i^* + \hat{x}_i^* - \hat{x}_{i-1}^*\right)\right]^2 \tag{14.35}$$

Taking the derivative $\tilde{W}(t)$ yields

$$\begin{aligned}
\dot{\tilde{W}}(t) &= \left[\sum_{i=1}^{N} 2\xi_i\left(\hat{x}_i^* - \hat{x}_{i-1}^*\right)\right]\left(\sum_{i=1}^{N} 2\xi_i\dot{\hat{x}}_i^*\right) + \left[\sum_{i=1}^{N}\xi_i\left(\hat{v}_i^* + \hat{x}_i^* - \hat{x}_{i-1}^*\right)\right]\left[\sum_{i=1}^{N}\xi_i\left(\dot{\hat{v}}_i^* + \hat{v}_i^* - \hat{v}_{i-1}^*\right)\right] \\
&= \left[\sum_{i=1}^{N} 2\xi_i\left(\hat{x}_i^* - \hat{x}_{i-1}^*\right)\right]\left(\sum_{i=1}^{N} 2\xi_i\hat{v}_i^*\right) + \left[\sum_{i=1}^{N}\xi_i\left(\hat{v}_i^* + \hat{x}_i^* - \hat{x}_{i-1}^*\right)\right]\left[\sum_{i=1}^{N}\xi_i\left(\phi_i + \hat{v}_i^* - \hat{v}_{i-1}^*\right)\right] \\
&= \left[\sum_{i=1}^{N} 2\xi_i\left(\hat{x}_i^* - \hat{x}_{i-1}^*\right)\right]\left(\sum_{i=1}^{N} 2\xi_i\hat{v}_i^*\right) + \left[\sum_{i=1}^{N}\xi_i\left(\hat{v}_i^* + \hat{x}_i^* - \hat{x}_{i-1}^*\right)\right]\left[\sum_{i=1}^{N} 4\xi_i\left(\hat{x}_i^* - \hat{x}_{i-1}^*\right)\right] \\
&= -\left(\sum_{i=1}^{N} 2\xi_i\left(\hat{x}_i^* - \hat{x}_{i-1}^*\right)\right)^2
\end{aligned}$$

Therefore, $\tilde{W}(t) \leq 0$ if and only if $\sum_{i=1}^{N} 2\xi_i\left(\hat{x}_i^* - \hat{x}_{i-1}^*\right) = 0$. Based on Barbalat's lemma [12], it is concluded that $\sum_{i=1}^{N} 2\xi_i\left(\hat{x}_i^* - \hat{x}_{i-1}^*\right)$ will converge to zero asymptomatically. In addition, the cost function (14.16) can be reduced for any i and satisfies $\hat{x}_i^* - \hat{x}_{i-1}^*$, $\hat{v}_i^* \rightarrow \hat{v}_{i-1}^*$. Therefore, for the optimization problem defined in (14.6), we can conclude that x_i^* and v_i^* are the unique solutions of the optimal trajectory. This completes the proof.

Remark 14.1 Compared with [1], a novel distributed trajectory optimization algorithm is proposed in this chapter to minimize the intervehicle spacing errors, which largely reduced the computation complexity of the algorithm.

14.4 Fixed-Time Tracking Control Design

This section aims to design a tracking controller for each vehicle to track the reference trajectories of the position and velocity generated by the first layer. As shown in Figure 14.3, a continuous sliding mode controller is introduced to achieve trajectory tracking and reject disturbance in fixed time with zero steady-state spacing errors.

Define the following tracking errors

$$\begin{aligned}
e_{1i} &= x_i - x_i^* \\
e_{2i} &= v_i - v_i^* \\
e_{ei} &= a_i - u_{i1}
\end{aligned} \tag{14.36}$$

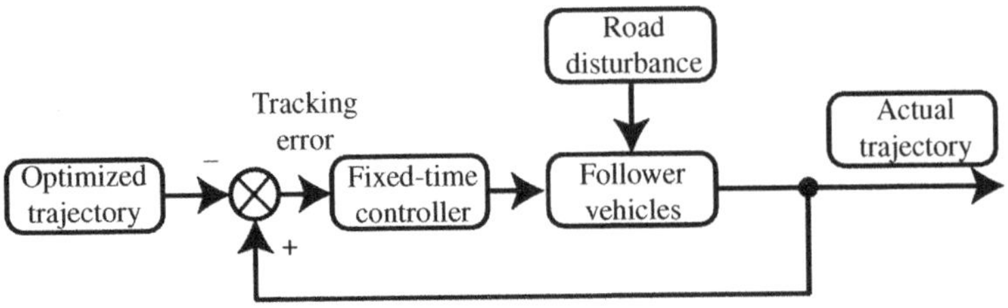

Figure 14.3 Fixed-time tracking control block diagram.

According to [13, 14], it can be obtained

$$\begin{aligned} \dot{e}_{1i} &= e_{2i} \\ \dot{e}_{2i} &= e_{3i} \\ \dot{e}_{3i} &= \frac{1}{\tau_i} u_i + \mu_i - \dot{u}_{i1} \end{aligned} \tag{14.37}$$

The design of a finite-time tracking controller based on terminal sliding mode is presented next. Introduce the following terminal sliding mode surface:

$$s_i = \dot{e}_{3i} + \left(c_1 \lceil e_{1i} \rfloor^{\eta_1} + c_2 \lceil e_{2i} \rfloor^{\eta_2} + c_3 \lceil e_{3i} \rfloor^{\eta_3}\right) + \left(k_1 \lceil e_{1i} \rfloor^{\theta_1} + k_2 \lceil e_{2i} \rfloor^{\theta_2} + k_3 \lceil e_{3i} \rfloor^{\theta_3}\right) \tag{14.38}$$

where c_i, k_i, η_i, θ_i, and $i = 1, 2, 3$ are positive constants. This surface is fixed-time convergent, provided that the following conditions are satisfied [15]:

1) $s^3 + c_3 s^2 + c_2 s + c_1$ is Hurwitz,
2) $s^3 + k_3 s^2 + k_2 s + b_1$ is Hurwitz,
3) $\theta_{i-1} = \dfrac{\theta_i \theta_{i+1}}{2\theta_{i+1} - \theta_i}$, $\eta_{i-1} = \dfrac{\eta_i \eta_{i+1}}{2\eta_{i+1} - \eta_i}$, $i = 2, 3$

where $\theta_4 = 1, 0 < \theta_3 < 1$, and $\eta_4 = 1, \eta_3 > 1$. Note here that $\eta_i \in (0, 1)$, $\theta_i \in (1, 1 + m)$, where $m > 0$ is sufficiently small.

For $s_i = 0$, we can easily get the dynamics of system (14.37) during sliding procedure

$$\begin{aligned} \dot{e}_{1i} &= e_{2i} \\ \dot{e}_{2i} &= e_{3i} \\ \dot{e}_{3i} &= -\left(c_1 \lceil e_{1i} \rfloor^{\eta_1} + c_2 \lceil e_{2i} \rfloor^{\eta_2} + c_3 \lceil e_{3i} \rfloor^{\eta_3}\right) - \left(k_1 \lceil e_{1i} \rfloor^{\theta_1} + k_2 \lceil e_{2i} \rfloor^{\theta_2} + k_3 \lceil e_{3i} \rfloor^{\theta_3}\right) \end{aligned} \tag{14.39}$$

Theorem 14.2 The sliding mode trajectory (14.39) can converge to the origin in a fixed time if parameters c_i, k_i, η_i, and θ_i satisfy the aforementioned constraints.

Proof: Define the following Lyapunov function candidates:

$$V_{i1} = x^{\mathrm{T}} P x \tag{14.40}$$

where $x = [e_{1i}, e_{2i}, e_{3i}]^T$. The symmetric positive definite matrix P must satisfy the Lyapunov equation as defined here:

$$PA + A^{\mathrm{T}} P = -Q \tag{14.41}$$

where Q is a positive definite symmetric matrix with A defined as:

$$A = \begin{bmatrix} 0 & 1 & 0 \\ 0 & 0 & 1 \\ -c_1 & -c_2 & -c_3 \end{bmatrix} \tag{14.42}$$

The time derivative of V_{i1} satisfies the following:

$$\dot{V}_{i1} = -x^T Q x < 0 \tag{14.43}$$

The above equality also holds true for $\eta_i \in (1 - m, 1)$, for sufficiently small $m > 0$. Moreover, the right-hand side of (14.39) is homogeneous vector field with degree [15] $n_1 = ((\eta_n - 1)/(\eta_n)) < 0$ with dilation $1/\eta_i$, where $i = 1, 2, 3$. Therefore, the following inequality holds leading to a fix-time convergence.

$$\dot{V}_1(\zeta) \leq -\Theta_1 V_1^{1+n_1}(\zeta) \tag{14.44}$$

where $\Theta_1 = (\lambda_{\min}(Q))/(\lambda_{\max}(P))$ with $\zeta = \left[e_{1i}^{\eta_1}, e_{2i}^{\eta_2}, e_{3i}^{\eta_3}\right]$. The convergence time for this differential equation can be found easily by solving it and is given by $T_1 \leq \left(\left(V_{i1}^{\delta\zeta(t_0)}\right)/(\Theta_1\delta)\right)$, where $\delta = (1 - \eta_3)/\eta_3$. By using Rayleigh inequality, we can show $T_1 \leq (\lambda_{\max}(P))/(\Theta_1\delta)$.

Now let us define another Lyapunov function for the system of (14.39) as:

$$V_{i2} = x^T P' x \tag{14.45}$$

The symmetric positive definite matrix P' must satisfy the following Lyapunov equation:

$$P'A' + A'^{\mathrm{T}} P' = -Q' \tag{14.46}$$

where

$$A' = \begin{bmatrix} 0 & 1 & 0 \\ 0 & 0 & 1 \\ -k_1 & -k_2 & -k_3 \end{bmatrix} \tag{14.47}$$

The time derivative of V_{i2} satisfies that

$$\dot{V}_{i2} = x^T Q' x < 0 \tag{14.48}$$

The proof is the same as earlier so it can be seen that V_{i2} decreases and reaches a value in a time no later than $T_2 = 1/(\Theta_2 n_2 \lambda_{\min}(P')^{n_2})$, where $n_2 = (\theta_3 - 1)/\theta_3 > 0$, $\Theta_2 = (\lambda_{\min}(Q'))/(\lambda_{\max}(P'))$ which is independent of initial conditions.

Based on the analysis given here, we now find that the total convergence time can be achieved in $T_{\max} \leq T_1 + T_2$. This completes the proof.

Remark 14.2 The existing methods [16, 17] can only guarantee convergence of the spacing errors to zero in finite time, but gives no hints regarding the settling time. The proposed method in this chapter gives a specific hint of the convergence time.

Theorem 14.3 For the tracking error system as described by (14.37), the terminal sliding mode-based control law is given here, which can enable the tracking error converge to zero within the fixed time that is independent of any initial condition.

$$
\begin{aligned}
u_i &= \tau_i\left(u_{i,eq} + u_{i,n}\right) \\
u_{i,eq} &= -\left(c_1\lceil e_{1i}\rfloor^{\eta_1} + c_2\lceil e_{2i}\rfloor^{\eta_2} + c_3\lceil e_{3i}\rfloor^{\eta_3}\right) - \left(k_1\lceil e_{1i}\rfloor^{\theta_1} + k_2\lceil e_{2i}\rfloor^{\theta_2} + k_3\lceil e_{3i}\rfloor^{\theta_3}\right) \\
u_{i,n} &= -\left(k_d + T\right)\operatorname{sgn}(s_i) - \varpi\lceil s_i\rfloor^{\vartheta} - \kappa\lceil s_i\rfloor^{\varepsilon}
\end{aligned}
\tag{14.49}
$$

where controller parameters θ_i, $\eta_i (i = 1, 2, 3)$ and k_d have been given previously, $T > 0$ is an adjustable constant, ϖ, κ are all positive constants, $\vartheta > 1$, and $0 < \varepsilon < 1$.

Proof: Substituting (14.49) into (14.38) yields,

$$
\begin{aligned}
s_i &= \dot{e}_{3i} + \left(c_1\lceil e_{1i}\rfloor^{\eta_1} + c_2\lceil e_{2i}\rfloor^{\eta_2} + c_3\lceil e_{3i}\rfloor^{\eta_3}\right) + \left(k_1\lceil e_{1i}\rfloor^{\theta_1} + k_2\lceil e_{2i}\rfloor^{\theta_2} + k_3\lceil e_{3i}\rfloor^{\theta_3}\right) \\
&= \mu_i - u_{i1} + \frac{1}{\tau_i}u_i + \left(c_1\lceil e_{1i}\rfloor^{\eta_1} + c_2\lceil e_{2i}\rfloor^{\eta_2} + c_3\lceil e_{3i}\rfloor^{\eta_3}\right) + \left(k_1\lceil e_{1i}\rfloor^{\theta_1} + k_2\lceil e_{2i}\rfloor^{\theta_2} + k_3\lceil e_{3i}\rfloor^{\theta_3}\right)
\end{aligned}
\tag{14.50}
$$

Involving (14.49), it is clear that

$$
s_i = \mu_i - u_{i1} + u_{i,eq} + u_{i,n} + \left(c_1\lceil e_{1i}\rfloor^{\eta_1} + c_2\lceil e_{2i}\rfloor^{\eta_2} + c_3\lceil e_{3i}\rfloor^{\eta_3}\right) + \left(k_1\lceil e_{1i}\rfloor^{\theta_1} + k_2\lceil e_{2i}\rfloor^{\theta_2} + k_3\lceil e_{3i}\rfloor^{\theta_3}\right)
\tag{14.51}
$$

whose time derivative can be obtained as:

$$
\dot{s}_i = \dot{\mu}_i + \dot{u}_n = \dot{\mu}_i - (k_d + \mathrm{T})\operatorname{sgn}(s_i) - \varpi\lceil s_i\rfloor^{\vartheta} - \kappa\lceil s_i\rfloor^{\varepsilon}
\tag{14.52}
$$

Consider the following Lyapunov function:

$$
V_{S,i}(s_i) = \frac{1}{2}s_i^2
\tag{14.53}
$$

The time derivative of $V_{S,i}(s_i)$ can be obtained:

$$
\begin{aligned}
\dot{V}_{S,i}(s_i) &= \left(s_i\dot{\mu}_i - (k_d + \mathrm{T})|s_i|\right) - \varpi s_i\lceil s_i\rfloor^{\vartheta} - \kappa s_i\lceil s_i\rfloor^{\varepsilon} \\
&\le -T|s_i| - \varpi|s_i|^{\vartheta+1} - \kappa|s_i|^{\varepsilon+1} \le -\varpi V_{S,i}^{\frac{\vartheta+1}{2}} - \kappa V_{S,i}^{\frac{\varepsilon+1}{2}}
\end{aligned}
\tag{14.54}
$$

Since $\vartheta > 1$, $\varepsilon \in (0, 1)$ we have $\dfrac{\vartheta+1}{2} > 1$, $\dfrac{\varepsilon+1}{2} \in (0, 1)$. From Lemma 14.3, the tracking error of vehicular platoon can arrive $s_i = 0$ with the bound in the convergence time is given as T_r, and $T_r < T_r^*(\varpi, \kappa, \vartheta, \varepsilon) = (2)/(\varpi(\vartheta - 1)) + 2/(\kappa(1 - \varepsilon))$. Combined with the above two results, the entire convergence time is:

$$
T \le T_1 + T_2 + T_r
\tag{14.55}
$$

Now, the string stability is conducted based on Definition 14.3. Choose the following global Lyapunov function candidate:

$$
V(s_i) = \sum_{i=1}^{N} V_{S,i}(s_i)
\tag{14.56}
$$

We can obtain $\dot{V}$ as follows:

$$
\dot{V} \le -\varpi\sum_{i=1}^{N}\left(V_{S,i}^{(\vartheta+1)/2}\right) - \kappa\sum_{i=1}^{N}\left(V_{S,i}^{(\vartheta+1)/2}\right) \le 0
\tag{14.57}
$$

Form (14.36), one has

$$
x_i = x_i^* + e_{1i}
\tag{14.58}
$$

The spacing error can be equivalently transformed by the following equation:

$$\delta_i(t) = x^*_{i-1} - x^*_i + e_{1i-1} - e_{1i} - d_{i-1,i} \tag{14.59}$$

Based on the results of Theorems 14.1 and 14.3, we can find this indicates that δ_i is uniformly ultimately bounded by choosing the appropriate parameters. By Definition 14.3, for the platoon system, string stability can be ensured.

Remark 14.3 Unlike the conventional finite-time sliding control methods whose convergence time depends on the initial state of vehicles, the terminal sliding mode control method here can always make the tracking errors of the collection of vehicles converge to zero within a bounded fixed time irrespective of the initial conditions.

Remark 14.4 The simple sliding mode surface and reaching law used in [1, 16] are subject to poor convergence and anti-chattering performance that is not good for string stability. However, a new terminal sliding mode control law is introduced, resulting in significantly better platoon control performance and anti-chattering capacity, as well as it also can ensure string stability.

14.5 Numerical Simulations

In this section, the proposed speed planning and tracking control algorithms are evaluated by MATLAB. Consider a heterogeneous platoon of a leader and four follower vehicles, wherein the vehicle length and safety distance $d_{i-1,i}$ of each vehicle are different, and the engine time constant τ_i is set to be different, whose parameters are listed in Table 14.1.

In order to better evaluate the control performance of the platoon system, the following cost functions are defined:

$$J_1(t) = \lim_{T\to\infty} \int_0^T \sum_{i=1}^{4} \left(x_i(t) - x_i^*(t)\right)^2 dt$$

$$J_2(t) = \lim_{T\to\infty} \int_0^T \sum_{i=1}^{4} (x_{i-1}(t) - x_i(t) - d_{i,i-1})^2 dt$$

First, the initial positions are set as follows: $x(0) = [0, -5, -11, -15, -20]$ m, the acceleration of the leader is set as 0, the velocity trajectories of the leader is given in Figure 14.4, which is usually considered the practical situation. Let the disturbance $\mu_i = \sin(t)$. This section considers the case for a platoon with nearest neighbor (NN) and two-NN (2NN) communication topologies that are given in Figure 14.5. For NN, the information between adjacent vehicles is able to be transmitted through onboard sensors. Table 14.2 gives a description of parameters and appropriate controller and sliding mode parameters are shown in Table 14.3.

Two layer control architecture of five-vehicle platoon under topologies NN and 2NN are given in Figures 14.6 and 14.7, respectively. In these figures, the optimization performance of the trajectory optimization layer is displayed in Figure 14.6a, b and 14.7a, b. From these results, one can observe that the first layer can guarantee better accuracy of the spacing errors and strengthen string stability. Results of the new terminal sliding surface tracking controller shown in Figure 14.6c, d and

Table 14.1 Model parameters.

$d_{0,1}$	$d_{1,2}$	$d_{2,3}$	$d_{3,4}$	τ_1	τ_2	τ_3	τ_4
5	6	4	5	2.5	2.6	2.7	2.8

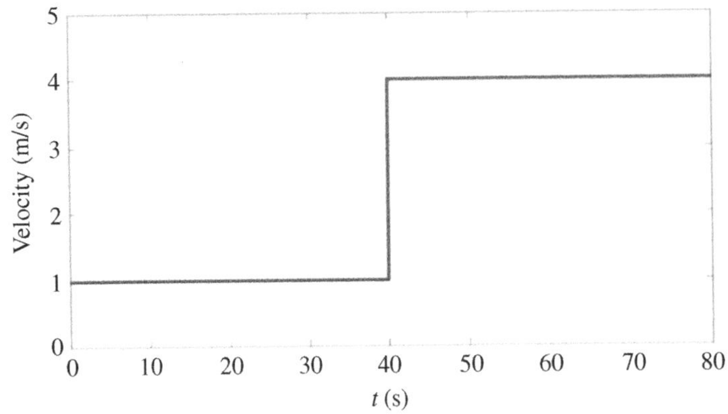

Figure 14.4 Velocity of the leader.

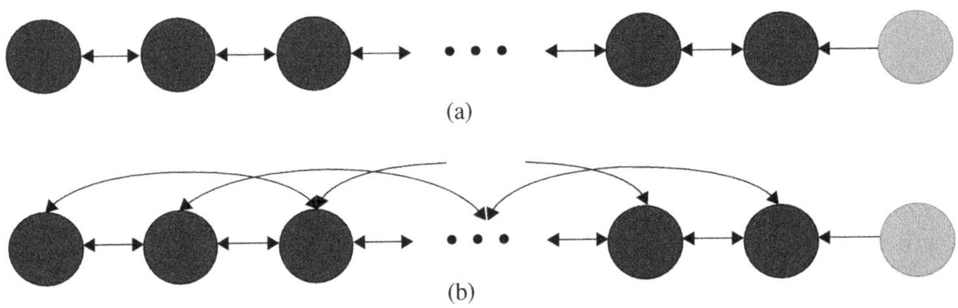

Figure 14.5 Types of bidirectional information flow topology. (a) NN. (b) 2NN.

Table 14.2 Description of parameters.

	Parameter	Description
1	d	Intervehicle spacing
2	τ	Engine time constant
3	μ	Unknow uncertainty
4	c, k, η, θ	Sliding surface coefficient
5	μ, ∂, β, λ, ζ	Distributed controller parameters
6	ϖ, κ, ϑ, ε, T	Tracking controller parameters

Table 14.3 Controller parameters.

c_1	c_2	c_3	k_1	k_2	k_3	η_1	η_2	η_3
15	66	80	15	66	80	$\frac{7}{10}$	$\frac{7}{13}$	$\frac{7}{16}$
θ_1	θ_2	θ_3	T	ϑ	ε	ϖ	κ	
$\frac{21}{20}$	$\frac{21}{19}$	$\frac{21}{18}$	2	8	0.01	2	0.5	

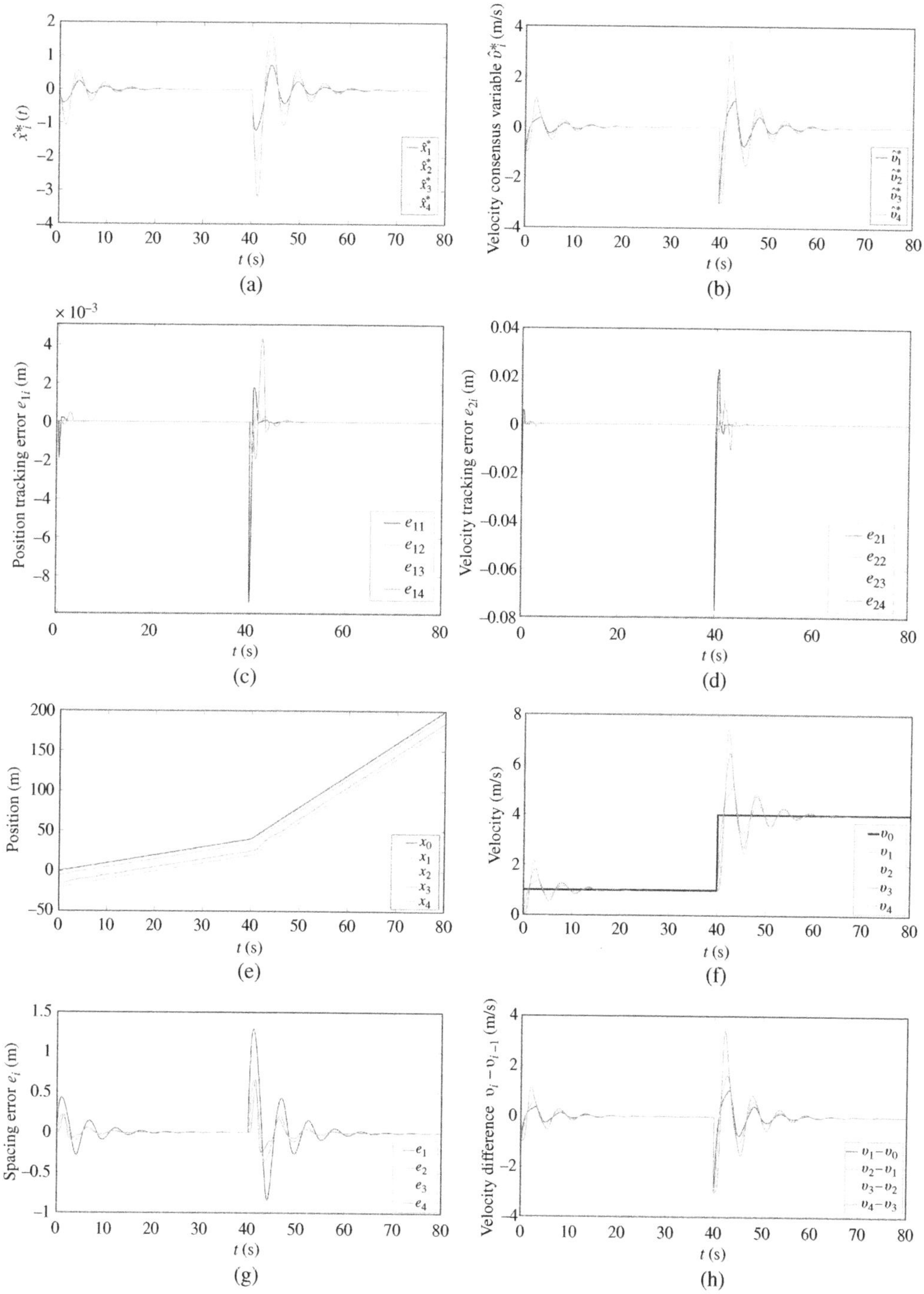

Figure 14.6 Nearest neighbor (NN) topology (a) Position consensus variable. (b) Velocity consensus variable. (c) Position tracking error. (d) Velocity tracking error. (e) Position. (f) Velocity. (g) Spacing error. (h) Velocity difference.

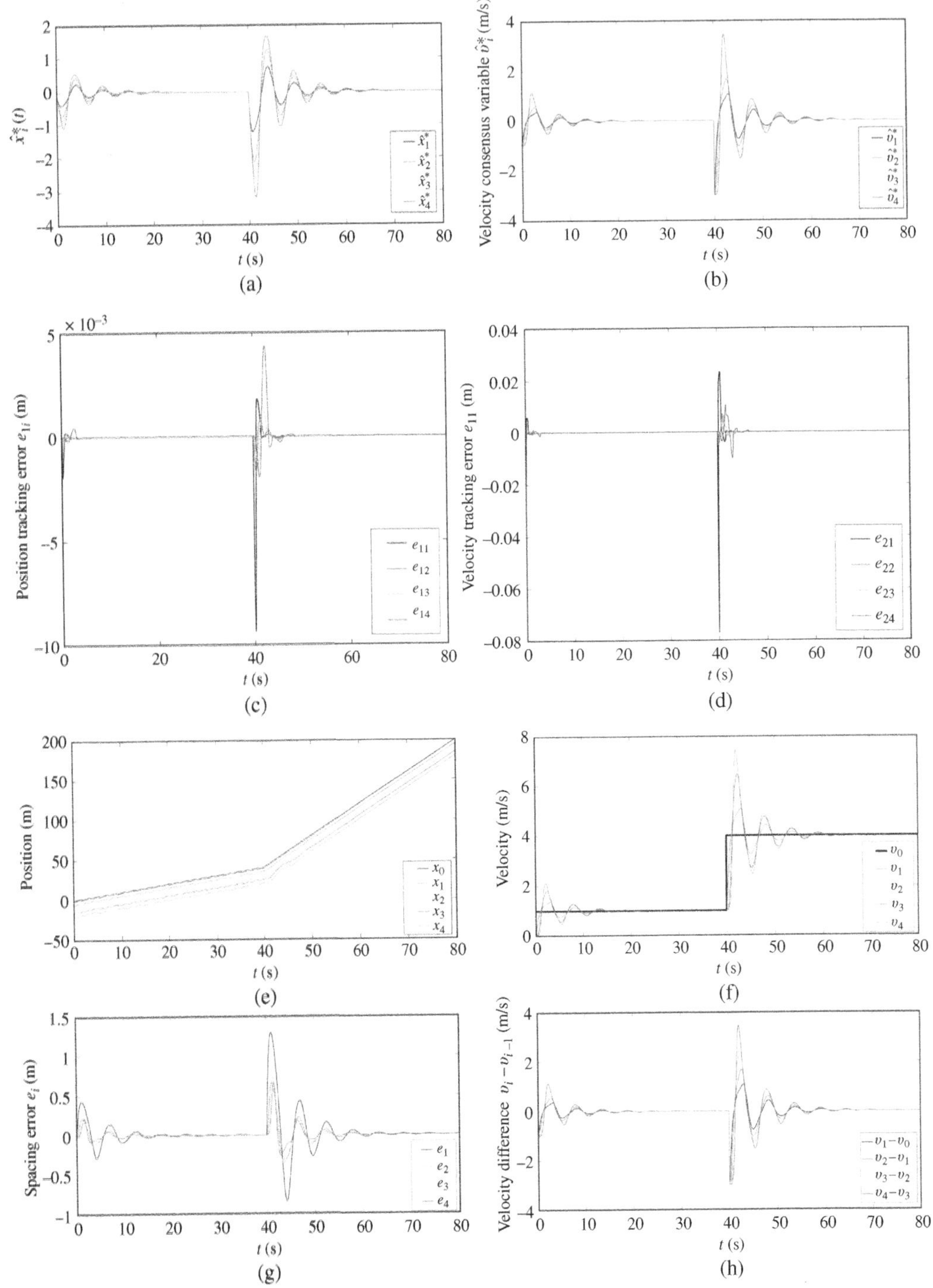

Figure 14.7 Two-NN topology. (a) Position consensus variable. (b) Velocity consensus variable. (c) Position tracking error. (d) Velocity tracking error. (e) Position. (f) Velocity. (g) Spacing error. (h) Velocity difference.

14.7c, d. Figures 14.6e, f and 14.7e, f show the position and velocity states of five vehicles. Figure 14.6g, h and 14.7g, h show that the spacing and velocity errors between neighboring vehicles can both converge to zero. By introducing a double-layer control algorithm, each following vehicle can maintain the desired safety distance and keep safe pace with the leader. At the same time, since spacing errors and velocity differences $v_i - v_{i-1}$ are not expanded in a platoon, then it can obviously guarantee platoon string stability subject to disturbance. The analysis shows the upper layer can reduce the spacing errors, moreover controller designed at the lower layer can track its optimization signal in a fixed time, at the same time, the tracking errors can converge to zero.

Figures 14.8 and 14.9 respectively show that the intervehicle spacing errors and speed tracking errors of the proposed algorithm are smaller and converge faster without obvious chattering.

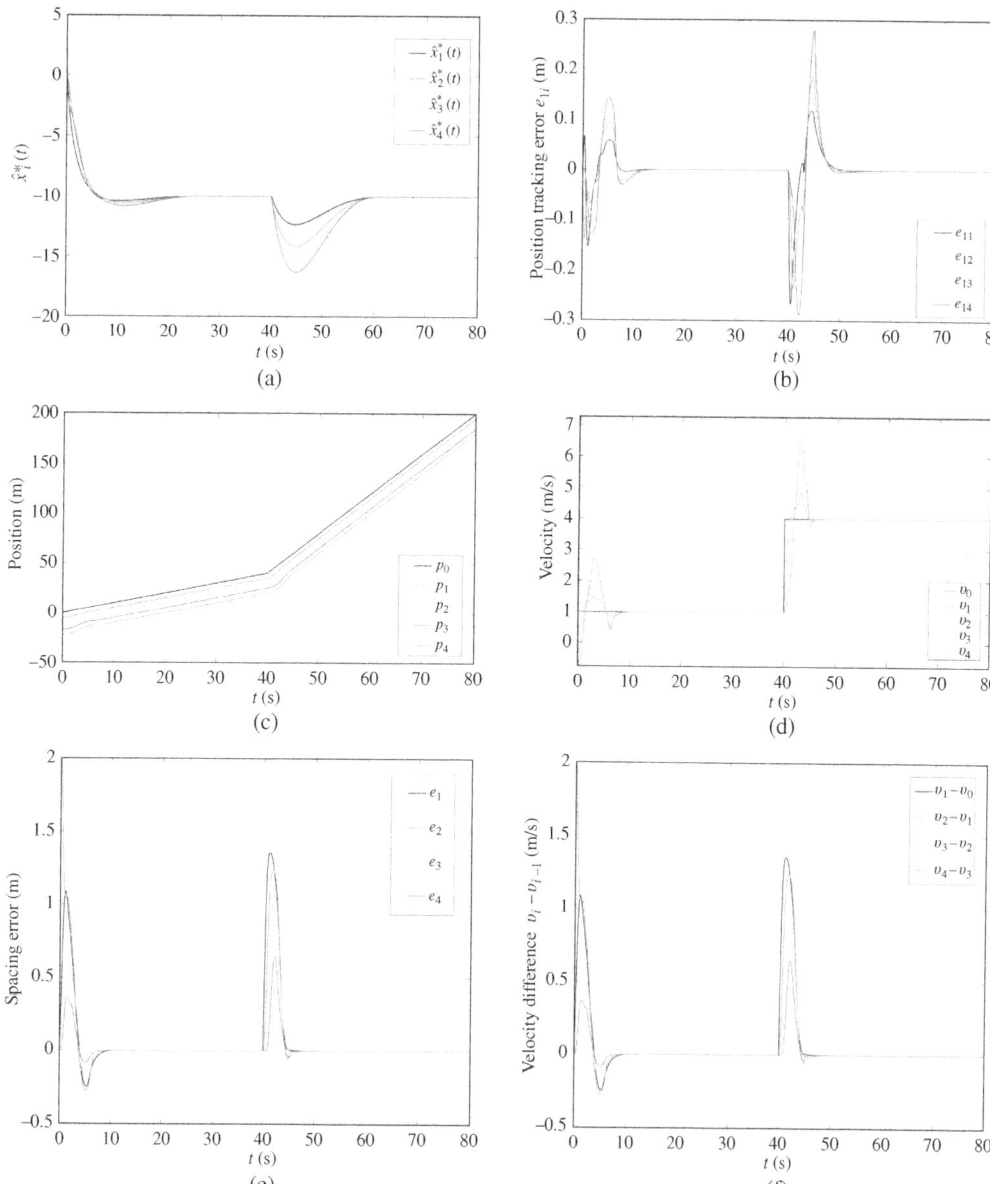

Figure 14.8 Comparison experiment [1] Nearest neighbor (NN) topology. (a) Position error variable. (b) Position tracking error. (c) Position. (d) Velocity. (e) Spacing error. (f) Velocity difference.

Firstly, compared with the distributed trajectory optimization algorithm designed in literature [1], from Figure 14.8a, b, we can obtain $J_1(t) = 24.7826$, it can be clearly shown that reference position errors and velocity errors in our paper are smaller than [1]. And at the same time, from Figure 14.8e, f, we can obtain $J_2(t) = 11.41$, compared to $J_2(t) = 6.4415$ in this chapter, we can find the interspacing errors and velocity differences between neighbors, from which, it can be found the proposed strategy is clearly much better to the method in literature [1].

At the same time, the hierarchical control method in this chapter and the literature [18] without optimization of the design of sliding mode control algorithm are compared, which adopt the same vehicle dynamics, CSP and NN communication topology structure. From Figure 14.9a, b, we can get spacing errors and velocity differences curve in [18]. We can find the superior performance of the proposed method in the trajectory optimization layer. According to the cost function defined, we can obtain $J_2(t) = 21.2658$ in [18]. Compared with an optimization algorithm of NN

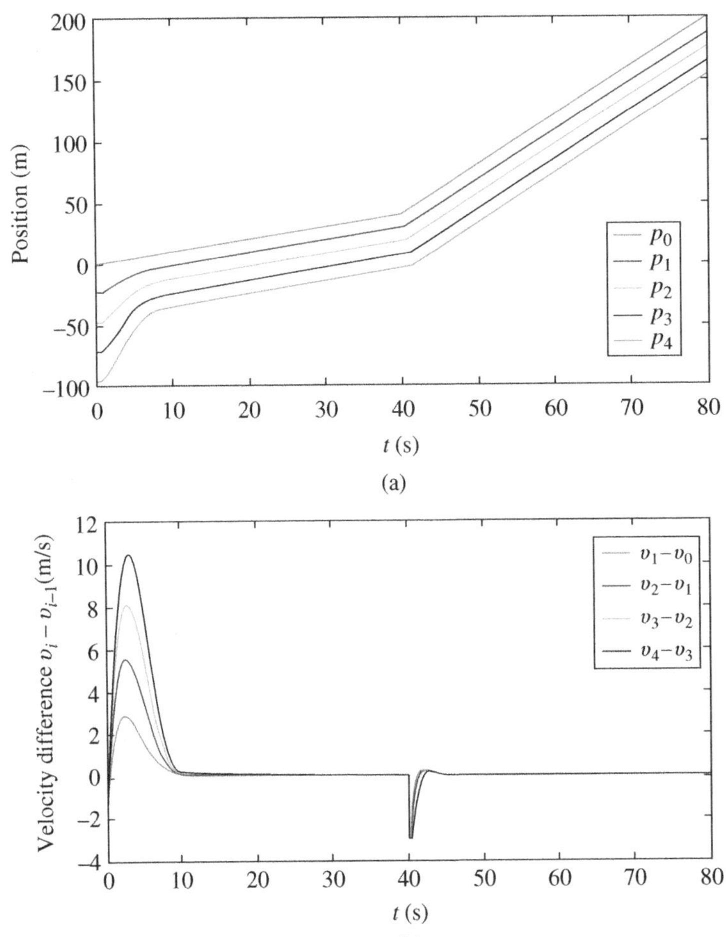

Figure 14.9 Comparison experiment [18] Nearest neighbor (NN) topology. (a) Spacing error. (b) Velocity difference.

communication topology in this chapter $J_2(t) = 6.4415$, it shows the superiority of the optimization method proposed in our paper. To sum up, the effectiveness of the proposed hierarchical control algorithm is well demonstrated, which can be shown that the intervehicle spacing errors and velocity differences are smaller and converging faster without obvious chattering with steady-state tracking performance.

14.6 Conclusions and Future Work

This chapter studied a novel distributed optimal platoon and fixed-time trajectory tracking control algorithm. The optimal trajectory for each following vehicle is derived in the context of distributed convex optimization based on a spacing error cost function. Then, a novel control law is proposed for the vehicles to track the optimal trajectory. Results show that the proposed hierarchical framework can minimize the spacing error and ensure the string stability of the platoon.

References

1 Wen, S. and Guo, G. (2021). Distributed trajectory optimization and sliding mode control of heterogenous vehicular platoons. *IEEE Trans. Intell. Transp. Syst.* 23: 1–16.

2 Rahili, S. and Ren, W. (2017). Distributed continuous-time convex optimization with time-varying cost functions. *IEEE Trans. Autom. Control* 62 (4): 1590–1605.

3 Tian, B., Zuo, Z., Yan, X., and Wang, H. (2017). A fixed-time output feedback control scheme for double integrator systems. *Automatica* 80: 17–24.

4 Hou, H., Yu, X., Xu, L. et al. (2020). Finite-time continuous terminal sliding mode control of servo motor systems. *IEEE Trans. Ind. Electron.* 67: 5647–5656.

5 Yu, W., Chen, G., Cao, M., and Kurths, J. (2010). Second-order consensus for multiagent systems with directed topologies and nonlinear dynamics. *IEEE Trans. Syst. Man Cybern. B Cybern.* 40 (3): 881–891.

6 Wen, G., Duan, Z., Yu, W., and Chen, G. (2012). Consensus in multi-agent systems with communication constraints. *IET Control Theory Appl.* 22 (2): 170–182.

7 Andrieu, V., Praly, L., and Astolfi, A. (2009). Homogeneous approximation, recursive observer design, and output feedback. *SIAM J. Control. Optim.* 47 (4): 1814–1850.

8 Huang, Y. and Jia, Y. (2017). Fixed-time consensus tracking control for second-order multi-agent systems with bounded input uncertainties via NFFTSM. *IET Control Theory Appl.* 11 (16): 2900–2909.

9 Guo, G. and Yue, W. (2014). Sampled-data cooperative adaptive cruise control of vehicles with sensor failures. *IEEE Trans. Intell. Transp. Syst.* 15 (6): 2404–2418.

10 Guo, G. and Li, D. (2019). Adaptive sliding mode control of vehicular platoons with prescribed tracking performance. *IEEE Trans. Intel. Transp. Syst.* 68 (8): 7511–7520.

11 Yaz, E. (1998). Linear matrix inequalities in system and control theory. *Proc. IEEE* 86 (12): 2473–2474.

12 Slotine, J. and Li, W. (1991). *Applied Nonlinear Control*. Upper Saddle River, NJ: Prentice-Hall.

13 Xiao, S., Ge, X., Han, Q., and Zhang, Y. (2021). Dynamic event-triggered platooning control of automated vehicles under random communication topologies and various spacing policies. *IEEE Trans. Cybern.* 52: 11477–11490.

14 Dunbar, W.B. and Caveney, D.S. (2012). Distributed receding horizon control of vehicle platoons: stability and string stability. *IEEE Trans. Autom. Control* 57 (3): 620–633.

15 Mishra, J.P., Yu, X., and Jalili, M. (2018). Arbitrary-order continuous finite-time sliding mode controller for fixed-time convergence. *IEEE Trans. Circ. Syst. II Express Briefs* 65 (12): 1988–1992.

16 Guo, G. and Li, D. (2018). Pmp-based set-point optimization and sliding-mode control of vehicular platoons. *IEEE Trans. Comput. Social Syst.* 5 (2): 553–562.

17 Guo, G., Kang, J., Lei, H., and Li, D. (2021). Finite-time stabilization of a collection of connected vehicles subject to communication interruptions. *IEEE Trans. Intel. Transp. Syst.* 23: 1–9.

18 Li, Y., Tang, C., Peeta, S., and Wang, Y. (2019). Integral-sliding-mode braking control for a connected vehicle platoon: theory and application. *IEEE Trans. Ind. Electron.* 66 (6): 4618–4628.

Index

a

acceleration 3–5, 7, 16–22, 27–28, 31, 35, 44, 46, 57–66, 68–69, 72–73, 86, 89–95, 98–101, 103, 107–110, 112–113, 115–119, 124, 126, 135, 137–138, 140–141, 144, 147, 149, 157–162, 166, 176, 178, 181–182, 190–191, 202–203, 208, 217–218, 226–227, 229–232, 237, 243, 246, 253–255, 257–258, 260, 270, 278–279, 285, 292, 301
accident 3
actuator delay 4, 9, 98, 266
actuator faults 123–126, 128–129, 134, 136–141, 143–145, 147, 149, 151, 156–158, 160, 162–163
adaptive control 34–37, 119, 123, 170
adaptive switching 33
ad hoc networks 3, 97
algorithm 4–6, 16, 23, 26–27, 34, 39, 43, 48–49, 51–57, 60, 66, 70, 98, 112–113, 117, 123, 165, 175, 201–202, 205, 207, 212–214, 216–217, 220–221, 223–226, 228, 231–232, 235–238, 240, 254–255, 272, 290, 297, 301, 305–307
analysis 4–5, 9, 23, 27, 31, 42, 48, 55, 60, 70, 72–74, 77, 79, 81, 91, 98, 106, 110, 124, 128–129, 131, 133, 144, 149, 151, 153, 166, 168, 173, 181, 185–186, 190, 202, 212, 226, 234, 247, 252, 284, 299, 305
Arduino cars 5, 21–23, 213, 220, 222
assignment 42–46, 48–52, 54, 56–58, 60, 62, 64, 66, 70

b

binary variable 7
block entry matrix 10, 77
braking delay 4, 6, 98, 100, 266
break input 166, 182, 292

c

capacity 3, 75, 127, 128, 148, 149, 155, 165, 268, 301
chattering 129, 136, 157, 186, 268, 283, 290, 301, 305, 306
closed-loop system 133, 154, 244, 273
codesign 26
communication channel 16, 26, 97, 166, 182
communication constraints 3, 7, 265
communication delay 4, 7, 31, 72, 76–77, 81–83, 89–92, 95, 265
communication interruption 26–30, 32–36, 39
communication link 32, 44, 46, 59–60, 66, 70, 74, 87, 243
communication network 26, 43, 97–98, 100, 103, 202
communication topology 4, 27, 42, 44–45, 48–49, 51, 53, 56, 66–67, 74–76, 82, 87, 244, 255–261, 264–265, 269, 283, 285, 306
computational burden 124, 129, 131, 144, 150
connected vehicles 3, 9, 26, 28–29, 31–34, 36, 39, 42, 70, 72–73, 76–77, 81–82, 123, 225
connection status 43, 244
control design 27, 98, 202, 234, 290, 297

Connected Vehicular Systems: Communication, Control, and Optimization, First Edition. Ge Guo and Shixi Wen.

controller 4–5, 7–9, 11–17, 19–20, 22–23, 27–40, 42–44, 46–51, 53–61, 63–64, 66–67, 69–70, 73–75, 82–86, 88–95, 98, 100, 102–105, 107–110, 112–113, 115–118, 124, 127–129, 131, 133–134, 136, 143–144, 148–149, 155, 157, 165, 167–171, 173–176, 186, 189–190, 201–202, 207–212, 214, 217, 220, 225–226, 228, 231, 236, 242–243, 245–247, 249–255, 260, 262–263, 266–268, 270–272, 275–276, 278, 285, 288, 290, 293–294, 297–298, 300–302, 305
controller gains 8, 46, 53, 66, 72, 75, 83–85, 100, 231
control performance 42, 59–60, 66, 70, 83, 93, 98, 107–108, 129, 134, 216, 243, 254–258, 260, 263–266, 268, 284, 286, 301
control system 4–5, 9, 12–13, 15–16, 19, 31, 48–49, 51–52, 58, 73, 76–77, 80, 82–83, 85–86, 89, 98–99, 101, 165, 206, 242, 244–245, 247–248, 265, 269, 280, 282
convex inequality 9, 106, 110
coordination 201
cost function 43–51, 48–51, 59–60, 66, 68, 70, 205–206, 227, 245, 250, 254, 260–263, 272, 274, 288, 297, 306

d

data transmission 7, 16, 21, 46, 75, 98, 101, 103, 108
dead-zone 143–147, 149, 156–157, 160, 162
density 126, 134, 143, 147, 155, 182, 203, 220, 227, 270, 274, 277, 278
difference 165, 167, 184, 210, 306
discrete-time 48, 50
distributed control 42, 242, 302
disturbance 4, 26–27, 34, 36, 39–40, 43–44, 47–48, 50, 59–64, 70, 72, 74, 76–78, 81–84, 86, 88–89, 93, 95, 98, 104–108, 113, 117–118, 124, 126, 128, 135, 144, 146, 149, 155, 166, 177, 181–183, 188, 192, 197, 203, 206, 217, 219–221, 225, 254–255, 263, 276, 292, 294, 297–298, 301, 305
disturbance attenuation level 76, 78, 81–83, 88, 98, 105, 107–108, 113, 117
driveline 26, 73, 99
driven 103

e

effect 3–5, 7, 15–17, 19, 23, 26, 34, 36, 39, 42–43, 46, 57, 59–60, 70, 73, 77, 82, 86, 89, 98, 118, 123, 125–126, 129, 135–136, 139, 143, 145–146, 155–157, 165, 168, 173, 175, 185, 188, 190–192, 197, 203, 211, 214, 219–220, 240, 243, 245, 250, 253–256, 260, 263–264, 266, 268, 293, 306
effectiveness 23, 34, 43, 57, 70, 73, 86, 98, 118, 123, 125–126, 135–136, 145–146, 155–156, 175, 190–192, 197, 220, 240, 243, 253–254, 256, 263, 268, 306
efficiency 58, 73, 86, 99, 101, 143, 203–204, 227, 240
energy density 227
equations 6, 15, 50, 53, 79, 81, 172, 185, 207–208, 211–212, 227, 231–232, 234–235, 294
estimation errors 128, 149, 151, 173
event-triggered communication 97, 105
event-triggered condition 98, 101, 103, 107
external disturbances 27, 34, 36, 39, 40, 43, 60, 74, 76, 124, 128, 144, 149, 155, 166, 181–183, 188, 197, 217

f

fault-tolerant control 123, 126, 128, 129, 135, 139, 143, 150, 157
feedback 5, 8, 9, 14–17, 30, 31, 42–44, 46–49, 51, 55, 56, 58, 60, 66, 70, 74, 75, 83–85, 88, 93, 98, 100, 104, 105, 107, 108, 110, 112, 113, 118, 129, 150, 181, 203, 234, 236, 244, 250, 253, 260, 264
final theorem 14–15
fixed-time 290–293, 297, 298, 307

follower 3–5, 7, 9, 13, 15, 17, 28, 31, 43–47, 55–57, 60, 66–67, 72–74, 89, 126, 135–136, 147, 156–157, 175, 185, 204, 232, 239–240, 242–246, 251, 254–255, 259–260, 263–264, 269–270, 275, 278, 291–292, 298, 301
force 202, 204, 208, 230–232
frequency domain 27, 28, 60, 72, 205, 235, 266
fuel consumption 3, 118, 201, 203–205, 213, 214, 216, 217, 225, 227, 228, 240

g
graph 4, 26, 44, 45, 74, 75, 87, 226, 243, 244, 255, 269, 291, 292

h
hybrid reference model 9

i
individual stability 165, 275
information feedback 98, 100, 250
information flow 72, 75, 165, 166, 185, 242, 244, 269, 291, 302
information topology 292
initial condition 8, 9, 29, 33, 34, 48, 53, 59, 60, 72, 76, 78, 79, 81, 83, 103, 105, 123, 144, 207, 252, 253, 299, 301
input quantization 143–145, 147, 149, 151, 156, 157, 162
integral 3, 128, 149, 276
intelligent transportation system 3, 204
inter-vehicle communication topology 1
inter-vehicle distance 193, 195, 203

j
jerk 31, 256–258, 260–262, 266, 285–286, 288

k
kinematics status 46

l
Laplace transformation 48, 53, 235, 252, 253
leader 3–7, 9, 11, 13, 16–17, 19, 21–22, 28, 42–43, 45, 47, 57, 59–60, 74–76, 86–87, 124, 126, 135–136, 144, 147, 155, 165–167, 175, 181–183, 185, 190–191, 201–202, 205, 207, 209, 211, 220, 222, 226, 228, 232, 234, 237, 238, 242–246, 254–255, 257–258, 260, 264, 268–270, 278–279, 285, 290–291, 301–302, 305
length 59, 100, 127, 148, 183, 190, 206, 270, 292
level 58, 72, 76–78, 81–83, 86, 88, 105, 107–108, 113, 117, 146, 201, 207, 225, 256–258, 260, 262, 266
linearization 44, 74, 100, 181, 202–203, 211, 234, 236, 244, 253
linear matrix inequality 16, 77, 252
longitudinal behavior 244
longitudinal model 270
longitudinal vehicle dynamic model 235
Lyapunov function 130, 151, 171, 173, 186, 188, 210, 235, 248–249, 251–252, 273–276, 291, 295, 297–299
Lyapunov stability 118, 123, 143, 235, 251, 291

m
Markovian 72, 77, 80–81, 83–84, 88–89, 91–92, 94–95
Markovian chain 72, 77, 80–81, 83–84
Markovian jumping system 72, 95
Markovian switching 77, 80, 83–84, 88–89, 91–92, 94
mass 58, 73, 86, 99, 119, 124–125, 135, 145, 156, 166, 182, 202–205, 213–214, 220, 227, 237, 253
matrix inequality 11, 16, 77, 243, 252, 269
matrix variables 85, 112
medium access constraints 7, 266
model 4–6, 16, 19–20, 23, 26–29, 31–34, 40, 42–45, 57–59, 70, 72–76, 93, 98–101, 124–125, 128, 135, 143–145, 147, 155, 166–167, 181–182, 184, 186, 202–205, 207–208, 213, 221, 226–228, 234–235, 237, 244, 253, 260, 270, 291, 301

n
neural network 123, 145
nonlinearity 143–145, 147, 149, 156–158, 160, 162, 181–182

nonlinear vehicle dynamics 4, 57–58, 123, 143, 202, 228, 253

o

objectives 28, 31, 47, 79, 85, 168, 225, 228, 243, 268, 270, 293
optimal control 43, 53, 59, 70, 207, 270
optimal IVCT 43, 48, 50, 52, 55–56, 70
optimization 118, 201, 204–206, 225, 227–228, 242–243, 245–247, 249–250, 254, 260, 262–263, 265–266, 268, 270–274, 278, 288, 290, 292–294, 297, 301, 305–307

p

packet dropout 4, 7, 21–22, 26, 43, 46–47, 59–64, 70, 265
parameter 156, 161, 162, 179–182
performance 3, 9, 15, 17–19, 28, 31, 33, 36, 39, 41–42, 59–66, 68–70, 83, 89–91, 93, 97–98, 105, 107–108, 110, 113, 118–119, 123, 127–129, 134, 136, 143, 147–148, 155–157, 165–166, 168–186, 188–192, 194, 196–198, 201, 209, 216, 218–219, 237–238, 243, 254–258, 260, 263–266, 268, 276, 278, 283–286, 288, 301, 306
period 3, 5, 31, 42–43, 46, 48, 51–52, 55–56, 59, 66–70, 75, 82, 86, 97, 101, 108, 182, 208, 220, 226, 229
platoon control 4–5, 9, 11–13, 16–18, 23, 26, 43–44, 49, 51–52, 58, 60, 70, 72–73, 82–87, 89, 99, 105, 123, 126, 128, 143, 147, 149, 165, 168, 175, 179, 183, 185, 190–191, 197, 201–203, 205–206, 212, 221, 223, 225, 228, 235–237, 242–243, 245, 247–248, 253, 260, 265, 268, 270, 278, 280, 282, 284, 286, 288, 301
platoon stability 29, 168
pollution ix
positive 5, 9, 34, 48–50, 52, 73–74, 77, 81–83, 99, 106, 126–129, 146–150, 167, 169–170, 183–184, 186, 210, 243, 246–248, 251–252, 269, 272–273, 275, 291, 294–295, 298–300
possibility 16, 26, 29–32, 35, 39, 44, 56–60, 86, 89, 93, 97–98, 101, 103, 105, 107–108, 110, 118–119, 123–124, 128–129, 131, 133–136, 139–141, 144, 146, 149–152, 154–157, 166, 170, 175–177, 181, 183, 186, 188–192, 197, 201, 203–204, 212–214, 217, 220, 222, 225, 231–232, 234, 236–237, 242, 245, 250–255, 270, 272, 276, 278–279, 292, 294, 298, 300–302

q

quantization 4, 143–152, 154, 156–158, 160, 162, 207

r

random 26, 35, 72, 242, 256–258, 260, 262, 266, 285–286, 288
reference trajectory 209, 238, 272, 278, 294
Riccati equation 42, 50, 52, 53
road capacity 3, 165
road traffic 3, 270, 277

s

sampled-data 43, 46–48, 58, 70, 72–73, 75, 95, 98, 100, 104–105, 107–108, 110, 112–113, 115, 117–118
sensor 3–5, 21, 28, 32, 66, 75, 98–101, 107, 119, 166, 168, 182, 202, 220, 226, 301
sequence 45, 74, 103, 201, 207, 243, 264–266, 269
shock wave effect 3
signal 7, 9, 12, 36–40, 46, 49, 89, 100, 103, 123, 155, 159, 173, 246–247, 250, 272, 305
single 98, 100, 165, 201, 205, 223
sliding mode control 149, 165, 179, 181, 186, 190, 225, 228, 234, 242, 290, 301, 306
spacing error 5, 9, 13–15, 17–20, 23, 26–28, 31, 36–38, 40, 46, 48, 57, 60–66, 68–69, 90–92, 98, 100, 108–110, 112–113, 115–118, 125, 127–128, 133–134, 136–141, 143–144, 147–149, 151, 154, 157–158, 160, 162, 183, 191, 203, 205, 211, 215–216, 218–221, 223, 227, 235–237, 240, 242, 244–247, 250, 252–253, 260–261, 263–264, 268, 270, 275–276, 278, 280–288, 290, 292, 294, 297, 299, 301, 303–307

spacing policy 4–5, 23, 26, 28, 30, 32, 36, 75, 123–124, 126–129, 134–135, 143–144, 147–149, 151, 155, 165, 167, 203, 245, 268, 270, 274, 277–278, 288, 292
sting stability 205
stochastic 43, 46–48, 53–55, 70
switched systems 10, 23, 265
switching IVCT 43, 51, 55–56, 66, 70
switching topology 72

t

terminal sliding mode 290, 292–294, 298–302
third-order dynamics 165, 260, 266
throttle input 166, 182, 292
throttle torque 54, 86
time derivative 106, 166–168, 170, 185–186, 188, 230, 232, 234–235, 271, 274–276, 296, 299–300, 248253
time interval 8–9, 16, 19, 46, 51, 59–60, 89–90, 104, 108, 110–111, 114, 264
time-invariant 43, 52, 244
tracking control 170, 179, 185, 187, 189–192, 195, 201–202, 204, 206–212, 214, 216, 218, 220, 222, 225–226, 228, 230–236, 238, 240, 242–243, 245–246, 250, 254–255, 263, 268, 270–272, 275, 278, 290, 292–293, 297–299, 301–302, 307
tracking error 17, 26–27, 32–34, 36, 44–49, 53, 56–57, 59–60, 72, 75–78, 81, 89, 98, 100, 104–105, 107, 113, 118, 165, 167–171, 173–178, 181, 183–186, 188–197, 208, 234–235, 243, 246–247, 250–252, 257–258, 260, 263–276, 280–283, 285, 293, 297–301, 303–305
traffic density 126, 134, 143, 147, 155, 274, 277, 278
traffic flow 123, 126, 128–129, 134–135, 139, 143, 147, 149, 151, 154–155, 157, 268, 270, 274–275, 277–278, 283, 288
traffic flow stability 123, 126, 128–129, 134–135, 139, 143, 147, 149, 151, 154–155, 157, 268, 270, 274–275, 277–278, 288
transformation 31, 48, 53, 60, 167, 174, 177, 181, 184–185, 208, 235, 252–253, 263, 295
transform function 15–16
transition rate 72, 75, 77, 80–81, 83–84, 88–89

u

unified framework 48, 118
unique solutions 274, 297
unknown dead-zone 145, 157
unknown disturbances 121
unknown parameters 231
unknown transition rate 72, 80
unmodeled dynamics 93, 166

v

vehicle dynamics 4, 26, 34, 44, 57–58, 73–74, 86, 99–100, 123, 128, 143, 149, 165, 181–183, 202–203, 208, 228, 233, 242, 245–246, 253, 260, 265, 306
vehicle platoon 4, 6, 8–9, 16–20, 27, 51, 58–59, 66, 70, 86–87, 89–97, 99, 126, 181, 185, 190, 197–198, 213, 215–216, 218–219, 254–258, 263–266, 276, 283, 285, 287–288, 292–293, 301
vehicular platoon 5, 11, 26–28, 42–43, 54, 57, 60, 72–73, 89, 125–126, 129, 134, 139, 143–144, 147, 149, 151, 154, 157, 165–166, 175, 179, 181, 188–190, 192, 201–202, 205, 214, 225–227, 229, 231, 235, 242, 246, 260, 272, 277, 300
velocity 5–7, 15–23, 26–27, 31, 39–40, 44, 46–47, 57, 59–66, 68–69, 72–73, 89–95, 98–101, 105, 107–110, 112–113, 115–119, 124, 126, 134, 137–138, 140–141, 144, 147, 154, 157–158, 160, 162, 165–168, 175–179, 183, 190–192, 195, 227, 238–239, 243, 245–247, 250, 254–255, 257, 260–261, 264, 269–272, 274–276, 278–287, 292–294, 297, 301–306
velocity error 7, 31, 46–47, 66, 100, 126, 147, 192, 195, 239, 246–247, 250, 255, 260–261, 264, 270–287, 274, 278, 305

virtual leader 121, 165, 166

W

weak infinitesimal operator 78, 81
weak string stability 147, 151, 154
wireless channel 7, 72
wireless communication 21, 28, 166, 202, 226

Z

zero-order holder 48
zero steady-state 4, 9, 13, 27

Printed and bound by CPI Group (UK) Ltd, Croydon, CR0 4YY
17/09/2026
14964315-0001